Numerical analysis pr̶ ̶ ̶ ̶ ̶ ̶ ̶ ̶ ̶ ̶ ̶ ̶ ̶aces to the world. For mathematicians
it is a b̶ ̶ ̶ ̶ ̶ ̶ ̶ ̶ ̶ ̶ ̶ ̶ ̶ ̶with an applicable flavour. For scientists
and en̶g̶ ̶ ̶ ̶ ̶ ̶ ̶ ̶ ̶ ̶ ̶ ̶ject, part of the standard repertoire
of mod̶ ̶ ̶ ̶ ̶ ̶ ̶ ̶ ̶ ̶ ̶ ̶s it is a theory on the interplay
of com̶ ̶ ̶ ̶ ̶ ̶and algorithms for real-number calculations. It is
the tension between these standpoints that is the driving force of this book,
which presents a rigorous account of the fundamentals of numerical analysis
of both ordinary and partial differential equations. The point of departure
is mathematical but the exposition strives to maintain a balance between
theoretical, algorithmic and applied aspects of the subject.

In detail, topics covered include numerical solution of ordinary differential equations by multistep and Runge–Kutta methods; finite difference and
finite element techniques for the Poisson equation; a variety of algorithms
to solve large, sparse algebraic systems; methods for parabolic and hyperbolic differential equations and techniques of their analysis. The book is
accompanied by an appendix that presents brief back-up in a number of
mathematical topics.

Dr Iserles concentrates on fundamentals: deriving methods from first
principles, analysing them with a variety of mathematical techniques and
occasionally discussing questions of implementation and applications. By
doing so, he is able to lead the reader to theoretical understanding of the
subject without neglecting its practical aspects. The outcome is a textbook
that is mathematically honest and rigorous and provides its target audience
with a wide range of skills in both ordinary and partial differential equations.

A First Course in the Numerical Analysis of Differential Equations

Cambridge Texts in Applied Mathematics

A First Course in the Numerical Analysis of Differential Equations

ARIEH ISERLES

Department of Applied Mathematics and Theoretical Physics
University of Cambridge

CAMBRIDGE
UNIVERSITY PRESS

Published by the Press Syndicate of the University of Cambridge
The Pitt Building, Trumpington Street, Cambridge CB2 1RP
40 West 20th Street, New York, NY 10011-4211, USA
10 Stamford Road, Oakleigh, Melbourne 3166, Australia

© Cambridge University Press 1996

First published 1996

Library of Congress cataloging in publication data available
British Library cataloging in publication data available

ISBN 0 521 55376 8 Hardback
ISBN 0 521 55655 4 Paperback

Printed by Bell and Bain Ltd., Glasgow

Contents

Preface

Books – so we are often told – should be born out of a sense of mission, a wish to share knowledge, experience and ideas, a penchant for beauty. This book has been born out of a sense of frustration.

For the last decade or so I have been teaching the numerical analysis of differential equations to mathematicians, in Cambridge and elsewhere. Examining this extensive period of trial and (frequent) error, two main conclusions come to mind and both have guided my choice of material and presentation in this volume.

Firstly, mathematicians are different from other varieties of *homo sapiens*. It may be observed that people study numerical analysis for various reasons. **Scientists** and **engineers** require it as a means to an end, a tool to investigate the subject matter that *really* interests them. Entirely justifiably, they wish to spend neither time nor intellectual effort on the finer points of mathematical analysis, typically preferring a style that combines a cook-book presentation of numerical methods with a leavening of intuitive and hand-waving explanations. **Computer scientists** adopt a different, more algorithmic, attitude. Their heart goes after the clever algorithm and its interaction with computer architecture. Differential equations and their likes are abandoned as soon as decency allows (or sooner). They are replaced by discrete models, which in turn are analysed by combinatorial techniques. **Mathematicians,** though, follow a different mode of reasoning. Typically, mathematics students are likely to participate in an advanced numerical analysis course in their final year of undergraduate studies, or perhaps in the first postgraduate year. Their studies until that point in time would have consisted, to a large extent, of a progression of formal reasoning, the familiar sequence of axiom $\Rightarrow$ theorem $\Rightarrow$ proof $\Rightarrow$ corollary $\Rightarrow$ .... Numerical analysis does not fit easily into this straitjacket, and this goes a long way toward explaining why many students of mathematics find it so unattractive.

Trying to teach numerical analysis to mathematicians, one is thus in a dilemma: should the subject be presented purely as a mathematical theory, intellectually pleasing but arid insofar as applications are concerned or, alternatively, should the audience be administered an application-oriented culture shock that might well cause it to vote with its feet?! The resolution is not very difficult, namely to present the material in a *bona fide* mathematical manner, occasionally veering toward issues of applications and algorithmics but never abandoning honesty and rigour. It is perfectly allowable to omit an occasional proof (which might well require material outside the scope of the presentation) and even to justify a numerical method on the grounds of plausibility and a good track record in applications. But plausibility, a good track record, intuition and old-fashioned hand-waving do not constitute an honest mathematical argument and should never be presented as such.

Secondly, students should be exposed in numerical analysis to both ordinary and partial differential equations, as well as to means of dealing with large sparse algebraic systems. The pressure of many mathematical subjects and sub-disciplines is such that only a modest proportion of undergraduates are likely to take part in more than a single advanced numerical analysis course. Many more will, in all likelihood, be faced with the need to solve differential equations numerically in the future course of their professional life. Therefore, the option of restricting the exposition to ordinary differential equations, say, or to finite elements, while having the obvious merit of cohesion and sharpness of focus is counterproductive in the long term.

To recapitulate, the ideal course in the numerical analysis of differential equations, directed toward mathematics students, should be mathematically honest and rigorous and provide its target audience with a wide range of skills in both ordinary and partial differential equations. For the last decade I have been desperately trying to find a textbook that can be used to my satisfaction in such a course – in vain. There are many fine textbooks on particular aspects of the subject: numerical methods for ordinary differential equations, finite elements, computation of sparse algebraic systems. There are several books that span the whole subject but, unfortunately, at a relatively low level of mathematical sophistication and rigour. But, to the best of my knowledge, no text addresses itself to the right mathematical agenda at the right level of maturity. Hence my frustration and hence the motivation behind this volume.

This is perhaps the place to review briefly the main features of this book.

★ We cover a broad range of material: the numerical solution of ordinary differential equations by multistep and Runge–Kutta methods; finite difference and finite element techniques for the Poisson equation; a variety of algorithms for solving the large systems of sparse algebraic equations that occur in the course of computing the solution of the Poisson equation; and, finally, methods for parabolic and hyperbolic differential equations and techniques for their analysis. There is probably enough material in this book for a one-year, fast-paced course and probably many lecturers will wish to cover only part of the material. To help them – and their students – to navigate in the numerical minefield, this preface is accompanied by a flowchart on page xvii that displays the 'connectivity' of this book's contents. The darker-shaded items along the centre form the core: no decent exposition of the subject can afford to avoid these topics. (Of course, it is entirely legitimate to pick and choose *within* each chapter!) The boxes corresponding to optional material are shaded lighter – the incorporation of these topics is a matter for individual choice. They all contain valuable material but, in their entirety, might well exceed the capacity and attention span of a typical advanced undergraduate course.

★ This is a textbook for mathematics students. By implication, it is not a textbook for computer scientists, engineers or natural scientists. As I have already argued, each group of students has different concerns and thought modes. Each assimilates knowledge differently. Hence, a textbook that attempts to be different things to different audiences is likely to disappoint them all. Nevertheless, non-mathematicians in need of numerical knowledge can benefit from this volume, but it is fair to observe that they should perhaps peruse it somewhat later in their careers, when in possession of the appropriate degree of motivation and background knowledge.

On an even more basic level of restriction, this is a textbook, not a monograph or a collection of recipes. Emphatically, our mission is *not* to bring the exposition to the state of the art or to highlight the most advanced developments. Likewise, it is not our intention to provide techniques that cater for all possible problems and eventualities.

★ An annoying feature of many numerical analysis texts is that they display inordinately long lists of methods and algorithms to solve any one problem. Thus, not just one Runge–Kutta method but twenty! The hapless reader is left with an arsenal of weapons but, all too often, without a clue which one to use and why. In this volume we adopt an alternative approach: methods are derived from underlying principles and these principles, rather than the algorithms themselves, are at the centre of our argument. As soon as the underlying principles are sorted out, algorithmic fireworks become the least challenging part of numerical analysis – the real intellectual effort goes into the mathematical analysis.

This is not to say that issues of software are not important or that they are somehow of a lesser scholarly pedigree. They receive our attention in Chapter 5 and I hasten to emphasize that good software design is just as challenging as theorem-proving. Indeed, the proper appreciation of difficulties in software and applications is enhanced by the understanding of the analytic aspects of numerical mathematics.

★ A truly exciting aspect of numerical analysis is the extensive use it makes of different mathematical disciplines. If you believe that numerics are a mathematical cop-out, a device for abandoning mathematics in favour of something 'softer', you are in for a shock. Numerical analysis is perhaps the most extensive and varied user of a very wide range of mathematical theories, from basic linear algebra and calculus all the way to functional analysis, differential topology, graph theory, analytic function theory, nonlinear dynamical systems, number theory, convexity theory – and the list goes on and on. Hardly any theme in modern mathematics fails to inspire and help numerical analysis. Hence, numerical analysts must be open-minded and ready to borrow from a wide range of mathematical skills – this is not a good bolt-hole for narrow specialists!

In this volume we emphasize the variety of mathematical themes that inspire and inform numerical analysis. This is not as easy as it might sound, since it is impossible to take for granted that students in different universities have a similar knowledge of pure mathematics. In other words, it is often necessary to devote a few pages to a topic which, in principle, has nothing to do with numerical analysis *per se* but which, nonetheless, is required in our exposition. I ask for the indulgence of those readers who are more knowledgeable in arcane mathematical matters – all they need is simply to skip few pages....

★ There is a major difference between recalling and understanding a mathematical concept. Reading mathematical texts I often come across concepts that are familiar and which I have certainly encountered in the past. Ask me, however, to recite their precise definition and I will probably flunk the test. The proper and virtuous course of action in such an instance is to pause, walk to the nearest mathematical library and consult the right source. To be frank, although sometimes I pursue this course of action, more often than not I simply go on reading. I have every reason to believe that I am not alone in this dubious practice.

In this volume I have attempted a partial remedy to the aforementioned phenomenon, by adding an appendix named 'Bluffer's guide to useful mathematics'. This appendix lists in a perfunctory manner definitions and major theorems in a range of topics – linear algebra, elementary functional analysis and approximation theory – to which students should have been exposed previously but which might have been forgotten. Its purpose is neither to substitute elementary mathematical courses nor to offer remedial teaching. If you flick too often to the end of the book in search of a definition then, my friend, perhaps you had better stop for a while and get to grips with the underlying subject, using a proper textbook. Likewise, if you always pursue a virtuous course of action, consulting a proper source in each and every case of doubt, please do not allow me to tempt you off the straight and narrow.

★ Part of the etiquette of writing mathematics is to attribute material and to refer to primary sources. This is important not just to quench the vanity of one's colleagues but also to set the record straight, as well as allowing an interested reader access to more advanced material. Having said this, I entertain serious doubts with regard to the practice of sprinkling each and every paragraph in a *textbook* with copious references. The scenario is presumably that, having read the sentence '... suppose that $x \in \mathbb{U}$, where $\mathbb{U}$ is a foliated widget [37]', the reader will look up the references, identify '[37]' with a paper of J. Bloggs in *Proc. SDW,* recognize the latter as *Proceedings of the Society of Differentiable Widgets,* walk to the library, locate the journal (which will be actually on the shelf, rather than on loan, misplaced or stolen).... All this might not be far-fetched as far as advanced mathematics monographs are concerned but makes very little sense in an undergraduate context. Therefore I have adopted a practice whereby there are no references in the text proper. Instead, each chapter is followed by a section of 'Comments and bibliography', where we survey briefly further literature that might be beneficial to students (and lecturers).

Such sections serve a further important purpose. Some students – am I too optimistic? – might be interested and inspired by the material of the chapter. For their benefit I have given in each 'Comments and bibliography' section a brief discussion of further developments, algorithms, methods of analysis and connections with other mathematical disciplines.

★ Clarity of exposition often hinges on transparency of notation. Thus, throughout this book we use the following convention.

- Lower-case lightface sloping letters $(a, b, c, \alpha, \beta, \gamma, \ldots)$ represent scalars;
- Lower-case boldface sloping letters $(\boldsymbol{a}, \boldsymbol{b}, \boldsymbol{c}, \boldsymbol{\alpha}, \boldsymbol{\beta}, \boldsymbol{\gamma}, \ldots)$ represent vectors;
- Upper-case lightface letters $(A, B, C, \Theta, \Phi, \ldots)$ represent matrices;
- Letters in calligraphic font $(\mathcal{A}, \mathcal{B}, \mathcal{C}, \ldots)$ represent operators;
- Shell capitals $(\mathbb{A}, \mathbb{B}, \mathbb{C} \ldots)$ represent sets.

Mathematical constants like $\mathrm{i} = \sqrt{-1}$ and e, the base of natural logarithms, are denoted by roman, rather than italic letters. This follows British typesetting convention and helps to identify the different components of a mathematical formula.

As with any principle, our notational convention has its exceptions. For example, in Section 3.1 we refer to Legendre and Chebyshev polynomials by the conventional notation, P_n and T_n: any other course of action would have caused utter confusion. And, again as with any principle, grey areas and ambiguities abound. I have tried to eliminate them by applying common sense but this, needless to say, is a highly subjective criterion.

This book started out life as two sets of condensed lecture notes – one for students of Part II (the last year of undergraduate mathematics in Cambridge) and the other for students of Part III (the Cambridge advanced degree course in mathematics). The task of expanding lecture notes to a full-scale book is, unfortunately, more complicated than producing a cup of hot soup from concentrate by adding boiling water, stirring and simmering for a short while. Ultimately, it has taken the better part of a year, shared with the usual commitments of academic life. The main portion of the manuscript was written in Autumn 1994, during a sabbatical leave at the California Institute of Technology (Caltech). It is my pleasant duty to acknowledge the hospitality of my many good friends there and the perfect working environment in Pasadena.

A familiar computer proverb states that while the first 90% of a programming job takes 90% of the time the remaining 10% also takes 90% of the time.... Writing a textbook follows similar rules and, back home in Cambridge, I have spent several months reading and rereading the manuscript. This is the place to thank a long list of friends and colleagues whose help has been truly crucial: Brad Baxter (Imperial College, London), Martin Buhmann (Swiss Institute of Technology, Zürich), Yu-Chung Chang (Caltech), Stephen Cowley (Cambridge), George Goodsell (Cambridge), Mike Holst (Caltech), Herb Keller (Caltech), Yorke Liu (Cambridge), Michelle Schatzman (Lyon), Andrew Stuart (Stanford), Stefan Vandewalle (Louven) and Antonella Zanna (Cambridge). Some have read the manuscript and offered their comments. Some provided software well beyond my own meagre programming skills and helped with the figures and with computational examples. Some have experimented with the manuscript upon their students and listened to their complaints. Some contributed insight and occasionally saved me from embarrassing blunders. All have been helpful, encouraging and patient to a fault with my foibles and idiosyncrasies. None is responsible for blunders, errors, mistakes, misprints and infelicities that, in spite of my sincerest efforts, are bound to persist in this volume.

This is perhaps the place to extend thanks to two 'friends' that have made the process of writing this book considerably easier: the TEX typesetting system and the `Matlab` package. These days we take mathematical typesetting for granted but it is often forgotten that just a decade ago a mathematical manuscript would have been hand-written, then typed and retyped and, finally, typeset by publishers – each stage requiring laborious proofreading. In turn, `Matlab` allows us a unique opportunity to turn our office into a computational-cum-graphic laboratory, to bounce ideas off the computer screen and produce informative figures and graphic displays. Not since the discovery of coffee have any inanimate objects caused so much pleasure to so many mathematicians!

The editorial staff of Cambridge University Press, in particular Alan Harvey, David Tranah and Roger Astley, went well beyond the call of duty in being helpful, friendly and cooperative. Susan Parkinson, the copy editor, has worked to the highest standards. Her professionalism, diligence and good taste have done wonders in sparing the readers numerous blunders and the more questionable examples of my hopeless wit. This is a pleasant opportunity to thank them all.

Last but never the least, my wife and best friend, Dganit. Her encouragement, advice and support cannot be quantified in conventional mathematical terms. Thank you!

I wish to dedicate this book to my parents, Gisella and Israel. They are not mathematicians, yet I have learnt from them all the really important things that have motivated me as a mathematician: love of scholarship and admiration for beauty and art.

Arieh Iserles
August 1995

Flowchart of contents

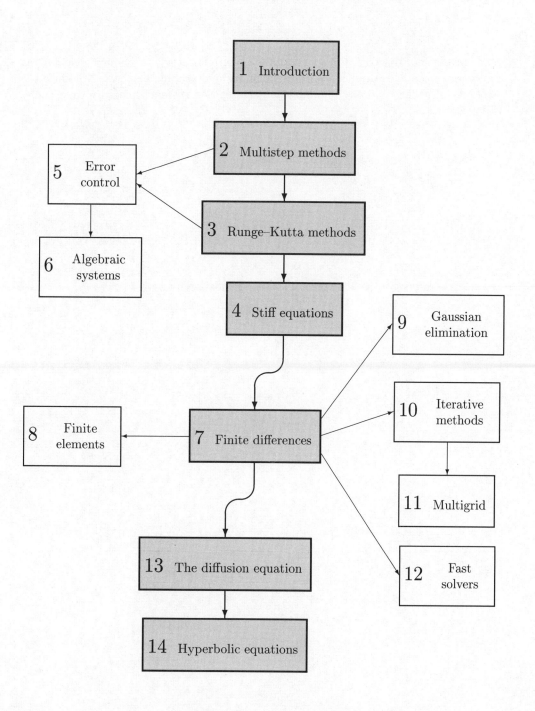

1 Introduction

2 Multistep methods

5 Error control

6 Algebraic systems

3 Runge–Kutta methods

4 Stiff equations

9 Gaussian elimination

8 Finite elements

7 Finite differences

10 Iterative methods

11 Multigrid

13 The diffusion equation

12 Fast solvers

14 Hyperbolic equations

PART I

Ordinary differential equations

1

Euler's method and beyond

1.1 Ordinary differential equations and the Lipschitz condition

We commence our exposition of the computational aspects of differential equations by a close – yet extensive – examination of numerical methods for *ordinary differential equations (ODEs)*. This is important because of the central role of ODEs in a multitude of applications. Not less crucial is the critical part that numerical ODEs play in the design and analysis of computational methods for *partial differential equations (PDEs)*.

Our goal is to approximate the solution of the problem

$$y' = f(t, y), \quad t \geq t_0, \qquad y(t_0) = y_0. \tag{1.1}$$

Here f is a sufficiently well-behaved function that maps $[t_0, \infty) \times \mathbb{R}^d$ to $\mathbb{R}^d$ and the initial condition $y_0 \in \mathbb{R}^d$ is a given vector. $\mathbb{R}^d$ denotes here – and elsewhere in this book – the d-dimensional real Euclidean space.

The 'niceness' of f may span a whole range of desirable attributes. At the very least, we insist on f obeying, in a given vector norm $\| \cdot \|$, the *Lipschitz condition*

$$\|f(t, x) - f(t, y)\| \leq \lambda \|x - y\| \quad \text{for all} \quad x, y \in \mathbb{R}^d, \ t \geq t_0. \tag{1.2}$$

Here $\lambda > 0$ is a real constant that is independent of the choice of x and y – a *Lipschitz constant*. Subject to (1.2), it is possible to prove that the ODE system (1.1) possesses a unique solution.[1] Taking a stronger requirement, we may stipulate that f is an *analytic* function – in other words, that the Taylor series of f about every $(t, y_0) \in [0, \infty) \times \mathbb{R}^d$ has a positive radius of convergence. It is then possible to prove that the solution y itself is analytic. Analyticity comes in handy, since much of our investigation of numerical methods is based on Taylor expansions, but it is often an excessive requirement and excludes many ODEs of practical importance.

In this volume we strive to steer a middle course between the complementary vices of mathematical nitpicking and of hand-waving. We solemnly undertake to avoid any needless mention of exotic function spaces that present the theory in its most general form, whilst desisting from woolly and inexact statements. Thus, we *always* assume that f is Lipschitz and, as necessary, may explicitly stipulate that it is analytic. An intelligent reader could, if the need arose, easily weaken many of our 'analytic' statements so that they are applicable also to sufficiently-differentiable functions.

---

[1] We refer to the Appendix for a brief refresher course on norms, existence and uniqueness theorems for ODEs and other useful odds and ends of mathematics.

1.2   Euler's method

Let us ponder briefly the meaning of the ODE (1.1). We possess two items of information: we know the value of y at a single point $t = t_0$ and, given any function value $y \in \mathbb{R}^d$ and time $t \geq t_0$, we can tell the slope from the differential equation. The purpose of the exercise being to guess the value of y at a new point, the most elementary approach is to use a linear interpolant. In other words, we estimate $y(t)$ by making the approximation $f(t, y(t)) \approx f(t_0, y(t_0))$ for $t \in [t_0, t_0 + h]$, where $h > 0$ is sufficiently small. Integrating (1.1),

$$y(t) = y(t_0) + \int_{t_0}^{t} f(\tau, y(\tau))\, \mathrm{d}\tau \approx y_0 + (t - t_0)f(t_0, y_0). \tag{1.3}$$

Given a sequence t_0, $t_1 = t_0 + h$, $t_2 = t_0 + 2h, \ldots$, where $h > 0$ is the *time step*, we denote by y_n a numerical estimate of the exact solution $y(t_n)$, $n = 0, 1, \ldots$. Motivated by (1.3), we choose

$$y_1 = y_0 + hf(t_0, y_0).$$

This procedure can be continued to produce approximants at t_2, t_3 and so on. In general, we obtain the recursive scheme

$$y_{n+1} = y_n + hf(t_n, y_n), \qquad n = 0, 1, \ldots, \tag{1.4}$$

the celebrated *Euler method*.

Euler's method is not only the most elementary computational scheme for ODEs and, simplicity notwithstanding, of enduring practical importance. It is also the cornerstone of the numerical analysis of differential equations of evolution. In a deep and profound sense, all the fancy multistep and Runge–Kutta schemes that we shall discuss in the sequel are nothing but a generalization of the basic paradigm (1.4).

$\diamond$ **Graphic interpretation**   Euler's method can be illustrated pictorially. Consider, for example, the scalar *logistic equation* $y' = y(1 - y)$, $y(0) = \frac{1}{10}$.

Figure 1.1 displays the first few steps of Euler's method, with a grotesquely large step $h = 1$. For each step we show the exact solution with initial condition $y(t_n) = y_n$ in the vicinity of $t_n = nh$ (solid line) and the linear interpolation via Euler's method (1.4) (dotted line).

The initial condition being, by definition, exact, so is the slope at t_0. However, instead of following a curved trajectory, the numerical solution is piecewise-linear. Having reached t_1, say, we have moved to a wrong trajectory (i.e., corresponding to a different initial condition). The slope at t_1 is wrong – or, rather, it is the correct slope of a wrong solution! Advancing further, we might well stray even more from the original trajectory.

A realistic goal of numerical solution is not, however, to avoid errors altogether; after all, we approximate since we do not know the exact solution in the first place! An error-generating mechanism exists in every algorithm for numerical ODEs and our purpose is to understand it and to ensure that, in

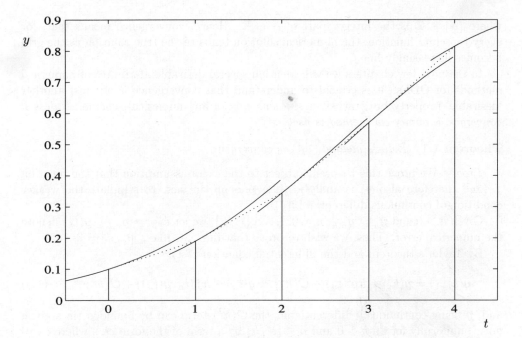

Figure 1.1   Euler's method, as applied to the equation $y' = y(1 - y)$, $y(0) = \frac{1}{10}$.

a given implementation, errors do not accumulate beyond a specified tolerance. Remarkably, even the excessive step $h = 1$ leads in Figure 1.1 to relatively modest local error. $\diamond$

Euler's method can be easily extended to cater for variable steps. Thus, for a general monotone sequence $t_0 < t_1 < t_2 < \cdots$ we approximate as follows:

$$\boldsymbol{y}(t_{n+1}) \approx \boldsymbol{y}_{n+1} = \boldsymbol{y}_n + h_n \boldsymbol{f}(t_n, \boldsymbol{y}_n),$$

where $h_n = t_{n+1} - t_n$, $n = 0, 1, \ldots$. However, for the time being we restrict ourselves to constant steps.

How good is Euler's method in approximating (1.1)? Before we even attempt to answer this question, we need to formulate it with considerably more rigour. Thus, suppose that we wish to compute a numerical solution of (1.1) in the compact interval $[t_0, t_0 + t^*]$ with some time-stepping numerical method, not necessarily Euler's scheme. In other words, we cover the interval by an equidistant grid and employ the time-stepping procedure to produce a numerical solution. Each grid is associated with a different numerical sequence and the critical question is whether, as $h \to 0$ and the grid is being refined, the numerical solution tends to the exact solution of (1.1). More formally, we express the dependence of the numerical solution upon the step size by the notation $\boldsymbol{y}_n = \boldsymbol{y}_{n,h}$, $n = 0, 1, \ldots, \lfloor t^*/h \rfloor$. A method is said to be *convergent* if, for every ODE (1.1) with a Lipschitz function $\boldsymbol{f}$ and every $t^* > 0$ it is true that

$$\lim_{h \to 0+} \max_{n = 0, 1, \ldots, \lfloor t^*/h \rfloor} \|\boldsymbol{y}_{n,h} - \boldsymbol{y}(t_n)\| = 0,$$

where $\lfloor \alpha \rfloor \in \mathbb{Z}$ is the integer part of $\alpha \in \mathbb{R}$. Hence, convergence means that, for every Lipschitz function, the numerical solution tends to the true solution as the grid becomes increasingly fine.[2]

In the next few chapters we will mention several desirable attributes of numerical methods for ODEs. It is crucial to understand that convergence is not just another 'desirable' property but, rather, a *sine qua non* of any numerical scheme. *Unless it converges, a numerical method is useless!*

Theorem 1.1   *Euler's method* (1.4) *is convergent.*

Proof   We prove this theorem subject to the extra assumption that the function $\boldsymbol{f}$ (and therefore also $\boldsymbol{y}$) is analytic (it is enough, in fact, to stipulate the weaker condition of continuous differentiability).

Given $h > 0$ and $\boldsymbol{y}_n = \boldsymbol{y}_{n,h}$, $n = 0, 1, \ldots, \lfloor t^*/h \rfloor$, we let $\boldsymbol{e}_{n,h} = \boldsymbol{y}_{n,h} - \boldsymbol{y}(t_n)$ denote the numerical error. Thus, we wish to prove that $\lim_{h \to 0+} \max_n \|\boldsymbol{e}_{n,h}\| = 0$.

By Taylor's theorem and the differential equation (1.1),

$$\boldsymbol{y}(t_{n+1}) = \boldsymbol{y}(t_n) + h\boldsymbol{y}'(t_n) + \mathcal{O}(h^2) = \boldsymbol{y}(t_n) + h\boldsymbol{f}(t_n, \boldsymbol{y}(t_n)) + \mathcal{O}(h^2), \qquad (1.5)$$

and, $\boldsymbol{y}$ being continuously differentiable, the $\mathcal{O}(h^2)$ term can be bounded (in a given norm) uniformly for all $h > 0$ and $n \leq \lfloor t^*/h \rfloor$ by a term of the form ch^2, where $c > 0$ is a constant. We subtract (1.5) from (1.4), giving

$$\boldsymbol{e}_{n+1,h} = \boldsymbol{e}_{n,h} + h[\boldsymbol{f}(t_n, \boldsymbol{y}(t_n) + \boldsymbol{e}_{n,h}) - \boldsymbol{f}(t_n, \boldsymbol{y}(t_n))] + \mathcal{O}(h^2).$$

Thus, it follows by the triangle inequality from the Lipschitz condition and the afore-mentioned bound on the $\mathcal{O}(h^2)$ reminder term that

$$\begin{aligned} \|\boldsymbol{e}_{n+1,h}\| &\leq \|\boldsymbol{e}_{n,h}\| + h\|\boldsymbol{f}(t_n, \boldsymbol{y}(t_n) + \boldsymbol{e}_{n,h}) - \boldsymbol{f}(t_n, \boldsymbol{y}(t_n))\| + ch^2 \\ &\leq (1 + h\lambda)\|\boldsymbol{e}_{n,h}\| + ch^2, \qquad n = 0, 1, \ldots, \lfloor t^*/h \rfloor - 1. \end{aligned} \qquad (1.6)$$

We now claim that

$$\|\boldsymbol{e}_{n,h}\| \leq \frac{c}{\lambda} h \left[(1 + h\lambda)^n - 1 \right], \qquad n = 0, 1, \ldots. \qquad (1.7)$$

The proof is by induction on n. When $n = 0$ we need to prove that $\|\boldsymbol{e}_{0,h}\| \leq 0$ and hence that $\boldsymbol{e}_{0,h} = \boldsymbol{0}$. This is certainly true, since at t_0 the numerical solution matches the initial condition and the error is zero.

For general $n \geq 0$ we assume that (1.7) is true up to n and use (1.6) to argue that

$$\|\boldsymbol{e}_{n+1,h}\| \leq (1 + h\lambda)\frac{c}{\lambda} h \left[(1 + h\lambda)^n - 1 \right] + ch^2 = \frac{c}{\lambda} h \left[(1 + h\lambda)^{n+1} - 1 \right].$$

This advances the inductive argument from n to $n+1$ and proves that (1.7) is true. The constant $h\lambda$ is positive, therefore $1 + h\lambda < e^{h\lambda}$, and we deduce that $(1 + h\lambda)^n < e^{nh\lambda}$.

---

[2]We have just introduced a norm through the back door. This, however, should cause no worry, since all norms are equivalent in finite-dimensional spaces. In other words, if a method is convergent in one norm, it converges in all....

The index n is allowed to range in $\{0, 1, \ldots, \lfloor t^*/h \rfloor\}$, hence $(1 + h\lambda)^n < e^{\lfloor t^*/h \rfloor h\lambda} \leq e^{t^*\lambda}$. Substituting into (1.7), we obtain the inequality

$$\|e_{n,h}\| \leq \frac{c}{\lambda}(e^{t^*\lambda} - 1)h, \qquad n = 0, 1, \ldots, \lfloor t^*/h \rfloor.$$

Since $c(e^{t^*\lambda} - 1)/\lambda$ is independent of h, it follows that

$$\lim_{\substack{h \to 0 \\ 0 \leq nh \leq t^*}} \|e_{n,h}\| = 0.$$

In other words, Euler's method is convergent. ∎

◇ **Health warning** At first sight, it might appear that there is more to the last theorem than meets the eye – not just a proof of convergence but also an upper bound on the error. In principle it is perfectly true and the error of Euler's method is indeed always bounded by $hce^{t^*\lambda}/\lambda$. Moreover, with very little effort it is possible to demonstrate, e.g. by using the *Peano kernel theorem*, that a reasonable choice is $c = \max_{t \in [t_0, t_0 + t^*]} \|\boldsymbol{y}''(t)\|$. The problem with this bound, unfortunately, is that in an overwhelming majority of practical cases it is too large by many orders of magnitude. It falls into the broad category of statements like 'the distance between London and New York is less than 47 light years' which, although manifestly true, fail to contribute significantly to the sum total of human knowledge.

The problem is not with the proof *per se* but with the insensitivity of a Lipschitz constant. A trivial example is the scalar linear equation $y' = -100y$, $y(0) = 1$. Therefore $\lambda = 100$ and, since $y(t) = e^{-100t}$, $c = \lambda^2$. We thus derive the upper bound of $100h(e^{100t^*} - 1)$. Letting $t^* = 1$, say, we have

$$|y_n - y(nh)| \leq 2.69 \times 10^{45} h. \tag{1.8}$$

It is easy, however, to show that $y_n = (1 - 100h)^n$, hence to derive the exact expression

$$|y_n - y(nh)| = |(1 - 100h)^n - e^{-100nh}|$$

which is smaller by many orders of magnitude than (1.8) (note that, unless nh is very small, to all intents and purposes $e^{-100nh} \approx 0$).

The moral of our discussion is simple. *The bound from the proof of Theorem 1.1 must not be used in practical estimation of numerical error!* ◇

Euler's method can be rewritten in the form $\boldsymbol{y}_{n+1} - [\boldsymbol{y}_n + h\boldsymbol{f}(t_n, \boldsymbol{y}_n)] = \boldsymbol{0}$. Replacing $\boldsymbol{y}_m$ by the exact solution $\boldsymbol{y}(t_k)$, $k = n, n+1$, and expanding into the first few terms of the Talyor series about $t = t_0 + nh$, we obtain

$$\boldsymbol{y}(t_{n+1}) - [\boldsymbol{y}(t_n) + h\boldsymbol{f}(t_n, \boldsymbol{y}(t_n))]$$
$$= [\boldsymbol{y}(t_n) + h\boldsymbol{y}'(t_n) + \mathcal{O}(h^2)] - [\boldsymbol{y}(t_n) + h\boldsymbol{y}'(t_n)] = \mathcal{O}(h^2).$$

We say the Euler's method (1.4) is of *order* one. In general, given an arbitrary time-stepping method

$$\boldsymbol{y}_{n+1} = \boldsymbol{\mathcal{Y}}_n(\boldsymbol{f}, h, \boldsymbol{y}_0, \boldsymbol{y}_1, \ldots, \boldsymbol{y}_n), \qquad n = 0, 1, \ldots,$$

for the ODE (1.1), we say that it is of *order p* if

$$y(t_{n+1}) - \mathcal{Y}_n(f, h, y(t_0), y(t_1), \ldots, y(t_n)) = \mathcal{O}(h^{p+1})$$

for every analytic f and $n = 0, 1, \ldots$. Alternatively, a method is of order p if it recovers *exactly* every polynomial solution of degree p or less.

The order of a numerical method provides us with information about its *local behaviour* – advancing from t_n to t_{n+1}, where $h > 0$ is sufficiently small, we are incurring an error of $\mathcal{O}(h^{p+1})$. Our main interest, however, is in not the local but the *global* behaviour of the method: how well is it doing in a fixed bounded interval of integration as $h \to 0$? Does it converge to the true solution? How fast? Since the local error decays as $\mathcal{O}(h^{p+1})$, the number of steps increases as $\mathcal{O}(h^{-1})$. The naive expectation is that the global error decreases as $\mathcal{O}(h^p)$, but – as we will see in Chapter 2 – it cannot be taken for granted for each and every numerical method without an additional condition. As far as Euler's method is concerned, Theorem 1.1 demonstrates that all is well and the error indeed decays as $\mathcal{O}(h)$.

1.3   The trapezoidal rule

Euler's method approximates the derivative by a constant in $[t_n, t_{n+1}]$, namely by its value at t_n (again, we denote $t_k = t_0 + kh$, $k = 0, 1, \ldots$). Clearly, the 'cantilevering' approximation is not very good and it makes more sense to make the constant approximation of the derivative equal to the average of its values at the end-points. Bearing in mind that derivatives are given by the differential equation, we thus obtain an expression similar to (1.3):

$$
\begin{aligned}
y(t) &= y(t_n) + \int_{t_n}^{t} f(\tau, y(\tau)) \, d\tau \\
&\approx y(t_n) + \tfrac{1}{2}(t - t_n)\{f(t_n, y(t_n)) + f(t_{n+1}, y(t_{n+1}))\}.
\end{aligned}
$$

This is the motivation behind the *trapezoidal rule*

$$y_{n+1} = y_n + \tfrac{1}{2}h[f(t_n, y_n) + f(t_{n+1}, y_{n+1})]. \tag{1.9}$$

To obtain the order of (1.9), we substitute the exact solution,

$$
\begin{aligned}
&y(t_{n+1}) - \left\{ y(t_n) + \tfrac{1}{2}h[f(t_n, y(t_n)) + f(t_{n+1}, y(t_{n+1}))] \right\} \\
&= \left[y(t_n) + hy'(t_n) + \tfrac{1}{2}h^2 y''(t_n) + \mathcal{O}(h^3) \right] \\
&\quad - \left(y(t_n) + \tfrac{1}{2}h \left\{ y'(t_n) + \left[y'(t_n) + hy''(t_n) + \mathcal{O}(h^2) \right] \right\} \right) = \mathcal{O}(h^3).
\end{aligned}
$$

Therefore the trapezoidal rule is of order two.

Forewarned of the shortcomings of local analysis, we should not jump to conclusions. Before we infer that the error decays globally as $\mathcal{O}(h^2)$, we must first prove that the method is convergent. Fortunately, this can be accomplished by a straightforward generalization of the method of proof of Theorem 1.1.

Theorem 1.2   *The trapezoidal rule* (1.9) *is convergent.*

Proof Subtracting

$$y(t_{n+1}) = y(t_n) + \tfrac{1}{2}h\left[f(t_n, y(t_n)) + f(t_{n+1}, y_{n+1})\right] + \mathcal{O}(h^3)$$

from (1.9), we obtain

$$
\begin{aligned}
e_{n+1,h} = \; & e_{n,h} + \tfrac{1}{2}h\left\{[f(t_n, y_n) - f(t_n, y(t_n))]\right. \\
& \left. + \left[f(t_{n+1}, y_{n+1}) - f(t_{n+1}, y(t_{n+1}))\right]\right\} + \mathcal{O}(h^3).
\end{aligned}
$$

For analytic f we may bound the $\mathcal{O}(h^3)$ term by ch^3 for some $c > 0$, and this upper bound is valid uniformly throughout $[t_0, t_0 + t^*]$. Therefore, it follows from the Lipschitz condition (1.2) and the triangle inequality that

$$\|e_{n+1,h}\| \leq \|e_{n,h}\| + \tfrac{1}{2}h\lambda \left\{\|e_{n,h}\| + \|e_{n+1,h}\|\right\} + ch^3.$$

Since we are ultimately interested in letting $h \to 0$, there is no harm in assuming that $h\lambda < 2$, and we can thus deduce that

$$\|e_{n+1,h}\| \leq \left(\frac{1 + \tfrac{1}{2}h\lambda}{1 - \tfrac{1}{2}h\lambda}\right)\|e_{n,h}\| + \left(\frac{c}{1 - \tfrac{1}{2}h\lambda}\right)h^3. \tag{1.10}$$

Our next step closely parallels the derivation of inequality (1.7). We thus argue that

$$\|e_{n,h}\| \leq \frac{c}{\lambda}\left[\left(\frac{1 + \tfrac{1}{2}h\lambda}{1 - \tfrac{1}{2}h\lambda}\right)^n - 1\right]h^2. \tag{1.11}$$

This follows by induction on n from (1.10) and is left as an exercise to the reader.

Since $0 < h\lambda < 2$, it is true that

$$\frac{1 + \tfrac{1}{2}h\lambda}{1 - \tfrac{1}{2}h\lambda} = 1 + \frac{h\lambda}{1 - \tfrac{1}{2}h\lambda} \leq \sum_{\ell=0}^{\infty}\frac{1}{\ell!}\left(\frac{h\lambda}{1 - \tfrac{1}{2}h\lambda}\right)^{\ell} = \exp\left(\frac{h\lambda}{1 - \tfrac{1}{2}h\lambda}\right).$$

Consequently, (1.11) yields

$$\|e_{n,h}\| \leq \frac{ch^2}{\lambda}\left(\frac{1 + \tfrac{1}{2}h\lambda}{1 - \tfrac{1}{2}h\lambda}\right)^n \leq \frac{ch^2}{\lambda}\exp\left(\frac{nh\lambda}{1 - \tfrac{1}{2}h\lambda}\right).$$

This bound is true for every nonnegative integer n such that $nh \leq t^*$. Therefore

$$\|e_{n,h}\| \leq \frac{ch^2}{\lambda}\exp\left(\frac{t^*\lambda}{1 - \tfrac{1}{2}h\lambda}\right)$$

and we deduce that

$$\lim_{\substack{h \to 0 \\ 0 \leq nh \leq t^*}} \|e_{n,h}\| = 0.$$

In other words, the trapezoidal rule converges. ∎

The number $ch^2 \exp[t^*\lambda/(1 - \tfrac{1}{2}h\lambda)]/\lambda$ is, again, of absolutely no use in practical error bounds. However, a significant difference from Theorem 1.1 is that for the trapezoidal rule the error decays *globally* as $\mathcal{O}(h^2)$. This is to be expected from a second-order method if its convergence has been established.

Another difference between the trapezoidal rule and Euler's method is of an entirely different character. Whereas Euler's method (1.4) can be executed explicitly – knowing $\boldsymbol{y}_n$ we can produce $\boldsymbol{y}_{n+1}$ by computing a value of $\boldsymbol{f}$ and making a few arithmetic operations – this is not the case with (1.9). The vector $\boldsymbol{v} = \boldsymbol{y}_n + \frac{1}{2}h\boldsymbol{f}(t_n, \boldsymbol{y}_n)$ can be evaluated from known data, but that leaves us in each step with the task of finding $\boldsymbol{y}_{n+1}$ as the solution of the system of algebraic equations

$$\boldsymbol{y}_{n+1} - \tfrac{1}{2}h\boldsymbol{f}(t_{n+1}, \boldsymbol{y}_{n+1}) = \boldsymbol{v}.$$

The trapezoidal rule is thus said to be *implicit*, to distinguish it from the *explicit* Euler's method and its ilk.

Solving nonlinear equations is hardly a mission impossible, but we cannot take it for granted either. Only in texts on pure mathematics are we allowed to wave a magic wand, exclaim 'let $\boldsymbol{y}_{n+1}$ be a solution of ...' and assume that all our problems are over. As soon as we come to deal with actual computation, we had better specify how we plan (or our computer plans) to undertake the task of evaluating $\boldsymbol{y}_{n+1}$. This will be one of the themes of Chapter 6, which deals with the implementation of ODE methods. It suffices to state now that the cost of numerically solving nonlinear equations does not rule out the trapezoidal rule (and other implicit methods) as viable computational instruments. Implicitness is just one attribute of a numerical method and we must weigh it alongside other features.

⋄ **A 'good' example**   Figure 1.2 displays the (natural) logarithm of the error in the numerical solution of the scalar linear equation $y' = -y + 2\mathrm{e}^{-t}\cos 2t$, $y(0) = 0$ for (in descending order) $h = \frac{1}{2}$, $h = \frac{1}{10}$ and $h = \frac{1}{50}$.

How well does the plot illustrate our main distinction between Euler's method and the trapezoidal rule, namely faster decay of the error for the latter? As often in life, information is somewhat obscured by extraneous 'noise'; in the present case the error oscillates. This can be easily explained by the periodic component of the exact solution $y(t) = \mathrm{e}^{-t}\sin 2t$. Another observation is that, for both Euler's method and the trapezoidal rule, the error, twists and turns notwithstanding, does decay. This, on the face of it, can be explained by the decay of the exact solution, but is an important piece of news nonetheless. Our most pessimistic assumption is that errors might accumulate from step to step but, as can be seen from this example, this prophecy of doom is often misplaced. This is a highly nontrivial point which will be debated at greater length throughout Chapter 4.

Factoring out oscillations and decay, we observe that errors indeed decrease with h. More careful examination verifies that they increase at roughly the rate predicted by order considerations. Specifically, for a convergent method of order p we have $\|\boldsymbol{e}\| \approx ch^p$, hence $\ln\|\boldsymbol{e}\| \approx \ln c + p\ln h$. Denoting by $\boldsymbol{e}^{(1)}$ and $\boldsymbol{e}^{(2)}$ errors corresponding to step sizes $h^{(1)}$ and $h^{(2)}$ respectively, it follows that $\ln\|\boldsymbol{e}^{(2)}\| \approx \ln\|\boldsymbol{e}^{(1)}\| - p\ln(h^{(2)}/h^{(1)})$. The ratio of consecutive step sizes in Figure 1.2 being five, we expect the error to decay by (at least) a constant multiple of $\ln 5 \approx 1.6094$ and $2\ln 5 \approx 3.2189$ for Euler and the trapezoidal rule respectively. The actual error decays if anything slightly faster than this. ⋄

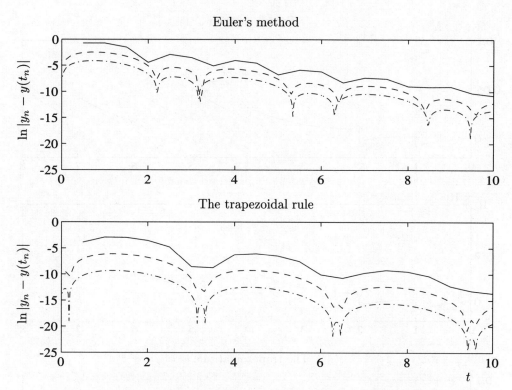

Figure 1.2 Euler's method and the trapezoidal rule, as applied to $y' = -y + 2e^{-t}\cos 2t$, $y(0) = 0$. The logarithm of the error is displayed for $h = \frac{1}{2}$ (solid line), $h = \frac{1}{10}$ (broken line) and $h = \frac{1}{50}$ (broken-and-dotted line).

$\diamond$ **A 'bad' example** Theorems 1.1 and 1.2 and, indeed, the whole numerical ODE theory, rest upon the assumption that (1.1) obeys the Lipschitz condition. We can expect numerical methods to underperform in the absence of (1.2), and this is vindicated by experiment. In Figure 1.3 we display the numerical solution of the equation $y' = \ln 3 \left(y - \lfloor y \rfloor - \frac{1}{2} \right)$, $y(0) = 0$. It is easy to verify that the exact solution is

$$y(t) = -n + \frac{1}{2}\left(1 - 3^{t-n}\right), \qquad n \le t \le n+1, \quad n = 0, 1, \ldots.$$

However, the equation fails the Lipschitz condition. In order to demonstrate this, we let $m \ge 1$ be an integer and set $x = m+\varepsilon$, $z = m-\varepsilon$, where $\varepsilon \in \left(0, \frac{1}{4}\right)$. Then

$$\left| \left(x - \lfloor x \rfloor - \tfrac{1}{2}\right) - \left(z - \lfloor z \rfloor - \tfrac{1}{2}\right) \right| = \frac{1 - 4\varepsilon}{2\varepsilon}|x - z|$$

and, since ε can be arbitrarily small, we see that inequality (1.2) cannot be satisfied with a finite λ.

Figure 1.3 displays the error for $h = \frac{1}{100}$ and $h = \frac{1}{1000}$. We observe that, although the error decreases with h, the rate of decay is just $\mathcal{O}(h)$ and, for the trapezoidal rule, falls short of what can be expected in a Lipschitz case.

Figure 1.3 The error using Euler's method and using the trapezoidal rule for $y' = \ln 3\left(y - \lfloor y \rfloor - \frac{1}{2}\right)$, $y(0) = 0$. For each method the upper figure corresponds to $h = \frac{1}{100}$ and the lower to $h = \frac{1}{1000}$.

Moreover, Euler's method outperforms the trapezoidal rule – but when the ODE is not Lipschitz, all bets are off! ◇

Two assumptions have led us to the trapezoidal rule. Firstly, for sufficiently small h, it is a good idea to approximate the derivative by a constant and, secondly, in choosing the constant we should not 'discriminate' between the endpoints – hence the average

$$\boldsymbol{y}'(t) \approx \tfrac{1}{2}[\boldsymbol{f}(t_n, \boldsymbol{y}_n) + \boldsymbol{f}(t_{n+1}, \boldsymbol{y}_{n+1})]$$

is a sensible choice. Similar reasoning leads, however, to an alternative approximant,

$$\boldsymbol{y}'(t) \approx \boldsymbol{f}\left(t_n + \tfrac{1}{2}h, \tfrac{1}{2}(\boldsymbol{y}_n + \boldsymbol{y}_{n+1})\right), \qquad t \in [t_n, t_{n+1}],$$

and to the *implicit midpoint rule*

$$\boldsymbol{y}_{n+1} = \boldsymbol{y}_n + h\boldsymbol{f}\left(t_n + \tfrac{1}{2}h, \tfrac{1}{2}(\boldsymbol{y}_n + \boldsymbol{y}_{n+1})\right). \tag{1.12}$$

It is easy to prove that (1.12) is second order and that it converges. This is left to the reader in Exercise 1.1.

The implicit midpoint rule is a special case of the *Runge–Kutta* method. We defer the discussion of such methods to Chapter 3.

1.4 The theta method

Both Euler's method and the trapezoidal rule fit the general pattern

$$\boldsymbol{y}_{n+1} = \boldsymbol{y}_n + h[\theta\boldsymbol{f}(t_n, \boldsymbol{y}_n) + (1 - \theta)\boldsymbol{f}(t_{n+1}, \boldsymbol{y}_{n+1})], \qquad n = 0, 1, \ldots, \tag{1.13}$$

with $\theta = 0$ and $\theta = \tfrac{1}{2}$ respectively. We may contemplate using (1.13) for any fixed value of $\theta \in [0, 1]$ and this, appropriately enough, is called a *theta method*. It is explicit for $\theta = 0$, otherwise implicit.

Although we can interpret (1.13) geometrically – the slope of the solution is assumed to be piecewise constant and provided by a linear combination of derivatives at the endpoints of each interval – we prefer the formal route of a Taylor expansion. Thus, substituting the exact solution $\boldsymbol{y}(t)$,

$$\boldsymbol{y}(t_{n+1}) - \boldsymbol{y}(t_n) - h[\theta\boldsymbol{f}(t_n, \boldsymbol{y}(t_n)) + (1 - \theta)\boldsymbol{f}(t_{n+1}, \boldsymbol{y}(t_{n+1}))]$$
$$= \boldsymbol{y}(t_{n+1}) - \boldsymbol{y}(t_n) - h[\theta\boldsymbol{y}'(t_n) + (1 - \theta)\boldsymbol{y}'(t_{n+1})]$$
$$= \left[\boldsymbol{y}(t_n) + h\boldsymbol{y}'(t_n) + \tfrac{1}{2}h^2\boldsymbol{y}''(t_n) + \tfrac{1}{6}h^3\boldsymbol{y}'''(t_n)\right] - \boldsymbol{y}(t_n)$$
$$\quad - h\left\{\theta\boldsymbol{y}'(t_n) + (1 - \theta)\left[\boldsymbol{y}'(t_n) + h\boldsymbol{y}''(t_n) + \tfrac{1}{2}h^2\boldsymbol{y}'''(t_n)\right]\right\} + \mathcal{O}(h^4)$$
$$= \left(\theta - \tfrac{1}{2}\right)h^2\boldsymbol{y}''(t_n) + \left(\tfrac{1}{2}\theta - \tfrac{1}{3}\right)h^3\boldsymbol{y}'''(t_n) + \mathcal{O}(h^4). \tag{1.14}$$

Therefore the method is of order two for $\theta = \tfrac{1}{2}$ (the trapezoidal rule) and otherwise of order one. Moreover, by expanding further than strictly required by order considerations, we can extract from (1.14) an extra morsel of information. Thus, subtracting the last expression from

$$\boldsymbol{y}_{n+1} - \boldsymbol{y}_n - h\left[\theta\boldsymbol{f}(t_n, \boldsymbol{y}_n) + (1 - \theta)\boldsymbol{f}(t_{n+1}, \boldsymbol{y}_{n+1})\right] = 0$$

we obtain for sufficiently small $h > 0$,

$$
\begin{aligned}
\boldsymbol{e}_{n+1} = \; & \boldsymbol{e}_n + \theta h [\boldsymbol{f}(t_n, \boldsymbol{y}(t_n) + \boldsymbol{e}_n) - \boldsymbol{f}(t_n, \boldsymbol{y}(t_n))] \\
& + (1 - \theta) h [\boldsymbol{f}(t_{n+1}, \boldsymbol{y}(t_{n+1}) + \boldsymbol{e}_{n+1}) \\
& \qquad - \boldsymbol{f}(t_{n+1}, \boldsymbol{y}(t_{n+1}))]
\begin{cases}
-\frac{1}{12} h^3 \boldsymbol{y}'''(t_n) + \mathcal{O}(h^4), & \theta = \frac{1}{2}, \\[2mm]
+ \left(\theta - \frac{1}{2}\right) h^2 \boldsymbol{y}''(t_n) + \mathcal{O}(h^3), & \theta \neq \frac{1}{2}.
\end{cases}
\end{aligned}
$$

Considering $\boldsymbol{e}_{n+1}$ as an unknown, we apply the *implicit function theorem* – this is allowed since $\boldsymbol{f}$ is analytic and, for sufficiently small $h > 0$, the matrix

$$
I - (1 - \theta) h \frac{\partial \boldsymbol{f}(t_{n+1}, \boldsymbol{y}(t_{n+1}))}{\partial \boldsymbol{y}}
$$

is nonsingular. The conclusion is that

$$
\boldsymbol{e}_{n+1} = \boldsymbol{e}_n
\begin{cases}
-\frac{1}{12} h^3 \boldsymbol{y}'''(t_n) + \mathcal{O}(h^4), & \theta = \frac{1}{2}, \\[2mm]
+ \left(\theta - \frac{1}{2}\right) h^2 \boldsymbol{y}''(t_n) + \mathcal{O}(h^3), & \theta \neq \frac{1}{2}.
\end{cases}
$$

The theta method is convergent for every $\theta \in [0, 1]$, as can be verified with ease by generalizing the proofs of Theorems 1.1 and 1.2. This is is the subject of Exercise 1.1.

Why, a vigilant reader might ask, bother with the theta method except for the special values of $\theta = 0$ and $\theta = \frac{1}{2}$? After all, the first is unique in conferring explicitness and the second is the only second-order theta method. The reasons are threefold. Firstly, the whole concept of order is based on the assumption that the numerical error is concentrated mainly in the leading term of its Taylor expansion. This is true as $h \to 0$, except that the step length, when implemented on a real computer, never actually tends to zero. ... Thus, in very special circumstances we might wish to annihilate higher-order terms in the error expansion – for example, letting $\theta = \frac{2}{3}$ gets rid of the $\mathcal{O}(h^3)$ term while retaining the $\mathcal{O}(h^2)$ component. Secondly, the theta method is our first example of a more general approach to the design of numerical algorithms, whereby simple geometric intuition is replaced by a more formal approach based on a Taylor expansion and the implicit function theorem. Its study is a good preparation for the material of Chapters 2 and 3. Finally, the choice $\theta = 1$ is of great practical relevance. The first-order implicit method

$$
\boldsymbol{y}_{n+1} = \boldsymbol{y}_n + h \boldsymbol{f}(t_{n+1}, \boldsymbol{y}_{n+1}), \qquad n = 0, 1, \ldots, \tag{1.15}
$$

is called the *backward Euler's method* and is one of the favourite algorithms for the solution of *stiff* ODEs. We defer the discussion of stiff equations to Chapter 4, where the merits of the backward Euler's method and similar schemes will become clear.

Comments and bibliography

We assume very little knowledge of the analytic (as distinguished from the numerical) theory of ODEs throughout this volume: just the concepts of existence, uniqueness, the Lipschitz condition and (mainly in Chapter 4) explicit solution of linear initial value systems. A brief résumé of essential knowledge is reviewed in the Appendix at Section A.2.3, but a diligent reader will do well to refresh her memory with a thorough look at a reputable textbook, for example Birkhoff & Rota (1978) or Boyce & DiPrima (1986).

Euler's method, the grandaddy of all numerical schemes for differential equations, is introduced in just about every relevant textbook (e.g. Conte & de Boor, 1990; Hairer *et al.*, 1991; Isaacson & Keller, 1966; Lambert, 1991), as is the trapezoidal rule. More traditional books have devoted considerable effort toward proving, with the Euler–Maclaurin formula (Ralston, 1965), that the error of the trapezoidal rule can be expanded in odd powers of h (cf. Exercise 1.8), but it seems that nowadays hardly anybody cares much about this observation, except for its applications to Richardson's extrapolation (Isaacson & Keller, 1966).

We have mentioned in Section 1.2 the Peano kernel theorem. Its knowledge is marginal to the subject matter of this book. However, if you want to understand mathematics and learn a simple, yet beautiful, result in approximation theory, we refer to A.2.2.6 and A.2.2.7 and references therein.

Birkhoff, G. and Rota, G.-C. (1978), *Ordinary Differential Equations* (3rd ed.), Wiley, New York.

Boyce, W.E. and DiPrima, R.C. (1986), *Elementary Differential Equations and Boundary Value Problems* (4th ed.), Wiley, New York.

Conte, S.D. and de Boor, C. (1980), *Elementary Numerical Analysis: An Algorithmic Approach* (3rd ed.), McGraw-Hill Kōgakusha, Tokyo.

Hairer, E, Nørsett, S.P. and Wanner, G. (1991), *Solving Ordinary Differential Equations I: Nonstiff Problems* (2nd ed.) Springer-Verlag, Berlin.

Isaacson, E. and Keller, H.B. (1966), *Analysis of Numerical Methods*, Wiley, New York.

Lambert, J.D. (1991), *Numerical Methods for Ordinary Differential Systems*, Wiley, London.

Ralston, A. (1965), *A First Course in Numerical Analysis*, McGraw-Hill Kōgakusha, New York.

Exercises

1.1 Apply the method of proof of Theorems 1.1 and 1.2 to prove convergence of the implicit midpoint rule (1.12) and of the theta method (1.13).

1.2 The linear system $y' = Ay$, $y(0) = y_0$, where A is a symmetric matrix, is solved by Euler's method.

a Letting $e_n = y_n - y(nh)$, $n = 0, 1, \ldots$, prove that

$$\|e_n\|_2 \leq \|y_0\|_2 \max_{\lambda \in \sigma(A)} \left|(1 + h\lambda)^n - e^{nh\lambda}\right|,$$

where $\sigma(A)$ is the set of eigenvalues of A and $\|\cdot\|_2$ is the Euclidean matrix norm (cf. A.1.3.3).

b Demonstrate that for every $-1 \ll x \leq 0$ and $n = 0, 1, \ldots$ it is true that

$$e^{nx} - \tfrac{1}{2}nx^2 e^{(n-1)x} \leq (1 + x)^n \leq e^{nx}.$$

[*Hint: Prove first that $1 + x \leq e^x$, $1 + x + \frac{1}{2}x^2 \geq e^x$ for all $x \leq 0$, and then argue that, provided $|a - 1|$ and $|b|$ are small, it is true that $(a - b)^n \geq a^n - na^{n-1}b$.*]

c Suppose that the maximal eigenvalue of A is $\lambda_{\max} < 0$. Prove that, as $h \to 0$ and $nh \to t \in [0, t^*]$,

$$\|\boldsymbol{e}_n\|_2 \leq \tfrac{1}{2}t\lambda_{\max}^2 e^{\lambda_{\max}t}\|\boldsymbol{y}_0\|_2 h \leq \tfrac{1}{2}t^*\lambda_{\max}^2\|\boldsymbol{y}_0\|_2 h.$$

d Compare the order of magnitude of this bound with the upper bound from Theorem 1.1 in the case

$$A = \begin{bmatrix} -2 & 1 \\ 1 & -2 \end{bmatrix}, \quad t^* = 10.$$

1.3   We solve the scalar linear system $y' = ay$, $y(0) = 1$.

a Show that the 'continuous output' method

$$u(t) = \frac{1 + \frac{1}{2}a(t - nh)}{1 - \frac{1}{2}a(t - nh)}y_n, \quad nh \leq t \leq (n + 1)h, \quad n = 0, 1, \dots,$$

is consistent with the values of y_n and y_{n+1} which are obtained by the trapezoidal rule.

b Demonstrate that u obeys the perturbed ODE

$$u'(t) = au(t) + \frac{\frac{1}{4}a^3(t - nh)^2}{1 - \frac{1}{2}a(t - nh)}y_n, \quad t \in [nh, (n + 1)h],$$

with the initial condition $u(nh) = y_n$. Thus, prove that

$$u((n + 1)h) = e^{ha}\left(1 + \tfrac{1}{4}a^3 \int_0^h \frac{e^{-\tau a}\tau^2 \, d\tau}{1 - \frac{1}{2}a\tau}\right)y_n.$$

c Let $e_n = y_n - y(nh)$, $n = 0, 1, \dots$. Show that

$$e_{n+1} = e^{ha}\left(1 + \tfrac{1}{4}a^3 \int_0^h \frac{e^{-\tau a}\tau^2 \, d\tau}{1 - \frac{1}{2}a\tau}\right)e_n + \tfrac{1}{4}a^3 e^{(n+1)ha} \int_0^h \frac{e^{-\tau a}\tau^2 \, d\tau}{1 - \frac{1}{2}a\tau}.$$

In particular, deduce that $a < 0$ implies that the error propagates subject to the inequality

$$|e_{n+1}| \leq e^{ha}\left(1 + \tfrac{1}{4}|a|^3 \int_0^h e^{-\tau a}\tau^2 \, d\tau\right)|e_n| + \tfrac{1}{4}|a|^3 e^{(n+1)ha} \int_0^h e^{-\tau a}\tau^2 \, d\tau.$$

1.4   Given $\theta \in [0, 1]$, find the order of the method

$$\boldsymbol{y}_{n+1} = \boldsymbol{y}_n + h\boldsymbol{f}\left(t_n + (1 - \theta)h, \theta\boldsymbol{y}_n + (1 - \theta)\boldsymbol{y}_{n+1}\right).$$

1.5 Provided that f is analytic, it is possible to obtain from $y' = f(t, y)$ an expression for the second derivative of y, namely $y'' = g(t, y)$, where

$$g(t, y) = \frac{\partial f(t, y)}{\partial t} + \frac{\partial f(t, y)}{\partial y} f(t, y).$$

Find the order of the methods

$$y_{n+1} = y_n + h f(t_n, y_n) + \tfrac{1}{2} h^2 g(t_n, y_n)$$

and

$$y_{n+1} = y_n + \tfrac{1}{2} [f(t_n, y_n) + f(t_{n+1}, y_{n+1})] + \tfrac{1}{12} h^2 [g(t_n, y_n) - g(t_{n+1}, y_{n+1})].$$

1.6* Assuming that g is Lipschitz, prove that both methods from Exercise 1.5 converge.

1.7 Repeated differentiation of the ODE (1.1), for an analytic f, yields explicit expressions for functions g_m such that

$$\frac{\mathrm{d}^m y(t)}{\mathrm{d}t^m} = g_m(t, y(t)), \qquad m = 0, 1, \ldots;$$

hence $g_0(t, y) = y$ and $g_1(t, y) = f(t, y)$, whereas g_2 has been already defined in Exercise 1.5 as g.

a Assuming for simplicity that $f = f(y)$ (i.e., that the ODE system (1.1) is *autonomous*), derive g_3.

b Prove that the mth *Taylor method*

$$y_{n+1} = \sum_{k=0}^{m} \frac{1}{k!} h^k g_k(t_n, y_n), \qquad n = 0, 1, \ldots,$$

is of order m for $m = 1, 2, \ldots$.

c Let $f(y) = \Lambda y + b$, where the matrix Λ and the vector b are independent of t. Find the explicit form of g_m for $m = 0, 1, \ldots$ and thereby prove that the mth Taylor method reduces to the recurrence

$$y_{n+1} = \left(\sum_{k=0}^{m} \frac{1}{k!} h^k \Lambda^k \right) y_n + \left(\sum_{k=1}^{m} \frac{1}{k!} h^k \Lambda^{k-1} \right) b, \qquad n = 0, 1, \ldots.$$

1.8 Let f be analytic. Prove that, for sufficiently small $h > 0$ and an analytic function x, the function

$$x(t + h) - x(t - h) - h f \left(\tfrac{1}{2} (x(t - h) + x(t + h)) \right)$$

can be expanded into power series in *odd* powers of h. Deduce that the error in the implicit midpoint rule (1.13), when applied to *autonomous* ODEs $y' = f(y)$ also admits an expansion in odd powers of h. [*Hint: First try to prove the statement for a scalar function f. Once you solve this problem, a generalization should present no difficulties.*]

2

Multistep methods

2.1   The Adams method

A typical numerical method for an initial-value ODE system computes the solution on a step-by-step basis. Thus, the Euler method advances the solution from t_0 to t_1 using $\boldsymbol{y}_0$ as an initial value. Next, to advance from t_1 to t_2, we *discard* $\boldsymbol{y}_0$ and employ $\boldsymbol{y}_1$ as the new initial value.

Numerical analysts, however, are thrifty by nature. Why discard a potentially valuable vector $\boldsymbol{y}_0$? With greater generality, why not make the solution depend on several past values, provided that these values are available?

There is one perfectly good reason why not – the exact solution of

$$\boldsymbol{y}' = \boldsymbol{f}(t, \boldsymbol{y}), \quad t \geq t_0, \qquad \boldsymbol{y}(t_0) = \boldsymbol{y}_0 \tag{2.1}$$

is uniquely determined ($\boldsymbol{f}$ being Lipschitz) by a single initial condition. Any attempt to pin the solution down at more than one point is mathematically nonsensical or, at best, redundant. This, however, is valid only with regard to the true solution of (2.1). When it comes to computation, this redundancy becomes our friend and past values of $\boldsymbol{y}$ can be put to a very good use – provided, however, that we are very careful indeed.

Let us thus suppose again that $\boldsymbol{y}_n$ is the numerical solution at $t_n = t_0 + nh$, where $h > 0$ is the step-size, and let us attempt to derive an algorithm that intelligently exploits past values. To that end, we assume that

$$\boldsymbol{y}_m = \boldsymbol{y}(t_m) + \mathcal{O}\big(h^{s+1}\big), \qquad m = 0, 1, \ldots, n + s - 1, \tag{2.2}$$

where $s \geq 1$ is a given integer. Our wish being to advance the solution from t_{n-s+1} to t_{n+s}, we commence from the trivial identity

$$\boldsymbol{y}(t_{n+s}) = \boldsymbol{y}(t_{n+s-1}) + \int_{t_{n+s-1}}^{t_{n+s}} \boldsymbol{y}'(\tau)\,\mathrm{d}\tau = \boldsymbol{y}(t_{n+s-1}) + \int_{t_{n+s-1}}^{t_{n+s}} \boldsymbol{f}(\tau, \boldsymbol{y}(\tau))\,\mathrm{d}\tau. \tag{2.3}$$

Wishing to exploit (2.3) for computational ends, we note that the integral on the right incorporates $\boldsymbol{y}$ not just at the grid points – where approximants are available – but throughout the interval $[t_{n+s-1}, t_{n+s}]$. The main idea of an *Adams method* is to use past values of the solution to approximate $\boldsymbol{y}'$ in the interval of integration. Thus, let $\boldsymbol{p}$ be an interpolation polynomial (cf. A.2.2.1–A.2.2.5) that matches $\boldsymbol{f}(t_m, \boldsymbol{y}_m)$ for $m = n, n+1, , \ldots, n+s-1$. Explicitly,

$$\boldsymbol{p}(t) = \sum_{m=0}^{s-1} p_m(t) \boldsymbol{f}(t_{n+m}, \boldsymbol{y}_{n+m}),$$

where the functions

$$p_m(t) = \prod_{\substack{\ell=0 \\ \ell \neq m}}^{s-1} \frac{t - t_{n+\ell}}{t_{n+m} - t_{n+\ell}} = \frac{(-1)^{s-1-m}}{m!(s-1-m)!} \prod_{\substack{\ell=0 \\ \ell \neq m}}^{s-1} \left(\frac{t - t_n}{h} - \ell \right), \tag{2.4}$$

for every $m = 0, 1, \ldots, s - 1$, are *Lagrange interpolation polynomials*. It is an easy exercise to verify that indeed $p(t_m) = f(t_m, y_m)$ for all $m = n, n+1, \ldots, n+s-1$. Hence, (2.2) implies that $p(t_m) = y'(t_m) + \mathcal{O}(h^s)$ for this range of m. We now use interpolation theory from A.2.2.2 to argue that, y being sufficiently smooth,

$$p(t) = y'(t) + \mathcal{O}(h^s), \qquad t \in [t_{n+s-1}, t_{n+s}].$$

We next substitute p in the integrand of (2.3), replace $y(t_{n+s-1})$ by y_{n+s-1} there and, having integrated along an interval of length h, we incurr an error of $\mathcal{O}(h^{s+1})$. In other words, the method

$$y_{n+s} = y_{n+s-1} + h \sum_{m=0}^{s-1} b_m f(t_{n+m}, y_{n+m}), \tag{2.5}$$

where

$$b_m = h^{-1} \int_{t_{n+s-1}}^{t_{n+s}} p_m(\tau) \, d\tau = h^{-1} \int_0^h p_m(t_{n+s-1} + \tau) \, d\tau, \qquad m = 0, 1, \ldots, s-1,$$

is of order $p = s$. Note from (2.4) that the coefficients $b_0, b_1, \ldots, b_{s-1}$ are independent of n and of h – thus we can subsequently use them to advance the iteration from t_{n+s} to t_{n+s+1} and so on.

The scheme (2.5) is called the s-step *Adams–Bashforth* method.

Having derived explicit expressions, it is easy to state Adams–Bashforth methods for moderate values of s. Thus, for $s = 1$ we encounter our old friend, the Euler method, whereas $s = 2$ gives

$$y_{n+2} = y_{n+1} + h \left[\tfrac{3}{2} f(t_{n+1}, y_{n+1}) - \tfrac{1}{2} f(t_n, y_n) \right] \tag{2.6}$$

and $s = 3$ gives

$$y_{n+3} = y_{n+2} + h \left[\tfrac{23}{12} f(t_{n+2}, y_{n+2}) - \tfrac{4}{3} f(t_{n+1}, y_{n+1}) + \tfrac{5}{12} f(t_n, y_n) \right]. \tag{2.7}$$

Figure 2.1 displays the logarithm of the error in the solution of $y' = -y^2$, $y(0) = 1$, by Euler's method and the schemes (2.6) and (2.7). The important information can be read off the y-scale: when h is halved, Euler's error decreases linearly, the error of (2.6) decays quadratically and (2.7) displays cubic decay. This is hardly surprising, since the order of the s-step Adams–Bashforth method is, after all, s and the global error decays as $\mathcal{O}(h^s)$.

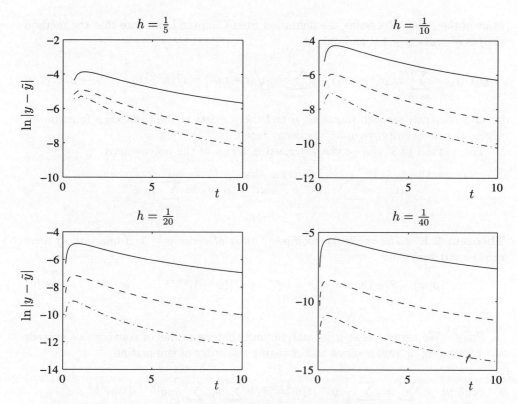

Figure 2.1 The first three Adams–Bashforth methods, as applied to the equation $y' = -y^2$, $y(0) = 1$. Euler's method, (2.6) and (2.7) correspond to the solid, broken and broken-and-dotted lines respectively.

Adams–Bashforth methods are just one instance of multistep methods. In the remainder of this chapter we will encounter several other families of such schemes. Later in this book we will learn that different multistep methods are suitable in different situations. First, however, we need to study the general theory of order and convergence.

2.2 Order and convergence of multistep methods

We write a general s-step method in the form

$$\sum_{m=0}^{s} a_m \boldsymbol{y}_{n+m} = h \sum_{m=0}^{s} b_m \boldsymbol{f}(t_{n+m}, \boldsymbol{y}_{n+m}), \qquad n = 0, 1, \ldots, \qquad (2.8)$$

where a_m, b_m, $m = 0, 1, \ldots, s$, are given constants, independent of h, n and the underlying differential equation. It is conventional to normalize (2.8) by letting $a_s = 1$. When $b_s = 0$ (as is the case with the Adams–Bashforth method) the method is said to be *explicit*; otherwise it is *implicit*.

Since we are about to encounter several criteria that play an important role in choosing the coefficients a_m and b_m, a central consideration is to obtain a reasonable value of the order. Recasting the definition from Chapter 1, we note that the method (2.8) is of *order* $p \geq 1$ if and only if

$$\psi(t, \boldsymbol{y}) := \sum_{m=0}^{s} a_m \boldsymbol{y}(t + mh) - h \sum_{m=0}^{s} b_m \boldsymbol{y}'(t + mh) = \mathcal{O}(h^{p+1}), \qquad h \to 0, \quad (2.9)$$

for all sufficiently smooth functions $\boldsymbol{y}$ and there exists at least one such function for which we cannot improve upon the decay rate $\mathcal{O}(h^{p+1})$.

The method (2.8) can be characterized in terms of the polynomials

$$\rho(w) := \sum_{m=0}^{s} a_m w^m \qquad \text{and} \qquad \sigma(w) := \sum_{m=0}^{s} b_m w^m.$$

Theorem 2.1   *The multistep method* (2.8) *is of order* $p \geq 1$ *if and only if there exists* $c \neq 0$ *such that*

$$\rho(w) - \sigma(w) \ln w = c(w-1)^{p+1} + \mathcal{O}(|w-1|^{p+2}), \qquad w \to 1. \qquad (2.10)$$

Proof   We assume that $\boldsymbol{y}$ is analytic and that its radius of convergence exceeds sh. Expanding in Taylor series and changing the order of summation,

$$\psi(t, \boldsymbol{y}) = \sum_{m=0}^{s} a_m \sum_{k=0}^{\infty} \frac{1}{k!} \boldsymbol{y}^{(k)}(t) m^k h^k - h \sum_{m=0}^{s} b_m \sum_{k=0}^{\infty} \frac{1}{k!} \boldsymbol{y}^{(k+1)}(t) m^k h^k$$

$$= \left(\sum_{m=0}^{s} a_m \right) \boldsymbol{y}(t) + \sum_{k=1}^{\infty} \frac{1}{k!} \left(\sum_{m=0}^{s} m^k a_m - k \sum_{m=0}^{s} m^{k-1} b_m \right) h^k \boldsymbol{y}^{(k)}(t).$$

Thus, to obtain order p it is neccesary and sufficient that

$$\sum_{m=0}^{s} a_m = 0, \qquad \sum_{m=0}^{s} m^k a_m = k \sum_{m=0}^{s} m^{k-1} b_m, \qquad k = 1, 2, \ldots, p.$$

$$\sum_{m=0}^{s} m^{p+1} a_m \neq (p+1) \sum_{m=0}^{s} m^p b_m. \qquad (2.11)$$

Let $w = \mathrm{e}^z$; therefore $w \to 1$ corresponds to $z \to 0$. Expanding again in a Taylor series,

$$\rho(\mathrm{e}^z) - z\sigma(\mathrm{e}^z) = \sum_{m=0}^{s} a_m \mathrm{e}^{mz} - z \sum_{m=0}^{s} b_m \mathrm{e}^{mz}$$

$$= \sum_{m=0}^{s} a_m \left(\sum_{k=0}^{\infty} \frac{1}{k!} m^k z^k \right) - z \sum_{m=0}^{s} b_m \left(\sum_{k=0}^{\infty} \frac{1}{k!} m^k z^k \right)$$

$$= \sum_{k=0}^{\infty} \frac{1}{k!} \left(\sum_{m=0}^{s} m^k a_m \right) z^k - \sum_{k=1}^{\infty} \frac{1}{(k-1)!} \left(\sum_{m=0}^{s} m^{k-1} b_m \right) z^k.$$

Therefore

$$\rho(\mathrm{e}^z) - z\sigma(\mathrm{e}^z) = ch^{p+1} + \mathcal{O}\!\left(h^{p+2}\right)$$

for some $c \neq 0$ if and only if (2.11) is true. The theorem follows by restoring $w = \mathrm{e}^z$. ∎

An alternative derivation of the order conditions (2.11) assists in our understanding of them. The map $\boldsymbol{y} \mapsto \boldsymbol{\psi}(t, \boldsymbol{y})$ is linear, consequently $\boldsymbol{\psi}(t, \boldsymbol{y}) = \mathcal{O}\!\left(h^{p+1}\right)$ if and only if $\boldsymbol{\psi}(t, q) = 0$ for every polynomial q of degree p. Because of linearity, this is equivalent to

$$\boldsymbol{\psi}(t, q_k) = 0, \qquad k = 0, 1, \ldots, p,$$

where $\{q_0, q_1, \ldots, q_p\}$ is a basis of the $(p+1)$-dimensional space of p-degree polynomials (cf. A.2.1.2–A.2.1.3). Setting $q_k(t) = (t - mh)^k$ for $k = 0, 1, \ldots, p$ we immediately obtain (2.11).

◇ **Adams–Bashforth revisited. . .**   Theorem 2.1 obviates the need for 'special tricks' such as used in our derivation of the Adams–Bashforth methods in Section 2.1. Given any multistep scheme (2.8), we can verify its order by a fairly painless expansion into series. It is convenient to express everything in the currency $\xi := w - 1$. For example, (2.6) results in

$$\rho(w) - \sigma(w)\ln w = (\xi + \xi^2) - \left(1 + \tfrac{3}{2}\xi\right)\left(\xi - \tfrac{1}{2}\xi^2 + \tfrac{1}{3}\xi^3 + \cdots\right) = \tfrac{5}{12}\xi^3 + \mathcal{O}\!\left(\xi^4\right);$$

thus order two is validated. Likewise, we can check that (2.7) is indeed of order three from the expansion

$$
\begin{aligned}
\rho(w) - \sigma(w)\ln w &= \xi + 2\xi^2 + \xi^3 \\
&\quad - \left(1 + \tfrac{5}{2}\xi + \tfrac{23}{12}\xi^2\right)\left(\xi - \tfrac{1}{2}\xi^2 + \tfrac{1}{3}\xi^3 - \tfrac{1}{4}\xi^4 + \cdots\right) \\
&= \tfrac{3}{8}\xi^4 + \mathcal{O}\!\left(\xi^5\right).
\end{aligned}
$$

◇

Nothing, unfortunately, could be further from good numerical practice than to assess a multistep method solely – or primarily – in terms of its order. Thus, let us consider the two-step implicit scheme

$$\boldsymbol{y}_{n+2} - 3\boldsymbol{y}_{n+1} + 2\boldsymbol{y}_n = h\left[\tfrac{13}{12}\boldsymbol{f}(t_{n+2}, \boldsymbol{y}_{n+2}) - \tfrac{5}{3}\boldsymbol{f}(t_{n+1}, \boldsymbol{y}_{n+1}) - \tfrac{5}{12}\boldsymbol{f}(t_n, \boldsymbol{y}_n)\right]. \quad (2.12)$$

It is easy to ascertain that the order of (2.12) is two. Encouraged by this – and not being very ambitious – we attempt to use this method to solve numerically the exceedingly simple equation $y' \equiv 0$, $y(0) = 1$. A single step reads $y_{n+2} - 3y_{n+1} + 2y_n = 0$, a recurrence relation whose general solution is $y_n = c_1 + c_2 2^n$, $n = 0, 1, \ldots$, where $c_1, c_2 \in \mathbb{R}$ are arbitrary. Suppose that $c_2 \neq 0$; we need both y_0 and y_1 to launch time-stepping and it is trivial to verify that $c_2 \neq 0$ is equivalent to $y_1 \neq y_0$. It is easy to prove that the method fails to converge. Thus, choose $t > 0$ and let $h \to 0$ so that $nh \to t$. Obviously, $n \to \infty$, and this implies that $|y_n| \to \infty$, which is far from the exact value $y(t) \equiv 1$.

Figure 2.2   The breakdown in the numerical solution of $y' = -y$, $y(0) = 1$, by a nonconvergent numerical scheme, showing how the situation worsens with decreasing step-size.

The failure in convergence does not require, realistically, that $c_2 \neq 0$ be induced by y_1. Any calculation on a real computer introduces roundoff error which, sooner or later, is bound to render $c_2 \neq 0$ and bring about a geometric growth in the error.

Needless to say, a method that cannot integrate the simplest possible ODE with any measure of reliability should not be used for more substantial computational ends. Nontrivial order is not sufficient to ensure convergence! The need thus arises for a criterion that allows us to discard bad methods and narrow the field down to convergent multistep schemes.

$\diamond$ **Failure to converge**   Suppose that the linear equation $y' = -y$, $y(0) = 1$, is solved by a two-step method with $\rho(w) = w^2 - 2.01w + 1.01$, $\sigma(w) = 0.995w - 1.005$. As will be soon evident, this method also fails the convergence criterion, although not by a wide margin! Figure 2.2 displays three solution trajectories, for progressively decreasing step-sizes $h = \frac{1}{10}, \frac{1}{20}, \frac{1}{40}$. In all instances, in its early stages the solution perfectly resembles the decaying exponential, but after a while small perturbations grow at an increasing pace and render the computation meaningless. It is a characteristic of nonconvergent methods that decreasing the step-size actually makes matters worse! $\diamond$

We say that a polynomial obeys the *root condition* if all its zeros reside in the closed complex unit disc and all its zeros of unit modulus are simple.

Theorem 2.2 (The Dahlquist equivalence theorem)   *Suppose that the error in the starting values $\boldsymbol{y}_1, \boldsymbol{y}_2, \ldots, \boldsymbol{y}_{s-1}$ tends to zero as $h \to 0+$. The multistep method*

(2.8) *is convergent if and only if it is of order $p \geq 1$ and the polynomial ρ obeys the root condition.*   ∎

It is important to make crystal clear that convergence is not simply another attribute of a numerical method, to be weighed alongside its other features. If a method is not convergent – and regardless of how attractive it may look – **do not use it!**

Theorem 2.2 allows us to discard method (2.12) without further ado, since $\rho(w) = (w-1)(w-2)$ violates the root condition. Of course, this method is contrived and, even were it convergent, it is doubtful whether it would have been of much interest. However, more 'respectable' methods fail the convergence test. For example, the method

$$\boldsymbol{y}_{n+3} + \tfrac{27}{11}\boldsymbol{y}_{n+2} - \tfrac{27}{11}\boldsymbol{y}_{n+1} - \boldsymbol{y}_n$$
$$= h\left[\tfrac{3}{11}\boldsymbol{f}(t_{n+3},\boldsymbol{y}_{n+3}) + \tfrac{27}{11}\boldsymbol{f}(t_{n+2},\boldsymbol{y}_{n+2}) + \tfrac{27}{11}\boldsymbol{f}(t_{n+1},\boldsymbol{y}_{n+1}) + \tfrac{3}{11}\boldsymbol{f}(t_n,\boldsymbol{y}_n)\right]$$

is of order six; it is the *only* three-step method that attains this order! Unfortunately,

$$\rho(w) = (w-1)\left(w + \frac{19+4\sqrt{15}}{11}\right)\left(w + \frac{19-4\sqrt{15}}{11}\right)$$

and the root condition fails. On the other hand, note that Adams–Bashforth methods are safe for all $s \geq 1$, since $\rho(w) = w^{s-1}(w-1)$.

◇ **Analysis and algebraic conditions**   Theorem 2.2 demonstrates a state of affairs that prevails throughout mathematical analysis. Thus, we desire to investigate an *analytic* condition, e.g. whether a differential equation has a solution, whether a continuous dynamical system is asymptotically stable, whether a numerical method converges. By their very nature, analytic concepts involve infinite processes and continua, hence one can expect analytic conditions to be difficult to verify, to the point of unmanageability. For all we know, the human brain (exactly like a digital computer) might be essentially an algebraic machine. It is thus an important goal in mathematical analysis to search for equivalent *algebraic* conditions. The Dahlquist equivalence theorem is a remarkable example of this: everything essentially reduces to the determination whether zeros of a polynomial reside in a unit disc, and this can be determined in a finite number of algebraic operations! In the course of this book we will encounter numerous other examples of this state of affairs. Cast your mind back to basic infinitesimal calculus and you are bound to recall further instances of analytic problems being rendered into an algebraic language.   ◇

The multistep method (2.8) has $2s+1$ parameters. Had order been the sole consideration, we could have utilized all the available degrees of freedom to maximize it. The outcome, an (implicit) s-step method of order $2s$, is unfortunately not convergent for $s \geq 3$ (we have already seen the case $s = 3$). In general, it is possible to prove that the maximal order of a convergent s-step method (2.8) is at most $2\lfloor (s+2)/2 \rfloor$ for implicit schemes and just s for explicit ones; this is known as *the Dahlquist first barrier*.

The usual practice is to employ orders $s + 1$ and s for s-step implicit and explicit methods respectively. An easy procedure for constructing such schemes is as follows. Choose an arbitrary s-degree polynomial ρ that obeys the root condition and such that $\rho(1) = 0$ (according to (2.11), $\rho(1) = \sum a_m = 0$ is necessary for order $p \geq 1$). Dividing the order condition (2.10) by $\ln w$ we obtain

$$\sigma(w) = \frac{\rho(w)}{\ln w} + \mathcal{O}(|w - 1|^p). \tag{2.13}$$

(Note that division by $\ln w$ shaves off a power of $|w - 1|$ and that the singularity at $w = 1$ in the numerator and the denominator is removable.) Suppose first that $p = s + 1$ and no restrictions are placed on σ. We expand the fraction in (2.13) into a Taylor series about $w = 1$ and let σ be the s-degree polynomial that matches the series up to $\mathcal{O}(|w - 1|^{s+1})$. The outcome is a convergent, s-step method of order $s + 1$. Likewise, to obtain an explicit method of order s, we let σ be an $(s - 1)$-degree polynomial (to force $b_m = 0$) that matches the series up to $\mathcal{O}(|w - 1|^s)$.

Let us, for example, choose $s = 2$ and $\rho(w) = w^2 - w$. Letting, as before, $\xi = w - 1$, we have

$$
\begin{aligned}
\frac{\rho(w)}{\ln w} &= \frac{\xi + \xi^2}{\xi - \frac{1}{2}\xi^2 + \frac{1}{3}\xi^3 + \mathcal{O}(\xi^4)} = \frac{1 + \xi}{1 - \frac{1}{2}\xi + \frac{1}{3}\xi^2} + \mathcal{O}(\xi^3) \\
&= (1 + \xi)\left(1 + \frac{1}{2}\xi - \frac{1}{12}\xi^2\right) + \mathcal{O}(\xi^3) = 1 + \frac{3}{2}\xi + \frac{5}{12}\xi^2 + \mathcal{O}(\xi^3).
\end{aligned}
$$

Thus, for quadratic σ and order 3 we truncate, obtaining

$$\sigma(w) = 1 + \frac{3}{2}(w - 1) + \frac{5}{12}(w - 1)^2 = -\frac{1}{12} + \frac{2}{3}w + \frac{5}{12}w^2,$$

whereas in the explicit case σ is linear, $p = 2$, and we recover, unsurprisingly, the Adams–Bashforth scheme (2.6).

The choice $\rho(w) = w^{s-1}(w - 1)$ is associated with *Adams* methods. We have already seen the explicit Adams–Bashforth schemes, whilst their implicit counterparts bear the name of Adams–Moulton methods. However, provided that we wish to maximize the order subject to convergence, without placing any extra constraints on the multistep method, Adams schemes are the most reasonable choice. After all, if – as implied in the statement of Theorem 2.2 – large zeros of ρ are bad, it makes perfect sense to drive as many zeros as we can to the origin!

2.3   Backward differentiation formulae

Classical texts in numerical analysis present several distinct families of multistep methods. For example, letting $\rho(w) = w^{s-2}(w^2 - 1)$ leads to s-order explicit *Nystrom* methods and and to implicit *Milne* methods of order $s + 1$ (cf. Exercise 2.3). However, in a well-defined yet important situation, certain multistep methods are significantly better than other schemes of the type (2.8). These are the *backward differentiation formulae* (BDFs).

An s-order, s-step method is said to be a BDF if $\sigma(w) = \beta w^s$ for some $\beta \in \mathbb{R} \setminus \{0\}$.

Lemma 2.3   *For a BDF we have*

$$\beta = \left(\sum_{m=1}^{s} \frac{1}{m} \right)^{-1} \qquad and \qquad \rho(w) = \beta \sum_{m=1}^{s} \frac{1}{m} w^{s-m}(w-1)^m. \tag{2.14}$$

Proof   The order being $p = s$, (2.10) implies that

$$\rho(w) - \beta w^s \ln w = \mathcal{O}\left(|w - 1|^{s+1}\right), \qquad w \to 1.$$

We substitute $v = w^{-1}$, hence

$$v^s \rho(v^{-1}) = -\beta \ln v + \mathcal{O}\left(|v - 1|^{s+1}\right), \qquad v \to 1.$$

Since

$$\ln v = \ln[1 + (v - 1)] = \sum_{m=1}^{s} \frac{(-1)^{m-1}}{m}(v-1)^m + \mathcal{O}\left(|v - 1|^{s+1}\right),$$

we deduce that

$$v^s \rho(v^{-1}) = \beta \sum_{m=1}^{s} \frac{(-1)^m}{m}(v-1)^m.$$

Therefore

$$\begin{aligned}
\rho(w) &= \beta v^{-s} \sum_{m=1}^{s} \frac{(-1)^m}{m}(v-1)^m = \beta \sum_{m=1}^{s} \frac{(-1)^m}{m} w^s(w^{-1} - 1)^m \\
&= \beta \sum_{m=1}^{s} \frac{1}{m} w^{s-m}(w-1)^m.
\end{aligned}$$

To complete the proof of (2.14), we need only to derive the explicit form of β. It follows at once by imposing the normalization condition $a_s = 1$ on the polynomial ρ. ∎

The simplest BDF has been already encountered in Chapter 1: when $s = 1$ we recover the backward Euler method (1.15). The next two BDFs are

$$s = 2: \qquad \boldsymbol{y}_{n+2} - \tfrac{4}{3}\boldsymbol{y}_{n+1} + \tfrac{1}{3}\boldsymbol{y}_n = \tfrac{2}{3}h\boldsymbol{f}(t_{n+2}, \boldsymbol{y}_{n+2}), \tag{2.15}$$

$$s = 3: \qquad \boldsymbol{y}_{n+3} - \tfrac{18}{11}\boldsymbol{y}_{n+2} + \tfrac{9}{11}\boldsymbol{y}_{n+1} - \tfrac{2}{11}\boldsymbol{y}_n = \tfrac{6}{11}h\boldsymbol{f}(t_{n+3}, \boldsymbol{y}_{n+3}). \tag{2.16}$$

Their derivation is trivial; for example, (2.16) follows by letting $s = 3$ in (2.14). Therefore

$$\beta = \frac{1}{1 + \frac{1}{2} + \frac{1}{3}} = \frac{6}{11}$$

and

$$\rho(w) = \tfrac{6}{11}\left[w^2(w-1) + \tfrac{1}{2}w(w-1)^2 + \tfrac{1}{3}(w-1)^3 \right] = w^3 - \tfrac{18}{11}w^2 + \tfrac{9}{11}w - \tfrac{2}{11}.$$

Figure 2.3   The norm of the numerical solution of (2.17) by the Adams–Bashforth method (2.6) for $h = 0.027$ (solid line) and $h = 0.0275$ (broken line).

Since BDFs are derived by specifying σ, we cannot be assured that the polynomial ρ of (2.14) obeys the root condition. In fact, the root condition fails for all but a few such methods.

Theorem 2.4   *The polynomial* (2.14) *obeys the root condition and the underlying BDF method is convergent if and only if* $1 \leq s \leq 6$.   ∎

Fortunately, the 'good' range of s is sufficient for all practical considerations.

Underscoring the importance of BDFs, we present a simple example that demonstrates the limitations of Adams schemes; we hasten to emphasize that this is by way of a trailer for our discussion of *stiff* ODEs in Chapter 4.

Let us consider the linear ODE system

$$\boldsymbol{y}' = \begin{bmatrix} -20 & 10 & 0 & \cdots & 0 \\ 10 & -20 & \ddots & \ddots & \vdots \\ 0 & \ddots & \ddots & \ddots & 0 \\ \vdots & \ddots & \ddots & -20 & 10 \\ 0 & \cdots & 0 & 10 & -20 \end{bmatrix} \boldsymbol{y}, \qquad \boldsymbol{y}(0) = \begin{bmatrix} 1 \\ 1 \\ \vdots \\ 1 \end{bmatrix}. \qquad (2.17)$$

We will encounter in this book numerous instances of similar systems; (2.17) is a handy paradigm for many linear ODEs that occur in the context of discretization of the partial differential equations of evolution.

Figure 2.3 displays the Euclidean norm of the solution of (2.17) by the second-order Adams-Bashforth method (2.6), with two (slightly) different step-sizes, $h = 0.027$ (the solid line) and $h = 0.0275$ (the broken line). The solid line is indistinguishable in the figure from the norm of the true solution, which approaches zero as $t \to \infty$. Not so the norm for $h = 0.0275$: initially, it shadows pretty well the correct value but, after a while, it runs away. The whole qualitative picture is utterly false! And, by the way, things rapidly get considerably worse when h is increased: for $h = 0.028$ the norm reaches 2.5×10^4, while for $h = 0.029$ it shoots to 1.3×10^{11}.

What is the mechanism that degrades the numerical solution and renders it so sensitive to small changes in h? At the moment it suffices to state that the quality of local approximation (which we have quantified in the concept of 'order') is not to blame; taking the third-order scheme (2.7) in place of the current method would have only made matters worse. On the other hand, were we to attempt the solution of this ODE with (2.15), say, and with any $h > 0$, the norm would have tended to zero in tandem with the exact solution. In other words, methods such as BDFs are singled out by a favourable property that makes them the methods of choice for important classes of ODEs. Much more will be said about this in Chapter 4.

Comments and bibliography

There are several ways of introducing the theory of multistep methods. Traditional texts have emphasized the derivation of schemes by various interpolation formulae. The approach of Section 2.1 harks back to this approach, as does the name 'backward differentiation formula'. Other books derive order conditions by sheer brute force, requiring that the multistep formula (2.8) be exact for all polynomials of degree p, since this is equivalent to order p. This can be expressed as a linear system of $p + 1$ equations in the $2s + 1$ unknowns $a_0, a_1, \ldots, a_{s-1}, b_0, b_1, \ldots, b_s$. A solution of this system yields a multistep method of requisite order (of course, we must check it for convergence!), although this procedure does not add much to our understanding of such methods.[1] Linking order with an approximation of the logarithm, *pace* Theorem 2.1, elucidates matters on a considerably more profound level. This can be shown by the following hand-waving argument.

Given an analytic function g, say, and a number $h > 0$, we denote $g_n^{(k)} = g^{(k)}(t_0 + hn)$, $k, n = 0, 1, \ldots$, and define two operators that map such 'grid functions' into themselves, the *shift* operator $\mathcal{E} g_n^{(k)} := g_{n+1}^{(k)}$ and the *differential* operator $\mathcal{D} g_n^{(k)} := g_n^{(k+1)}$, $k, n = 0, 1, \ldots$ (cf. Section 7.1). Expanding in a Taylor series about $t_0 + nh$,

$$\mathcal{E} g_n^{(k)} = \sum_{\ell=0}^{\infty} \frac{1}{\ell!} g_n^{(k+\ell)} h^\ell = \left(\sum_{\ell=0}^{\infty} \frac{1}{\ell!} (h\mathcal{D})^\ell \right) g_n^{(k)}, \qquad k, n = 0, 1, \ldots.$$

Since this is true for every analytic g with a radius of convergence exceeding h, it follows that, at least formally, $\mathcal{E} = \exp(h\mathcal{D})$. The exponential of the operator, exactly like the more familiar matrix exponential, is defined by a Taylor series.

The above argument can be tightened at the price of some mathematical sophistication. The main problem with naively defining $\mathcal{E}$ as the exponential of $h\mathcal{D}$ is that, in the standard spaces beloved by mathematicians, $\mathcal{D}$ is not a *bounded* linear operator. To recover boundedness we need to resort to a more exotic space.

Let $\mathbb{U} \subseteq \mathbb{C}$ be an open connected set and denote by $\mathcal{A}(\mathbb{U})$ the vector space of analytic functions defined in $\mathbb{U}$. The sequence $\{f_n\}_{n=0}^{\infty}$, where $f_n \in \mathcal{A}(\mathbb{U})$, $n = 0, 1, \ldots$, is said to converge to f *locally uniformly* in $\mathcal{A}(\mathbb{U})$ if $f_n \to f$ uniformly in every compact (i.e., closed and

---

[1]Though low on insight and beauty, brute force techniques are occasionally useful in mathematics just as in more pedestrian walks of life.

bounded) subset of $\mathbb{U}$. It is possible to prove that there exists a metric (a 'distance function') on $\mathcal{A}(\mathbb{U})$ that is consistent with locally uniform convergence and to demonstrate, using the Cauchy integral formula, that the operator $\mathcal{D}$ is a bounded linear operator on $\mathcal{A}(\mathbb{U})$. Hence, so is $\mathcal{E} = \exp(h\mathcal{D})$ and we can justify a definition of the exponential via a Taylor series.

The correspondence between the shift operator and the differential operator is fundamental to the numerical solution of ODEs – after all, a differential equation provides us with the action of $\mathcal{D}$, as well as with a function value at a single point, and the act of numerical solution is concerned with (repeatedly) approximating the action of $\mathcal{E}$. Equipped with our new-found knowledge, we should realize that approximation of the exponential plays (often behind the scenes) a crucial role in designing numerical methods. Later, in Chapter 4, approximants of the exponential, this time with a matrix argument, will be crucial to our understanding of important stability issues, whereas the present theme forms the basis for our exposition of finite differences in Chapter 7.

Applying the operatorial approach to multistep methods, we note at once that

$$\sum_{m=0}^{s} a_m \boldsymbol{y}(t_{n+m}) - h \sum_{m=0}^{s} b_m \boldsymbol{y}'(t_{n+m}) = \left(\sum_{m=0}^{s} a_m \mathcal{E}^m - h\mathcal{D} \sum_{m=0}^{s} b_m \mathcal{E}^m \right) \boldsymbol{y}(t_n)$$

$$= [\rho(\mathcal{E}) - h\mathcal{D}\sigma(\mathcal{E})]\, \boldsymbol{y}(t_n).$$

Note that $\mathcal{E}$ and $\mathcal{D}$ commute (since $\mathcal{E}$ is given in terms of a power series in $\mathcal{D}$), and this justifies the above formula. Moreover, $\mathcal{E} = \exp(h\mathcal{D})$ means that $h\mathcal{D} = \ln \mathcal{E}$, where the logarithm, again, is defined by means of a Taylor expansion (about the identity operator $\mathcal{I}$). This, in tandem with the observation that $\lim_{h \to 0+} \mathcal{E} = \mathcal{I}$, is the basis to an alternative 'proof' of Theorem 2.1 – a proof that can be made completely rigorous with little effort by employing the implicit function theorem.

The proof of the equivalence theorem (Theorem 2.2) and the establishment of the first barrier (cf. Section 2.2) by Germund Dahlquist, in 1956 and 1959 respectively, were important milestones in the history of numerical analysis. Not only are these results of a great intrinsic impact but they were also instrumental in establishing numerical analysis as a *bona fide* mathematical discipline and imparting a much-needed rigour to numerical thinking. It goes without saying that numerical analysis is not *just* mathematics. It is much more! Numerical analysis is first and foremost about the computation of mathematical models originating in science and engineering. It employs mathematics – and computer science – to an end. Quite often we use a computational algorithm because, although it lacks formal mathematical justification, our experience and intuition tell us that it is efficient and (hopefully) provides the correct answer. There is nothing wrong with this! However, as always in applied mathematics, we must bear in mind the important goal of casting our intuition and experience into a rigorous mathematical framework. Intuition is fallible and experience attempts to infer from incomplete data – mathematics is still the best tool of a computational scientist!

Modern texts in the numerical analysis of ODEs highlight the importance of a structured mathematical approach. The classic monograph of Henrici (1962) is still a model of clear and beautiful exposition and includes an easily digestible proof of the Dahlquist first barrier. Hairer *et al.* (1991) and Lambert (1991) are also highly recommended. In general, books on numerical ODEs fall into two categories: pre-Dahlquist and post-Dahlquist. The first set is nowadays of mainly historical and antiquarian significance.

We will encounter multistep methods again in Chapter 4. As has been already seen in Section 2.3, convergence and reasonable order are far from sufficient for the successful computation of many ODEs. The solution of such equations, commonly termed *stiff*, requires numerical methods with superior stability properties.

Much of the discussion of multistep methods centres upon their implementation. The present chapter avoids any talk of implementation issues – solution of the (mostly nonlinear) algebraic equations associated with implicit methods, error and step-size control, the choice of the starting values $y_1, y_2, \ldots, y_{s-1}$. Our purpose has been an introduction to multistep schemes and their main properties (convergence, order), as well as a brief survey of the most distinguished members of the multistep methods menagerie. We defer the discussion of implementation issues to Chapters 5 and 6.

Hairer, E., Nørsett, S.P. and Wanner, G. (1991), *Solving Ordinary Differential Equations I: Nonstiff Problems* (2nd ed.), Springer-Verlag, Berlin.

Henrici, P. (1962), *Discrete Variable Methods in Ordinary Differential Equations*, Wiley, New York.

Lambert, J.D. (1991), *Numerical Methods for Ordinary Differential Systems*, Wiley, London.

Exercises

2.1 Derive explicitly the three-step and four-step Adams–Moulton methods and the three-step Adams-Bashforth method.

2.2 Let $\eta(z, w) = \rho(w) - z\sigma(w)$.

a Demonstrate that the multistep method (2.8) is of order p if and only if
$$\eta(z, e^z) = cz^{p+1} + \mathcal{O}(z^{p+2}), \qquad z \to 0,$$
for some $c \in \mathbb{R} \setminus \{0\}$.

b Prove that, subject to $\partial\eta(0, 1)/\partial w \neq 0$, there exists in a neighbourhood of the origin an analytic function $w_1(z)$ such that $\eta(z, w_1(z)) = 0$ and
$$w_1(z) = e^z - c\left(\frac{\partial\eta(0, 1)}{\partial w}\right)^{-1} z^{p+1} + \mathcal{O}(z^{p+2}), \qquad z \to 0. \tag{2.18}$$

c Show that (2.18) is true if the underlying method is convergent. [*Hint: Express $\partial\eta(0, 1)/\partial w$ in terms of the polynomial ρ.*]

2.3 Instead of (2.3), consider the identity
$$y(t_{n+s}) = y(t_{n+s-2}) + \int_{t_{n+s-2}}^{t_{n+s}} f(\tau, y(\tau))\, d\tau.$$

a Replace $f(\tau, y(\tau))$ by the interpolating polynomial p from Section 2.1 and substitute y_{n+s-2} in place of $y(t_{n+s-2})$. Prove that the resultant explicit *Nystrom* method is of order $p = s$.

b Derive the two-step Nystrom method in a closed form by using the above approach.

c Find the coefficients of the two-step and three-step Nystrom methods by noticing that $\rho(w) = w^{s-2}(w^2 - 1)$ and evaluating σ from (2.13).

d Derive the two-step, third-order implicit *Milne* method, again letting $\rho(w) = w^{s-2}(w^2 - 1)$ but allowing σ to be of degree s.

2.4   Determine the order of the three-step method

$$\boldsymbol{y}_{n+3} - \boldsymbol{y}_n = h\left[\tfrac{3}{8}\boldsymbol{f}(t_{n+3}, \boldsymbol{y}_{n+3}) + \tfrac{9}{8}\boldsymbol{f}(t_{n+2}, \boldsymbol{y}_{n+2}) + \tfrac{9}{8}\boldsymbol{f}(t_{n+1}, \boldsymbol{y}_{n+1}) \right.$$
$$\left. + \tfrac{3}{8}\boldsymbol{f}(t_n, \boldsymbol{y}_n)\right],$$

the *three-eighths* scheme. Is it convergent?

2.5*   By solving a three-term recurrence relation, calculate analytically the sequence of values $y_2, y_3, y_4, \ldots$ that is generated by the *midpoint rule*

$$\boldsymbol{y}_{n+2} = \boldsymbol{y}_n + 2h\boldsymbol{f}(t_{n+1}, \boldsymbol{y}_{n+1})$$

when it is applied to the differential equation $y' = -y$. Starting from the values $y_0 = 1$, $y_1 = 1-h$, show that the sequence diverges as $n \to \infty$. Recall, however, from Theorem 2.1 that the root condition, in tandem with order $p \geq 1$ and suitable starting conditions, imply convergence to the true solution in a *finite* interval as $h \to 0+$. Prove that this implementation of the midpoint rule is consistent with the above theorem. [*Hint: Express the roots of the characteristic polynomial of the recurrence relation as* $\exp\{\pm \sinh^{-1} h\}$.]

2.6   Show that the multistep method

$$\boldsymbol{y}_{n+3} + \alpha_2 \boldsymbol{y}_{n+2} + \alpha_1 \boldsymbol{y}_{n+1} + \alpha_0 \boldsymbol{y}_n = h[\beta_2 \boldsymbol{f}(t_{n+2}, \boldsymbol{y}_{n+2})$$
$$+ \beta_1 \boldsymbol{f}(t_{n+1}, \boldsymbol{y}_{n+1}) + \beta_0 \boldsymbol{f}(t_n, \boldsymbol{y}_n)]$$

is fourth order only if $\alpha_0 + \alpha_2 = 8$ and $\alpha_1 = -9$. Hence deduce that this method cannot be both fourth order and convergent.

2.7   Prove that the BDFs (2.15) and (2.16) are convergent.

2.8   Find the explicit form of the BDF for $s = 4$.

2.9   An s-step method with $\sigma(w) = w^{s-1}(w + 1)$ and order s might be superior to a BDF in certain situations.

a Find a general formula for ρ and β, along the lines of (2.14).

b Derive explicitly such methods for $s = 2$ and $s = 3$.

c Are the last two methods convergent?

3

Runge–Kutta methods

3.1 Gaussian quadrature

The exact solution of the trivial differential equation

$$y' = f(t), \quad t \geq t_0, \qquad y(t_0) = y_0,$$

whose right-hand side is independent of y, is $y_0 + \int_{t_0}^{t} f(\tau)\, d\tau$. Since a very rich theory and powerful methods exist to compute integrals numerically, it is only natural to wish to utilize them in the numerical solution of general ODEs

$$\boldsymbol{y}' = \boldsymbol{f}(t, \boldsymbol{y}), \quad t \geq t_0, \qquad \boldsymbol{y}(t_0) = \boldsymbol{y}_0, \tag{3.1}$$

and this is the rationale behind *Runge–Kutta* methods. Before we debate such methods, it is thus fit and proper to devote some attention to the numerical calculation of integrals.

It is usual to replace an integral with a finite sum, a procedure known as *quadrature*. Specifically, let ω be a nonnegative function acting in the interval (a, b), such that

$$0 < \int_a^b \omega(\tau)\, d\tau < \infty, \qquad \left| \int_a^b \tau^j \omega(\tau)\, d\tau \right| < \infty, \quad j = 1, 2, \ldots;$$

ω is dubbed the *weight function*. We approximate as follows:

$$\int_a^b f(\tau)\omega(\tau)\, d\tau \approx \sum_{j=1}^{\nu} b_j f(c_j), \tag{3.2}$$

where the numbers $b_1, b_2, \ldots, b_\nu$ and $c_1, c_2, \ldots, c_\nu$, which are independent of the function f (but, in general, depend upon ω, a and b), are called the quadrature *weights* and *nodes,* respectively. Note that we do not require a and b in (3.2) to be bounded; the choices $a = -\infty$ or $b = +\infty$ are perfectly acceptable. Of course, we stipulate $a < b$.

How good is the approximation (3.2)? Suppose that the quadrature matches the integral exactly whenever f is an arbitrary polynomial of degree $p - 1$. It is then easy to prove, e.g. by using the Peano kernel theorem (cf. A.2.2.6), that, for every function f with p smooth derivatives,

$$\left| \int_a^b f(\tau)\omega(\tau)\, d\tau - \sum_{j=1}^{\nu} b_j f(c_j) \right| \leq c \max_{a \leq t \leq b} \left| f^{(p)}(t) \right|,$$

where the constant $c > 0$ is independent of f. Such a quadrature formula is said to be of *order p*.

We denote the set of all real polynomials of degree m by $\mathbb{P}_m$. Thus, (3.2) is of order p if it is exact for every $f \in \mathbb{P}_{p-1}$.

Lemma 3.1   *Given any distinct set of nodes $c_1, c_2, \ldots, c_\nu$, it is possible to find a unique set of weights $b_1, b_2, \ldots, b_\nu$ such that the quadrature formula (3.2) is of order $p \geq \nu$.*

Proof   Since $\mathbb{P}_{\nu-1}$ is a linear space, it is necessary and sufficient for order ν that (3.2) is exact for elements of an arbitrary basis of $\mathbb{P}_{\nu-1}$. We choose the simplest such basis, namely $\{1, t, t^2, \ldots, t^{\nu-1}\}$, and the order conditions then read

$$\sum_{j=1}^{\nu} b_j c_j^m = \int_a^b \tau^m \omega(\tau) \, d\tau, \qquad m = 0, 1, \ldots, \nu - 1. \tag{3.3}$$

This is a system of ν equations in the ν unknowns $b_1, b_2, \ldots, b_\nu$, whose matrix, the nodes being distinct, is a nonsingular *Vandermonde* matrix (cf. A.1.2.5). Thus, the system possesses a unique solution and we recover a quadrature of order $p \geq \nu$.   ∎

The weights $b_1, b_2, \ldots, b_\nu$ can be derived explicitly with little extra effort and we make use of this in the sequel in (3.14). Let

$$p_j(t) = \prod_{\substack{k=1 \\ k \neq j}}^{\nu} \frac{t - c_k}{c_j - c_k}, \qquad j = 1, 2, \ldots, \nu,$$

be Lagrange polynomials (cf. A.2.2.3). Because

$$\sum_{j=1}^{\nu} p_j(t) g(c_j) = g(t)$$

for every polynomial g of degree ν, it follows that

$$\sum_{j=1}^{\nu} \int_0^1 p_j(\tau) \omega(\tau) \, d\tau \, c_j^m = \int_0^1 \left[\sum_{j=1}^{\nu} p_j(\tau) c_j^m \right] \omega(\tau) \, d\tau = \int_0^1 \tau^m \omega(\tau) \, d\tau$$

for every $m = 0, 1, \ldots, \nu - 1$. Therefore

$$b_j = \int_0^1 p_j(\tau) \omega(\tau) \, d\tau, \qquad j = 1, 2, \ldots, \nu,$$

is the solution of (3.3).

A natural inclination is to choose quadrature nodes that are equispaced in $[a, b]$, and this leads to the so-called *Newton–Cotes* methods. This procedure, however, falls far short of the optimal; by making an adroit choice of $c_1, c_2, \ldots, c_\nu$, we can, in fact, double the order to 2ν.

Each weight function ω determines an *inner product* (cf. A.1.3.1) in the interval (a, b), namely

$$\langle f, g \rangle := \int_a^b f(\tau)g(\tau)\omega(\tau)\,\mathrm{d}\tau,$$

whose domain is the set of all functions f, g such that

$$\int_a^b [f(\tau)]^2 \omega(\tau)\,\mathrm{d}\tau, \int_a^b [g(\tau)]^2 \omega(\tau)\,\mathrm{d}\tau < \infty.$$

We say that $p_m \in \mathbb{P}_m$, $p_m \not\equiv 0$, is an *m*th *orthogonal polynomial* (with respect to the weight function ω) if

$$\langle p_m, \hat{p} \rangle = 0, \quad \text{for every} \quad \hat{p} \in \mathbb{P}_{m-1}. \tag{3.4}$$

Orthogonal polynomials are not unique, since we can always multiply p_m by a nonzero constant without violating (3.4). However, it is easy to demonstrate that *monic* orthogonal polynomials are unique. (The coefficient of the highest power of t in a monic polynomial equals one.) Suppose that both p_m and $\tilde{p}_m$ are monic m-degree orthogonal polynomials with respect to the same weight function. Then $p_m - \tilde{p}_m \in \mathbb{P}_{m-1}$ and, by (3.4), $\langle p_m, p_m - \tilde{p}_m \rangle = \langle \tilde{p}_m, p_m - \tilde{p}_m \rangle = 0$. We thus deduce from the linearity of the inner product that $\langle p_m - \tilde{p}_m, p_m - \tilde{p}_m \rangle = 0$, and this is possible, *pace* Appendix subsection A.1.3.1, only if $\tilde{p}_m = p_m$.

Orthogonal polynomials occur in many areas of mathematics; a brief list includes approximation theory, statistics, representation of groups, theory of (ordinary and partial) differential equations, functional analysis, quantum groups, coding theory, mathematical physics and, last but not least, numerical analysis.

◇ **Classical orthogonal polynomials**   Three families of weights give rise to *classical orthogonal polynomials*.

Let $a = -1$, $b = 1$ and $\omega(t) = (1-t)^\alpha(1+t)^\beta$, where $\alpha, \beta > -1$. The underlying orthogonal polynomials are known as *Jacobi* polynomials $P_m^{(\alpha,\beta)}$. We single out for special attention the *Legendre* polynomials P_m, which correspond to $\alpha = \beta = 0$, and the *Chebyshev* polynomials T_m, associated with the choice $\alpha = \beta = -\frac{1}{2}$. Note that for $\min\{\alpha, \beta\} < 0$ the weight function has a singularity at the endpoints ± 1. There is nothing wrong with that, provided ω is integrable in $[0, 1]$; but this is exactly the reason we require $\alpha, \beta > -1$.

The other two 'classics' are the *Laguerre* and *Hermite* polynomials. The Laguerre polynomial $L_m^{(\alpha)}$ is orthogonal with the weight function $\omega(t) = t^\alpha \mathrm{e}^{-t}$, $(a, b) = (0, \infty)$, $\alpha > -1$, whereas the Hermite polynomial H_m is orthogonal in $(a, b) = \mathbb{R}$ with the weight function $\omega(t) = \mathrm{e}^{-t^2}$.

Why are classical orthogonal polynomials so named? Firstly, they have been very extensively studied and occur in a very wide range of applications. Secondly, it is possible to prove that they are singled out by several properties that, in a well-defined sense, render them the 'simplest' orthogonal polynomials. For example – and do not try to prove this on your own! – $P_m^{(\alpha,\beta)}$, $L_m^{(\alpha)}$ and H_m are the only orthogonal polynomials whose derivatives are also orthogonal with some weight function.   ◇

The theory of orthogonal polynomials is replete with beautiful results which, perhaps regrettably, we do not require in this volume. However, one morsel of information, germane to the understanding of quadrature, is about the location of zeros of orthogonal polynomials.

Lemma 3.2   *All m zeros of an orthogonal polynomial p_m reside in the interval (a, b) and they are simple.*

Proof   Since

$$\int_a^b p_m(\tau)\omega(\tau)\,\mathrm{d}\tau = \langle p_m, 1\rangle = 0$$

and $\omega \geq 0$, it follows that p_m changes sign at least once in (a, b). Let us thus denote by $x_1, x_2, \ldots, x_k$ all the points in (a, b) where p_m changes sign. We already know that $k \geq 1$. Let us assume that $k \leq m - 1$ and set

$$q(t) := \prod_{j=1}^k (t - x_j) = \sum_{i=0}^k q_i t^i.$$

Therefore p_m changes sign in (a, b) at exactly the same points as q and the product $p_m q$ does not change sign there at all. The weight function being nonnegative and $p_m q \not\equiv 0$, we deduce that

$$\int_a^b p_m(\tau)q(\tau)\omega(\tau)\,\mathrm{d}\tau \neq 0.$$

On the other hand, the orthogonality condition (3.4) and the linearity of the inner product imply that

$$\int_a^b p_m(\tau)q(\tau)\,\mathrm{d}\tau = \sum_{i=0}^k q_i\langle p_m, t^i\rangle = 0,$$

because $k \leq m - 1$. This is a contradiction and we conclude that $k \geq m$. Since each sign-change of p_m is a zero of the polynomial and, according to the fundamental theorem of algebra, each $\hat{p} \in \mathbb{P}_m \setminus \mathbb{P}_{m-1}$ has exactly m zeros in $\mathbb{C}$, we deduce that p_m has exactly m simple zeros in (a, b). ∎

Theorem 3.3   *Let $c_1, c_2, \ldots, c_\nu$ be the zeros of p_ν and let $b_1, b_2, \ldots, b_\nu$ be the solution of the Vandermonde system (3.3). Then*

(i)   *The quadrature method (3.2) is of order 2ν;*

(ii)   *No other quadrature can exceed this order.*

Proof   Let $\hat{p} \in \mathbb{P}_{2\nu-1}$. Applying the Euclidean algorithm to the pair $\{\hat{p}, p_\nu\}$ we deduce that there exist $q, r \in \mathbb{P}_{\nu-1}$ such that $\hat{p} = p_\nu q + r$. Therefore, according to (3.4),

$$\int_a^b \hat{p}(\tau)\omega(\tau)\,\mathrm{d}\tau = \langle p_\nu, q\rangle + \int_a^b r(\tau)\omega(\tau)\,\mathrm{d}\tau = \int_a^b r(\tau)\omega(\tau)\,\mathrm{d}\tau;$$

we recall that $\deg q \leq \nu - 1$. Moreover,

$$\sum_{j=1}^\nu b_j\hat{p}(c_j) = \sum_{j=1}^\nu b_j p_\nu(c_j)q(c_j) + \sum_{j=1}^\nu b_j r(c_j) = \sum_{j=1}^\nu b_j r(c_j)$$

because $p_\nu(c_j) = 0$, $j = 1, 2, \ldots, \nu$. Finally, $r \in \mathbb{P}_{\nu-1}$ and Lemma 3.1 imply

$$\int_a^b r(\tau)\omega(\tau)\,\mathrm{d}\tau = \sum_{j=1}^\nu b_j r(c_j).$$

We thus deduce that

$$\int_a^b \hat{p}(\tau)\omega(\tau)\,\mathrm{d}\tau = \sum_{j=1}^\nu b_j \hat{p}(c_j), \qquad \hat{p} \in \mathbb{P}_{2\nu-1},$$

and the quadrature formula is of order $p \geq 2\nu$.

To prove (ii) (and, incidentally, to affirm that $p = 2\nu$, thereby completing the proof of (i)) we assume that, for some choice of weights $b_1, b_2, \ldots, b_\nu$ and nodes $c_1, c_2, \ldots, c_\nu$, the quadrature formula (3.2) is of order $p \geq 2\nu+1$. In particular, it must then integrate exactly the polynomial

$$\hat{p}(t) := \prod_{i=1}^\nu (t - c_i)^2, \qquad \hat{p} \in \mathbb{P}_{2\nu}.$$

This, however, is impossible, since

$$\int_a^b \hat{p}(\tau)\omega(\tau)\,\mathrm{d}\tau = \int_a^b \left[\prod_{i=1}^\nu (\tau - c_i)\right]^2 \omega(\tau)\,\mathrm{d}\tau > 0,$$

while

$$\sum_{j=1}^\nu b_j \hat{p}(c_j) = \sum_{j=1}^\nu b_j \prod_{i=1}^\nu (c_j - c_i)^2 = 0.$$

The proof is complete. ∎

The optimal methods of the last theorem are commonly known as *Gaussian quadrature* formulae.

In the sequel we require a generalization of Theorem 3.3. Its proof is left as an exercise to the reader.

Theorem 3.4   *Let $r \in \mathbb{P}_\nu$ obey the orthogonality conditions*

$$\langle r, \hat{p} \rangle = 0 \quad \text{for every} \quad \hat{p} \in \mathbb{P}_{m-1}, \qquad \langle r, t^m \rangle \neq 0,$$

for some $m \in \{0, 1, \ldots, \nu\}$. We let $c_1, c_2, \ldots, c_\nu$ be the zeros of the polynomial r and choose $b_1, b_2, \ldots, b_\nu$ consistently with (3.3). The quadrature formula (3.2) has order $p = \nu + m$. ∎

3.2   Explicit Runge–Kutta schemes

How do we extend a quadrature formula to the ODE (3.1)? The obvious approach is to integrate from t_n to $t_{n+1} = t_n + h$:

$$\boldsymbol{y}(t_{n+1}) = \boldsymbol{y}(t_n) + \int_{t_n}^{t_{n+1}} \boldsymbol{f}(\tau, \boldsymbol{y}(\tau))\,\mathrm{d}\tau = \boldsymbol{y}(t_n) + h \int_0^1 \boldsymbol{f}(t_n + h\tau, \boldsymbol{y}(t_n + h\tau))\,\mathrm{d}\tau,$$

and to replace the second integral by a quadrature. The outcome might have been the 'method'

$$y_{n+1} = y_n + h \sum_{j=1}^{\nu} b_j f(t_n + c_j h, \, y(t_n + c_j h)), \qquad n = 0, 1, \ldots,$$

except that we do not know the value of y at the nodes $t_n + c_1 h, t_n + c_2, \ldots, t_n + c_\nu h$. We must resort to an approximation!

We denote our approximant of $y(t_n + c_j h)$ by $\boldsymbol{\xi}_j$, $j = 1, 2, \ldots, \nu$. To start with, we let $c_1 = 0$, since then the approximant is already provided by the former step of the numerical method, $\boldsymbol{\xi}_1 = y_n$. The idea behind *explicit Runge–Kutta (ERK)* methods is to express each $\boldsymbol{\xi}_j$, $j = 2, 3, \ldots, \nu$, by updating y_n with a linear combination of $f(t_n, \boldsymbol{\xi}_1)$, $f(t_n + hc_2, \boldsymbol{\xi}_2)$, $\ldots$, $f(t_n + c_{j-1} h, \boldsymbol{\xi}_{j-1})$. Specifically, we let

$$
\begin{aligned}
\boldsymbol{\xi}_1 &= y_n \\
\boldsymbol{\xi}_2 &= y_n + h a_{2,1} f(t_n, \boldsymbol{\xi}_1) \\
\boldsymbol{\xi}_3 &= y_n + h a_{3,1} f(t_n, \boldsymbol{\xi}_1) + h a_{3,2} f(t_n + c_2 h, \boldsymbol{\xi}_2) \\
&\;\;\vdots \\
\boldsymbol{\xi}_\nu &= y_n + h \sum_{i=1}^{\nu-1} a_{\nu,i} f(t_n + c_i h, \boldsymbol{\xi}_i) \\
y_{n+1} &= y_n + h \sum_{j=1}^{\nu} b_j f(t_n + c_j h, \boldsymbol{\xi}_j).
\end{aligned}
\tag{3.5}
$$

The matrix $A = (a_{j,i})_{j,i=1,2,\ldots,\nu}$, where missing elements are defined to be zero, is called the *RK matrix*, while

$$
\boldsymbol{b} = \begin{bmatrix} b_1 \\ b_2 \\ \vdots \\ b_\nu \end{bmatrix} \qquad \text{and} \qquad \boldsymbol{c} = \begin{bmatrix} c_1 \\ c_2 \\ \vdots \\ c_\nu \end{bmatrix}
$$

are the *RK weights* and *RK nodes* respectively. We say that (3.5) has ν *stages*. Confusingly, sometimes the $\boldsymbol{\xi}_j$ are called 'RK stages'; elsewhere this name is reserved for $f(t_n + c_j h, \boldsymbol{\xi}_j)$, $j = 1, 2, \ldots, s$. To avoid confusion, we henceforth desist from using the phrase 'RK stages'.

How should we choose the RK matrix? The most obvious way consists of expanding everything in sight into Taylor series about (t_n, y_n); but, in a naive rendition, this is of strictly limited utility. For example, let us consider the simplest nontrivial case, $\nu = 2$. Assuming sufficient smoothness of the vector function f, we have

$$
\begin{aligned}
f(t_n + c_2 h, \boldsymbol{\xi}_2) &= f(t_n + c_2 h, \, y_n + a_{2,1} h f(t_n, y_n)) \\
&= f(t_n, y_n) + h \left[c_2 \frac{\partial f(t_n, y_n)}{\partial t} + a_{2,1} \frac{\partial f(t_n, y_n)}{\partial y} f(t_n, y_n) \right] + \mathcal{O}(h^2) ;
\end{aligned}
$$

therefore the last equation in (3.5) becomes

$$\boldsymbol{y}_{n+1} = \boldsymbol{y}_n + h(b_1 + b_2)\boldsymbol{f}(t_n, \boldsymbol{y}_n) \tag{3.6}$$
$$+ h^2 b_2 \left[c_2 \frac{\partial \boldsymbol{f}(t_n, \boldsymbol{y}_n)}{\partial t} + a_{2,1} \frac{\partial \boldsymbol{f}(t_n, \boldsymbol{y}_n)}{\partial \boldsymbol{y}} \boldsymbol{f}(t_n, \boldsymbol{y}_n) \right] + \mathcal{O}(h^3).$$

We need to compare (3.6) with the Taylor expansion of the exact solution about the same point $(t_n, \boldsymbol{y}_n)$. The first derivative is provided by the ODE, whereas we can obtain $\boldsymbol{y}''$ by differentiating (3.1) with respect to t:

$$\boldsymbol{y}'' = \frac{\partial \boldsymbol{f}(t, \boldsymbol{y})}{\partial t} + \frac{\partial \boldsymbol{f}(t, \boldsymbol{y})}{\partial \boldsymbol{y}} \boldsymbol{f}(t, \boldsymbol{y}).$$

We denote the exact solution at t_{n+1}, subject to the initial condition $\boldsymbol{y}_n$ at t_n, by $\tilde{\boldsymbol{y}}$. Therefore

$$\tilde{\boldsymbol{y}}(t_{n+1}) = \boldsymbol{y}_n + h\boldsymbol{f}(t_n, \boldsymbol{y}_n) + \tfrac{1}{2}h^2 \left[\frac{\partial \boldsymbol{f}(t_n, \boldsymbol{y}_n)}{\partial t} + \frac{\partial \boldsymbol{f}(t_n, \boldsymbol{y}_n)}{\partial \boldsymbol{y}} \boldsymbol{f}(t_n, \boldsymbol{y}_n) \right] + \mathcal{O}(h^3).$$

Comparison with (3.6) gives us the condition for order $p \geq 2$:

$$b_1 + b_2 = 1, \qquad b_2 c_2 = \tfrac{1}{2}, \qquad a_{2,1} = c_2. \tag{3.7}$$

It is easy to verify that the order cannot exceed two, e.g. by applying the ERK method to the scalar equation $y' = y$.

The conditions (3.7) do not define a two-stage ERK uniquely. Popular choices of parameters are displayed in the *RK tableaux*

$$
\begin{array}{c|cc}
0 & & \\
\frac{1}{2} & \frac{1}{2} & \\
\hline
 & 0 & 1
\end{array}, \qquad
\begin{array}{c|cc}
0 & & \\
\frac{2}{3} & \frac{2}{3} & \\
\hline
 & \frac{1}{4} & \frac{3}{4}
\end{array} \qquad \text{and} \qquad
\begin{array}{c|cc}
0 & & \\
1 & 1 & \\
\hline
 & \frac{1}{2} & \frac{1}{2}
\end{array}.
$$

which are of the following form:

$$
\begin{array}{c|c}
\boldsymbol{c} & A \\
\hline
 & \boldsymbol{b}^T
\end{array}.
$$

A naive expansion can be carried out (with substantially greater effort) for $\nu = 3$, whereby we can obtain third-order schemes. However, this is clearly not a serious contender in the technique-of-the-month competition. Fortunately, there are substantially more powerful and easier means of analysing the order of Runge–Kutta methods. We commence by observing that the condition

$$\sum_{i=1}^{j-1} a_{j,i} = c_j, \qquad j = 2, 3, \ldots, \nu,$$

is necessary for order one – otherwise we cannot recover the solution of $y' \equiv 1$. The simplest device, which unfortunately is valid *only* for $p \leq 3$, consists of verifying the order for the scalar *autonomous* equation

$$y' = f(y), \quad t \geq t_0, \qquad y(t_0) = y_0, \tag{3.8}$$

rather than for (3.1). We do not intend here to justify the above assertion bur merely to demonstrate its efficacy in the case $\nu = 3$. We henceforth adopt the 'local convention' that, unless indicated otherwise, all the quantities are evaluated at t_n, e.g. $y \sim y_n$, $f \sim f(y_n)$ etc. Subscripts denote derivatives.

$$\xi_1 = y$$
$$\Rightarrow \quad f(\xi_1) = f;$$
$$\xi_2 = y_h c_2 f$$
$$\Rightarrow \quad f(\xi_2) = f(y + hc_2 f) = f + hc_2 f_y f + \tfrac{1}{2} h^2 c_2 f_{yy} f + \mathcal{O}(h^3) ;$$
$$\xi_3 = y + h(c_3 - a_{3,2})f(\xi_1) + ha_{3,2}f(\xi_2)$$
$$= y + (c_3 - a_{3,2})f + ha_{3,2}f(y + hc_2 f) + \mathcal{O}(h^3)$$
$$= y + hc_3 f + h^2 a_{3,2} c_2 f_y f + \mathcal{O}(h^3)$$
$$\Rightarrow \quad f(\xi_3) = f(y + hc_3 f + h^2 a_{3,2} c_2 f_y f) + \mathcal{O}(h^3)$$
$$= f + hc_3 f_y f + h^2 \left(\tfrac{1}{2} c_3^2 f_{yy} f^2 + a_{3,2} c_2 f_y^2 f \right) + \mathcal{O}(h^3) .$$

Therefore

$$y_{n+1} = y + hb_1 f + hb_2 \left(f + hc_2 f_y f + \tfrac{1}{2} h^2 c_2^2 f_{yy} f^2 \right)$$
$$+ hb_3 \left[f + hc_3 f_y f + h^2 \left(\tfrac{1}{2} c_3^2 f_{yy} f^2 + a_{3,2} c_2 f_y^2 f \right) \right] + \mathcal{O}(h^4)$$
$$= y_n + h(b_1 + b_2 + b_3)f + h^2 (c_2 b_2 + c_3 b_3) f_y f + h^3 \left[\tfrac{1}{2} (b_2 c_2^2 + b_3 c_3^2) f_{yy} f^2 \right.$$
$$\left. + b_3 a_{3,2} c_2 f_y^2 f \right] + \mathcal{O}(h^4) .$$

Since

$$\tilde{y}' = f, \qquad \tilde{y}'' = f_y f, \qquad \tilde{y}''' = f_{yy} f^2 + f_y^2 f$$

the expansion of $\tilde{y}$ reads

$$\tilde{y}_{n+1} = y + hf + \tfrac{1}{2} h^2 f_y f + \tfrac{1}{6} h^3 \left(f_{yy} f^2 + f_y^2 f \right) + \mathcal{O}(h^4) .$$

Comparison of the powers of h leads to third-order conditions, namely

$$b_1 + b_2 + b_3 = 1, \qquad b_2 c_2 + b_3 c_3 = \tfrac{1}{2}, \qquad b_2 c_2^2 + b_3 c_3^2 = \tfrac{1}{3}, \qquad b_3 a_{3,2} c_2 = \tfrac{1}{6}.$$

Some instances of third-order, three-stage ERK methods are important enough to bear an individual name, for example the *classical* RK method

$$
\begin{array}{c|ccc}
0 & & & \\
\frac{1}{2} & \frac{1}{2} & & \\
1 & -1 & 2 & \\
\hline
 & \frac{1}{6} & \frac{2}{3} & \frac{1}{6}
\end{array}
$$

and the *Nystrom* scheme

$$
\begin{array}{c|ccc}
0 & & & \\
\frac{2}{3} & \frac{2}{3} & & \\
\frac{2}{3} & 0 & \frac{2}{3} & \\
\hline
 & \frac{1}{4} & \frac{3}{8} & \frac{3}{8}
\end{array} .
$$

Fourth order is not beyond the capabilities of a Taylor expansion, although a great deal of persistence and care (or, alternatively, a good symbolic manipulator) are required. The best-known fourth-order, four-stage ERK method is

$$
\begin{array}{c|cccc}
0 & & & & \\
\frac{1}{2} & \frac{1}{2} & & & \\
\frac{1}{2} & 0 & \frac{1}{2} & & \\
1 & 0 & 0 & 1 & \\
\hline
& \frac{1}{6} & \frac{1}{3} & \frac{1}{3} & \frac{1}{6}
\end{array}
\ .
$$

The derivation of higher-order ERK methods requires a substantially more advanced technique based upon graph theory. It is well beyond the scope of this volume (but see the comments at the end of this chapter). The analysis is further complicated by the fact that ν-stage ERKs of order ν exist only for $\nu \leq 4$. To obtain order five we need six stages and matters become considerably worse for higher orders.

3.3 Implicit Runge–Kutta schemes

The idea behind *implicit Runge–Kutta* (IRK) methods is to allow the vector functions $\boldsymbol{\xi}_1, \boldsymbol{\xi}_2, \ldots, \boldsymbol{\xi}_\nu$ to depend upon each other in a more general manner than that of (3.5). Thus, let us consider the scheme

$$
\begin{aligned}
\boldsymbol{\xi}_j &= \boldsymbol{y}_n + h \sum_{i=1}^{\nu} a_{j,i} \boldsymbol{f}(t_n + c_i h, \boldsymbol{\xi}_i), \qquad j = 1, 2, \ldots, \nu \\
\boldsymbol{y}_{n+1} &= \boldsymbol{y}_n + h \sum_{j=1}^{\nu} b_j \boldsymbol{f}(t_n + c_j h, \boldsymbol{\xi}_j).
\end{aligned}
\tag{3.9}
$$

Here $A = (a_{j,i})_{j,i=1,2,\ldots,\nu}$ is an arbitrary matrix, whereas in (3.5) it was strictly lower triangular. We impose the convention

$$
\sum_{i=1}^{\nu} a_{j,i} = c_j, \qquad j = 1, 2, \ldots, \nu,
$$

which is necessary for the method to be of nontrivial order. The ERK terminology – RK nodes, RK weights etc. – stays in place.

For general RK matrix A, the algorithm (3.9) is a system of νd coupled algebraic equations, where $\boldsymbol{y} \in \mathbb{R}^d$. Hence, its calculation faces us with an ordeal altogether of a different kind than that for the explicit method (3.5). However, IRK schemes possess important advantages; in particular they may exhibit superior stability properties. Moreover, as will be apparent in Section 3.4, there exists for every $\nu \geq 1$ a unique IRK method of order 2ν, a natural extension of the Gaussian quadrature formulae of Theorem 3.3.

◇ **A two-stage IRK method** Let us consider the method

$$
\begin{aligned}
\boldsymbol{\xi}_1 &= \boldsymbol{y}_n + \tfrac{1}{4} h \left[\boldsymbol{f}(t_n, \boldsymbol{\xi}_1) - \boldsymbol{f}(t_n + \tfrac{2}{3} h, \boldsymbol{\xi}_2) \right], \\
\boldsymbol{\xi}_2 &= \boldsymbol{y}_n + \tfrac{1}{12} h \left[3\boldsymbol{f}(t_n, \boldsymbol{\xi}_1) + 5\boldsymbol{f}(t_n + \tfrac{2}{3} h, \boldsymbol{\xi}_2) \right], \\
\boldsymbol{y}_{n+1} &= \boldsymbol{y}_n + \tfrac{1}{4} h \left[\boldsymbol{f}(t_n, \boldsymbol{\xi}_1) + 3\boldsymbol{f}(t_n + \tfrac{2}{3} h, \boldsymbol{\xi}_2) \right].
\end{aligned}
\tag{3.10}
$$

In tableau notation it reads

$$
\begin{array}{c|cc}
0 & \frac{1}{4} & -\frac{1}{4} \\
\frac{2}{3} & \frac{1}{4} & \frac{5}{12} \\
\hline
& \frac{1}{4} & \frac{3}{4}
\end{array} \; .
$$

To investigate the order of (3.10), we again assume that the underlying ODE is scalar and autonomous – a procedure that is justified since we do not intend to exceed third order. As before, the convention is that each quantity, unless explicitly stated to the contrary, is evaluated at y_n. Let $k_1 := f(\xi_1)$ and $k_2 := f(\xi_2)$. Expanding about y_n,

$$
\begin{aligned}
k_1 &= f + \tfrac{1}{4}hf_y(k_1 - k_2) + \tfrac{1}{32}h^2 f_{yy}(k_1 - k_2)^2 + \mathcal{O}(h^3)\,, \\
k_2 &= f + \tfrac{1}{12}hf_y(3k_1 + 5k_2) + \tfrac{1}{288}h^2 f_{yy}(3k_1 + 5k_2)^2 + \mathcal{O}(h^3)\,,
\end{aligned}
$$

therefore $k_1, k_2 = f + \mathcal{O}(h)$. Substituting this on the right-hand side of the above equations yields $k_1 = f + \mathcal{O}(h^2)$, $k_2 = f + \tfrac{2}{3}hf_y f + \mathcal{O}(h^2)$. Substituting again these enhanced estimates, we finally obtain

$$
\begin{aligned}
k_1 &= f - \tfrac{1}{6}h^2 f_y^2 f + \mathcal{O}(h^3)\,, \\
k_2 &= f + \tfrac{2}{3}hf_y f + h^2 \left(\tfrac{5}{18}f_y^2 f + \tfrac{2}{9}f_{yy}f^2\right) + \mathcal{O}(h^3)\,.
\end{aligned}
$$

Consequently,

$$
\begin{aligned}
y_{n+1} &= y_n + h(b_1 k_1 + b_2 k_2) \\
&= y + hf + \tfrac{1}{2}h^2 f_y f + \tfrac{1}{6}h^3(f_y^2 f + f_{yy}f^2) + \mathcal{O}(h^4)\,. \qquad (3.11)
\end{aligned}
$$

On the other hand, $y' = f$, $y'' = f_y f$, $y''' = f_y^2 f^2 + f_{yy}f^2$ and the exact expansion is

$$
\tilde{y}_{n+1} = y + hf + \tfrac{1}{2}h^2 f_y f + \tfrac{1}{6}h^3(f_y^2 f + f_{yy}f^2) + \mathcal{O}(h^4)\,,
$$

and this matches (3.11). We thus deduce that the method (3.10) is of order at least three. It is, actually, of order exactly three, and this can be demonstrated by applying (3.10) to the linear equation $y' = y$. $\quad\diamond$

It is perfectly possible to derive IRK methods of higher order by employing the graph-theoretic technique mentioned at the end of Section 3.2. However, an important subset of implicit Runge–Kutta schemes can be investigated very easily and without any cumbersome expansions by an entirely different approach. This will be the theme of the next section.

3.4   Collocation and IRK methods

Let us abandon Runge–Kutta methods for a little while and consider instead an alternative approach to the numerical solution of the ODE (3.1). As before, we assume that the integration has been already carried out up to $(t_n, \boldsymbol{y}_n)$ and we seek a recipe to advance it to $(t_{n+1}, \boldsymbol{y}_{n+1})$, where $t_{n+1} = t_n + h$. To this end we choose ν distinct *collocation parameters* $c_1, c_2, \ldots, c_\nu$ (preferably in $[0,1]$, although this is not essential to our argument) and seek a νth-degree polynomial $\boldsymbol{u}$ (with vector coefficients) such that

$$
\begin{aligned}
\boldsymbol{u}(t_n) &= \boldsymbol{y}_n, \\
\boldsymbol{u}'(t_n + c_j h) &= \boldsymbol{f}(t_n + c_j h, \boldsymbol{u}(t_n + c_j h)), \qquad j = 1, 2, \ldots, \nu.
\end{aligned}
\tag{3.12}
$$

In other words, $\boldsymbol{u}$ obeys the initial condition and satisfies the differential equation (3.1) exactly at ν distinct points. A *collocation* method consists of finding such a $\boldsymbol{u}$ and setting

$$
\boldsymbol{y}_{n+1} = \boldsymbol{u}(t_{n+1}).
$$

The collocation method sounds eminently plausible. Yet, you will search for it in vain in most expositions of ODE methods. The reason is that we have not been entirely sincere at the beginning of this section: collocation is nothing other than a Runge–Kutta method in disguise.

Lemma 3.5   *Set*

$$
q(t) := \prod_{j=1}^{\nu}(t - c_j), \qquad q_\ell(t) := \frac{q(t)}{t - c_\ell}, \qquad \ell = 1, 2, \ldots, \nu,
$$

and let

$$
a_{j,i} := \int_0^{c_j} \frac{q_i(\tau)}{q_i(c_i)} \, \mathrm{d}\tau, \qquad j, i = 1, 2, \ldots, \nu,
\tag{3.13}
$$

$$
b_j := \int_0^1 \frac{q_j(\tau)}{q_j(c_j)} \, \mathrm{d}\tau, \qquad j = 1, 2, \ldots, \nu.
\tag{3.14}
$$

The collocation method (3.12) is identical to the IRK method

$$
\begin{array}{c|c}
\boldsymbol{c} & A \\
\hline
& \boldsymbol{b}^T
\end{array}.
$$

Proof   According to Appendix subsection A.2.2.3, the *Lagrange interpolation polynomial*

$$
\boldsymbol{q}(t) := \sum_{\ell=1}^{\nu} \frac{q_\ell((t - t_n)/h)}{q_\ell(c_\ell)} \boldsymbol{w}_\ell
$$

satisfies $\boldsymbol{q}(c_\ell) = \boldsymbol{w}_\ell$, $\ell = 1, 2, \ldots, \nu$. Let us choose $\boldsymbol{w}_\ell = \boldsymbol{u}'(t_n + c_\ell h)$, $\ell = 1, 2, \ldots, \nu$. The two $(\nu - 1)$th-degree polynomials $\boldsymbol{q}$ and $\boldsymbol{u}'$ coincide at ν points and we thus conclude that $\boldsymbol{q} \equiv \boldsymbol{u}'$. Therefore, invoking (3.12),

$$
\boldsymbol{u}'(t) = \sum_{\ell=1}^{\nu} \frac{q_\ell((t - t_n)/h)}{q_\ell(c_\ell)} \boldsymbol{u}'(t_n + c_\ell h) = \sum_{\ell=1}^{\nu} \frac{q_\ell((t - t_n)/h)}{q_\ell(c_\ell)} \boldsymbol{f}(t_n + c_\ell h, \boldsymbol{u}(t_n + c_\ell h)).
$$

We will integrate the last expression. Since $u(t_n) = y_n$, the outcome is

$$u(t) = y_n + \int_{t_n}^{t} \sum_{\ell=1}^{\nu} f(t_n + c_\ell h, u(t_n + c_\ell h)) \frac{q_\ell((\tau - t_n)/h)}{q_\ell(c_\ell)} \, d\tau$$

$$= y_n + h \sum_{\ell=1}^{\nu} f(t_n + c_\ell h, u(t_n + c_\ell h)) \int_{0}^{(t-t_n)/h} \frac{q_\ell(\tau)}{q_\ell(c_\ell)} \, d\tau. \qquad (3.15)$$

We set $\xi_j := u(t_n + c_j h)$, $j = 1, 2, \ldots, \nu$. Letting $t = t_n + c_j h$ in (3.15), the definition (3.13) implies that

$$\xi_j = y_n + h \sum_{i=1}^{\nu} a_{j,i} f(t_n + c_i h, \xi_i), \qquad j = 1, 2, \ldots, \nu,$$

whereas $t = t_{n+1}$ and (3.14) yield

$$y_{n+1} = u(t_{n+1}) = y_n + \sum_{j=1}^{\nu} b_j f(t_n + c_j h, \xi_j).$$

Thus, we recover the definition (3.9) and conclude that the collocation method (3.12) is an IRK method. ∎

◇ **Not every Runge–Kutta method originates in collocation**   Let $\nu = 2$, $c_1 = 0$ and $c_2 = \frac{2}{3}$. Therefore

$$q(t) = t \left(t - \tfrac{2}{3}\right), \qquad q_1(t) = t - \tfrac{2}{3}, \qquad q_2(t) = t$$

and (3.13), (3.14) yield the IRK method with tableau

$$
\begin{array}{c|cc}
0 & 0 & 0 \\
\frac{2}{3} & \frac{1}{3} & \frac{1}{3} \\
\hline
 & \frac{1}{4} & \frac{3}{4}
\end{array}.
$$

Given that every choice of collocation points corresponds to a unique collocation method, we deduce that the IRK method (3.10) (again, with $\nu = 2$, $c_1 = 0$ and $c_2 = \frac{2}{3}$) has no collocation counterpart. There is nothing wrong in this, except that we cannot use the remainder of this section to elucidate the order of (3.10).   ◇

Not only are collocation methods a special case of IRK but, as far as actual computation is concerned, to all intents and purposes the IRK formulation (3.9) is preferable. The one advantage of (3.12) is that it lends itself very conveniently to analysis and obviates the need for cumbersome expansions. In a sense, collocation methods are the true inheritors of the quadrature formulae.

Before we can reap the benefits of the formulation (3.12), we need first to present (without proof) an important result on the estimation of error in a numerical solution. It is frequently the case that we possess a smoothly differentiable 'candidate solution'

v, say, to the ODE (3.1). Typically, such a solution can be produced by any of a myriad of approximation or perturbation techniques, by extending (e.g. by interpolation) a numerical solution from a grid to the whole interval of interest or by formulating 'continuous' numerical methods – the collocation (3.12) is a case in point.

Given such a function v, we can calculate the *defect*

$$d(t, v) := v'(t) - f(t, v(t)).$$

Clearly, there is a connection between the magnitude of the defect and the error $v(t) - y(t)$: since $d(t, y) \equiv 0$, we can expect a small value of $\|d(t, v)\|$ to imply that the error is small. Such a connection is important, since, unlike the error, we can evaluate the defect without knowing the exact solution y.

Matters are simple for linear equations. Thus, suppose that

$$y' = \Lambda y, \qquad y(t_0) = y_0. \tag{3.16}$$

We have $d(t) = v' - \Lambda v$ and therefore the linear inhomogeneous ODE

$$v' = \Lambda v + d(t), \quad t \geq t_0, \qquad v(t_0) \text{ given.}$$

The exact solution is provided by the familiar variation-of-constants formula,

$$v(t) = e^{(t-t_0)\Lambda} + \int_{t_0}^{t} e^{(t-\tau)\Lambda} d(\tau) \, d\tau, \qquad t \geq t_0,$$

while the solution of (3.16) is, of course,

$$y(t) = e^{(t-t_0)\Lambda} y_0 \qquad t \geq t_0.$$

We deduce that

$$v(t) - y(t) = e^{(t-t_0)\Lambda}(v_0 - y_0) + \int_{t_0}^{t} e^{(t-\tau)\Lambda} d(\tau) \, d\tau, \qquad t \geq t_0;$$

thus the error can be expressed completely in terms of the 'observables' $v_0 - y_0$ and d.

It is perhaps not very surprising that we can establish a connection between the error and the defect for the linear equation (3.16) since, after all, its exact solution is known. Remarkably, the variation-of-constants formula can be rendered, albeit in a somewhat weaker form, in a nonlinear setting.

Theorem 3.6 (The Alekseev–Gröbner lemma) *Let v be a smoothly differentiable function that obeys the initial condition $v(t_0) = y_0$. Then*

$$v(t) - y(t) = \int_{t_0}^{t} \Phi(t, \tau, v(\tau)) d(\tau, v) \, d\tau, \qquad t \geq t_0, \tag{3.17}$$

where Φ is the matrix of partial derivatives of the solution of the ODE $w' = f(t, w)$, $w(\tau) = v(\tau)$, with respect to $v(\tau)$. ∎

The matrix Φ is, in general, unknown. It can be estimated quite efficiently, a practice which is useful in error control, but this ranges well beyond the scope of this book. Fortunately, we do not need to know Φ for the application that we have in mind!

Theorem 3.7   *Suppose that*

$$\int_0^1 q(\tau)\tau^j \, d\tau = 0, \qquad j = 0, 1, \ldots, m - 1, \tag{3.18}$$

for some $m \in \{0, 1, \ldots, \nu\}$. (The polynomial $q(t) = \prod_{\ell=1}^{\nu}(t - c_\ell)$ has been already defined in the proof of Lemma 3.5.) Then the collocation method (3.12) is of order $\nu + m$.[1]

Proof   We express the error of the collocation method by using the Alekseev–Gröbner formula (3.17) (with t_0 replaced by t_n and, of course, the collocation solution u playing the role of v; we recall that $u(t_n) = y_n$, hence the conditions of Theorem 3.6 are satisfied). Thus

$$y_{n+1} - \tilde{y}(t_{n+1}) = \int_{t_n}^{t_{n+1}} \Phi(t_{n+1}, \tau, u(\tau)) d(\tau, u) \, d\tau.$$

(We recall that $\tilde{y}$ denotes the exact solution of the ODE for the initial condition $\tilde{y}(t_n) = y_n$.) We next replace the integral by the quadrature formula with respect to the weight function $\omega(t) \equiv 1$, $t_n < t < t_{n+1}$, with the quadrature nodes $t_n + c_1 h$, $t_n + c_2 h$, ..., $t_n + c_\nu h$. Therefore

$$y_{n+1} - \tilde{y}(t_{n+1}) = \sum_{j=1}^{\nu} b_j d(t_n + c_j h, u) + \text{the error of the quadrature.} \tag{3.19}$$

However, according to the definition (3.12) of collocation,

$$d(t_n + c_j h, u) = u'(t_n + c_j h) - f(t_n + c_j h, u(t_n + c_j h)) = 0, \qquad j = 1, 2, \ldots, \nu.$$

According to Theorem 3.4, the order of quadrature with the weight function $\omega(t) \equiv 1$, $0 \leq t \leq 1$, with nodes $c_1, c_2, \ldots, c_\nu$, is $m + \nu$. Therefore, translating linearly from $[0, 1]$ to $[t_n, t_{n+1}]$ and paying heed to the length of the latter interval, $t_{n+1} - t_n = h$, it follows that the error of the quadrature in (3.19) is $\mathcal{O}(h^{\nu+m+1})$. We thus deduce that $y_{n+1} - \tilde{y}(t_{n+1}) = \mathcal{O}(h^{\nu+m+1})$ and prove the theorem.[2]   ∎

Corollary   *Let $c_1, c_2, \ldots, c_\nu$ be the zeros of the polynomial $\tilde{P}_\nu \in \mathbb{P}_\nu$ that is orthogonal with the weight function $\omega(t) \equiv 1$, $0 \leq t \leq 1$. Then the underlying collocation method (3.12) has order 2ν.*

Proof   The corollary is a straightforward consequence of the last theorem, since the definition of orthogonality (3.4) implies in the present context that (3.18) is satisfied by $m = \nu$.   ∎

---

[1] If $m = 0$ this means that (3.18) does not hold for any value of j and the theorem claims that the underlying collocation method is then of order ν.

[2] Strictly speaking, we have only proved that the error is at least $\nu + m$. However, if m is the largest integer such that (3.18) holds, then it is trivial to prove that the order cannot exceed $\nu + m$; for example, apply the collocation to the equation $y' = (\nu + m + 1)t^{\nu+m}$, $y(0) = 0$.

◇ **Gauss–Legendre methods** The ν-stage, order-2ν methods from the last corollary are called *Gauss–Legendre* (Runge–Kutta) methods. Note that, according to Lemma 3.2, $c_1, c_2, \ldots, c_\nu \in (0,1)$ and, as necessary for collocation, are distinct. The polynomial $\tilde{P}_\nu$ can be obtained explicitly, e.g. by linearly transforming the more familiar *Legendre* polynomial P_ν, which is orthogonal with the weight function $\omega(t) \equiv 1$, $-1 < t < 1$. The (monic) outcome is

$$\tilde{P}_\nu(t) = \frac{(\nu!)^2}{(2\nu)!} \sum_{k=0}^{\nu} (-1)^{\nu-k} \binom{\nu}{k} \binom{\nu+k}{k} t^k.$$

For $\nu = 1$ we obtain $\tilde{P}_1(t) = t - \frac{1}{2}$, hence $c_1 = \frac{1}{2}$. The method, which can be written in a tableau form as

$$
\begin{array}{c|c}
\frac{1}{2} & \frac{1}{2} \\
\hline
& 1
\end{array},
$$

is the familiar *implicit midpoint rule* (1.12). In the case $\nu = 2$ we have $\tilde{P}_2(t) = t^2 - t + \frac{1}{6}$, therefore $c_1 = \frac{1}{2} - \frac{\sqrt{3}}{6}$, $c_2 = \frac{1}{2} + \frac{\sqrt{3}}{6}$. The formulae (3.13), (3.14) lead to the two-stage order-four IRK method

$$
\begin{array}{c|cc}
\frac{1}{2} - \frac{\sqrt{3}}{6} & \frac{1}{4} & \frac{1}{4} - \frac{\sqrt{3}}{6} \\
\frac{1}{2} + \frac{\sqrt{3}}{6} & \frac{1}{4} + \frac{\sqrt{3}}{6} & \frac{1}{4} \\
\hline
& \frac{1}{2} & \frac{1}{2}
\end{array}.
$$

The computation of nonlinear algebraic systems that originate in IRK methods with large ν is expensive but this is compensated by the increase in order. It is impossible to lay down firm rules, and the exact point whereby the law of diminishing returns compels us to choose a lower-order method changes from equation to equation. It is fair to remark, however, that the three-stage Gauss–Legendre is probably the largest that is consistent with reasonable implementation costs:

$$
\begin{array}{c|ccc}
\frac{1}{2} - \frac{\sqrt{15}}{10} & \frac{5}{36} & \frac{2}{9} - \frac{\sqrt{15}}{15} & \frac{5}{36} - \frac{\sqrt{15}}{30} \\
\frac{1}{2} & \frac{5}{36} + \frac{\sqrt{15}}{24} & \frac{2}{9} & \frac{5}{36} - \frac{\sqrt{15}}{24} \\
\frac{1}{2} + \frac{\sqrt{15}}{10} & \frac{5}{36} + \frac{\sqrt{15}}{30} & \frac{2}{9} + \frac{\sqrt{15}}{15} & \frac{5}{36} \\
\hline
& \frac{5}{18} & \frac{4}{9} & \frac{5}{18}
\end{array}.
$$

◇

Comments and bibliography

A standard text on numerical integration is Davis & Rabinowitz (1967), while highly readable accounts of orthogonal polynomials can be found in Chihara (1978) and Rainville (1967). We emphasize that although the theory of orthogonal polynomials is of tangential importance to the subject matter of this volume, it is well worth studying for its intrinsic beauty, as well as its numerous applications.

Runge–Kutta methods have been known for a long time; Runge himself produced the main idea in 1895.[3] Their theoretical understanding, however, is much more recent and associated mainly with the work of John Butcher. As is often the case with progress in computational science, an improved theory has spawned new and better algorithms, these in turn have led to further theoretical comprehension and so on. Lambert's textbook (1991) presents a readable account of Runge–Kutta methods, based upon a relatively modest theoretical base. More advanced accounts can be found in Butcher (1987) and Hairer *et al.* (1991).

Let us present in a nutshell the main idea behind the graph-theoretical approach of Butcher to the derivation of the order of Runge–Kutta methods. The few examples of expansion in Sections 3.2–3.3 already demonstrate that the main difficulty rests in the need to differentiate repeatedly composite functions. For expositional reasons only, we henceforth restrict our attention to scalar, autonomous equations.[4] Thus,

$$y' = f(y)$$
$$\Rightarrow \quad y'' = f_y(y)f(y)$$
$$\Rightarrow \quad y''' = f_{yy}(y)[f(y)]^2 + [f_y(y)]^2 f(y)$$
$$\Rightarrow \quad y^{(iv)} = f_{yyy}(y)[f(y)]^3 + 4f_{yy}(y)f_y(y)[f(y)]^2 + [f_y(y)]^3 f(y)$$

and so on. Although it cannot yet be seen from the above, the number of terms increases *exponentially*. This should not deter us from exploring high-order methods, since there is a great deal of redundancy in the order conditions (recall from the corollary to Theorem 2.7 that it is possible to attain order 2ν with a ν-stage method!), but we need an intelligent mechanism to express the increasingly more complicated derivatives in a compact form.

Such a mechanism is provided by *graph theory*. Briefly, a *graph* is a collection of *vertices* and *edges:* it is usual to render the vertices pictorially as solid circles, while the edges are the lines joining them. For example, two simple five-vertex graphs are

The *order* of a graph is the number of vertices therein: both graphs above are of order five. We say that a graph is a *tree* if each two vertices are joined by a single path of edges. Thus, the second graph is a tree, whereas the first is not. Finally, in a tree, we single out one vertex and call it the *root*. This imposes a partial ordering on a *rooted tree:* the root is the lowest, its children (i.e., all vertices that are joined to the root by a single edge) are next in line, then its children's children and so on. We adopt in our pictures the (obvious) convention that the root is always at the bottom. Two rooted trees of the same order are said to be *equivalent* if each exhibits the same pattern of paths from its 'top' to its root – the following picture of three equivalent rooted trees should clarify this concept: the graphs

---

[3]The monograph of Collatz (1966), and in particular its copious footnotes, is an excellent source on the life of many past heroes of numerical analysis.

[4]This restriction leads to loss of generality. A comprehensive order analysis should be done for systems of equations.

are all equivalent. We keep just one representative of each equivalence class and, hopefully without much confusion, refer to members of this reduced set as 'rooted trees'. We denote by $\gamma(\hat{t})$ the product of the order of the tree $\hat{t}$ and the orders of all possible trees that occur upon the consecutive removal of the roots of $\hat{t}$. For example, for the above tree we have

(an open circle denotes a vertex that has been removed) and $\gamma(\hat{t}) = 5 \times (2 \times 1 \times 1) \times 1 = 10$.

As we have seen above, the derivatives of y can be expressed as linear combinations of products of derivatives of f. The latter are called *elementary differentials* and they can be assigned to rooted trees according to the following rule: to each vertex of a rooted tree corresponds $f_{yy\ldots y}$, where the suffix occurs the same number of times as the number of children of the vertex, and the elementary differential corresponding to the whole tree is a product of these terms. For example,

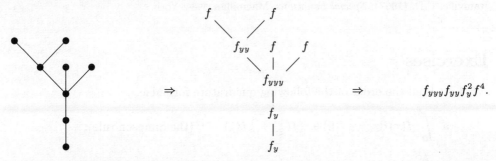

To every rooted tree there corresponds an order condition, which we can express in terms of the RK matrix A and the RK weights $\boldsymbol{b}$. This is best demonstrated by an example. We assign an index to every vertex of a tree $\hat{t}$, e.g. the tree

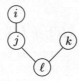

corresponds to the condition

$$\sum_{\ell,j,i,k=1}^{\nu} b_\ell a_{\ell,j} a_{\ell,k} a_{j,i} = \frac{1}{\gamma(\hat{t})} = \frac{1}{8}.$$

The general rule is clear – we multiply b_ℓ by all components $a_{q,r}$, where q and r are the indices of a parent and a child respectively, sum up for all indices ranging in $\{1, 2, \ldots, \nu\}$ and equate to the reciprocal of $\gamma(\hat{t})$. The main result linking rooted trees and Runge–Kutta methods is that the scheme (3.9) (or, for that matter, (3.5)) is of order p if and only if the above order conditions are satisfied for all rooted trees of order less than or equal to p.

The graph-theoretical technique is the standard tool in the construction of Runge–Kutta schemes. The alternative approach of collocation is less well known, although it is presented in more recent texts, e.g. Hairer *et al.* (1991). Of course, only a subset of all Runge–Kutta methods are equivalent to collocation and the technique is of little value for ERK schemes. However, it is possible to generalize the concept of collocation to cater for *all* Runge–Kutta methods.

Butcher, J.C. (1987), *The Numerical Analysis of Ordinary Differential Equations*, John Wiley, New York.

Chihara, T.S. (1978), *An Introduction to Orthogonal Polynomials*, Gordon & Breach, New York.

Collatz, L. (1966), *The Numerical Treatment of Differential Equations* (3rd ed.), Springer-Verlag, Berlin.

Davis, P.J. and Rabinowitz, P. (1967), *Numerical Integration*, Blaisdell, London.

Hairer, E., Nørsett, S.P. and Wanner, G. (1991), *Solving Ordinary Differential Equations I: Nonstiff Problems* (2nd ed.), Springer-Verlag, Berlin.

Lambert, J.D. (1991), *Numerical Methods for Ordinary Differential Systems*, Wiley, London.

Rainville, E.D. (1967), *Special Functions*, Macmillan, New York.

Exercises

3.1   Find the order of the following quadrature formulae.

   a $\displaystyle\int_0^1 f(\tau)\,\mathrm{d}\tau = \tfrac{1}{6}f(0) + \tfrac{2}{3}f(\tfrac{1}{2}) + \tfrac{1}{6}f(1)$ (the Simpson rule);

   b $\displaystyle\int_0^1 f(\tau)\,\mathrm{d}\tau = \tfrac{1}{8}f(0) + \tfrac{3}{8}f(\tfrac{1}{3}) + \tfrac{3}{8}f(\tfrac{2}{3}) + \tfrac{1}{8}f(1)$ (the three-eighths rule);

   c $\displaystyle\int_0^1 f(\tau)\,\mathrm{d}\tau = \tfrac{2}{3}f(\tfrac{1}{4}) - \tfrac{1}{3}f(\tfrac{1}{2}) + \tfrac{2}{3}f(\tfrac{3}{4});$

   d $\displaystyle\int_0^\infty f(\tau)\mathrm{e}^{-\tau}\,\mathrm{d}\tau = \tfrac{5}{3}f(1) - \tfrac{3}{2}f(2) + f(3) - \tfrac{1}{6}f(4).$

3.2   Let us define

$$T_n(\cos\theta) := \cos n\theta, \qquad n = 0, 1, 2, \ldots, \qquad -\pi \le \theta \le \pi.$$

   a Show that each T_n is a polynomial of degree n and that the T_n satisfy the three–term recurrence relation

$$T_{n+1}(t) = 2t\,T_n(t) - T_{n-1}(t), \qquad n = 1, 2, \ldots.$$

b Prove that T_n is an nth orthogonal polynomial with respect to the weight function $\omega(t) = (1-t)^{-\frac{1}{2}}$, $-1 < t < 1$.

c Find the explicit values of the zeros of T_n, thereby verifying the statement of Lemma 3.2, namely that all the zeros of an orthogonal polynomial reside in the open support of the weight function.

d Find b_1, b_2, c_1, c_2 such that the order of the quadrature

$$\int_{-1}^{1} f(\tau) \frac{\mathrm{d}\tau}{\sqrt{1-\tau^2}} \approx b_1 f(c_1) + b_2 f(c_2)$$

is four.

[*The T_ns are known as* Chebyshev polynomials *and they have many applications in mathematical analysis.*]

3.3 Construct the Gaussian quadrature formulae for the weight function $\omega(t) \equiv 1$, $0 \le t \le 1$, of orders two, four and six.

3.4 Restricting your attention to scalar autonomous equations $y' = f(y)$, prove that the ERK method with the tableau

$$
\begin{array}{c|cccc}
0 \\
\frac{1}{2} & \frac{1}{2} \\
\frac{1}{2} & 0 & \frac{1}{2} \\
1 & 0 & 0 & 1 \\
\hline
 & \frac{1}{6} & \frac{1}{3} & \frac{1}{3} & \frac{1}{6}
\end{array}
$$

is of order four.

3.5 Suppose that an ν-stage ERK method of order ν is applied to the linear scalar equation $y' = \lambda y$. Prove that

$$y_n = \left[\sum_{k=0}^{\nu} \frac{1}{k!} (h\lambda)^k \right]^n y_0, \qquad n = 0, 1, \ldots.$$

3.6 Determine all choices of $\boldsymbol{b}$, $\boldsymbol{c}$ and A such that the two-stage IRK method

$$
\begin{array}{c|c}
\boldsymbol{c} & A \\
\hline
 & \boldsymbol{b}^T
\end{array}
$$

is of order $p \ge 3$.

3.7 Write the theta method, (1.13), as a Runge–Kutta method.

3.8 Derive the three-stage Runge–Kutta method that corresponds to the collocation points $c_1 = \frac{1}{4}$, $c_2 = \frac{1}{2}$, $c_3 = \frac{3}{4}$ and determine its order.

3.9   Let $\kappa \in \mathbb{R}\backslash\{0\}$ be a given constant. We choose collocation nodes $c_1, c_2, \ldots, c_\nu$ as zeros of the polynomial $\tilde{P}_\nu + \kappa\tilde{P}_{\nu-1}$. [$\tilde{P}_m$ *is the mth-degree Legendre polynomial, shifted to the interval* $(0,1)$. *In other words, it is orthogonal there with the weight function* $\omega(t) \equiv 1$.]

a Prove that the collocation method (3.12) is of order $2\nu - 1$.

b Let $\kappa = -1$ and find explicitly the corresponding IRK method for $\nu = 2$.

4

Stiff equations

4.1   What are stiff ODEs?

Let us try to solve the seemingly innocent linear ODE

$$\boldsymbol{y}' = \Lambda \boldsymbol{y}, \qquad \boldsymbol{y}(0) = \boldsymbol{y}_0, \qquad \text{where} \qquad \Lambda = \begin{bmatrix} -100 & 1 \\ 0 & -\frac{1}{10} \end{bmatrix}, \qquad (4.1)$$

by Euler's method (1.4). We obtain

$$\boldsymbol{y}_1 = \boldsymbol{y}_0 + h\Lambda \boldsymbol{y}_0 = (I + h\Lambda)\boldsymbol{y}_0, \qquad \boldsymbol{y}_2 = \boldsymbol{y}_1 + h\Lambda \boldsymbol{y}_1 = (I + h\Lambda)\boldsymbol{y}_1 = (I + h\Lambda)^2 \boldsymbol{y}_0$$

(where I is the identity matrix) and, in general, it is easy to prove by elementary induction that

$$\boldsymbol{y}_n = (I + h\Lambda)^n \boldsymbol{y}_0, \qquad n = 0, 1, 2, \ldots. \qquad (4.2)$$

Since the *spectral factorization* (A.1.5.4) of Λ is

$$\Lambda = VDV^{-1}, \qquad \text{where} \qquad V = \begin{bmatrix} 1 & 1 \\ 0 & \frac{999}{10} \end{bmatrix} \qquad \text{and} \qquad D = \begin{bmatrix} -100 & 0 \\ 0 & -\frac{1}{10} \end{bmatrix},$$

we deduce that the exact solution of (4.1) is

$$\boldsymbol{y}(t) = \mathrm{e}^{t\Lambda} = V\mathrm{e}^{tD}V^{-1}\boldsymbol{y}_0, \qquad t \geq 0, \qquad \text{where} \qquad \mathrm{e}^{tD} = \begin{bmatrix} \mathrm{e}^{-100t} & 0 \\ 0 & \mathrm{e}^{-t/10} \end{bmatrix}.$$

In other words, there exist two vectors $\boldsymbol{x}_1$ and $\boldsymbol{x}_2$, say, dependent on $\boldsymbol{y}_0$ but not on t, such that

$$\boldsymbol{y}(t) = \mathrm{e}^{-100t}\boldsymbol{x}_1 + \mathrm{e}^{-t/10}\boldsymbol{x}_2, \qquad t \geq 0. \qquad (4.3)$$

The function $g(t) = \mathrm{e}^{-100t}$ decays exceedingly fast: $g\left(\frac{1}{10}\right) \approx 4.54 \times 10^{-5}$ and $g(1) \approx 3.72 \times 10^{-44}$, while the decay of $\mathrm{e}^{-t/10}$ is a thousandfold more sedate. Thus, even for small $t > 0$, the contribution of $\boldsymbol{x}_1$ is nil to all intents and purposes and $\boldsymbol{y}(t) \approx \mathrm{e}^{-t/10}\boldsymbol{x}_2$. What about the Euler solution $\{\boldsymbol{y}_n\}_{n=0}^{\infty}$, though? It follows from (4.2) that

$$\boldsymbol{y}_n = V(I + hD)^n V^{-1}\boldsymbol{y}_0, \qquad n = 0, 1, \ldots$$

and, since

$$(I + hD)^n = \begin{bmatrix} (1 - 100h)^n & 0 \\ 0 & (1 - \frac{1}{10}h)^n \end{bmatrix},$$

Figure 4.1   The logarithm of the Euclidean norm $\|\boldsymbol{y}_n\|$ of the Euler steps, as applied to the equation (4.1) with $h = \frac{1}{10}$ and an initial condition identical to the second (i.e., the 'stable') eigenvector. The divergence is entirely due to roundoff error!

it follows that

$$\boldsymbol{y}_n = (1 - 100h)^n\boldsymbol{x}_1 + (1 - \tfrac{1}{10}h)^n\boldsymbol{x}_2, \qquad n = 0, 1, \ldots \qquad (4.4)$$

(it is left to the reader in Exercise 4.1 to prove that the constant vectors $\boldsymbol{x}_1$ and $\boldsymbol{x}_2$ are the same in (4.3) and (4.4)). Suppose that $h > \frac{1}{50}$. Then $|1 - 100h| > 1$ and it is a consequence of (4.4) that, for sufficiently large n, *the Euler iterates grow geometrically in magnitude,* in contrast with the asymptotic behaviour of the true solution.

Suppose that we choose an initial condition identical to an eigenvector corresponding to the eigenvalue -0.1, for example

$$\boldsymbol{y}_0 = \left[\begin{array}{c} 1 \\ \frac{999}{10} \end{array} \right].$$

Then, in exact arithmetic, $\boldsymbol{x}_1 = \boldsymbol{0}$, $\boldsymbol{x}_2 = \boldsymbol{y}_0$ and $\boldsymbol{y}_n = \left(1 - \frac{1}{10}h\right)^n\boldsymbol{y}_0$, $n = 0, 1, \ldots$. The latter converges to $\boldsymbol{0}$ as $n \to \infty$ for all reasonable values of $h > 0$ (specifically, for $h < 20$). Hence, we might hope that all will be well with the Euler method. Not so! Real computers produce roundoff errors and, unless $h < \frac{1}{50}$, sooner or later these are bound to attribute a nonzero contribution to an eigenvector corresponding to the eigenvalue -100. As soon as this occurs, the unstable component grows geometrically, as $(1 - 100h)^n$, and rapidly overwhelms the true solution.

Figure 4.1 displays $\ln\|\boldsymbol{y}_n\|$, $n = 0, 1, \ldots, 25$, with the above initial condition and the time step $h = \frac{1}{10}$. The calculation has been performed on a computer equipped with the ubiquitous IEEE arithmetic,[1] which is correct (in a single algebraic operation) to about fifteen decimal digits. The norm of the first seventeen steps decreases at the right pace, dictated by $\left(1 - \frac{1}{10}h\right)^n = \left(\frac{99}{100}\right)^n$. However, everything then breaks down and, after just two steps, the norm increases geometrically, as $|1 - 100h|^n = 9^n$. The reader is welcome to check that the slope of the curve in Figure 4.1 is indeed $\ln\frac{99}{100} \approx -0.0101$ initially, but becomes $\ln 9 \approx 2.1972$ in the second, unstable regime.

The choice of $\boldsymbol{y}_0$ as a 'stable' eigenvector is not contrived. Faced with an equation like (4.1) (with an arbitrary initial condition) we are likely to employ a small step size in the initial *transient* regime, in which the contribution of the 'unstable' eigenvector is still significant. However, as soon as this has disappeared and the solution is completely described by the 'stable' eigenvector, it is tempting to increase h. This must be resisted: like a malign version of the Cheshire cat, the rogue eigenvector might seem to have disappeared, but its hideous grin stays and is bound to thwart our endeavours.

It is important to understand that this behaviour has nothing to do with the local error of the numerical method; the step size is depressed not by accuracy considerations (to which we should be always willing to pay heed) but by instability.

Not every numerical method displays a similar breakdown in stability. Thus, solving (4.1) with the trapezoidal rule (1.9), we obtain

$$\boldsymbol{y}_1 = \left(\frac{I + \frac{1}{2}h\Lambda}{I - \frac{1}{2}h\Lambda}\right)\boldsymbol{y}_0, \qquad \boldsymbol{y}_2 = \left(\frac{I + \frac{1}{2}h\Lambda}{I - \frac{1}{2}h\Lambda}\right)\boldsymbol{y}_1 = \left(\frac{I + \frac{1}{2}h\Lambda}{I - \frac{1}{2}h\Lambda}\right)^2\boldsymbol{y}_0,$$

noting that since $(I - \frac{1}{2}h\Lambda)^{-1}$ and $(I + \frac{1}{2}h\Lambda)$ commute, the order of multiplication does not matter, and, in general,

$$\boldsymbol{y}_n = \left(\frac{I + \frac{1}{2}h\Lambda}{I - \frac{1}{2}h\Lambda}\right)^n \boldsymbol{y}_0, \qquad n = 0, 1, \ldots. \tag{4.5}$$

Substituting for Λ from (4.1) and factorizing, we deduce, in the same way as for (4.4), that

$$\boldsymbol{y}_n = \left(\frac{1 - 50h}{1 + 50h}\right)^n \boldsymbol{x}_1 + \left(\frac{1 - \frac{1}{20}h}{1 + \frac{1}{20}h}\right)^n \boldsymbol{x}_2, \qquad n = 0, 1, \ldots.$$

Thus, since

$$\left|\frac{1 - 50h}{1 + 50h}\right|, \left|\frac{1 - \frac{1}{20}h}{1 + \frac{1}{20}h}\right| < 1$$

for every $h > 0$, we deduce that $\lim_{n\to\infty}\boldsymbol{y}_n = \boldsymbol{0}$. This recovers the correct asymptotic behaviour of the ODE (4.1) (cf. (4.3)) regardless of the size of h.

In other words, the trapezoidal rule does not require any restriction in the step size to avoid instability. We hasten to say that this does not mean, of course, that *any* h is suitable. It is necessary to choose $h > 0$ small enough to ensure that the local error is within reasonable bounds and the exact solution is adequately approximated. However, there is no need to decrease h to a minuscule size to prevent rogue components of the solution growing out of control.

---

[1] The current standard of computer arithmetic on workstations.

The equation (4.1) is an example of a *stiff ODE*. Several attempts at a rigorous definition of stiffness appear in the literature, but it is perhaps more informative to adopt an operative (and slightly vague) designation. Thus, we say that an ODE system

$$\boldsymbol{y}' = \boldsymbol{f}(t, \boldsymbol{y}), \quad t \geq t_0, \qquad \boldsymbol{y}(t_0) = \boldsymbol{y}_0, \tag{4.6}$$

is *stiff* if its numerical solution by some methods requires (perhaps in a portion of the solution interval) a significant depression of the step-size to avoid instability. Needless to say, this is not a proper mathematical definition, but then we are not aiming to prove theorems of the sort 'if a system is stiff then...'. The main importance of the above concept is in helping us to choose and implement numerical methods – a procedure that, anyway, is far from an exact science!

We have already seen the most ubiquitous mechanism that generates stiffness, namely, that modes with vastly different scales and 'lifetimes' are present in the solution. It is sometimes the practice to designate the quotient of the largest and the smallest (in modulus) eigenvalues of a linear system (and, for a general system (4.6), the eigenvalues of the Jacobian matrix) as the *stiffness ratio:* the stiffness ratio of (4.1) is 10^3. This concept is helpful in elucidating the behaviour of many ODE systems and, in general, it is a safe bet that if (4.6) has a large stiffness ratio than it is stiff. Having said this, it is also valuable to stress the shortcomings of linear analysis and emphasize that the stiffness ratio might fail to elucidate the behaviour of a nonlinear ODE system.

A large proportion of the ODEs that occur in real life (or for whatever passes for 'real life' in an academic environment) are stiff. Whenever equations model several processes with vastly different rates of evolution, stiffness is not far away. For example, the differential equations of chemical kinetics describe reactions that often proceed at very different time scales; a stiffness ratio of 10^{17} is quite typical. Other popular sources of stiffness are control theory, reactor kinetics, weather prediction, mathematical biology and electronics: they all abound with phenomena that display variation at significantly different time scales. The world record, to the author's knowledge, is held, unsurprisingly perhaps, by the equations that describe the cosmological Big Bang: the stiffness ratio is 10^{31}.

One of the main sources of stiff equations is numerical analysis itself. As we will see in Chapter 13, parabolic partial differential equations are often approximated by large systems of stiff ODEs.

4.2   The linear stability domain and A-stability

Let us suppose that a given numerical method is applied with a constant step size $h > 0$ to the scalar linear equation

$$y' = \lambda y, \quad t \geq 0, \qquad y(0) = 1, \tag{4.7}$$

where $\lambda \in \mathbb{C}$. The exact solution of (4.7) is, of course, $y(t) = \mathrm{e}^{\lambda t}$, hence $\lim_{t \to \infty} y(t) = 0$ if and only if $\mathrm{Re}\, \lambda < 0$. We say that the *linear stability domain* $\mathcal{D}$ of the underlying numerical method is the set of all numbers $h\lambda \in \mathbb{C}$ such that $\lim_{n \to \infty} y_n = 0$. In other words, $\mathcal{D}$ is the set of all $h\lambda$ for which the correct asymptotic behaviour of (4.7) is recovered, provided that the latter equation is stable.[2]

Let us commence with Euler's method (1.4). We obtain the solution sequence identically to the derivation of (4.2),

$$y_n = (1 + h\lambda)^n, \qquad n = 0, 1, \ldots. \tag{4.8}$$

Therefore $\{y_n\}_{n=0,1,\ldots}$ is a geometric sequence and $\lim_{n\to\infty} y_n = 0$ if and only if $|1 + h\lambda| < 1$. We thus conclude that

$$\mathcal{D}_{\text{Euler}} = \{z \in \mathbb{C} : |1 + z| < 1\}$$

is the interior of a complex disc of unit radius, centred at $z = -1$ (see Figure 4.2).

Before we proceed any further, let us ponder briefly the rationale behind this sudden interest in a humble scalar linear equation. After all, we do not need numerical analysis to solve (4.7)! However, for Euler's method and for all other methods that have been the theme of Chapters 1–3 we can extrapolate from scalar linear equations to linear ODE systems. Thus, suppose that we solve (4.1) *with an arbitrary $d \times d$ matrix* Λ. The solution sequence is given by (4.2). Suppose that Λ has a full set of eigenvectors and hence the spectral factorization $\Lambda = VDV^{-1}$, where V is a nonsingular matrix of eigenvectors and $D = \text{diag}(\lambda_1, \lambda_2, \ldots, \lambda_d)$ contains the eigenvalues of Λ. Exactly as in (4.4), we can prove that there exist vectors $\boldsymbol{x}_1, \boldsymbol{x}_2, \ldots, \boldsymbol{x}_d \in \mathbb{C}^d$, dependent only on $\boldsymbol{y}_0$, not on n, such that

$$\boldsymbol{y}_n = \sum_{k=1}^{d} (1 + h\lambda_k)^n \, \boldsymbol{x}_k, \qquad n = 0, 1, \ldots. \tag{4.9}$$

Let us suppose that the exact solution of the linear system is asymptotically stable. This happens if and only if $\text{Re}\, \lambda_k < 0$ for all $k = 1, 2, \ldots, d$. To mimic this behaviour with Euler's method, we deduce from (4.9) that the step-size $h > 0$ must be such that $|1 + h\lambda_k| < 1$, $k = 1, 2, \ldots, d$: all the products $h\lambda_1, h\lambda_2, \ldots, h\lambda_d$ must lie in $\mathcal{D}_{\text{Euler}}$. This means in practice that the step-size is determined by the stiffest component of the system!

The restriction to systems with a full set of eigenvectors is made for ease of exposition only. In general, we may use a *Jordan factorization* (A.1.5.6) in place of a spectral factorization; see Exercise 4.2 for a simple example. Moreover, the analysis can be easily extended to inhomogeneous systems $\boldsymbol{y}' = \Lambda\boldsymbol{y} + \boldsymbol{a}$, and this is illustrated by Exercise 4.3.

The importance of $\mathcal{D}$ ranges well beyond linear systems. Given a nonlinear ODE system

$$\boldsymbol{y}' = \boldsymbol{f}(t, \boldsymbol{y}), \quad t \geq t_0, \qquad \boldsymbol{y}(t_0) = \boldsymbol{y}_0,$$

---

[2]The interest in (4.7) with $\text{Re}\,\lambda > 0$ is limited, since the exact solution rapidly becomes very large. However, for nonlinear equations there exists an intense interest, which we will not pursue in this volume, in those equations for which a counterpart of λ, namely the Liapunov exponent, is positive.

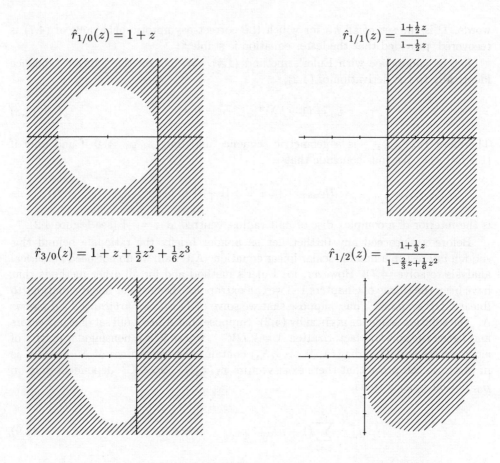

$$\hat{r}_{1/0}(z) = 1 + z \qquad\qquad \hat{r}_{1/1}(z) = \frac{1+\frac{1}{2}z}{1-\frac{1}{2}z}$$

$$\hat{r}_{3/0}(z) = 1 + z + \tfrac{1}{2}z^2 + \tfrac{1}{6}z^3 \qquad\qquad \hat{r}_{1/2}(z) = \frac{1+\frac{1}{3}z}{1-\frac{2}{3}z+\frac{1}{6}z^2}$$

Figure 4.2 Stability domains (unshaded areas) for various rational approximants. Note that $\hat{r}_{1/0}$ corresponds to the Euler method, while $\hat{r}_{1/1}$ corresponds both to the trapezoidal rule and the implicit midpoint rule. The $\hat{r}_{\alpha/\beta}$ notation is introduced in Section 4.3.

where $\boldsymbol{f}$ is differentiable with respect to $\boldsymbol{y}$, it is usual to require that in the nth step

$$h\lambda_{n,1},\ h\lambda_{n,2},\ \ldots,\ h\lambda_{n,d} \in \mathcal{D},$$

where the complex numbers $\lambda_{n,1},\ \lambda_{n,2},\ \ldots,\ \lambda_{n,d}$ are the eigenvalues of the *Jacobian matrix* $J_n := \partial \boldsymbol{f}(t_n, \boldsymbol{y}_n)/\partial\boldsymbol{y}$. This is based on the assumption that the local behaviour of the ODE is modelled well by the variational equation $\boldsymbol{y}' = \boldsymbol{y}_n + J_n(\boldsymbol{y} - \boldsymbol{y}_n)$. We hasten to emphasize that this practice is far from exact. Naive translation of any linear theory to a nonlinear setting can be dangerous and a more correct approach is to embrace a nonlinear approach from the outset. Although this ranges well beyond the material of this book, we provide a few pointers to modern nonlinear stability theory in the comments following this chapter.

Let us continue our investigation of linear stability domains. Replacing Λ by λ

from (4.7) in (4.5) and bearing in mind that $y_0 = 1$, we obtain

$$y_n = \left(\frac{1 + \frac{1}{2}h\lambda}{1 - \frac{1}{2}h\lambda}\right)^n, \qquad n = 0, 1, \ldots. \tag{4.10}$$

Again, $\{y_n\}_{n=0,1,\ldots}$ is a geometric sequence. Therefore, we obtain for the linear stability domain in the case of the trapezoidal rule,

$$\mathcal{D}_{\mathrm{TR}} = \left\{ z \in \mathbb{C} : \left|\frac{1 + \frac{1}{2}z}{1 - \frac{1}{2}z}\right| < 1 \right\}.$$

It is trivial to verify that the inequality within the braces is *identical* to $\mathrm{Re}\, z < 0$. In other words, *the trapezoidal rule mimics the asymptotic stability of linear ODE systems without any need to decrease the step-size*, a property that we have already noticed in a special example in Section 4.1.

The latter feature is of sufficient importance to deserve a name of its own. We say that a method is *A-stable* if

$$\mathbb{C}^- := \{ z \in \mathbb{C} : \mathrm{Re}\, z < 0 \} \subseteq \mathcal{D}.$$

In other words, whenever a method is A-stable, we can choose the step-size h (at least, for linear systems) on accuracy considerations only, without paying heed to stability constraints.

The trapezoidal rule is A-stable, whilst Euler's method is not. As is evident from Figure 4.2, the graph labelled $\hat{r}_{1/2}(z)$ – but not the one labelled $\hat{r}_{3/0}(z)$ – corresponds to an A-stable method. It is left to the reader to ascertain in Exercise 4.4 that the theta method (1.13) is A-stable if and only if $0 \leq \theta \leq \frac{1}{2}$.

4.3   A-stability of Runge–Kutta methods

Applying the Runge–Kutta method (3.9) to the linear equation (4.7), we obtain

$$\xi_j = y_n + h\lambda \sum_{i=1}^{\nu} a_{j,i}\xi_i, \qquad j = 1, 2, \ldots, \nu.$$

Denote

$$\boldsymbol{\xi} := \begin{bmatrix} \xi_1 \\ \xi_2 \\ \vdots \\ \xi_\nu \end{bmatrix}, \qquad \mathbf{1} := \begin{bmatrix} 1 \\ 1 \\ \vdots \\ 1 \end{bmatrix} \in \mathbb{R}^\nu,$$

then $\boldsymbol{\xi} = \mathbf{1}y_n + h\lambda A\boldsymbol{\xi}$ and the exact solution of this linear algebraic system is

$$\boldsymbol{\xi} = (I - h\lambda A)^{-1}\mathbf{1}y_n.$$

Therefore

$$y_{n+1} = y_n + h\sum_{j=1}^{\nu} b_j\xi_j = \left[1 + h\lambda \boldsymbol{b}^T(I - h\lambda A)^{-1}\mathbf{1}\right]y_n, \qquad n = 0, 1, \ldots. \tag{4.11}$$

We denote by $\mathbb{P}_{\alpha/\beta}$ the set of all rational functions $\hat{p}/\hat{q}$, where $\hat{p} \in \mathbb{P}_\alpha$ and $\hat{q} \in \mathbb{P}_\beta$.

Lemma 4.1   *For every Runge–Kutta method (3.9) there exists $r \in \mathbb{P}_{\nu/\nu}$ such that*

$$y_n = [r(h\lambda)]^n, \qquad n = 0, 1, \ldots. \tag{4.12}$$

Moreover, if the Runge–Kutta method is explicit then $r \in \mathbb{P}_\nu$.

Proof   It follows at once from (4.11) that (4.12) is valid with

$$r(z) := 1 + z\boldsymbol{b}^T (I - zA)^{-1}\mathbf{1}, \qquad z \in \mathbb{C}, \tag{4.13}$$

and it remains to verify that r is indeed a rational function (a polynomial for an explicit scheme) of the stipulated type.

We represent the inverse of $I - zA$ using a familiar formula from linear algebra,

$$(I - zA)^{-1} = \frac{\text{adj}\,(I - zA)}{\det(I - zA)},$$

where $\text{adj}\,C$ is the *adjunct* of the $\nu \times \nu$ matrix C: the (i, j)th entry of the adjunct is the determinant of the (j, i)th principal minor, multiplied by $(-1)^{i+j}$. Since each entry of $I - zA$ is linear in z, we deduce that each element of $\text{adj}\,(I - zA)$, being (up to a sign) a determinant of a $(\nu - 1) \times (\nu - 1)$ matrix, is in $\mathbb{P}_{\nu-1}$. We thus conclude that

$$\boldsymbol{b}^T \text{adj}\,(I - zA)\mathbf{1} \in \mathbb{P}_{\nu-1},$$

therefore $\det(I - zA) \in \mathbb{P}_\nu$ implies $r \in \mathbb{P}_{\nu/\nu}$.

Finally, if the method is explicit then A is strictly lower triangular and $I - zA$ is, regardless of $z \in \mathbb{C}$, a lower triangular matrix with ones along the diagonal. Therefore $\det(I - zA) \equiv 1$ and r is a polynomial.   ∎

Lemma 4.2   *Suppose that an application of a numerical method to the linear equation (4.7) produces a geometric solution sequence, $y_n = [r(h\lambda)]^n$, $n = 0, 1, \ldots$, where r is an arbitrary function. Then*

$$\mathcal{D} = \{z \in \mathbb{C} : |r(z)| < 1\}. \tag{4.14}$$

Proof   This follows at once from the definition of the set $\mathcal{D}$.   ∎

Corollary   *No explicit Runge–Kutta (ERK) method (3.5) may be A-stable.*

Proof   Given an ERK method, Lemma 4.1 states that the function r is a polynomial and (4.13) implies that $r(0) = 1$. No polynomial, except for the constant function $r(z) \equiv c \in (-1, 1)$, may be uniformly bounded by the value unity in $\mathbb{C}^-$, and this excludes A-stability.   ∎

For both Euler's method and the trapezoidal rule we have already observed that the solution sequence obeys the conditions of Lemma 4.2. This is hardly surprising, since both methods can be written in a Runge–Kutta formalism.

◇ **The function r for specific IRK schemes**   Let us consider the methods

$$
\begin{array}{c|cc}
0 & \frac{1}{4} & -\frac{1}{4} \\
\frac{2}{3} & \frac{1}{4} & \frac{5}{12} \\
\hline
 & \frac{1}{4} & \frac{3}{4}
\end{array}
\qquad \text{and} \qquad
\begin{array}{c|cc}
0 & 0 & 0 \\
\frac{2}{3} & \frac{1}{3} & \frac{1}{3} \\
\hline
 & \frac{1}{4} & \frac{3}{4}
\end{array}.
$$

We have already encountered both in Chapter 3: the first is (3.10), whereas the second corresponds to collocation at $c_1 = 0$, $c_2 = \frac{2}{3}$.

Substitution into (4.13) confirms that the function r is identical for the two methods:

$$
r(z) = \frac{1 + \frac{1}{3}z}{1 - \frac{2}{3}z + \frac{1}{6}z^2}. \tag{4.15}
$$

To check A-stability we employ (4.14). Representing $z \in \mathbb{C}$ in polar coordinates, $z = \rho e^{i\theta}$, where $\rho > 0$ and $|\theta + \pi| < \frac{1}{2}\pi$, we query whether $|r(\rho e^{i\theta})| < 1$. This is equivalent to

$$
\left| 1 + \tfrac{1}{3}\rho e^{i\theta} \right|^2 < \left| 1 - \tfrac{2}{3}\rho e^{i\theta} + \tfrac{1}{6}\rho^2 e^{2i\theta} \right|^2,
$$

and hence to

$$
1 + \tfrac{2}{3}\rho\cos\theta + \tfrac{1}{9}\rho^2 < 1 - \tfrac{4}{3}\rho\cos\theta + \rho^2 \left(\tfrac{1}{3}\cos 2\theta + \tfrac{4}{9} \right) - \tfrac{2}{9}\rho^3 \cos\theta + \tfrac{1}{36}\rho^4.
$$

Rearranging terms, the condition for $\rho e^{i\theta} \in \mathcal{D}$ becomes

$$
2\rho \left(1 + \tfrac{1}{9}\rho^2 \right)\cos\theta < \tfrac{1}{3}\rho^2(1 + \cos 2\theta) + \tfrac{1}{36}\rho^4 = \tfrac{2}{3}\rho^2\cos^2\theta + \tfrac{1}{36}\rho^4,
$$

and this is obeyed for all $z \in \mathbb{C}^-$ since $\cos\theta < 0$ for all such z. Both methods are therefore A-stable.

A similar analysis can be applied to the Gauss–Legendre methods of Section 3.4, but the calculations become increasingly labour intensive for large values of ν. Fortunately, we are just about to identify a few shortcuts that render this job significantly easier.   ◇

Our first observation is that there is no need to check every $z \in \mathbb{C}^-$ to verify that a given rational function r originates in an A-stable method (such an r is called *A-acceptable*).

Lemma 4.3   *Let r be an arbitrary rational function that is not a constant. Then $|r(z)| < 1$ for all $z \in \mathbb{C}^-$ if and only if all the poles of r have positive real parts and $|r(it)| \leq 1$ for all $t \in \mathbb{R}$.*

Proof   If $|r(z)| < 1$ for all $z \in \mathbb{C}^-$ then, by continuity, $|r(z)| \leq 1$ for all $z \in \mathrm{cl}\,\mathbb{C}^-$. In particular, r is not allowed poles in the closed left half-plane and $|r(it)| \leq 1$, $t \in \mathbb{R}$.

To prove the converse we note that, provided its poles reside to the right of $i\mathbb{R}$, the rational function r is analytic in the closed set $\mathrm{cl}\,\mathbb{C}^-$. Therefore, and since r is not constant, it attains its maximum along the boundary. In other words, $|r(it)| \leq 1$, $t \in \mathbb{R}$, implies $|r(z)| < 1$, $z \in \mathbb{C}^-$, and the proof is complete.   ∎

The benefits of the lemma are apparent in the case of the function (4.15): the poles reside at $2 \pm i\sqrt{2}$, hence at the open right half-plane. Moreover, $|r(it)| \leq 1$, $t \in \mathbb{R}$, is equivalent to

$$\left|1 + \tfrac{1}{3}\mathrm{i}t\right|^2 \le \left|1 - \tfrac{2}{3}\mathrm{i}t - \tfrac{1}{6}t^2\right|^2, \qquad t \in \mathbb{R},$$

and hence to

$$1 + \tfrac{1}{9}t^2 \le 1 + \tfrac{1}{9}t^2 + \tfrac{1}{36}t^4, \qquad t \in \mathbb{R}.$$

The gain is even more spectacular for the two-stage Gauss–Legendre method, since in this case

$$r(z) = \frac{1 + \tfrac{1}{2}z + \tfrac{1}{12}z^2}{1 - \tfrac{1}{2}z + \tfrac{1}{12}z^2}$$

(although it is possible to evaluate this from the RK tableau in Section 3.4, a considerably easier derivation follows from the proof of the corollary to Theorem 4.6). Since the poles $3 \pm \mathrm{i}\sqrt{3}$ are in the open right half-plane and $|r(\mathrm{i}t)| \equiv 1$, $t \in \mathbb{R}$, the method is A-stable.

Our next result demonstrates that not every rational function r is likely to feature in (4.12).

Lemma 4.4   *Suppose that the solution sequence $\{y_n\}_{n=0}^{\infty}$, which is produced by applying a method of order p to the linear equation (4.7) with a constant step-size, obeys (4.12). Then necessarily*

$$r(z) = \mathrm{e}^z + \mathcal{O}\big(z^{p+1}\big), \qquad z \to 0. \tag{4.16}$$

Proof   Since $y_{n+1} = r(h\lambda)y_n$ and the exact solution, subject to the initial condition $y(t_n) = y_n$, is $\mathrm{e}^{h\lambda}y_n$, the relation (4.16) follows from the definition of order. ∎

We say that a function r that obeys (4.16) is of *order* p. This should not be confused with the order of a numerical method: it is easy to construct p-order methods with a function r whose order exceeds p, in other words, methods that exhibit superior order when applied to linear equations.

The lemma narrows down considerably the field of rational functions r that might occur in A-stability analysis. The most important functions exploit all available degrees of freedom to increase the order.

Theorem 4.5   *Given any integers $\alpha, \beta \ge 0$, there exists a unique function $\hat{r}_{\alpha/\beta} \in \mathbb{P}_{\alpha/\beta}$ such that*

$$\hat{r}_{\alpha/\beta} = \frac{\hat{p}_{\alpha/\beta}}{\hat{q}_{\alpha/\beta}}, \qquad \hat{q}_{\alpha/\beta}(0) = 1$$

and $\hat{r}_{\alpha/\beta}$ is of order $\alpha + \beta$. The explicit forms of the numerator and the denominator are respectively

$$\hat{p}_{\alpha/\beta}(z) = \sum_{k=0}^{\alpha} \binom{\alpha}{k} \frac{(\alpha+\beta)!}{(\alpha+\beta-k)!} z^k,$$

$$\hat{q}_{\alpha/\beta}(z) = \sum_{k=0}^{\beta} \binom{\beta}{k} \frac{(\alpha+\beta)!}{(\alpha+\beta-k)!} (-z)^k = \hat{p}_{\beta/\alpha}(-z). \tag{4.17}$$

Moreover, $\hat{r}_{\alpha/\beta}$ is (up to a rescaling of the numerator and the denominator by a nonzero multiplicative constant) the only member of $\mathbb{P}_{\alpha/\beta}$ of order $\alpha + \beta$ and no function in $\mathbb{P}_{\alpha/\beta}$ may exceed this order. ∎

The functions $\hat{r}_{\alpha/\beta}$ are called *Padé approximants* to the exponential. Most of the functions r that have been encountered so far are of this kind; thus (compare with (4.8), (4.10) and (4.15))

$$\hat{r}_{1/0}(z) = 1 + z, \qquad \hat{r}_{1/1} = \frac{1 + \frac{1}{2}z}{1 - \frac{1}{2}z}, \qquad \hat{r}_{1/2}(z) = \frac{1 + \frac{1}{3}z}{1 - \frac{2}{3}z + \frac{1}{6}z^2}.$$

Padé approximants can be classified according to whether they are A-acceptable. Obviously, we need $\alpha \le \beta$, otherwise $\hat{r}_{\alpha/\beta}$ cannot be bounded in $\mathbb{C}^-$. Surprisingly, the latter condition is not sufficient. It is not difficult to prove, for example, that $\hat{r}_{0/3}$ is not A-acceptable!

Theorem 4.6 (The Wanner–Hairer–Nørsett theorem) *The Padé approximant $\hat{r}_{\alpha/\beta}$ is A-acceptable if and only if $\alpha \le \beta \le \alpha + 2$.* ∎

Corollary *The Gauss–Legendre IRK methods are A-stable for every $\nu \ge 1$.*

Proof We know from Section 3.4 that a ν-stage Gauss–Legendre method is of order 2ν. By Lemma 4.1 the underlying function r belongs to $\mathbb{P}_{\nu/\nu}$ and, by Lemma 4.4, it approximates the exponential function to order 2ν. Therefore, according to Theorem 4.5, $r = \hat{r}_{\nu/\nu}$, a function that is A-acceptable by Theorem 4.6. It follows that the Gauss–Legendre method is A-stable. ∎

4.4 A-stability of multistep methods

Attempting to extend the definition of A-stability to the multistep method (2.8), we are faced with a problem: the implementation of an s-step method requires the provision of s values and only one of these is supplied by the initial condition. We will see in Chapter 6 how such values are derived in realistic computation. Here we adopt the attitude that a stable solution of the linear equation (4.7) is required *for all possible values of* $y_1, y_2, \ldots, y_{s-1}$. The justification of this pessimistic approach is that otherwise, even were we somehow to choose 'good' starting values, a small perturbation (e.g., a roundoff error) might well divert the solution trajectory toward instability. The reasons are similar to those already discussed in Section 4.1 in the context of the Euler method.

Let us suppose that the method (2.8) is applied to the solution of (4.7). The outcome is

$$\sum_{m=0}^{s} a_m y_{n+m} = h\lambda \sum_{m=0}^{s} b_m y_{n+m}, \qquad n = 0, 1, \ldots,$$

which we write in the form

$$\sum_{m=0}^{s} (a_m - h\lambda b_m)y_{n+m} = 0, \qquad n = 0, 1, \ldots. \tag{4.18}$$

The equation (4.18) is an example of a *linear difference equation*

$$\sum_{m=0}^{s} g_m x_{n+m} = 0, \qquad n = 0, 1, \ldots, \tag{4.19}$$

and it can be solved similarly to the more familiar linear differential equation

$$\sum_{m=0}^{s} g_m x^{(m)} = 0, \qquad t \geq t_0,$$

where the superscript indicates differentiation m times. Specifically, we form the characteristic polynomial

$$\eta(w) := \sum_{m=0}^{s} g_m w^m.$$

Let the zeros of η be $w_1, w_2, \ldots, w_q$, say, with multiplicities $k_1, k_2, \ldots, k_q$ respectively, where $\sum_{i=1}^{q} k_i = s$. Then the general solution of (4.19) is

$$x_n = \sum_{i=1}^{q} \left(\sum_{j=0}^{k_i-1} c_{i,j} n^j \right) w_i^n, \qquad n = 0, 1, \ldots. \tag{4.20}$$

The s constants $c_{i,j}$ are uniquely determined by the s starting values $x_0, x_1, \ldots, x_{s-1}$.

Lemma 4.7 *Let us suppose that the zeros (as a function of w) of*

$$\eta(z, w) := \sum_{m=0}^{s} (a_m - b_m z) w^m, \qquad z \in \mathbb{C},$$

are $w_1(z), w_2(z), \ldots, w_{q(z)}(z)$, *while their multiplicities are* $k_1(z), k_2(z), \ldots, k_{q(z)}(z)$ *respectively. The multistep method* (2.8) *is A-stable if and only if*

$$|w_i(z)| < 1, \quad i = 1, 2, \ldots, q(z) \qquad \text{for every} \quad z \in \mathbb{C}^-. \tag{4.21}$$

Proof As for (4.20), the behaviour of y_n is determined by the magnitude of the numbers $w_i(h\lambda)$, $i = 1, 2, \ldots, q(h\lambda)$. If all reside inside the complex unit disc then their powers decay faster than any polynomial in n, therefore $y_n \to 0$. Hence, (4.21) is sufficient for A-stability.

On the other hand, if $|w_1(h\lambda)| \geq 1$, say, then there exist starting values such that $c_{1,0} \neq 0$; therefore it is impossible for y_n to tend to zero as $n \to \infty$. We deduce that (4.21) is necessary for A-stability and conclude the proof. ∎

Instead of a single geometric component in (4.11), we have now a linear combination of several (in general, s) components to reckon with. This is the *quid pro quo* for using $s - 1$ starting values in addition to the initial condition, a practice whose perils have been already highlighted in the introduction to Chapter 2. According to Exercise 2.2, if a method is convergent then one of these components approximates

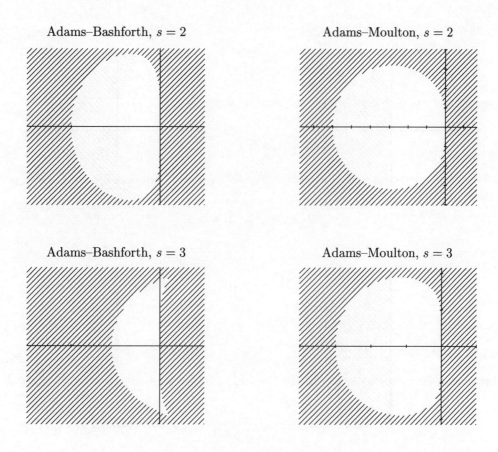

Figure 4.3   Linear stability domains $\mathcal{D}$ of Adams methods, explicit on the left and implicit on the right.

the exponential function to the same order as the order of the method: this is similar to Lemma 4.4. However, the remaining zeros are purely parasitic: we can attribute no meaning to them so far as approximation is concerned.

Figure 4.3 displays the linear stability domains of Adams methods, all drawn to the same scale. Notice first how small they are and that they are becoming progressively smaller with s. Next, pay attention to the difference between the explicit Adams–Bashforth and the implicit Adams–Moulton. In the latter case the stability domain, although not very impressive compared with those for other methods of Section 4.3, is substantially larger than for the explicit counterpart. This goes some way toward explaining the interest in implicit Adams methods, but more important reasons will be presented in Chapter 5.

However, as already mentioned in Chapter 2, Adams methods were never intended to cope with stiff equations. After all, this was the motivation for the introduction of backward differentiation formulae in Section 2.3. We turn therefore to Figure 4.4, which displays linear stability domains for BDF – and are disappointed.... True, the set $\mathcal{D}$ is larger than was the case with, say, the Adams–Moulton method. However,

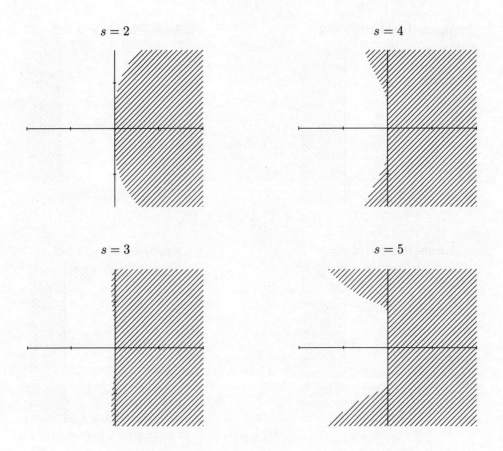

Figure 4.4   Linear stability domains $\mathcal{D}$ of BDF methods of orders $s = 2, 3, 4, 5$,
drawn to the same scale. Note that only $s = 2$ is A-stable.

Let us commence with the good news: the BDF is indeed A-stable in the case
$s = 2$. To demonstrate this we require two technical lemmas, which will be presented
with a comment in lieu of a complete proof.

Lemma 4.8   *The multistep method* (2.8) *is A-stable if and only if* $b_s > 0$ *and*

$$|w_1(\mathrm{i}t)|, \; |w_2(\mathrm{i}t)|, \; \ldots, \; |w_{q(\mathrm{i}t)}(\mathrm{i}t)| \leq 1, \qquad t \in \mathbb{R},$$

where $w_1, w_2, \ldots, w_{q(z)}$ *are the zeros of* $\eta(z, \cdot)$ *from Lemma 4.7.*

Proof   On the face of it, this is an exact counterpart of Lemma 4.3: $b_s > 0$ implies
analyticity in $\mathrm{cl}\,\mathbb{C}^-$ and the condition on the moduli of zeros extends the inequality on
$|r(z)|$. This is deceptive, since the zeros of $\eta(z, \cdot)$ do not reside in the complex plane
but in an s-sheeted *Riemann surface* over $\mathbb{C}$. This does not preclude the application
of the maximum principle, except that somewhat more sophisticated mathematical
machinery is required.   ∎

Lemma 4.9 (The Cohn–Schur criterion) *Both zeros of the quadratic* $\alpha w^2 + \beta w + \gamma$, *where* $\alpha, \beta, \gamma \in \mathbb{C}$, $\alpha \neq 0$, *reside in the closed complex unit disc if and only if*

$$|\alpha| \geq |\gamma|, \qquad \left||\alpha|^2 - |\gamma|^2\right| \geq \left|\alpha\bar{\beta} - \beta\bar{\gamma}\right|. \tag{4.22}$$

Proof This is a special case of a more general result, the *Cohn–Lehmer–Schur* criterion. The latter provides a finite algorithm to check whether a given complex polynomial (of any degree) has all its zeros in any closed disc in $\mathbb{C}$. ∎

Theorem 4.10 *The two-step BDF* (2.15) *is A-stable*

Proof We have
$$\eta(z, w) = (1 - \tfrac{2}{3}z)w^2 - \tfrac{4}{3}w + \tfrac{1}{3}.$$

Therefore $b_2 = \tfrac{2}{3}$ and the first A-stability condition of Lemma 4.8 is satisfied. To verify the second condition we choose $t \in \mathbb{R}$ and use Lemma 4.9 to ascertain that neither of the moduli of the zeros of $\eta(it, \cdot)$ exceeds unity. Consequently $\alpha = 1 - \tfrac{2}{3}it$, $\beta = -\tfrac{4}{3}$, $\gamma = \tfrac{1}{3}$ and we obtain

$$|\alpha|^2 - |\gamma|^2 = \tfrac{4}{9}(2 + t^2) > 0$$

and

$$(|\alpha|^2 - |\gamma|^2)^2 - |\alpha\bar{\beta} - \beta\bar{\gamma}|^2 = \tfrac{16}{81}t^4 \geq 0.$$

Consequently, (4.22) is satisfied and we deduce A-stability. ∎

Unfortunately, not only the 'positive' deduction from Figure 4.4 is true. The absence of A-stability in the BDF for $s \geq 2$ (of course, $s \leq 6$, otherwise the method is not convergent and we never use it!) is a consequence of a more general and fundamental result.

Theorem 4.11 (The Dahlquist second barrier) *The highest order of an A-stable multistep method* (2.8) *is two.* ∎

Comparing the Dahlquist second barrier with the corollary to Theorem 4.6, it is difficult to escape the impression that multistep methods are inferior to Runge–Kutta methods when it comes to A-stability. This, however, does not mean that they should not be used with stiff equations! Let us again look at Figure 4.4. Although $s = 3, 4, 5$ fail A-stability, it is apparent that for each stability domain $\mathcal{D}$ there exists $\alpha \in (0, \pi]$ such that the infinite wedge

$$\mathcal{V}_\alpha := \left\{ \rho e^{i\theta} : \rho > 0, \ |\theta + \pi| < \alpha \right\} \subseteq \mathbb{C}^-,$$

belongs to $\mathcal{D}$. In other words, provided all the eigenvalues of a linear ODE system reside in $\mathcal{V}_\alpha$, no matter how far away they are from the origin, there is no need to depress the step-size in response to stability restrictions. Methods with $\mathcal{V}_\alpha \subseteq \mathcal{D}$ are called A(α)-stable.[3] All BDF for $s \leq 6$ are A(α)-stable: in particular $s = 3$ corresponds to $\alpha = 86°2'$ – as Figure 4.4 implies, almost all of $\mathbb{C}^-$ resides in the linear stability domain.

---

[3] Numerical analysts, being (mostly) human, tend to express α in degrees rather than radians.

Comments and bibliography

Different aspects of stiff equations and A-stability form the theme of several monographs of varying degrees of sophistication and detail. Gear (1971) and Lambert (1991) are the most elementary, whereas Hairer & Wanner (1991) is a compendium of just about everything known in the subject area *circa* 1991. (No text, however, for obvious reasons, abbreviates the phrase 'linear stability domain'....)

Before we comment on a few themes connected with stability analysis, let us mention briefly two topics which, while tangential to the subject matter of this chapter, deserve proper reference. Firstly, the functions $\hat{r}_{\alpha/\beta}$, which have played a substantial role in Section 4.3, are a special case of general Padé approximants. Let f be an arbitrary function that is analytic in the neighbourhood of the origin. The function $\hat{r} \in \mathbb{P}_{\alpha/\beta}$ is said to be an $[\alpha/\beta]$ *Padé approximant* of f if

$$\hat{r}(z) = f(z) + \mathcal{O}\left(z^{\alpha+\beta+1}\right), \qquad z \to 0.$$

Padé approximants possess beautiful theory and have numerous applications, not just in the more obvious fields – approximation of functions, numerical analysis etc. – but also in analytic number theory: they are a powerful tool in many transcendentality proofs. Baker & Graves-Morris (1981) present an up-to-date account of the Padé theory. Secondly, the Cohn–Schur criterion (Lemma 4.9) is a special case of a substantially more general body of knowledge that allows us to locate the zeros of polynomials in specific portions of the complex plane by a finite number of operations on the coefficients (Marden, 1966). A familiar example is the *Routh–Hurwitz* criterion, which tests whether all the zeros reside in $\mathbb{C}^-$ and is an important tool in control theory.

The characterization of all A-acceptable Padé approximants to the exponential was the subject of a long-standing conjecture. Its resolution in 1978 by Gerhard Wanner, Ernst Hairer and Syvert Nørsett introduced the novel technique of *order stars* and was one of the great heroic tales of modern numerical mathematics. This technique can be also used to prove a far-reaching generalization of Theorem 4.11, as well as many other interesting results in the numerical analysis of differential equations. A comprehensive account of order stars features in Iserles & Nørsett (1991).

As far as A-stability for multistep equations is concerned, Theorem 4.11 implies that not much can be done. One obvious alternative, which has been mentioned in Section 4.4, is to relax the stability requirement, in which case the order barrier disappears altogether. Another possibility is to combine the multistep rationale with the Runge–Kutta approach and possibly to incorporate higher derivatives as well. The outcome, a *general linear method* (Butcher, 1987), circumvents the barrier of Theorem 4.11 but is subjected to a less severe restriction. As long as A-stability is required, we cannot improve the order by ranging beyond one-step methods!

We have mentioned in Section 4.2 that the justification of the linear model, which has led us into the concept of A-stability, is open to question when it comes to nonlinear equations. It is, however, a convenient starting point. The stability analysis of discretized nonlinear ODEs is these days a thriving industry!

Nonlinear ODEs come in many varieties and the set pattern of nonlinear stability studies typically commences by choosing a subset of equations $y' = f(t, y)$ that is broad enough to include many interesting specimens, yet sufficiently narrow to be subjected to a particular brand of analysis. The subset of earliest interest comprises the *monotone equations*, defined by the inequality

$$\langle x - y, f(t, x) - f(t, y) \rangle \leq 0, \qquad t \geq t_0, \tag{4.23}$$

for every $x, y \in \mathbb{R}^d$. We refer to Appendix section A.1.3 for vector norms and inner products.

Any ODE that obeys the inequality (4.23) is *dissipative*: given any two solutions x and y, where $x(t_0) = x_0$ and $y(t_0) = y_0$, it is true that $\|x(t) - y(t)\|$ is a monotonically decreasing function for all $t \geq 0$ ($\|\cdot\|$ is the vector norm induced by the inner product $\langle \cdot, \cdot \rangle$). In other words, different trajectories of the solution bunch together with time. This is an important piece of asymptotic information, which we might wish to retain in a numerical scheme. It has been proved by Germund Dahlquist that, essentially, the A-stability of multistep methods is necessary and sufficient to reproduce this behaviour correctly (Hairer & Wanner, 1991). The situation is somewhat more complicated with regard to Runge–Kutta methods and A-stability here falls short of being sufficient. As proved by John Butcher, a sufficient condition for a dissipative Runge–Kutta solution for a monotone system is that $b_1, b_2, \ldots, b_\nu \geq 0$ and the matrix $M = (m_{i,j})_{i,j=1,2,\ldots,\nu}$, where

$$m_{i,j} = b_i a_{i,j} + b_j a_{j,i} - b_i b_j, \qquad i,j = 1, 2, \ldots, \nu,$$

is nonnegative definite. See Dekker & Verwer (1984) for a comprehensive review of this subject.

Many other sets of ODEs have been singled out for attention by numerical stability specialists. A case in point is the family of autonomous equations $y' = f(y)$ where

$$\langle x, f(x) \rangle \leq \alpha - \beta \|x\|^2, \qquad x \in \mathbb{R}^d, \tag{4.24}$$

α and β being positive constants. Thus, if $g(t) := \|y(t)\|^2$ is small, the inequality allows $g'(t)$ to grow; note that, by (4.24),

$$\tfrac{1}{2}g'(t) = \frac{\mathrm{d}}{\mathrm{d}t}\langle y(t), y(t) \rangle = \langle y(t), f(y(t)) \rangle \leq \alpha - \beta \|y(t)\|^2, \qquad t \geq t_0.$$

However, when g is sufficiently large then (4.24) implies that it must eventually decrease. In other words, the solution lives forever in a compact ball but it is allowed a great deal of latitude there – unlike the case of monotone equations, it need not tend to a fixed point. This brings into the scope of stability analysis a whole range of fashionable phenomena: periodic orbits, motion in invariant manifolds, and even chaotic solutions. Ordinary differential equations that obey (4.24), as well as other families of nonlinear ODEs with interesting dynamics, are analysed nowadays by techiques borrowed from the theory of nonlinear dynamical systems (Stuart, 1994). Favourable features of Runge–Kutta methods can be often phrased in terms of the matrix M and the vector b.

Hamiltonian systems

$$p' = \frac{\partial H(p, q)}{\partial q}, \qquad q' = -\frac{\partial H(p, q)}{\partial p},$$

where H is the *Hamiltonian function*, form a family of ODEs of truly singular importance. They are ubiquitous in mathematical physics and quantum chemistry and replete with features that are, on the face of it, profoundly difficult to model numerically: invariant manifolds and a large variety of conserved integrals. Remarkably, it is possible to prove that certain Runge–Kutta methods retain the most important attributes of Hamiltonian systems under discretization – all we need is $M = O$ (Sanz-Serna & Calvo, 1994). The famous matrix M again!

Baker, G.A. and Graves-Morris, P. (1981), *Padé Approximants*, Addison–Wesley, Reading, MA.

Butcher, J.C. (1987), *The Numerical Analysis of Ordinary Differential Equations*, John Wiley, New York.

Dekker, K. and Verwer, J.G. (1984), *Stability of Runge–Kutta Methods for Stiff Nonlinear Differential Equations,* North-Holland, Amsterdam.

Gear, C.W. (1971), *Numerical Initial Value Problems in Ordinary Differential Equations,* Prentice–Hall, Englewood Cliffs, NJ.

Hairer, E. and Wanner, G. (1991), *Solving Ordinary Differential Equations II: Stiff Problems and Differential-algebraic Equations,* Springer-Verlag, Berlin.

Iserles, A. and Nørsett, S.P. (1991), *Order Stars,* Chapman & Hall, London.

Lambert, J.D. (1991), *Numerical Methods for Ordinary Differential Systems,* Wiley, London.

Marden, M. (1966), *Geometry of Polynomials,* American Mathematical Society, Providence, RI.

Sanz-Serna, J.M. and Calvo, M.P. (1994), *Numerical Hamiltonian Problems,* Chapman & Hall, London.

Stuart, A.M. (1994), Numerical analysis of dynamical systems, *Acta Numerica* **3**, 467–572.

Exercises

4.1 Let $y' = \Lambda y$, $y(t_0) = y_0$, be solved (with a constant step-size $h > 0$) by a one-step method with a function r that obeys the relation (4.12). Suppose that a nonsingular matrix V and a diagonal matrix D exist such that $\Lambda = VDV^{-1}$. Prove that there exist vectors $x_1, x_2, \ldots, x_d \in \mathbb{R}^d$ such that

$$y(t_n) = \sum_{j=1}^{d} e^{t_n \lambda_j} x_j, \qquad n = 0, 1, \ldots,$$

and

$$y_n = \sum_{j=1}^{d} [r(h\lambda)]^n x_j, \qquad n = 0, 1, \ldots,$$

where $\lambda_1, \lambda_2, \ldots, \lambda_d$ are the eigenvalues of Λ. Therefore, deduce that the values of x_1, and of x_2, given in (4.3) and (4.4) are identical.

4.2* Consider the solution of $y' = \Lambda y$ where

$$\Lambda = \begin{bmatrix} \lambda & 1 \\ 0 & \lambda \end{bmatrix}, \qquad \lambda \in \mathbb{C}^-.$$

a Prove that

$$\Lambda^n = \begin{bmatrix} \lambda^n & n\lambda^{n-1} \\ 0 & \lambda^n \end{bmatrix}, \qquad n = 0, 1, \ldots.$$

b Let g be an arbitrary function that is analytic about the origin. The 2×2 matrix $g(\Lambda)$ can be defined by substituting powers of Λ into the Taylor expansion of g. Prove that

$$g(t\Lambda) = \begin{bmatrix} g(t\lambda) & tg'(t\lambda) \\ 0 & g(t\lambda) \end{bmatrix}.$$

c By letting $g(z) = e^z$ prove that $\lim_{t\to\infty} y(t) = 0$.

d Suppose that $y' = \Lambda y$ is solved with a Runge–Kutta method, using a constant step $h > 0$. Let r be the function from Lemma 4.1. Letting $g = r$, obtain the explicit form of $[r(h\Lambda)]^n$, $n = 0, 1, \ldots$.

e Prove that if $h\lambda \in \mathcal{D}$, where $\mathcal{D}$ is the linear stability domain of the Runge–Kutta method, then $\lim_{n\to\infty} y_n = 0$.

4.3* This question is concerned with the relevance of the linear stability domain to the numerical solution of *inhomogeneous* linear systems.

a Let Λ be a nonsingular matrix. Prove that the solution of $y' = \Lambda y + a$, $y(t_0) = y_0$, is

$$y(t) = e^{(t-t_0)\Lambda} y_0 + \Lambda^{-1}[e^{(t-t_0)\Lambda} - I]a, \qquad t \geq t_0.$$

Thus, deduce that if Λ has a full set of eigenvectors and all its the eigenvalues reside in $\mathbb{C}^-$ then $\lim_{t\to\infty} y(t) = -\Lambda^{-1}a$.

b Assuming for simplicity's sake that the underlying equation is scalar, i.e. $y' = \lambda y + a$, $y(t_0) = y_0$, prove that a single step of the Runge–Kutta method (3.9) results in

$$y_{n+1} = r(h\lambda)y_n + q(h\lambda), \qquad n = 0, 1, \ldots,$$

where r is given by (4.13) and

$$q(z) := hab^T(I - zA)^{-1}c \in \mathbb{P}_{(\nu-1)/\nu}, \qquad z \in \mathbb{C}.$$

c Deduce, by induction or otherwise, that

$$y_n = [r(h\lambda)]^n y_0 + \left\{ \frac{[r(h\lambda)]^n - 1}{r(h\lambda) - 1} \right\} q(h\lambda), \qquad n = 0, 1, \ldots.$$

d Assuming that $h\lambda \in \mathcal{D}$, prove that $\lim_{n\to\infty} y_n$ exists and is bounded.

4.4 Determine all values of θ such that the theta method (1.13) is A-stable.

4.5 Prove that for every ν-stage explicit Runge–Kutta method (3.5) of order ν it is true that

$$r(z) = \sum_{k=0}^{\nu} \frac{1}{k!} z^k, \qquad z \in \mathbb{C}.$$

4.6 Evaluate explicitly the function r for the following Runge–Kutta methods.

a
$$\begin{array}{c|cc}
0 & 0 & 0 \\
\frac{2}{3} & \frac{1}{3} & \frac{1}{3} \\
\hline
 & \frac{1}{4} & \frac{3}{4}
\end{array},$$

b
$$\begin{array}{c|cc}
\frac{1}{6} & \frac{1}{6} & 0 \\
\frac{5}{6} & \frac{2}{3} & \frac{1}{6} \\
\hline
 & \frac{1}{2} & \frac{1}{2}
\end{array},$$

c
$$\begin{array}{c|ccc}
0 & 0 & 0 & 0 \\
\frac{1}{2} & \frac{1}{4} & \frac{1}{4} & 0 \\
1 & 0 & 1 & 0 \\
\hline
 & \frac{1}{6} & \frac{2}{3} & \frac{1}{6}
\end{array}.$$

Are these methods A-stable?

4.7   Prove that the Padé approximant $\hat{r}_{0/3}$ is not A-acceptable.

4.8   Determine the order of the two-step method

$$\boldsymbol{y}_{n+2} - \boldsymbol{y}_n = \tfrac{2}{3}h\left[\boldsymbol{f}(t_{n+2}, \boldsymbol{y}_{n+2}) + \boldsymbol{f}(t_{n+1}, \boldsymbol{y}_{n+1}) + \boldsymbol{f}(t_n, \boldsymbol{y}_n)\right], \quad n = 0, 1, \ldots$$

Is it A-stable?

4.9   The two-step method

$$\boldsymbol{y}_{n+2} - \boldsymbol{y}_n = 2h\boldsymbol{f}(t_{n+1}, \boldsymbol{y}_{n+1}), \qquad n = 0, 1, \ldots \qquad (4.25)$$

is called the *explicit midpoint rule*.

a Denoting by $w_1(z)$ and $w_2(z)$ the zeros of the underlying function $\eta(z, \cdot)$, prove that $w_1(z)w_2(z) \equiv -1$ for all $z \in \mathbb{C}$.

b Show that $\mathcal{D} = \emptyset$.

c We say that $\tilde{\mathcal{D}}$ is a *weak linear stability domain* of a numerical method if, when applied to the scalar linear test equation, it produces a uniformly bounded solution sequence. (It is easy to see that $\tilde{\mathcal{D}} = \mathrm{cl}\,\mathcal{D}$ for most methods of interest.) Determine explicitly $\tilde{\mathcal{D}}$ for the method (4.25).

The method (4.25) will feature again in Chapters 13–14, in the guise of the *leapfrog* scheme.

4.10   Prove that if the multistep method (2.8) is convergent then $0 \in \partial\mathcal{D}$.

4.11*   We say that the autonomous ODE system $\boldsymbol{y}' = \boldsymbol{f}(\boldsymbol{y})$ obeys a *quadratic conservation law* if there exists a symmetric, positive-definite matrix S such that for every initial condition $\boldsymbol{y}(t_0) = \boldsymbol{y}_0$ it is true that $[\boldsymbol{y}(t)]^T S\boldsymbol{y}(t) \equiv \boldsymbol{y}_0^T S\boldsymbol{y}_0$, $t \geq t_0$.

a Show that the d-dimensional linear system $\boldsymbol{y}' = \Lambda\boldsymbol{y}$ obeys a quadratic conservation law if d is even and Λ is of the form

$$\Lambda = \begin{bmatrix} O & \Theta \\ -\Theta & O \end{bmatrix},$$

where Θ is a symmetric, positive-definite $\lfloor d/2 \rfloor \times \lfloor d/2 \rfloor$ matrix. [*Hint: Try the matrix*

$$S = \begin{bmatrix} \Theta^{-1} & O \\ O & \Theta^{-1} \end{bmatrix},$$

having first proved that it is positive definite.]

b Prove that an autonomous ODE obeys a quadratic conservation law if and only if $\boldsymbol{x}^T S\boldsymbol{f}(\boldsymbol{x}) \equiv \boldsymbol{0}$ for every $\boldsymbol{x} \in \mathbb{R}^d$.

c Suppose that an autonomous ODE that obeys a quadratic conservation law is solved with the implicit midpoint rule (1.12), using a constant step-size. Prove that $\boldsymbol{y}_n^T S\boldsymbol{y}_n \equiv \boldsymbol{y}_0^T S\boldsymbol{y}_0$, $n = 0, 1, \ldots$.

5

Error control

5.1 Numerical software *vs* numerical mathematics

There comes a point in every exposition of numerical analysis when the theme shifts from the familiar mathematical progression of definitions, theorems and proofs to the actual ways and means whereby computational algorithms are implemented. This point is sometimes accompanied by an air of anguish and perhaps disdain: we abandon the palace of the Queen of Sciences for the lowly shop floor of a software engineer. *Nothing can be further from the truth!* Devising an algorithm that fulfils its goal accurately, robustly and economically is an intellectual challenge equal to the best in mathematical research.

In Chapters 1–4 we have seen a multitude of methods for the numerical solution of the ODE system

$$y' = f(t, y), \quad t \geq t_0, \qquad y(t_0) = y_0. \tag{5.1}$$

In the present chapter we are about to study how to incorporate a method into a *computational package*. It is important to grasp that, when it comes to software design, a time-stepping method is just one – albeit very important – component. A good analogy is the design of a motor car. The time-stepping method is like the engine: it powers the vehicle along. A car without an engine is just about useless: a multitude of other components – wheels, chassis, transmission – are essential for an orderly operation. Different parts of the system should not be optimized on their own, but as a part of an integrated plan; there is little point in fitting a Formula 1 racing car engine into a family saloon. Moreover, the very goal of optimization is problem-dependent: do we want to optimize for speed? economy? reliability? marketability? In a well-designed car the right components are combined in such a way that they operate together as required, reliably and smoothly. The same is true for a computational package.

A user of a software package for ODEs, say, typically does not (and should not!) care about the particular choice of method or, for that matter, for the other 'operating parts' – error and step-size control, solution of nonlinear algebraic equations, choice of starting values and of the initial step, visualization of the numerical solution etc. As far as a user is concerned, a computational package is simply a tool.

The tool designer – be it a numerical analyst or a software engineer – must adopt a more discerning view. The package is no longer a black box but an integrated system, which can be represented in the following flowchart:

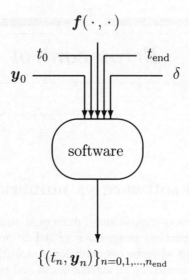

$$\{(t_n, \boldsymbol{y}_n)\}_{n=0,1,\ldots,n_{\mathrm{end}}}$$

The inputs are not just the function $\boldsymbol{f}$, the starting point t_0, the initial value $\boldsymbol{y}_0$ and the end-point t_{end}, but also the error *tolerance* $\delta > 0$; we wish the numerical error in, say, the Euclidean norm to be within δ. The output is the computed solution sequence at the points $t_0 < t_1 < \cdots < t_{\mathrm{end}}$, which, of course, are not equi-spaced.

We hasten to say that the above is actually the simplest possible model for a computational package, but it will do for expositional purposes. In general, the user might be expected to specify whether (5.1) is stiff and to express a range of preferences with regard to the form of the output. An increasingly important component in the design of a modern software package is *visualization* – it is difficult to absorb information from long arrays of numbers and its display in the form of time series, phase diagrams, Poincaré sections etc. often makes a great deal of difference. Writing, debugging, testing and documenting modern, advanced, broad-purpose software for ODEs is a highly professional and time-demanding enterprise.

In the present chapter we plan to elaborate a major component of any computational package, *the mechanism whereby numerical error is estimated* in the course of solution *and controlled by means of step-size changes.* Chapter 6 is devoted to another aspect, namely solution of the nonlinear algebraic systems that occur whenever implicit methods are applied to the system (5.1).

We will describe a number of different devices for the estimation of the *local error,* i.e. the error incurred when we integrate from t_n to t_{n+1} under the assumption that $\boldsymbol{y}_n$ is 'exact'. This should not be confused with the *global error,* namely the difference between $\boldsymbol{y}_n$ and $\boldsymbol{y}(t_n)$ for all $n = 0, 1, \ldots, n_{\mathrm{end}}$. Clever procedures for the estimation of global error are fast becoming standard in modern software packages.

The error-control devices of this chapter are applied to three relatively simple systems (5.1): the *van der Pol* equation

$$
\begin{aligned}
y_1' &= y_2, \\
y_2' &= (1 - y_1^2)y_2 - y_1,
\end{aligned}
\qquad 0 \le t \le 25,
\qquad
\begin{aligned}
y_1(0) &= \tfrac{1}{2}, \\
y_2(0) &= \tfrac{1}{2};
\end{aligned}
\qquad (5.2)
$$

the *Mathieu* equation

$$
\begin{aligned}
y_1' &= y_2, \\
y_2' &= -(2 - \cos 2t)y_1,
\end{aligned}
\qquad 0 \le t \le 30,
\qquad
\begin{aligned}
y_1(0) &= 1, \\
y_2(0) &= 0;
\end{aligned}
\tag{5.3}
$$

and the *Curtiss–Hirschfelder* equation

$$
y' = -50(y - \cos t), \qquad 0 \le t \le 10, \qquad y(0) = 1.
\tag{5.4}
$$

The first two equations are not stiff, while (5.4) is moderately so. The solution of the van der Pol equation (which is more commonly written as a second-order equation $y'' - \varepsilon(1 - y^2)y' + y = 0$; here we take $\varepsilon = 1$) models electrical circuits connected with triode oscillators. It is well known that, for every initial value, the solution tends to a periodic curve (see Figure 5.1). The Mathieu equation (which, likewise, is usually written in the second-order form $y'' + (a - b\cos 2t)y = 0$, here with $a = 2$ and $b = 1$) arises in the analysis of the vibrations of an elliptic membrane, and also in celestial mechanics. Its (nonperiodic) solution remains forever bounded, without approaching a fixed point (see Figure 5.2). Finally, the solution of the Curtiss–Hirschfelder equation (which has no known significance except as a good test case for computational algorithms) is

$$
y(t) = \tfrac{2500}{2501}\cos t + \tfrac{50}{2501}\sin t + \tfrac{1}{2501}\,\mathrm{e}^{-50t}, \qquad t \ge 0,
$$

and approaches a periodic curve at an exponential speed (see Figure 5.3).

We assume throughout this chapter that $\boldsymbol{f}$ is as smooth as required.

5.2 The Milne device

Let

$$
\sum_{m=0}^{s} a_m \boldsymbol{y}_{n+m} = h \sum_{m=0}^{s} b_m \boldsymbol{f}(t_{n+m}, \boldsymbol{y}_{n+m}), \qquad n = 0, 1, \ldots, \qquad a_s = 1,
\tag{5.5}
$$

be a given convergent multistep method of order p. The goal of assessing the local error in (5.5) is attained by employing another convergent multistep method of the same order, which we write in the form

$$
\sum_{m=q}^{s} \tilde{a}_m \boldsymbol{x}_{n+m} = h \sum_{m=q}^{s} \tilde{b}_m \boldsymbol{f}(t_{n+m}, \boldsymbol{x}_{n+m}), \qquad n \ge \max\{0, -q\} \qquad \tilde{a}_s = 1.
\tag{5.6}
$$

Here $q \le s - 1$ is an integer, which might be of either sign; the main reason for allowing negative q is that we wish to align both methods so that they approximate at the same point t_{n+s} in the nth step. Of course, $\boldsymbol{x}_{n+m} = \boldsymbol{y}_{n+m}$ for $m = \min\{0, q\}, \min\{0, q\} + 1, \ldots, s - 1$.

According to Theorem 2.1, the method (5.5), say, is of order p if and only if

$$
\rho(w) - \sigma(w)\ln w = c(w - 1)^{p+1} + \mathcal{O}\big(|w - 1|^{p+2}\big), \qquad w \to 1,
$$

where $c \neq 0$ and

$$\rho(w) = \sum_{m=0}^{s} a_m w^m, \qquad \sigma(w) = \sum_{m=0}^{s} b_m w^m.$$

By expanding $\boldsymbol{\psi}$ one term further in the proof of Theorem 2.1, it is easy to demonstrate that, provided $\boldsymbol{y}_n, \boldsymbol{y}_{n+1}, \ldots, \boldsymbol{y}_{n+s-1}$ are assumed to be error free, it is true that

$$\boldsymbol{y}(t_{n+s}) - \boldsymbol{y}_{n+s} = ch^{p+1}\boldsymbol{y}^{(p+1)}(t_{n+s}) + \mathcal{O}\big(h^{p+2}\big), \qquad h \to 0. \tag{5.7}$$

The number c is termed the *(local) error constant* of the method (5.5).[1] Let $\tilde{c}$ be the error constant of (5.6) and assume that we have selected the method in such a way that $\tilde{c} \neq c$. Therefore

$$\boldsymbol{y}(t_{n+s}) - \boldsymbol{x}_{n+s} = \tilde{c}h^{p+1}\boldsymbol{y}^{(p+1)}(t_{n+s}) + \mathcal{O}\big(h^{p+2}\big), \qquad h \to 0.$$

We subtract this expression from (5.7) and disregard the $\mathcal{O}\big(h^{p+2}\big)$ terms. The outcome is $\boldsymbol{x}_{n+s} - \boldsymbol{y}_{n+s} \approx (c - \tilde{c})h^{p+1}\boldsymbol{y}^{(p+1)}(t_{n+s})$, hence

$$h^{p+1}\boldsymbol{y}^{(p+1)}(t_{n+s}) \approx \frac{1}{c - \tilde{c}}(\boldsymbol{x}_{n+s} - \boldsymbol{y}_{n+s}).$$

Substitution into (5.7) yields an estimate of the local error, namely

$$\boldsymbol{y}(t_{n+s}) - \boldsymbol{y}_{n+s} \approx \frac{c}{c - \tilde{c}}(\boldsymbol{x}_{n+s} - \boldsymbol{y}_{n+s}). \tag{5.8}$$

This method of assessing the local error is known as the *Milne device*. Recall that our critical requirement is to maintain the local error at less than the tolerance δ. A naive approach is *error control per step*, namely to require that the *local error* κ satisfies

$$\kappa \leq \delta,$$

where

$$\kappa = \left| \frac{c}{c - \tilde{c}} \right| \|\boldsymbol{x}_{n+s} - \boldsymbol{y}_{n+s}\|$$

originates in (5.8). A better requirement, *error control per unit step*, incorporates a crude global consideration into the local estimate. It is based on the assumption that the accumulation of global error occurs roughly at a constant pace and is allied to our observation in Chapter 1 that the global error behaves like $\mathcal{O}(h^p)$. Therefore, the smaller the step-size, the more stringent requirement must we place upon κ; the right inequality is

$$\kappa \leq h\delta. \tag{5.9}$$

This is the criterion that we adopt in the remainder of this exposition.

Suppose that we have executed a single time step, thereby computing a candidate solution $\boldsymbol{y}_{n+s}$. We use (5.9), where κ has been evaluated by the Milne device (or by other means), to decide whether $\boldsymbol{y}_{n+s}$ is an acceptable approximation to $\boldsymbol{y}(t_{n+s})$. If

---

[1]The *global* error constant is defined as $c/\rho'(1)$, for reasons that are related to the theme of Exercise 2.2, but are outside the scope of our exposition.

not, the time step is rejected: we go back to t_{n+s-1}, halve h and resume time-stepping. If, however, (5.9) holds, the new value is acceptable and we advance to t_{n+s}. Moreover, if κ is significantly smaller that $h\delta$ – for example, if $\kappa < \frac{1}{10}h\delta$ – we take this as an indication that the time step is too small (hence, wasteful) and double it. A simple scheme of this kind might be conveniently represented in flowchart form:

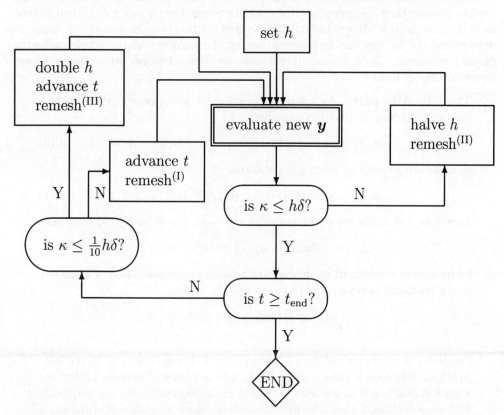

except that each box in the flowchart hides a multitude of sins! In particular, observe the need to 'remesh' the variables: multistep methods require that starting values for each step are provided on an equally spaced grid and this is no longer true when h is amended. We need then to approximate starting values, typically by polynomial interpolation (A.2.2.3–A.2.2.5). In each iteration we need $\hat{s} := s + \max\{0, q\}$ vectors $\boldsymbol{y}_{n+\min\{0,-q\}}, \ldots, \boldsymbol{y}_{n+s-1}$, which we rename $\boldsymbol{w}_1, \boldsymbol{w}_2, \ldots, \boldsymbol{w}_{\hat{s}}$ respectively. There are three possible cases:

(1) **h is unamended** We let $\boldsymbol{w}_j^{\text{new}} = \boldsymbol{w}_{j+1}$, $j = 1, 2, \ldots, \hat{s} - 1$, and $\boldsymbol{w}_{\hat{s}}^{\text{new}} = \boldsymbol{y}_{n+s}$.

(2) **h is halved** The values $\boldsymbol{w}_{\hat{s}-2j+1}^{\text{new}} = \boldsymbol{w}_{\hat{s}-j}$, $j = 1, 2, \ldots, \lfloor \hat{s}/2 \rfloor$, survive, while the rest, which approximate values at midpoints of the old grid, need to be computed by interpolation.

(3) **h is doubled** Here $\boldsymbol{w}_{\hat{s}}^{\text{new}} = \boldsymbol{y}_{n+s}$ and $\boldsymbol{w}_{\hat{s}-j}^{\text{new}} = \boldsymbol{w}_{\hat{s}-2j}$, $j = 1, 2, \ldots, \hat{s} - 1$. This requires an extra $\hat{s} - 1$ vectors, $\boldsymbol{w}_{-\hat{s}+2}, \ldots, \boldsymbol{w}_0$, that have not been defined

above. The remedy is simple, at least in principle: we need to carry forward in the previous two remeshings at least $2\hat{s} - 1$ vectors to allow the scope for step-size doubling. This procedure may impose a restriction on consecutive step-size doublings.

A glance at the flowchart affirms our claim in Section 5.1 that the specific method used to advance the time-stepping (the 'evaluate new y' box) is just a single instrument in a large orchestra. It might well be the first violin, but the quality of music is determined not by any one instrument but by the harmony of the whole orchestra playing in unison. It is the conductor, not the first violinist, whose name looms largest on the billboard!

◇ **The TR–AB2 pair**   As a simple example, let us suppose that we employ the two-step Adams–Bashforth method

$$x_{n+1} - x_n = \tfrac{1}{2}h[3f(t_n, x_n) - f(t_{n-1}, x_{n-1})] \tag{5.10}$$

to monitor the error of the trapezoidal rule

$$y_{n+1} - y_n = \tfrac{1}{2}h[f(t_{n+1}, y_{n+1}) + f(t_n, y_n)]. \tag{5.11}$$

Therefore $\tilde{s} = 2$, the error constants are $c = -\tfrac{1}{12}$, $\tilde{c} = \tfrac{5}{12}$ and (5.9) becomes

$$\|x_{n+1} - y_{n+1}\| \le 4h\delta.$$

Interpolation is required upon step-size halving at a single value of t, namely at the midpoint between (the old values of) t_n and t_{n+1}:

$$w_1^{\text{new}} = \tfrac{1}{8}(3w_2 + 6w_1 - w_0).$$

Figure 5.1 displays the solution of the van der Pol equation (5.2) by the TR–AB2 pair with tolerances $\delta = 10^{-3}, 10^{-4}$. The sequence of step-sizes attests to a marked reluctance on the part of the algorithm to experiment too frequently with step-doubling. This is healthy behaviour, since an excess of optimism is bound to breach the inequality (5.9) and is wasteful. Note, by the way, how strongly the step-size sequence correlates with the size of y_2', which, for the van der Pol equation, measures the 'awkwardness' of the solution – it is easy to explain this feature from the familiar phase portrait of (5.2).

The global error, as displayed in the bottom two graphs, is of the right order of magnitude and, as we might expect, slowly accumulates with time. This is typical of non-stiff problems like (5.2).

Similar lessons can be drawn from the Mathieu equation (5.3) (see Figure 5.2). It is perhaps more difficult to find a single characteristic of the solution that accounts for the variation in h, but it is striking how closely the step sequences for $\delta = 10^{-3}$ and $\delta = 10^{-4}$ correlate. The global error accumulates markedly faster but is still within what can be expected from the general theory and the accepted wisdom. The goal being to incur an error of at most δ in a unit-length interval, a final error of $(t_{\text{end}} - t_0)\delta$ is to be expected at the right-hand endpoint of the interval.

Figure 5.1   The top figure displays the two solution components of the
van der Pol equation (5.2) in the interval [0, 25]. The other figures are concerned with the Milne device, applied with the pair (5.10), (5.11). The second and third figures each feature the sequence of step-sizes for tolerances
δ equal to 10^{-3} and 10^{-4} respectively, while the lowest two figures show
the (exact) global error for these values of δ.

Figure 5.2 The top figure displays the two solution components of the
Mathieu equation (5.3) in the interval $[0, 30]$. The other figures are concerned with the Milne device, applied with the pair (5.10), (5.11). The
second and the third figures each feature the sequence of step-sizes for tolerances δ equal to 10^{-3} and 10^{-4} respectively, while the lowest two figures
show the (exact) global error for these values of δ.

Finally, Figure 5.3 displays the behaviour of the TR–AB2 pair for the mildly stiff equation (5.4). The first interesting observation, looking at the step sequences, is that the step-sizes are quite large, at least for $\delta = 10^{-3}$. Had we tried to solve this equation with the Euler method, we would have needed to impose $h < \frac{1}{25}$ to prevent instabilities, whereas the trapezoidal rule chugs along happily with h occsionally exceeding $\frac{1}{3}$. This is an important point to note since the stability analysis of Chapter 4 has been restricted to constant steps. It is worthwile to record that, at least in a single computational example, A-stability allows the trapezoidal rule to exploit step sizes solely in pursuit of accuracy.

The accumulation of global errors displays a pattern characteristic of stiff equations. Provided the method is adequately stable, global error does not accumulate at all and is often significantly smaller than δ! (The occasional jumps in the error in Figure 5.3 are probably attributable to the increase in h and might well have been eliminated altogether with sufficient fine-tuning of the computational scheme.)

Note that, of course, (5.10) has exceedingly poor stability characteristics (see Figure 4.3). This is not a handicap, since the Adams–Bashforth method is used solely for local error control. To demonstrate this point, in Figure 5.4 we display the error for the Curtiss–Hirschfelder equation (5.4) when, in lieu of (5.10), we employ the A-stable backward differentiation formula method (2.15). Evidently, not much changes! $\diamond$

We conclude this section by remarking again that execution of a variable-step code requires a multitude of choices and a great deal of fine-tuning. The need for brevity prevents us from commenting, for example, about an appropriate procedure for the choice of the initial step-size and of the starting values $\boldsymbol{y}_1, \boldsymbol{y}_2, \ldots, \boldsymbol{y}_{s-1}$.

5.3 Embedded Runge–Kutta methods

The comfort of a single constant whose magnitude reflects (at least for small h) the local error κ is denied us in the case of Runge–Kutta methods, (3.9). In order to estimate κ we need to resort to a different device, which again is based upon running two methods in tandem – one, of order p, to provide a candidate for solution and the other, of order $\tilde{p} \geq p + 1$, to control the error.

In line with (5.5) and (5.6), we denote by $\boldsymbol{y}_{n+1}$ the candidate solution at t_{n+1} obtained from the p-order method, whereas the solution there obtained from the higher-order scheme is $\boldsymbol{x}_{n+1}$. We have

$$\boldsymbol{y}_{n+1} = \tilde{\boldsymbol{y}}(t_{n+1}) + \boldsymbol{\ell}h^{p+1} + \mathcal{O}(h^{p+2}), \tag{5.12}$$
$$h \to 0,$$
$$\boldsymbol{x}_{n+1} = \tilde{\boldsymbol{y}}(t_{n+1}) + \mathcal{O}(h^{p+2}), \tag{5.13}$$

where $\boldsymbol{\ell}$ is a vector that depends on the equation (5.1) (but not upon h) and $\tilde{\boldsymbol{y}}$ is the exact solution of (5.1) with the initial value $\tilde{\boldsymbol{y}}(t_n) = \boldsymbol{y}_n$. Subtracting (5.13) from

Figure 5.3 The top figure displays the solution of the Curtiss–Hirschfelder equation (5.4) in the interval $[0, 10]$. The other figures are concerned with the Milne device, applied with the pair (5.10), (5.11). The second to fourth figures feature the sequence of step-sizes for tolerances δ equal to $10^{-3}, 10^{-4}$ and 10^{-5} respectively, while the bottom figures show the (exact) global error for these values of δ.

Figure 5.4   Global errors for the numerical solution of the Curtiss–Hirschfelder equation (5.4) by the TR–BDF2 pair with $\delta = 10^{-3}, 10^{-4}, 10^{-5}$.

(5.12), we obtain $\ell h^{p+1} \approx \boldsymbol{y}_{n+1} - \boldsymbol{x}_{n+1}$, the outcome being the error estimate

$$\kappa = \|\boldsymbol{y}_{n+1} - \boldsymbol{x}_{n+1}\|. \tag{5.14}$$

As soon as κ is available, we may proceed as in Section 5.2, except that, having opted for one-step methods, we are spared all the awkward and time-consuming minutiae of remeshing each time h is changed.

A naive application of the above approach requires the doubling, at the very least, of the expense of calculation, since we need to compute both $\boldsymbol{y}_{n+1}$ and $\boldsymbol{x}_{n+1}$. This is unacceptable, since, as a rule, the cost of error control should be marginal in comparison to the cost of the main scheme.[2] However, when an ERK method (3.5) is used for the main scheme it is possible to choose the two methods in such a way that the extra expense is small. Let us thus denote by

$$\begin{array}{c|c} \boldsymbol{c} & A \\ \hline & \boldsymbol{b}^T \end{array} \qquad \text{and} \qquad \begin{array}{c|c} \tilde{\boldsymbol{c}} & \tilde{A} \\ \hline & \tilde{\boldsymbol{b}}^T \end{array}$$

the pth-order method and the higher-order method, of ν and $\tilde{\nu}$ stages, respectively. The main idea is to choose

$$\tilde{\boldsymbol{c}} = \begin{bmatrix} \boldsymbol{c} \\ \hat{\boldsymbol{c}} \end{bmatrix}, \qquad \tilde{A} = \begin{bmatrix} A & O \\ \hat{A} & \end{bmatrix},$$

where $\hat{\boldsymbol{c}} \in \mathbb{R}^{\tilde{\nu}-\nu}$ and $\hat{A}$ is a $(\tilde{\nu} - \nu) \times \nu$ matrix consistent with $\tilde{A}$'s being strictly lower

---

[2]Note that we have used in Section 5.2 an explicit method, the Adams–Bashforth scheme, to control the error of the implicit trapezoidal rule. This is consistent with the latter remark.

triangular. In this case the first ν vectors $\boldsymbol{\xi}_1, \boldsymbol{\xi}_2, \ldots, \boldsymbol{\xi}_\nu$ are the same in both methods and the cost of the error controller is virtually the same as the cost of the higher-order method. We say the the first method is *embedded* in the second and that, together, they form an *embedded Runge–Kutta pair*. The tableau notation is

$$\begin{array}{c|c} \tilde{c} & \tilde{A} \\ \hline & b^T \\ \hline & \tilde{b}^T \end{array}.$$

A well-known example of an embedded RK pair is the Fehlberg method, with $p = 4$, $\tilde{p} = 5$, $\nu = 5$, $\tilde{\nu} = 6$ and the tableau

$$
\begin{array}{c|cccccc}
0 \\
\frac{1}{4} & \frac{1}{4} \\
\frac{3}{8} & \frac{3}{32} & \frac{9}{32} \\
\frac{12}{13} & \frac{1932}{2197} & -\frac{7200}{2197} & \frac{7296}{2197} \\
1 & \frac{439}{216} & -8 & \frac{3680}{513} & -\frac{845}{4104} \\
\frac{1}{2} & -\frac{8}{27} & 2 & -\frac{3544}{2565} & \frac{1859}{4104} & -\frac{11}{40} \\
\hline
& \frac{25}{216} & 0 & \frac{1408}{2565} & \frac{2197}{4104} & -\frac{1}{5} \\
\hline
& \frac{16}{135} & 0 & \frac{6656}{12825} & \frac{28561}{56430} & -\frac{9}{50} & \frac{2}{55}
\end{array}
$$

$\diamond$ **A simple embedded RK pair**   The RK pair

$$
\begin{array}{c|ccc}
0 \\
\frac{2}{3} & \frac{2}{3} \\
\frac{2}{3} & 0 & \frac{2}{3} \\
\hline
& \frac{1}{4} & \frac{3}{4} \\
\hline
& \frac{1}{4} & \frac{3}{8} & \frac{3}{8}
\end{array}
\tag{5.15}
$$

has orders $p = 2$, $\tilde{p} = 3$. The local error estimate becomes simply

$$\kappa = \tfrac{3}{8} \left\| \boldsymbol{f}\left(t_n + \tfrac{2}{3}h, \boldsymbol{\xi}_3\right) - \boldsymbol{f}\left(t_n + \tfrac{2}{3}h, \boldsymbol{\xi}_2\right) \right\|.$$

We have applied a variable-step algorithm based on the above error controller to the problems (5.2)–(5.4), within the same framework as the computational experiments using the Milne device in Section 5.2. The results are reported in Figures 5.5–5.7.

Comparing Figures 5.1 and 5.5 demonstrates that, as far as error control is concerned, the performance of (5.15) is roughly similar to the Milne device for the TR–AB2 pair. A similar conclusion can be drawn by comparing Figures 5.2 and 5.6. On the face of it, this is also the case with our single stiff example from Figures 5.3 and 5.7. However, a brief comparison of the step sequences confirms that the precision for the embedded RK pair has been attained at the cost of employing minute values of h. Needless to say, the smaller the step-size, the longer the computation and the higher the expense.

Figure 5.5 Global errors for the numerical solution of the van der Pol equation (5.2) by the embedded RK pair (5.15) with $\delta = 10^{-3}, 10^{-4}$.

Figure 5.6 Global errors for the numerical solution of the Mathieu equation (5.3) by the embedded RK pair (5.15) with $\delta = 10^{-3}, 10^{-4}$.

Figure 5.7   The step sequences (in the top two graphs) and global errors
for the numerical solution of the Curtiss–Hirschfelder equation (5.4) by the
embedded RK pair (5.15) with $\delta = 10^{-3}, 10^{-4}$.

It is easy to apportion the blame. The poor performance of the embedded pair
(5.15) for the last example can be attributed to the poor stability properties
of the 'inner' method

$$
\begin{array}{c|cc}
0 & & \\
\frac{2}{3} & \frac{2}{3} & \\
\hline
& \frac{1}{4} & \frac{3}{4}
\end{array}.
$$

Therefore $r(z) = 1 + z + \frac{1}{2}z^2$ (cf. Exercise 3.5) and it is an easy exercise to
show that $\mathcal{D} \cap \mathbb{R} = (-2, 0)$. In constant-step implementation we would have
thus needed $50h < 2$, hence $h < \frac{1}{25}$, to avoid instabilities. A bound of roughly
similar magnitude is consistent with Figure 5.7. The error-control mechanism
can cope with stiffness, but it does so at the price of drastically depressing
the step-size.   ◇

Exercise 5.5 demonstrates that the technique of embedded RK pairs can be generalized, at least to some extent, to cater for implicit methods, thereby rendering it
more suitable for stiff equations. It is fair, though, to point out that, in practical
implementations, the use of embedded RK pairs is almost always restricted to explicit
Runge–Kutta schemes.

Comments and bibliography

A classical approach to error control is to integrate once with step h and to integrate again,
along the same interval, with two steps of $\frac{1}{2}h$. Comparison of the two candidate solutions

at the new point yields an estimate of κ. This is clearly inferior to the methods of this chapter within the narrow framework of error control. However, the 'one step, two half-steps' technique is a rudimentary example of extrapolation, which can be used *both* to monitor the error and to improve locally the quality of the solution (cf. Exercise 5.6). See Hairer *et al.* (1991) for an extensive description of extrapolation techniques.

We wish to mention two further means whereby local error can be monitored. The first is a general technique due to Zadunaisky, which can ride piggyback on any time-stepping method

$$\boldsymbol{y}_{n+1} = \boldsymbol{\mathcal{Y}}(\boldsymbol{f}; h; (t_0, \boldsymbol{y}_0), (t_1, \boldsymbol{y}_1), \ldots, (t_n, \boldsymbol{y}_n))$$

of order p (Zadunaisky, 1976).

We are solving numerically the ODE system (5.1) and assume the availability of $p+1$ past values, $\boldsymbol{y}_{n-p}, \boldsymbol{y}_{n-p+1}, \ldots, \boldsymbol{y}_n$. They need not correspond to equally spaced points. Let us form a pth degree interpolating polynomial $\boldsymbol{p}$ such that $\boldsymbol{p}(t_i) = \boldsymbol{y}_i$, $i = n-p, n-p+1, \ldots, n$, and consider the ODE

$$\boldsymbol{z}' = \boldsymbol{f}(t, \boldsymbol{z}) + \left[\boldsymbol{p}'(t) - \boldsymbol{f}(t, \boldsymbol{p}(t))\right], \quad t \geq t_n, \qquad \boldsymbol{z}(t_n) = \boldsymbol{y}_n. \tag{5.16}$$

Two observations are crucial. Firstly, (5.16) is merely a small perturbation of the original system (5.1): since the numerical method is of order p and $\boldsymbol{p}$ interpolates at $p+1$ points, it follows that $\boldsymbol{p}(t) = \boldsymbol{y}(t) + \mathcal{O}\left(h^{p+1}\right)$, therefore, as long as $\boldsymbol{f}$ is sufficiently smooth, $\boldsymbol{p}'(t) - \boldsymbol{f}(t, \boldsymbol{p}(t)) = \mathcal{O}(h^p)$. Secondly, the *exact* solution of (5.16) is nothing other than $\boldsymbol{z} = \boldsymbol{p}$; this is verified at once by substitution.

We use the underlying numerical method to approximate the solution of (5.16) at t_{n+1}, using *exactly* the same ingredients that have been used in the computation of $\boldsymbol{y}_{n+1}$: the same starting values, the same approach to solving nonlinear algebraic systems, an identical stopping criterion... The outcome is

$$\boldsymbol{z}_{n+1} = \boldsymbol{\mathcal{Y}}(\boldsymbol{g}; h; (t_0, \boldsymbol{y}_0), (t_1, \boldsymbol{y}_1), \ldots, (t_n, \boldsymbol{y}_n)),$$

where $\boldsymbol{g}(t, \boldsymbol{z}) = \boldsymbol{f}(t, \boldsymbol{z}) + [\boldsymbol{p}'(t) - \boldsymbol{f}(t, \boldsymbol{p}(t))]$. Since $\boldsymbol{g} \approx \boldsymbol{f}$, we act upon the assumption (which can be firmed up mathematically) that the error in $\boldsymbol{z}_{n+1}$ is similar to the error in $\boldsymbol{y}_{n+1}$, and this motivates the estimate

$$\kappa = \|\boldsymbol{p}(t_{n+1}) - \boldsymbol{z}_{n+1}\|.$$

Our second technique for error control, the *Gear automatic integration* approach, is much more than simply a device to assess the local error. It is an integrated approach to the implementation of multistep methods that not only controls the growth of the error but also helps us to choose (on a local basis) the best multistep formula out of a given range.

The actual estimate of κ in Gear's method is probably the least striking detail. Recalling from (5.7) that the principal error term is of the form $ch^{p+1}\boldsymbol{y}^{(p+1)}(t_{n+s})$, we interpolate the $\boldsymbol{y}_i$ by a polynomial $\boldsymbol{p}$ of sufficiently high degree and replace $\boldsymbol{y}^{(p+1)}(t_{n+s})$ by $\boldsymbol{p}^{(p+1)}(t_{n+s})$. This, however, is only the beginning! Suppose, for example, that the underlying method is the pth-order Adams–Moulton. We subsequently form similar local-error estimates for its neighbours, Adams–Moulton methods of orders $p \pm 1$. Instead of doubling (or, if the error estimate for the pth method falls short of the tolerance, halving) the step-size, we ask ourselves which of the three methods would have attained, on the basis of our estimates, the requisite tolerance δ with the largest value of h? We then switch to this method and this step-size, advancing if the present step is acceptable, otherwise resuming the integration from the former point t_{n+s-1}.

This brief explanation does no justice to a complicated and sophisticated assembly of techniques, rules and tricks that makes the Gear approach the method of choice in many leading computational packages. The reader is referred to Gear (1971) and to Shampine & Gordon (1975) for details. Here we just comment on two important features. Firstly, a tremendous simplification of the tedious minutiae of interpolation and remeshing occurs if, instead of storing past values of the solution, the program deals with their finite differences, which, in effect, approximate the derivatives of y at t_{n+s-1}. This is called the *Nordsieck representation* of the multistep method (5.5). Secondly, the Gear automatic integration obviates the need for an independent (and tiresome) derivation of the requisite number of additional starting values, which is characteristic of other implementations of multistep methods (and which is often accomplished by a Runge–Kutta scheme). Instead, the integration can be commenced using a one-step method, allowing the algorithm to increase order only when enough information has accumulated for that purpose.

It might well be, gentle reader, that by this stage you are disenchanted with your prospects of programming a competitive computational package that can hold its own against the best in the field. If so, this exposition has achieved its purpose! Modern, high-quality software packages require years of planning, designing, programming, debugging, testing, debugging again, documenting, testing again, by whole teams of first class experts in numerical analysis and software engineering. It is neither a job for amateurs nor an easy alternative to proving theorems.

Good and reliable software for ODEs is available from commercial companies that specialize in numerical software, e.g. IMSL and NAG, as well as from *NetLib,* a depository of free software managed mainly by Oak Ridge National Laboratory (current `ftp` addresses are `netlib.ornl.gov` or `netlib.att.com`). General purpose software for partial differential equations is more problematic, for reasons that should be apparent later in this volume, but well-written, reliable and superbly documented packages exist for various families of such equations, often in a form suitable for specific applications such as computational fluid dynamics, electrical engineering etc.

A useful and (relatively) up-to-date guide to state-of-the-art mathematical software is available on the World-Wide Web at the site `http://gams.nist.gov/`, courtesy of the (American) National Institute of Standards and Technology. Another source of (often free) software is the `ftp` site `softlib.rice.edu` at Rice University (Houston, Texas). However, the number of `ftp` and World-Wide Web sites is expanding so fast as to render a more substantive list of little lasting value. Given the volume of traffic along the Information Superhighway, it is likely that the ideal program for your problem exists *somewhere*. It is a moot point, however, whether it is easier to locate it or to write one of your own....

Gear, C.W. (1971), *Numerical Initial Value Problems in Ordinary Differential Equations,* Prentice–Hall, Englewood Cliffs, NJ.

Hairer, E., Nørsett, S.P. and Wanner, G. (1991), *Solving Ordinary Differential Equations I: Nonstiff Problems* (2nd ed.), Springer-Verlag, Berlin.

Shampine, L.F. and Gordon, M.K. (1975), *Computer Solution of Ordinary Differential Equations,* W.H. Freeman, San Francisco.

Zadunaisky, P.E. (1976), On the estimation of errors propagated in the numerical integration of ordinary differential equations, *Numerische Mathematik* **27**, 21–39.

Exercises

5.1 Find the error constants for the Adams–Bashforth method (2.7) and for Adams–Moulton methods with $s = 2, 3$.

5.2* Prove that the error constant of the s-step backward differentiation formula is $-\beta/(s+1)$, where β has been defined in (2.14).

5.3 Instead of using (5.10) to estimate the error in the multistep method (5.11), we can use it to increase accuracy.

a Prove that the formula (5.7) yields

$$
\begin{aligned}
\boldsymbol{y}(t_{n+1}) - \boldsymbol{y}_{n+1} &= -\tfrac{1}{12}h^3\boldsymbol{y}'''(t_{n+1}) + \mathcal{O}(h^4)\,, \\
\boldsymbol{y}(t_{n+1}) - \boldsymbol{x}_{n+1} &= -\tfrac{5}{12}h^3\boldsymbol{y}'''(t_{n+1}) + \mathcal{O}(h^4)\,.
\end{aligned}
\tag{5.17}
$$

b Neglecting the $\mathcal{O}(h^4)$ terms, solve the two equations for the unknown $\boldsymbol{y}'''(t_n)$ (in contrast to the Milne device, where we solve for $\boldsymbol{y}(t_{n+1})$).

c Substituting the approximate expression back into (5.7) results in a two-step implicit multistep method. Derive it explicitly and determine its order. Is it convergent? Can you identify it?[3]

5.4 Prove that the embedded RK pair

$$
\begin{array}{c|ccc}
0 \\
\frac{1}{2} & \frac{1}{2} \\
1 & -1 & 2 \\
\hline
& 0 & 1 \\
\hline
& \frac{1}{6} & \frac{2}{3} & \frac{1}{6}
\end{array}
$$

combines a second-order and a third-order method.

5.5 Consider the embedded RK pair

$$
\begin{array}{c|ccc}
0 \\
1 & \frac{1}{2} & \frac{1}{2} \\
\frac{1}{2} & \frac{3}{2} & -1 \\
\hline
& \frac{1}{2} & \frac{1}{2} \\
\hline
& \frac{1}{6} & \frac{2}{3} & \frac{1}{6}
\end{array}\,.
$$

Note that the 'inner' two-stage method is implicit and that the third stage is explicit. This means that the added cost of error control is marginal.

a Prove that the 'inner' method is of order two, while the full three-stage method is of order three.

---

[3] It is always possible to use the method (5.6) to boost the order of (5.5), except when the outcome is not convergent, as is often the case.

b Show that the 'inner' method is A-stable. Can you identify it as a familiar method in disguise?

c Find the function r associated with the three-stage method and verify that, in line with Lemma 4.4, $r(z) = e^z + \mathcal{O}(z^4)$, $z \to 0$.

5.6 Let $y_{n+1} = \mathcal{Y}(f, h, y_n)$, $n = 0, 1, \ldots$, be a one-step method of order p. We assume (consistently with Runge–Kutta methods, cf. (5.12)) that there exists a vector ℓ_n, independent of h, such that

$$y_{n+1} = \tilde{y}(t_{n+1}) + \ell_n h^{p+1} + \mathcal{O}(h^{p+2}), \qquad h \to 0,$$

where $\tilde{y}$ is the exact solution of (5.1) with the initial condition $\tilde{y}(t_n) = y_n$. Let

$$x_{n+1} := \mathcal{Y}\left(f, \tfrac{1}{2}h, \mathcal{Y}\left(f, \tfrac{1}{2}h, y_n\right)\right).$$

Note that x_{n+1} is simply the result of traversing $[t_n, t_{n+1}]$ with the method $\mathcal{Y}$ in two equal steps of $\tfrac{1}{2}h$.

a Find a real constant α such that

$$x_{n+1} = \tilde{y}(t_{n+1}) + \alpha \ell_n h^{p+1} + \mathcal{O}(h^{p+2}), \qquad h \to 0.$$

b Determine a real constant β such that the linear combination $z_{n+1} := (1 - \beta)y_{n+1} + \beta x_{n+1}$ approximates $\tilde{y}(t_{n+1})$ up to $\mathcal{O}(h^{p+2})$. [*The procedure of using the enhanced value z_{n+1} as the approximant at t_{n+1} is known as* extrapolation. *It is capable of widespread generalization.*]

c Let $\mathcal{Y}$ correspond to the trapezoidal rule (1.9) and suppose that the above extrapolation procedure is applied to the scalar linear equation $y' = \lambda y$, $y(0) = 1$, with a constant step size h. Find a function r such that $z_{n+1} = r(h\lambda)z_n = [r(h\lambda)]^{n+1}$, $n = 0, 1, \ldots$. Is the new method A-stable?

6

Nonlinear algebraic systems

6.1 Functional iteration

From the point of view of a numerical mathematician, which we have adopted in Chapters 1–4, the solution of ordinary differential equations is all about analysis – i.e. convergence, order, stability and an endless progression of theorems and proofs. The outlook of Chapter 5 parallels that of a software engineer, and is concerned with the correct assembly of computational components and with choosing the step sequence dynamically. Computers, however, are engaged neither in analysis nor in algorithm design, but in the real work concerned with solving ODEs, and this consists in the main of the computation of (mostly nonlinear) algebraic systems of equations.

Why not – and this is a legitimate question – use explicit methods, whether multistep or Runge–Kutta, thereby dispensing altogether with the need to calculate algebraic systems? The main reason is computational cost. This is obvious in the case of stiff equations, since, for explicit time-stepping methods, stability considerations restrict the step-size to an extent that renders the scheme noncompetitive and downright ineffective. When stability questions are not at issue, it often makes very good sense to use explicit Runge–Kutta methods. The accepted wisdom is, however, that, as far as multistep methods are concerned, implicit methods should be used even for non-stiff equations since, as we will see in this chapter, the solution of the underlying algebraic systems can be approximated with relative ease.

Let us suppose that we wish to advance the (implicit) multistep method (2.8) by a single step. This entails solving the algebraic system

$$\boldsymbol{y}_{n+s} = hb_s \boldsymbol{f}(t_{n+s}, \boldsymbol{y}_{n+s}) + \boldsymbol{\gamma}, \tag{6.1}$$

where the vector

$$\boldsymbol{\gamma} = h \sum_{m=0}^{s-1} b_m \boldsymbol{f}(t_{n+m}, \boldsymbol{y}_{n+m}) - \sum_{m=0}^{s-1} a_m \boldsymbol{y}_{n+m}$$

is known. As far as implicit Runge–Kutta methods (3.9) are concerned, we need to solve at each step the system

$$\begin{aligned}
\boldsymbol{\xi}_1 &= \boldsymbol{y}_n + h[a_{1,1}\boldsymbol{f}(t_n + c_1 h, \boldsymbol{\xi}_1) + a_{1,2}\boldsymbol{f}(t_n + c_2 h, \boldsymbol{\xi}_2) + \cdots + a_{1,\nu}\boldsymbol{f}(t_n + c_\nu h, \boldsymbol{\xi}_\nu)], \\
\boldsymbol{\xi}_2 &= \boldsymbol{y}_n + h[a_{2,1}\boldsymbol{f}(t_n + c_1 h, \boldsymbol{\xi}_1) + a_{2,2}\boldsymbol{f}(t_n + c_2 h, \boldsymbol{\xi}_2) + \cdots + a_{2,\nu}\boldsymbol{f}(t_n + c_\nu h, \boldsymbol{\xi}_\nu)], \\
&\;\;\vdots \\
\boldsymbol{\xi}_\nu &= \boldsymbol{y}_n + h[a_{\nu,1}\boldsymbol{f}(t_n + c_1 h, \boldsymbol{\xi}_1) + a_{\nu,2}\boldsymbol{f}(t_n + c_2 h, \boldsymbol{\xi}_2) + \cdots + a_{\nu,\nu}\boldsymbol{f}(t_n + c_\nu h, \boldsymbol{\xi}_\nu)].
\end{aligned}$$

This system looks considerably more complicated than (6.1). Both, however, can be cast into a standard form, namely

$$w = hg(w) + \beta, \qquad w \in \mathbb{R}^{\tilde{d}}, \tag{6.2}$$

where the function g and the vector β are known. Obviously, for the multistep method in (6.1) $g(\cdot) = b_m f(t_{n+s}, \cdot)$, $\beta = \gamma$ and $\tilde{d} = d$. The solution of (6.2) then becomes y_{n+s}. The notation is slightly more complicated for Runge–Kutta methods, although it can be simplified a great deal by using *Kronecker products*. We provide an example for the case $\nu = 2$ where, mercifully, no Kronecker products are required. Thus, $\tilde{d} = 2d$ and

$$g(w) = \left[\begin{array}{c} a_{1,1} f(t_n + c_1 h, w_1) + a_{1,2} f(t_n + c_2 h, w_2) \\ a_{2,1} f(t_n + c_1 h, w_1) + a_{2,2} f(t_n + c_2 h, w_2) \end{array} \right] \qquad \text{where} \qquad w = \left[\begin{array}{c} w_1 \\ w_2 \end{array} \right].$$

Moreover,

$$\beta = \left[\begin{array}{c} y_n \\ y_n \end{array} \right].$$

Provided that the solution of (6.2) is known, we then set $\xi_j = w_j$, $j = 1, 2$, hence

$$y_{n+1} = y_n + h(b_1 w_1 + b_2 w_2).$$

Let us assume that g is nonlinear (if g were linear and the number of equations moderate, we could solve (6.2) by familiar *Gaussian elimination;* the numerical solution of large linear systems is discussed in Chapters 9–12). Our intention is to solve the algebraic system by iteration;[1] in other words, we need to make an *initial guess* $w^{[0]}$ and provide an algorithm

$$w^{[i+1]} = \mathcal{A}(w^{[i]}), \qquad i = 0, 1, \ldots, \tag{6.3}$$

such that

(1) $w^{[i]} \to \hat{w}$, the solution of (6.2);

(2)   the cost of each step (6.3) is small;   and

(3)   the progression to the limit is rapid.

The form (6.2) emphasizes two important aspects of this iterative procedure. Firstly, the vector β is known at the outset and there is no need to recalculate it in every iteration (6.3). Secondly, the step-size h is an important parameter and its magnitude is likely to determine central characteristics of the system (6.2): since the exact solution is obvious when $h = 0$, clearly the problem is likely to be easier for small $h > 0$. Moreover, although in principle (6.2) may possess many solutions, it follows at once from the implicit function theorem that, provided g is continuously differentiable, nonsingularity of the Jacobian matrix $I - h\partial g(\beta)/\partial w$ for $h \to 0$ implies the existence of a unique solution for sufficiently small $h > 0$.

---

[1]The reader will notice that two distinct iterative procedures are taking place: time-stepping, i.e. $y_{n+s-1} \mapsto y_{n+s}$, and the 'inner' iteration, $w^{[i]} \mapsto w^{[i+1]}$. To prevent confusion, we reserve the phrase 'iteration' for the latter.

The most elementary approach to the solution of (6.2) is the *functional iteration* $\mathcal{A}(\boldsymbol{w}) = hg(\boldsymbol{w}) + \beta$, which, using (6.3) can be expressed as

$$\boldsymbol{w}^{[i+1]} = hg(\boldsymbol{w}^{[i]}) + \beta, \qquad i = 0, 1, \ldots. \tag{6.4}$$

Much beautiful mathematics has been produced in the last decade in connection with functional iteration, concerned mainly with the fractal nature of basins of attraction in the complex case. For practical purposes, however, we resort to the tried and trusted *Banach fixed-point theorem*, which will now be stated and proved in a formalism appropriate for the recursion (6.4).

Given a vector norm $\| \cdot \|$ and $\boldsymbol{w} \in \mathbb{R}^{\tilde{d}}$, we denote by $\mathcal{B}_\rho(\boldsymbol{w})$ the closed ball of radius $\rho > 0$ centred at $\boldsymbol{w}$:

$$\mathcal{B}_\rho(\boldsymbol{w}) = \left\{ \boldsymbol{u} \in \mathbb{R}^{\tilde{d}} : \|\boldsymbol{u} - \boldsymbol{w}\| \leq \rho \right\}.$$

Theorem 6.1 *Let $h > 0$, $\boldsymbol{w}^{[0]} \in \mathbb{R}^{\tilde{d}}$, and suppose that there exist numbers $\lambda \in (0, 1)$ and $\rho > 0$ such that*

(i) $\|g(\boldsymbol{v}) - g(\boldsymbol{u})\| \leq \dfrac{\lambda}{h}\|\boldsymbol{v} - \boldsymbol{u}\|$ *for every $\boldsymbol{v}, \boldsymbol{u} \in \mathcal{B}_\rho(\boldsymbol{w}^{[0]})$;*

(ii) $\boldsymbol{w}^{[1]} \in \mathcal{B}_{(1-\lambda)\rho}(\boldsymbol{w}^{[0]}).$

Then

(a) $\boldsymbol{w}^{[i]} \in \mathcal{B}_\rho(\boldsymbol{w}^{[0]})$ *for every $i = 0, 1, \ldots$;*

(b) $\hat{\boldsymbol{w}} := \lim_{i \to \infty} \boldsymbol{w}^{[i]}$ *exists, obeys the equation (6.2) and $\hat{\boldsymbol{w}} \in \mathcal{B}_\rho(\boldsymbol{w}^{[0]})$;*

(c)   *no other point in $\mathcal{B}_\rho(\boldsymbol{w}^{[0]})$ is a solution of (6.2).*

Proof   We commence by using induction to prove that

$$\|\boldsymbol{w}^{[i+1]} - \boldsymbol{w}^{[i]}\| \leq \lambda^i (1 - \lambda)\rho \tag{6.5}$$

and that $\boldsymbol{w}^{[i+1]} \in \mathcal{B}_\rho(\boldsymbol{w}^{[0]})$ for all $i = 0, 1, \ldots$.

This is certainly true for $i = 0$ because of condition (ii) and the definition of $\mathcal{B}_\rho(\boldsymbol{w}^{[0]})$. Let us assume that the statement is true for all $m = 0, 1, \ldots, i - 1$. Then, by (6.2) and assumption (i),

$$\begin{aligned}
\|\boldsymbol{w}^{[i+1]} - \boldsymbol{w}^{[i]}\| &= \|[hg(\boldsymbol{w}^{[i]}) + \beta] - [hg(\boldsymbol{w}^{[i-1]}) + \beta]\| \\
&= h\|g(\boldsymbol{w}^{[i]}) - g(\boldsymbol{w}^{[i-1]})\| \leq \lambda\|\boldsymbol{w}^{[i]} - \boldsymbol{w}^{[i-1]}\|, \qquad i = 1, 2, \ldots.
\end{aligned}$$

This carries forward the induction for (6.5) from $i - 1$ to i.

The following sum,

$$\boldsymbol{w}^{[i+1]} - \boldsymbol{w}^{[0]} = \sum_{j=0}^{i} (\boldsymbol{w}^{[j+1]} - \boldsymbol{w}^{[j]}), \qquad i = 0, 1, \ldots,$$

telescopes; therefore, by the triangle inequality (A.1.3.3)

$$\|\boldsymbol{w}^{[i+1]} - \boldsymbol{w}^{[0]}\| = \left\| \sum_{j=0}^{i} (\boldsymbol{w}^{[j+1]} - \boldsymbol{w}^{[j]}) \right\| \leq \sum_{j=0}^{i} \|\boldsymbol{w}^{[j+1]} - \boldsymbol{w}^{[j]}\|, \qquad i = 0, 1, \ldots.$$

Exploiting (6.5) and summing the geometric series, we thus conclude that

$$\|\boldsymbol{w}^{[i+1]} - \boldsymbol{w}^{[0]}\| \leq \sum_{j=0}^{i} \lambda^i (1 - \lambda) \rho = (1 - \lambda^{i+1})\rho \leq \rho, \qquad i = 0, 1, \ldots.$$

Therefore $\boldsymbol{w}^{[i+1]} \in \mathcal{B}_\rho(\boldsymbol{w}^{[0]})$. This completes the inductive proof and we deduce that (a) is true.

Similarly telescoping series, the triangle inequality and (6.5) can be used to argue that

$$\|\boldsymbol{w}^{[i+k]} - \boldsymbol{w}^{[i]}\| = \left\| \sum_{j=0}^{k-1} (\boldsymbol{w}^{[i+j+1]} - \boldsymbol{w}^{[i+j]}) \right\| \leq \sum_{j=0}^{k-1} \|\boldsymbol{w}^{[i+j+1]} - \boldsymbol{w}^{[i+j]}\|$$

$$\leq \sum_{j=0}^{k-1} \lambda^{i+j} (1 - \lambda) \rho = \lambda^i (1 - \lambda^k) \rho, \qquad i = 0, 1, \ldots, \quad k = 1, 2, \ldots.$$

Therefore, $\lambda \in (0, 1)$ implies that for every $i = 0, 1, \ldots$ and $\varepsilon > 0$ we may choose k large enough that

$$\|\boldsymbol{w}^{[i+k]} - \boldsymbol{w}^{[i]}\| < \varepsilon.$$

In other words, $\{\boldsymbol{w}^{[i]}\}_{i=0,1,\ldots}$ is a *Cauchy sequence*.

The set $\mathcal{B}_\rho(\boldsymbol{w}^{[0]})$ being compact (i.e., closed and bounded), the Cauchy sequence $\{\boldsymbol{w}^{[i]}\}_{i=0,1,\ldots}$ converges to a limit within the set. This proves the existence of $\hat{\boldsymbol{w}} \in \mathcal{B}_\rho(\boldsymbol{w}^{[0]})$.

Finally, let us suppose that there exists $\boldsymbol{w}^\star \in \mathcal{B}_\rho(\boldsymbol{w}^{[0]})$, $\boldsymbol{w}^\star \neq \hat{\boldsymbol{w}}$, such that $\boldsymbol{w}^\star = h\boldsymbol{g}(\boldsymbol{w}^\star) + \boldsymbol{\beta}$. Therefore $\|\boldsymbol{w}^\star - \hat{\boldsymbol{w}}\| > 0$ implies that

$$\|\boldsymbol{w}^\star - \hat{\boldsymbol{w}}\| = \|[h\boldsymbol{g}(\boldsymbol{w}^\star) + \boldsymbol{\beta}] - [h\boldsymbol{g}(\hat{\boldsymbol{w}}) + \boldsymbol{\beta}]\| = h\|\boldsymbol{g}(\boldsymbol{w}^\star) - \boldsymbol{g}(\hat{\boldsymbol{w}})\|$$

$$\leq \lambda \|\boldsymbol{w}^\star - \hat{\boldsymbol{w}}\| < \|\boldsymbol{w}^\star - \hat{\boldsymbol{w}}\|.$$

This is impossible and we deduce that the fixed point $\hat{\boldsymbol{w}}$ is unique in $\mathcal{B}_\rho(\boldsymbol{w}^{[0]})$, thereby concluding the proof of the theorem. ∎

If $\boldsymbol{g}$ is smoothly differentiable then, by the mean value theorem, for every $\boldsymbol{v}$ and $\boldsymbol{u}$ there exists $\tau \in (0, 1)$ such that

$$\boldsymbol{g}(\boldsymbol{v}) - \boldsymbol{g}(\boldsymbol{u}) = \left[\frac{\partial \boldsymbol{g}(\tau \boldsymbol{v} + (1 - \tau)\boldsymbol{u})}{\partial \boldsymbol{w}} \right] (\boldsymbol{v} - \boldsymbol{u}).$$

Therefore, assumption (i) of Theorem 6.1 is nothing other than a statement on the magnitude of the step-size h in relation to $\|\partial \boldsymbol{g}/\partial \boldsymbol{w}\|$. In particular, if (6.2) originates in a multistep method, we need, in effect, $h|b_s| \times \|\partial \boldsymbol{f}(t_{n+s}, \boldsymbol{y}_{n+s})/\partial \boldsymbol{y}\| < 1$. (Similar inequality applies to Runge–Kutta methods.) The meaning of this restriction is phenomenologically similar to stiffness, as can be seen in the following example.

◇ **The trapezoidal rule and functional iteration**   The iterative scheme (6.4), as applied to the trapezoidal rule (1.9), reads

$$\boldsymbol{w}^{[i+1]} = \tfrac{1}{2}h\boldsymbol{f}(t_{n+1}, \boldsymbol{w}^{[i]}) + \left[\boldsymbol{y}_n + \tfrac{1}{2}h\boldsymbol{f}(t_n, \boldsymbol{y}_n)\right], \qquad i = 0, 1, \ldots.$$

Let us suppose that the underlying ODE is linear, i.e. of the form $\boldsymbol{y}' = \Lambda\boldsymbol{y}$, where Λ is symmetric. As long as we are employing the Euclidean norm, it is true that $\|\Lambda\| = \rho(\Lambda)$, the *spectral radius* of Λ (A.1.5.2). The outcome is the restriction $h\rho(\Lambda) < 2$, which imposes similar constraints to a stable implementation of the *Euler method* (1.4). Provided $\rho(\Lambda)$ is small and stiffness is not an issue, this makes little difference. However, to retain the A-stability of the trapezoidal rule for large $\rho(\Lambda)$ we must restrict $h > 0$ so drastically that all the benefits of A-stability are lost – we might just as well have used Adams–Bashforth, say, in the first place!   ◇

We conclude that a useful rule of a thumb is that we may use the functional iteration (6.4) for non-stiff problems but we need a different algorithm when stiffness becomes an issue.

6.2   The Newton–Raphson algorithm and its modification

Let us suppose that the function g is twice continuously differentiable. We expand (6.2) about a vector $\boldsymbol{w}^{[i]}$,

$$\boldsymbol{w} = \boldsymbol{\beta} + hg(\boldsymbol{w}^{[i]} + (\boldsymbol{w} - \boldsymbol{w}^{[i]})) = \boldsymbol{\beta} + hg(\boldsymbol{w}^{[i]}) + h\frac{\partial g(\boldsymbol{w}^{[0]})}{\partial \boldsymbol{w}}(\boldsymbol{w} - \boldsymbol{w}^{[i]}) + \mathcal{O}\left(\|\boldsymbol{w} - \boldsymbol{w}^{[i]}\|^2\right). \tag{6.6}$$

Disregarding the $\mathcal{O}(\|\boldsymbol{w} - \boldsymbol{w}^{[i]}\|^2)$ term, we solve (6.6) for $\boldsymbol{w} - \boldsymbol{w}^{[i]}$. The outcome,

$$\left[I - h\frac{\partial g(\boldsymbol{w}^{[i]})}{\partial \boldsymbol{w}}\right](\boldsymbol{w} - \boldsymbol{w}^{[i]}) \approx \boldsymbol{\beta} + hg(\boldsymbol{w}^{[i]}) - \boldsymbol{w}^{[i]},$$

motivates the iterative scheme

$$\boldsymbol{w}^{[i+1]} = \boldsymbol{w}^{[i]} - \left[I - \frac{\partial g(\boldsymbol{w}^{[i]})}{\partial \boldsymbol{w}}\right]^{-1}\left[\boldsymbol{w}^{[i]} - \boldsymbol{\beta} - hg(\boldsymbol{w}^{[i]})\right], \qquad i = 0, 1, \ldots. \tag{6.7}$$

This is (under a mild disguise) the celebrated *Newton–Raphson* algorithm.

The Newton–Raphson method has motivated several profound theories and attracted the attention of some of the towering mathematical minds of the twentieth century – Leonid Kantorowitz and Stephen Smale, to mention just two. We do not propose in this volume to delve into this issue, whose interest is tangential to our main theme. Instead, and without further ado, we merely comment on several features of the iterative scheme (6.7).

Firstly, as long as $h > 0$ is sufficiently small the rate of convergence of the Newton–Raphson algorithm is quadratic: it is possible to prove that there exists a constant $c > 0$ such that, for sufficiently large i,

$$\|\boldsymbol{w}^{[i+1]} - \hat{\boldsymbol{w}}\| \leq c\|\boldsymbol{w}^{[i]} - \hat{\boldsymbol{w}}\|^2,$$

where $\hat{w}$ is a solution of (6.2). This is already implicit in the fact that we have neglected an $\mathcal{O}(\|w - w^{[i]}\|^2)$ term in (6.6). It is important to comment that the 'sufficient smallness' of $h > 0$ is of a different order of magnitude to the minute values of $h > 0$ that are required when the functional iteration (6.4) is applied to stiff problems. It is easy to prove, for example, that (6.7) terminates in a single step when g is a linear function, regardless of any underlying stiffness (see Exercise 6.2).

Secondly, an implementation of (6.7) requires computation of the Jacobian matrix at every iteration. This is a formidable ordeal since, for a d-dimensional system, the Jacobian matrix has d^2 entries and its computation – even if all requisite formulae are available in an explicit form – is expensive.

Finally, each iteration requires the solution of a linear system of algebraic equations. It is highly unusual for such a system to be singular or ill conditioned (i.e., 'close' to singular) in a realistic computation, regardless of stiffness; the reasons, in the (simpler) case of multistep methods, are that $b_s > 0$ for all methods with reasonably large linear stability domains, the eigenvalues of $\partial f/\partial y$ reside in $\mathbb{C}^-$ and it is easy to prove that all the eigenvalues of the matrix in (6.7) are bounded away from zero. However, the solution of even a nonsingular, well-conditioned algebraic system is a nontrivial and potentially costly task.

Both shortcomings of Newton–Raphson – the computation of the Jacobian matrix and the need to solve linear systems in each iteration – can be alleviated by using the *modified Newton–Raphson* instead. The *quid pro quo,* however, is a significant slowing-down of the convergence rate. Before we introduce the modification of (6.7), let us comment briefly on an important special case when the 'full' Newton–Raphson can (and should) be used.

A significant proportion of stiff ODEs originate in the *semi-discretization* of parabolic partial differential equations by finite difference methods (Chapter 13). In these cases the Newton–Raphson method (6.7) is very effective indeed, since the Jacobian matrix is sparse (an overwhelming majority of its elements vanish): it has just $\mathcal{O}(d)$ nonzero components and usually can be computed with relative ease. Moreover, most methods for the solution of sparse algebraic systems confer no advantage to the special form of the modified equations, (6.9), an exception being the direct factorization algorithms of Chapter 9.

◇ **The reaction–diffusion equation**   A quasilinear parabolic partial differential equation with many applications in mathematical biology and in physics is the *reaction-diffusion* equation

$$\frac{\partial u}{\partial t} = \frac{\partial^2 u}{\partial x^2} + \varphi(u), \qquad 0 < x < 1, \quad t \geq 0, \tag{6.8}$$

where $u = u(x,t)$. It is given with the initial condition $u(x,0) = u_0(x)$, $0 < x < 1$, and (for simplicity) zero Dirichlet boundary conditions $u(0,t), u(1,t) \equiv 0$, $t \geq 0$.

Among the many applications of (6.8) we single out two for special mention. The choice $\varphi(u) = cu$, where $c > 0$, models the neutron density in an atom bomb (subject to the assumption that the latter is in the form of a thin uranium rod of unit length), whereas $\varphi(u) = \alpha u + \beta u^2$ (the *Fisher equation*) is used in population dynamics: the terms αu and βu^2 correspond respectively to the reproduction and interaction of a species while $\partial^2 u/\partial x^2$ models its diffusion in the underlying habitat.

A standard semi-discretization (that is, an approximation of a partial differential equation by an ODE system, see Chapter 13) of (6.8) is

$$y_k' = \frac{1}{(\Delta x)^2}(y_{k-1} - 2y_k + y_{k+1}) + \varphi(y_k), \qquad k = 1, 2, \ldots, d \quad t \geq 0,$$

where $\Delta x = 1/(d+1)$ and $y_0, y_{d+1} \equiv 0$. Suppose that φ' is easily available, e.g. that φ is a polynomial. The Jacobian matrix

$$\left(\frac{\partial f(t, y)}{\partial y}\right)_{k,\ell} = \begin{cases} -\dfrac{2}{(\Delta x)^2} + \varphi'(y_k), & k = \ell, \\ \dfrac{1}{(\Delta x)^2}, & |k - \ell| = 1, \\ 0 & \text{otherwise}, \end{cases} \quad k, \ell = 1, 2, \ldots, d,$$

is fairly easy to evaluate and store. Moreover, as apparent from the forthcoming discussion in Chapter 9, the solution of algebraic linear systems with tridiagonal matrices is very easy and fast. We conclude that in this case there is no need to trade off the superior speed of Newton–Raphson for 'easier' alternatives. $\diamond$

Unfortunately, most stiff systems do not share the features of the above example and for these we need to modify the Newton–Raphson iteration. This modification takes the form of a replacement of the matrix $\partial g(w^{[i]})/\partial w$ by another matrix, J, say, that does not vary with i. A typical choice might be

$$J = \frac{\partial g(w^{[0]})}{\partial w},$$

but it is not unusual, in fact, to retain the same matrix J for a number of time steps. In place of (6.7) we thus have

$$w^{[i+1]} = w^{[i]} - (I - hJ)^{-1}\left[w^{[i]} - \beta - hg(w^{[i]})\right], \qquad i = 0, 1, \ldots. \tag{6.9}$$

This *modified Newton–Raphson* scheme (6.9) confers two immediate advantages. The first is obvious: we need to calculate J only once per step (or perhaps per several steps). The second is realized when the underlying linear algebraic system is solved by Gaussian elimination – the method of choice for small or moderate d. In its LU formulation (A.1.4.5), Gaussian elimination of the linear system $Ax = b$, where $b \in \mathbb{R}^d$, consists of two stages. Firstly, the matrix A is factorized in the form LU, where L and U are lower triangular and upper triangular matrices respectively. Secondly, we solve $Lz = b$, followed by $Ux = z$. While factorization entails $\mathcal{O}(d^3)$ operations (for non-sparse matrices), the solution of two triangular $d \times d$ systems requires just $\mathcal{O}(d^2)$ operations and is considerably cheaper. In the case of the iterative scheme (6.9) it is enough to factorize $A = I - hJ$ just once per time step (or once per re-evaluation of

J and/or per change in h). Therefore, the cost of each single iteration goes down by an order of magnitude, as compared with the original Newton–Raphson scheme!

Of course, quadratic convergence is lost. As a matter of fact, modified Newton–Raphson is simply functional iteration except that, instead of $h\boldsymbol{g}(\boldsymbol{w}) + \boldsymbol{\beta}$, we iterate the new function $h\tilde{\boldsymbol{g}}(\boldsymbol{w}) + \tilde{\boldsymbol{\beta}}$, where

$$\tilde{\boldsymbol{g}}(\boldsymbol{w}) := (I - hJ)^{-1}[\boldsymbol{g}(\boldsymbol{w}) - J\boldsymbol{w}], \qquad \tilde{\boldsymbol{\beta}} := (I - hJ)^{-1}\boldsymbol{\beta}. \tag{6.10}$$

The proof is left to the reader in Exercise 6.3.

There is nothing to stop us from using Theorem 6.1 to explore the convergence of (6.9). It follows from (6.10) that

$$\tilde{\boldsymbol{g}}(\boldsymbol{v}) - \tilde{\boldsymbol{g}}(\boldsymbol{u}) = (I - hJ)^{-1}\{[\boldsymbol{g}(\boldsymbol{v}) - \boldsymbol{g}(\boldsymbol{u})] - J(\boldsymbol{v} - \boldsymbol{u})\}. \tag{6.11}$$

Recall, however, that, subject to the sufficient smoothness of $\boldsymbol{g}$, there exists a point $\boldsymbol{z}$ on the line segment joining $\boldsymbol{v}$ and $\boldsymbol{u}$ such that

$$\boldsymbol{g}(\boldsymbol{v}) - \boldsymbol{g}(\boldsymbol{u}) = \frac{\partial \boldsymbol{g}(\boldsymbol{z})}{\partial \boldsymbol{w}}(\boldsymbol{v} - \boldsymbol{u}).$$

Given that we have chosen

$$J = \frac{\partial \boldsymbol{g}(\tilde{\boldsymbol{w}})}{\partial \boldsymbol{w}}$$

(for example, $\tilde{\boldsymbol{w}} = \boldsymbol{w}^{[0]}$), it follows from (6.11) that

$$\frac{\|\tilde{\boldsymbol{g}}(\boldsymbol{v}) - \tilde{\boldsymbol{g}}(\boldsymbol{u})\|}{\|\boldsymbol{v} - \boldsymbol{u}\|} \leq \left\| \left[I - h\frac{\partial \boldsymbol{g}(\tilde{\boldsymbol{w}})}{\partial \boldsymbol{w}} \right]^{-1} \right\| \times \left\| \frac{\partial \boldsymbol{g}(\boldsymbol{z})}{\partial \boldsymbol{w}} - \frac{\partial \boldsymbol{g}(\tilde{\boldsymbol{w}})}{\partial \boldsymbol{w}} \right\|.$$

Unless the Jacobian matrix varies very considerably as a function of t, the second term on the right is likely to be small. Moreover, if all the eigenvalues of J are in $\mathbb{C}^-$, $\|(I - hJ)^{-1}\|$ is likely also to be small; stiffness is likely to help, not hinder, this estimate! Therefore, it is possible in general to satisfy assumption (i) of Theorem 6.1 with large $\rho > 0$.

6.3   Starting and stopping the iteration

Theorem 6.1 quantifies an important point that is equally valid for every iterative method for nonlinear algebraic equations (6.2), not just for the functional iteration (6.4) and the modified Newton–Raphson method (6.9): good performance hinges to a large extent on the quality of the starting value $\boldsymbol{w}^{[0]}$.

Provided that $\boldsymbol{g}$ is Lipschitz, condition (i) is always valid for a given $h > 0$ and sufficiently small $\rho > 0$. However, small ρ means that, to be consistent with condition (ii), we must choose an exceedingly good starting condition $\boldsymbol{w}^{[0]}$. Viewed in this light, the main purpose in replacing (6.4) by (6.9) is to allow convergence from imperfect starting values.

Even if the choice of an iterative scheme provides for a large *basin of attraction* of $\hat{\boldsymbol{w}}$ (the set of all $\boldsymbol{w}^{[0]} \in \mathbb{R}^{\tilde{d}}$ for which the iteration converges to $\hat{\boldsymbol{w}}$), it is important to commence the iteration with a good initial guess. This is true for every nonlinear algebraic system but our problem here is special – it originates in the use of a time-stepping method for ODEs. This is an important advantage.

Supposing for example that the underlying ODE method is a pth-order multistep scheme, let us recall the meaning of the solution of (6.2), namely $\boldsymbol{y}_{n+s}$. To paraphrase the last paragraph, it is an excellent policy to seek a starting condition $\boldsymbol{w}^{[0]}$ near to the vector $\boldsymbol{y}_{n+s}$. The latter is, of course, unknown, but we can obtain a good guess by using a different, *explicit* multistep method of order p. This multistep method, called the *predictor,* provides the platform upon which the iterative scheme seeks the solution of the implicit *corrector*.

It is not enough to start an iterative procedure, we must also provide a stopping criterion, which terminates the iterative process. This might appear as a relatively straightforward task: iterate until $\|\boldsymbol{w}^{[i+1]} - \boldsymbol{w}^{[i]}\| < \varepsilon$ for a given threshold value ε (distinct from, and probably significantly smaller than the tolerance δ that we employ in error control). However, this approach – perfectly sensible for the solution of general nonlinear algebraic systems – misses an important point: the origin of (6.2) is in a time-stepping computational method for ODEs, implemented with the step-size h. If convergence is slow we have two options, either to carry on iterating or to stop the procedure, abandon the current step-size and commence time-stepping with a smaller value of $h > 0$. In other words, there is nothing to prevent us from using the step-size both to control the error and to ensure rapid convergence of the iterative scheme.

The traditional attitude to iterative procedures, namely to proceed with perhaps thousands of iterations until convergence takes place (to a given threshold) is completely inadequate. Unless the process converges in a relatively small number of iterations - perhaps ten, perhaps fewer – the best course of action is to stop, decrease the step-size and recommence time-stepping. However, this does not exhaust the range of all possible choices. Let us remember that the goal is not to solve a nonlinear algebraic system *per se* but to compute a solution of an ODE system to a given tolerance. We thus have two options.

Firstly, we can iterate for $i = 0, 1, \ldots, i_{\text{end}}$, where $i_{\text{end}} = 10$, say. After each iteration we check for convergence. Unless it is attained (within the threshold ε) we decide that h is too large and abandon the current step. This is called *iteration to convergence.*

The second option is to identify the predictor–corrector pair with the two methods (5.6) and (5.5) that have been used in Chapter 5 to control the error by means of the *Milne device.* We perform just a single iteration of the corrector and substitute $\boldsymbol{w}^{[1]}$, instead of $\boldsymbol{y}_{n+s}$, into the error estimate (5.8). If $\kappa \le h\delta$ then all is well and we let $\boldsymbol{y}_{n+s} = \boldsymbol{w}^{[1]}$. Otherwise we abandon the step. Note that we are accepting a value of $\boldsymbol{y}_{n+s}$ that solves neither the nonlinear algebraic equation nor, as a matter of fact, the implicit multistep method. This, however, is of no consequence since our $\boldsymbol{y}_{n+s}$ passes the error test – and that is all that matters! This approach is called the *PECE iteration.*[2]

The choice between the PECE iteration and iteration to convergence hinges upon the relative cost of performing a single iteration and changing the step-size. If the cost of changing h is negligible, we might just as well abandon the iteration unless $\boldsymbol{w}^{[1]}$ (or

---

[2]Predict, Evaluate, Correct, Evaluate.

perhaps $w^{[2]}$, in which case we have a $PE(CE)^2$ procedure) satisfies the error criterion. If, though, this cost is large we should carry on with the iteration considerably longer. Another consideration is that a PECE iteration is likely to cause severe contraction of the linear stability domain of the corrector. In particular, no such procedure can be A-stable[3] (cf. Exercise 6.4).

We recall the dichotomy between stiff and non-stiff ODEs. If the ODE is non-stiff then we are likely to employ the functional iteration (6.2), which costs nothing to restart. The only cost of changing h is in remeshing which, although difficult to program, carries a very modest computational price tag. Since shrinkage of the linear stability domain is not an important issue for non-stiff ODEs, the clear conclusion is that the PECE approach is superior in this case. Moreover, if the equation is stiff, we should use the modified Newton–Raphson iteration (6.9). In order to change the step-size, we need to redo the LU factorization of $I - hJ$ (since h has changed). Moreover, it is a good policy to re-evaluate J as well, unless it has already been computed at y_{n+s-1}: it might well be that the failure of the iterative procedure follows from poor approximation of the Jacobian matrix. Finally, stability is definitely a crucial consideration and we should be unwilling to reconcile ourselves to a collapse in the size of the linear stability domain. All these reasons mean that the right approach is to iterate to convergence.

Comments and bibliography

The computation of nonlinear algebraic systems is as old as numerical analysis itself. This is not necessarily an advantage, since the theory has developed in many directions which, mostly, are irrelevant to the theme of this chapter.

The basic problem admits several equivalent formulations: firstly, we may regard it as finding a zero of the equation $h_1(x) = 0$; secondly, as computing a fixed point of the system $x = h_2(x)$, where $h_2 = x + \alpha h_1(x)$ for some $\alpha \in \mathbb{R} \setminus \{0\}$; thirdly, as minimizing $\|h_1(x)\|$. (With regard to the third formulation, minimization is often equivalent to the solution of a nonlinear system: provided the function $\psi : \mathbb{R}^d \to \mathbb{R}$ is continuously differentiable, the problem of finding the stationary values of ψ is equivalent to solving the system grad $\psi(x) = 0$.) Therefore, in a typical library nonlinear algebraic systems and their numerical analysis appear under several headings, probably on different shelves. Good sources are Ortega & Rheinboldt (1970) and Fletcher (1987) – the latter has a pronounced optimization flavour.

Modern numerical practice has moved a long way from the old days of functional iteration and the Newton–Raphson method and its modifications. The powerful algorithms of today owe much to tremendous advances in numerical optimization, as well as to the recent realization that certain acceleration schemes for linear systems can be applied with a telling effect to nonlinear problems (for example, the method of conjugate gradients, briefly mentioned in Chapter 10). However, it appears that, as far as the choice of nonlinear algebraic algorithms for practical implementation of ODE algorithms is concerned, not much has happened in the last three decades. The texts of Gear (1971) and of Shampine & Gordon (1975) represent, to a large extent, the state of the art today.

This conservatism is not necessarily a bad thing. After all, the test of the pudding is in the eating and, as far as we are aware, functional iteration (6.4) and modified Newton–Raphson

---

[3]in a formal sense. A-stability is defined only for constant steps, whereas the whole *raison d'être* of the PECE iteration is that it is operated within a variable-step procedure. However, experience tells us that the damage to the quality of the solution in an unstable situation is genuine.

(6.9), applied correctly and to the right problems, discharge their duty very well indeed. We cannot emphasize enough that the task in hand is not simply to solve an arbitrary nonlinear algebraic system, but to compute a problem that arises in the calculation of ODEs. This imposes a great deal of structure, highlights the crucial importance of the parameter h and, at each iteration, faces us with the question 'Should we continue to iterate or, rather, abandon the step and decrease h?'. The transplantation of modern methods for general nonlinear algebraic systems into this framework requires a great deal of work and fine-tuning. It might well be a worthwhile project, though: there are several good dissertations here, awaiting authors!

One aspect of functional iteration familiar to many readers (and, through the agency of the mass media, exhibitions and coffee-table volumes, to the general public) is the fractal sets that arise when complex functions are iterated. It is only fair to mention that, behind the façade of beautiful pictures, there lies some truly beautiful mathematics: complex dynamics, automorphic forms, Teichmüller spaces.... This, however, is largely irrelevant to the task in hand. It is a constant temptation of the wanderer in the mathematical garden to stray from the path and savour the sheer beauty and excitement of landscapes strange and wonderful. Although it may be a good idea occasionally to succumb to temptation, on this occasion we virtuously stay on the straight and narrow.

Fletcher, R. (1987), *Practical Methods of Optimization* (2nd ed.), Wiley, London.

Gear, C.W. (1971), *Numerical Initial Value Problems in Ordinary Differential Equations*, Prentice–Hall, Englewood Cliffs, NJ.

Ortega, J.M. and Rheinboldt, W.C. (1970), *Iterative Solution of Nonlinear Equations in Several Variables,* Academic Press, New York.

Shampine, L.F. and Gordon, M.K. (1975), *Computer Solution of Ordinary Differential Equations,* W.H. Freeman, San Francisco.

Exercises

6.1   Let $g(w) = \Omega w + a$, where Ω is a $d \times d$ matrix.

a Prove that the inequality (i) of Theorem 6.1 is satisfied for $\lambda = h\|\Omega\|$ and $\rho = \infty$. Deduce a condition on h that ensures that all the assumptions of the theorem are valid.

b Let $\| \cdot \|$ be the Euclidean norm (A.1.3.3). Show that the above value of λ is the best possible, in the following sense: there exist no $\rho > 0$ and $0 < \lambda < h\|\Omega\|$ such that

$$\|g(v) - g(u)\| \leq \frac{\lambda}{h}\|v - u\|$$

for all $v, u \in \mathcal{B}_\rho(w^{[0]})$. [*Hint: Recalling that*

$$\|\Omega\| = \max\left\{\frac{\|\Omega x\|}{\|x\|} : x \in \mathbb{R}^d,\ x \neq 0\right\},$$

prove that for every $\varepsilon > 0$ there exists $\boldsymbol{x}_\varepsilon$ such that $\|\Omega\| = \|\Omega \boldsymbol{x}_\varepsilon\| / \|\boldsymbol{x}_\varepsilon\|$ and $\|\boldsymbol{x}_\varepsilon\| = \varepsilon$. For any $\rho > 0$ choose $\boldsymbol{v} = \boldsymbol{w}^{[0]} + \boldsymbol{x}_\rho$ and $\boldsymbol{u} = \boldsymbol{w}^{[0]}$.]

6.2   Let $\boldsymbol{g}(\boldsymbol{w}) = \Omega \boldsymbol{w} + \boldsymbol{a}$, where Ω is a $d \times d$ matrix.

   a Prove that the Newton–Raphson method (6.7) converges (in exact arithmetic) in a single iteration.

   b Suppose that $J = \Omega$ in the modified Newton–Raphson method (6.9). Prove that also in this case just a single iteration is required.

6.3   Prove that the modified Newton–Raphson iteration (6.9) can be written as the functional iteration scheme

$$\boldsymbol{w}^{[i+1]} = h\tilde{\boldsymbol{g}}(\boldsymbol{w}^{[i]}) + \tilde{\boldsymbol{\beta}}, \qquad i = 0, 1, \ldots,$$

where $\tilde{\boldsymbol{g}}$ and $\tilde{\boldsymbol{\beta}}$ are given by (6.10).

6.4   Let the two-step Adams–Bashforth method (2.6) and the trapezoidal rule (1.9) be respectively the predictor and the corrector of a PECE scheme.

   a Applying the scheme with a constant step-size to the linear scalar equation $y' = \lambda y$, $y(0) = 1$, prove that

$$y_{n+1} - \left[1 + h\lambda + \tfrac{3}{4}(h\lambda)^2\right] y_n + \tfrac{1}{4}(h\lambda)^2 y_{n-1} = 0, \qquad n = 1, 2, \ldots. \quad (6.12)$$

   b Prove that, unlike the trapezoidal rule itself, the PECE scheme is not A-stable. [*Hint: Let $h|\lambda| \gg 1$. Prove that every solution of (6.12) is of the form $y_n \approx c \left[\tfrac{3}{4}(h\lambda)^2\right]^n$ for large n.*]

6.5   Consider the PECE iteration

$$\boldsymbol{x}_{n+3} = -\tfrac{1}{2}\boldsymbol{y}_n + 3\boldsymbol{y}_{n+1} - \tfrac{3}{2}\boldsymbol{y}_{n+2} + 3h\boldsymbol{f}(t_{n+2}, \boldsymbol{y}_{n+2}),$$

$$\boldsymbol{y}_{n+3} = \tfrac{1}{11}[2\boldsymbol{y}_n - 9\boldsymbol{y}_{n+1} + 18\boldsymbol{y}_{n+2} + 6h\boldsymbol{f}(t_{n+3}, \boldsymbol{x}_{n+3})].$$

   a Show that both methods are third-order and that the Milne device gives the estimate $\tfrac{6}{17}(\boldsymbol{x}_{n+3} - \boldsymbol{y}_{n+3})$ of the error of the corrector formula.

   b Let the method be applied to scalar equations, let the cubic polynomial p_{n+2} interpolate y_m at $m = n, n+1, n+2$ and let $p'_{n+2}(t_{n+2}) = f(t_{n+2}, y_{n+2})$. Verify that the predictor and corrector are equivalent to the formulae

$$\begin{aligned}
x_{n+1} &= p_{n+2}(t_{n+3}) = p_{n+2}(t_{n+2}) + hp'_{n+2}(t_{n+2}) + \tfrac{1}{2}h^2 p''_{n+2}(t_{n+2}) \\
&\quad + \tfrac{1}{6}h^3 p'''_{n+2}(t_{n+2}), \\
y_{n+1} &= p_{n+2}(t_{n+2}) + \tfrac{5}{11}hp'_{n+2}(t_{n+2}) - \tfrac{1}{22}h^2 p''_{n+2}(t_{n+2}) - \tfrac{7}{66}p'''_{n+2}(t_{n+2}) \\
&\quad + \tfrac{6}{11}hf(t_{n+3}, x_{n+3}).
\end{aligned}$$

respectively. These formulae make it easy to change the value of h at t_{n+2} if the Milne estimate is unacceptably large.

PART II
The Poisson equation

7

Finite difference schemes

7.1 Finite differences

The opening line of *Anna Karenina,* 'All happy families resemble one another, but each unhappy family is unhappy in its own way',[1] is a useful metaphor for the relationship between computational ordinary differential equations (ODEs) and computational *partial differential equations* (PDEs). ODEs are a happy family – perhaps they do not resemble each other, but, at the very least, we can treat them by a relatively small compendium of computational techniques. (True, upon closer examination, even ODEs are not the same: their classification into stiff and non-stiff is the most obvious example. However, how many happy families will survive the deconstructing attentions of a mathematician?)

PDEs are a huge and motley collection of problems, each unhappy in its own way. Most students of mathematics would be aware of the classification into elliptic, parabolic and hyperbolic equations, but this is only the first step in a long journey. As soon as nonlinear – or even quasilinear – PDEs are allowed, the subject is replete with an enormous number of different problems and each problem clamours for its own brand of numerics. No textbook can (or should) cover this enormous menagerie. Fortunately, however, it is possible to distil a small number of tools that allow for a well-informed numerical treatment of several important equations and form a sound basis for the understanding of the subject as a whole.

One such tool is the classical theory of *finite differences.* The main idea in the calculus of finite differences is to replace derivatives with linear combinations of discrete function values. Finite differences have the virtue of simplicity and they account for a large proportion of the numerical methods actually used in applications. This is perhaps a good place to stress that alternative approaches abound, each with its own virtue: finite elements, spectral and pseudospectral methods, boundary elements, spectral elements, particle methods.... Chapter 8 is devoted to the finite element method.

It is convenient to introduce finite differences in the context of real (or complex) sequences $z = \{z_k\}_{k=-\infty}^{\infty}$ indexed by all the integers. Everything can be translated to finite sequences in a straightforward manner, except that the notation becomes more cumbersome.

We commence by defining the following *finite difference operators,* which map the space $\mathbb{R}^{\mathbb{Z}}$ of all such sequences into itself. Each operator is defined in terms of its

---

[1] Leo Tolstoy, *Anna Karenina,* Translated by L. & A. Maude, Oxford University Press, London (1967).

action on individual elements of the sequence:

the *shift* operator,	$(\mathcal{E}z)_k = z_{k+1};$
the *forward difference* operator,	$(\Delta_+z)_k = z_{k+1} - z_k;$
the *backward difference* operator,	$(\Delta_-z)_k = z_k - z_{k-1};$
the *central difference* operator,	$(\Delta_0z)_k = z_{k+\frac{1}{2}} - z_{k-\frac{1}{2}};$
the *averaging* operator,	$(\Upsilon_0z)_k = \frac{1}{2}(z_{k-\frac{1}{2}} + z_{k+\frac{1}{2}}).$

The first three operations are defined for all $k = 0, \pm1, \pm2, \ldots$. Note, however, that the last two operators, Δ_0 and Υ_0, do not, as a matter of fact, map $\mathbb{R}^\mathbb{Z}$ into itself. After all, the values $z_{k+\frac{1}{2}}$ are meaningless for integer k. Having said this, we will soon see that, appropriately used, these operators can be perfectly well defined.

Let us further assume that the sequence z originates in the sampling of a function z, say, at equispaced points. In other words, $z_k = z(kh)$ for some $h > 0$. Stipulating (for the time being) that z is an entire function, we define

the *differential* operator,	$(\mathcal{D}z)_k = z'(kh).$

Our first observation is that all these operators are *linear:* given that

$$\mathcal{T} \in \{\mathcal{E}, \Delta_+, \Delta_-, \Delta_0, \Upsilon_0, \mathcal{D}\},$$

and that $w, z \in \mathbb{R}^\mathbb{Z}$, $a, b \in \mathbb{R}$, it is true that

$$\mathcal{T}(aw + bz) = a\mathcal{T}w + b\mathcal{T}z.$$

The superposition of finite difference operators is defined in an obvious manner, e.g.

$$\Delta_+\mathcal{E}^2 z_k = \Delta_+\left(\mathcal{E}(\mathcal{E}z_k)\right) = \Delta_+(\mathcal{E}z_{k+1}) = \Delta_+z_{k+2} = z_{k+3} - z_{k+2}.$$

Note that we have just introduced a notational shortcut: $\mathcal{T}z_k$ stands for $(\mathcal{T}z)_k$, where $\mathcal{T}$ is an arbitrary finite difference operator.

The purpose of the calculus of finite differences is, ultimately, to approximate derivatives by linear combinations of function values along a grid. We wish to get rid of $\mathcal{D}$ by expressing it in the currency of the other operators. This, however, requires us first to define formally general functions of finite difference operators.

Because of our assumption that $z_k = z(kh)$, $k = 0, \pm1, \pm2, \ldots$, finite difference operators depend upon the parameter h. Let $g(x) = \sum_{j=0}^\infty a_j x^j$ be an arbitrary analytic function, given in terms of its Taylor series. Noting that

$$\mathcal{E} - \mathcal{I}, \ \Upsilon_0 - \mathcal{I}, \ \Delta_+, \ \Delta_-, \ \Delta_0, \ h\mathcal{D} \ \overset{h\to0+}{\longrightarrow} \ O,$$

where $\mathcal{I}$ is the identity, we can formally expand g about $\mathcal{E} - \mathcal{I}$, $\Upsilon_0 - \mathcal{I}$, Δ_+ etc. For example,

$$g(\Delta_+)z = \left(\sum_{j=0}^\infty a_j\Delta_+^j\right)z = \sum_{j=0}^\infty a_j(\Delta_+^jz).$$

It is not our intention to argue here that the above expansions converge (although they do), but merely to use them in a formal manner to define functions of operators.

◇ **The operator** $\mathcal{E}^{1/2}$ What is the square root of the shift operator? One interpretation, which follows directly from the definition of $\mathcal{E}$, is that $\mathcal{E}^{1/2}$ is a 'half-shift', which takes z_k to $z_{k+\frac{1}{2}}$, which we can define as $z((k+\frac{1}{2})h)$. An alternative expression exploits the power series expansion

$$\sqrt{1+x} = 1 + \sum_{j=1}^{\infty} \left[\frac{(-1)^{j-1}}{2^{2j-1}} \right] \frac{(2j-2)!}{(j-1)!j!} x^j$$

to argue that

$$\mathcal{E}^{1/2} = \mathcal{I} - 2 \sum_{j=1}^{\infty} \frac{(2j-2)!}{(j-1)!j!} \left[-\tfrac{1}{4}(\mathcal{E} - \mathcal{I}) \right]^j.$$

Needless to say, the two definitions coincide, but the proof of this would proceed at a tangent to the theme of this chapter. Readers familiar with Newton's interpolation formula might seek a proof by interpolating $z(x + \frac{1}{2})$ on the set $\{x + jh\}_{j=0}^{\ell}$ and letting $\ell \to \infty$.   ◇

Recalling the purpose of our analysis, we next express all finite difference operators in a single currency, as functions of the shift operator $\mathcal{E}$. It is trivial that $\Delta_+ = \mathcal{E} - \mathcal{I}$ and $\Delta_- = \mathcal{I} - \mathcal{E}^{-1}$, while the interpretation of $\mathcal{E}^{1/2}$ as 'half-shift' implies that $\Delta_0 = \mathcal{E}^{1/2} - \mathcal{E}^{-1/2}$ and $\Upsilon_0 = \frac{1}{2}(\mathcal{E}^{-1/2} + \mathcal{E}^{1/2})$. Finally, to express $\mathcal{D}$ in terms of the shift operator, we recall the Taylor theorem: for any analytic function z it is true that

$$\mathcal{E}z(x) = z(x + h) = \sum_{j=0}^{\infty} \frac{1}{j!} \left[\frac{\mathrm{d}^j z(x)}{\mathrm{d}x^j} \right] h^j = \left[\sum_{j=0}^{\infty} \frac{1}{j!} (h\mathcal{D})^j \right] z(x) = \mathrm{e}^{h\mathcal{D}} z(x),$$

and we deduce $\mathcal{E} = \mathrm{e}^{h\mathcal{D}}$. Formal inversion yields

$$h\mathcal{D} = \ln \mathcal{E}. \tag{7.1}$$

We conclude that, each having been expressed in terms of $\mathcal{E}$, all six finite difference operators commute. This is a useful observation since it follows that we need not bother with the order of their action whenever they are superposed.

The above operator formulae can be (formally) inverted, thereby expressing $\mathcal{E}$ in terms of Δ_+ etc. It is easy to verify that $\mathcal{E} = \mathcal{I} + \Delta_+ = (I - \Delta_-)^{-1}$. The expression for Δ_0 is a quadratic equation for $\mathcal{E}^{1/2}$,

$$(\mathcal{E}^{1/2})^2 - \Delta_0 \mathcal{E}^{1/2} - \mathcal{I} = O,$$

with two solutions, $\frac{1}{2}\Delta_0 \pm \sqrt{\frac{1}{4}\Delta_0^2 + \mathcal{I}}$. Letting $h \to 0$, we deduce that the correct formula is

$$\mathcal{E} = \left(\tfrac{1}{2}\Delta_0 + \sqrt{\mathcal{I} + \tfrac{1}{4}\Delta_0^2} \right)^2.$$

We need not bother to express $\mathcal{E}$ in terms of Υ_0, since this serves no useful purpose.

Combining (7.1) with these expressions, we next write the differential operator in terms of other finite difference operators,

$$h\mathcal{D} = \ln(\mathcal{I} + \Delta_+) \tag{7.2}$$

$$h\mathcal{D} = -\ln(\mathcal{I} + \Delta_-) \tag{7.3}$$

$$h\mathcal{D} = 2\ln\left(\tfrac{1}{2}\Delta_0 + \sqrt{\mathcal{I} + \tfrac{1}{4}\Delta_0^2}\right). \tag{7.4}$$

Recall that the purpose of the exercise is to approximate the differential operator $\mathcal{D}$ and its powers (which, of course, correspond to higher derivatives). The formulae (7.2)–(7.4) are ideally suited to this purpose. For example, expanding (7.2) we obtain

$$\mathcal{D} = \frac{1}{h}\ln(\mathcal{I} + \Delta_+) = \frac{1}{h}\left[\Delta_+ - \tfrac{1}{2}\Delta_+^2 + \tfrac{1}{3}\Delta_+^3 + \mathcal{O}(\Delta_+^4)\right]$$

$$= \frac{1}{h}\left(\Delta_+ - \tfrac{1}{2}\Delta_+^2 - \tfrac{1}{3}\Delta_+^3\right) + \mathcal{O}(h^3), \qquad h \to 0,$$

where we exploit the estimate $\Delta_+ = \mathcal{O}(h)$, $h \to 0$. Operating s times, we obtain an expression for the sth derivative, $s = 1, 2, \ldots,$

$$\mathcal{D}^s = \frac{1}{h^s}\left[\Delta_+^s - \tfrac{1}{2}s\Delta_+^{s+1} + \tfrac{1}{24}s(3s + 5)\Delta_+^{s+2}\right] + \mathcal{O}(h^3), \qquad h \to 0. \tag{7.5}$$

The meaning of (7.5) is that the linear combination

$$\frac{1}{h^s}\left[\Delta_+^s - \tfrac{1}{2}s\Delta_+^{s+1} + \tfrac{1}{24}s(3s + 5)\Delta_+^{s+2}\right]z_k \tag{7.6}$$

of the $s + 3$ grid values $z_k, z_{k+1}, \ldots, z_{k+s+2}$ approximates $\mathrm{d}^s z(kh)/\mathrm{d}x^s$ up to $\mathcal{O}(h^3)$. Needless to say, truncating (7.5) a term earlier, for example, we obtain order $\mathcal{O}(h^2)$, whereas higher order can be obtained by expanding the logarithm further.

Similarly to (7.5), we can use (7.3) to express derivatives in terms of grid points wholly to the left,

$$\mathcal{D}^s = \frac{(-1)^s}{h^s}[\ln(I - \Delta_-)]^s = \frac{1}{h^s}\left[\Delta_-^s + \tfrac{1}{2}s\Delta_-^{s+1} + \tfrac{1}{24}s(3s + 5)\Delta_-^{s+2}\right] + \mathcal{O}(h^3), \quad h \to 0.$$

However, how much sense does it make to approximate derivatives solely in terms of grid points that all lie to one side? Sometimes we have little choice – more about this later – but in general it is a good policy to match the numbers of points to the left and to the right. The natural candidate for this task is the central finite difference operator Δ_0 except that now, having at last started to discuss approximation on a grid, not just operators in a formal framework, we can no longer loftily disregard the fact that $\Delta_0 z$ is not in the set $\mathbb{R}^{\mathbb{Z}}$ of grid sequences. The crucial observation is that *even powers* of Δ_0 map $\mathbb{R}^{\mathbb{Z}}$ to itself! Thus, $\Delta_0^2 z_n = z_{n-1} - 2z_n + z_{n+1}$ and the proof for all even power follows at once from the trivial observation that $\Delta_0^{2s} = (\Delta_0^2)^s$.

Recalling (7.4), we consider the Taylor expansion of the function $g(\xi) := \ln(\xi + \sqrt{1 + \xi^2})$. By the generalized binomial theorem,

$$g'(\xi) = \frac{1}{\sqrt{1 + \xi^2}} = \sum_{j=0}^{\infty}(-1)^j\binom{2j}{j}\left(\tfrac{1}{2}\xi\right)^{2j},$$

where $\binom{2j}{j}$ is a binomial coefficient equal to $(2j)!/(j!)^2$. Since $g(0) = 0$ and the Taylor series converges uniformly for $|\xi| < 1$, integration yields

$$g(\xi) = g(0) + \int_0^\xi g'(\tau)\,\mathrm{d}\tau = 2\sum_{j=0}^\infty \frac{(-1)^j}{2j+1}\binom{2j}{j}\left(\tfrac{1}{2}\xi\right)^{2j+1}.$$

Letting $\xi = \tfrac{1}{2}\Delta_0$, we thus deduce from (7.4) the formal expansion

$$\mathcal{D} = \frac{2}{h}\,g\!\left(\tfrac{1}{2}\Delta_0\right) = \frac{4}{h}\sum_{j=0}^\infty \frac{(-1)^j}{2j+1}\binom{2j}{j}\left(\tfrac{1}{4}\Delta_0\right)^{2j+1}. \tag{7.7}$$

Unfortunately, (7.7) is of exactly the wrong kind – all the powers of Δ_0 therein are odd! However, since even powers of odd powers are themselves even, raising (7.7) to an even power yields

$$\mathcal{D}^{2s} = \frac{1}{h^{2s}}\left[(\Delta_0^2)^s - \tfrac{1}{12}(\Delta_0^2)^{s+1} + \tfrac{1}{45}(\Delta_0^2)^{s+2}\right] + \mathcal{O}(h^6), \qquad h \to 0. \tag{7.8}$$

Thus, the linear combination

$$\frac{1}{h^{2s}}\left[(\Delta_0^2)^s - \tfrac{1}{12}(\Delta_0^2)^{s+1} + \tfrac{1}{45}(\Delta_0^2)^{s+2}\right] z_k \tag{7.9}$$

approximates $\mathrm{d}^{2s}z(kh)/\mathrm{d}x^{2s}$ to $\mathcal{O}(h^6)$.

How effective is (7.9) in comparison with (7.6)? To attain $\mathcal{O}(h^{2p})$, (7.6) requires $2s + 2p$ adjacent grid points and (7.9) just $2s + 2p - 1$, a relatively modest saving. Central difference operators, however, have smaller error constants (cf. Exercises 7.3 and 7.4). More importantly, they are more convenient to use and usually lead to more tractable linear algebraic systems (see Chapters 9–12).

The expansion (7.8) is valid only for even derivatives. To reap the benefits of central differencing for odd derivatives, we require a simple, yet clever, trick. Let us thus pay attention to the averaging operator Υ_0, which has until now played a silent role in the proceedings.

We express Υ_0 in terms of Δ_0. Since $\Upsilon_0 = \tfrac{1}{2}(\mathcal{E}^{1/2} + \mathcal{E}^{-1/2})$ and $\Delta_0 = \mathcal{E}^{1/2} - \mathcal{E}^{-1/2}$, it follows that

$$4\Upsilon_0^2 = \mathcal{E} + 2\mathcal{I} + \mathcal{E}^{-1},$$
$$\Delta_0^2 = \mathcal{E} - 2\mathcal{I} + \mathcal{E}^{-1}$$

and, subtracting, we deduce that $4\Upsilon_0 - \Delta_0^2 = 4\mathcal{I}$. We conclude that

$$\Upsilon_0 = (\mathcal{I} + \tfrac{1}{4}\Delta_0^2)^{1/2}. \tag{7.10}$$

The main idea now is to multiply (7.7) by the identity $\mathcal{I}$, which we craftily disguise by using (7.10),

$$\mathcal{I} = \Upsilon_0\,(\mathcal{I} + \tfrac{1}{4}\Delta_0^2)^{-1/2} = \Upsilon_0\sum_{j=0}^\infty (-1)^j \binom{2j}{j}\left(\tfrac{1}{16}\Delta_0^2\right)^j.$$

The outcome,

$$\mathcal{D} = \frac{1}{h}(\Upsilon_0 \Delta_0) \left[\sum_{j=0}^{\infty} (-1)^j \binom{2j}{j} \left(\tfrac{1}{16}\Delta_0^2\right)^j \right] \left[\sum_{i=0}^{\infty} \frac{(-1)^i}{2i+1} \binom{2i}{i} \left(\tfrac{1}{16}\Delta_0^2\right)^i \right], \qquad (7.11)$$

might look messy, but has one redeeming virtue: it is constructed exclusively from *even* powers of Δ_0 and $\Upsilon_0\Delta_0$. Since

$$\Upsilon_0 \Delta_0 z_k = \Upsilon_0 \left(z_{k+\frac{1}{2}} - z_{k-\frac{1}{2}} \right) = \tfrac{1}{2} \left(z_{k+1} - z_{k-1} \right),$$

we conclude that (7.11) is a linear combination of terms that reside on the grid.

The expansion (7.11) can be rised to a power, but this is not a good idea, since this procedure is wasteful in terms of grid points; an example is provided by

$$\mathcal{D}^2 = \frac{1}{h^2}(\Upsilon_0 \Delta_0)^2 \left(\mathcal{I} - \tfrac{1}{3}\Delta_0^2\right) + \mathcal{O}(h^4).$$

Since

$$(\Upsilon_0 \Delta_0)^2 \left(\mathcal{I} - \tfrac{1}{3}\Delta_0^2\right) z_k = \tfrac{1}{12}(-z_{k-3} + 5z_{k-2} + z_{k-1} - 10z_k + z_{k+1} + 5z_{k+2} - z_{k+3}),$$

we need seven points to attain $\mathcal{O}(h^4)$, while (7.9) requires just five points. In general, a considerably better idea is first to raise (7.7) to an odd power and then to multiply it by $\mathcal{I} = \Upsilon_0(\mathcal{I} + \tfrac{1}{4}\Delta_0^2)^{1/2}$. The outcome,

$$\begin{aligned}
\mathcal{D}^{2s+1} &= \frac{1}{h^{2s+1}}(\Upsilon_0 \Delta_0) \left[(\Delta_0^2)^s - \tfrac{1}{12}(s+2)(\Delta_0^2)^{s+1} \right. \\
&\quad \left. + \tfrac{1}{1440}(s+3)(5s+16)(\Delta_0^2)^{s+2} \right] + \mathcal{O}(h^5), \qquad h \to 0,
\end{aligned} \qquad (7.12)$$

lives on the grid and, other things being equal, is the recommended approximant of odd derivatives.

◇ **A simple example…**   Figure 7.1 displays the (natural) logarithm of the error in the approximation of the first derivative of $z(x) = x\,\mathrm{e}^x$. The first row corresponds to the forward difference approximants

$$\frac{1}{h}\Delta_+, \qquad \frac{1}{h}\left(\Delta_+ - \tfrac{1}{2}\Delta_+^2\right) \qquad \text{and} \qquad \frac{1}{h}\left(\Delta_+ - \tfrac{1}{2}\Delta_+^2 + \tfrac{1}{3}\Delta_+^3\right),$$

with $h = \tfrac{1}{10}$ and $h = \tfrac{1}{20}$ in the first and in the second column respectively. The second row displays the central difference approximants

$$\frac{1}{h}\Upsilon_0\Delta_0 \qquad \text{and} \qquad \frac{1}{h}\Upsilon_0\Delta_0\left(\mathcal{I} - \tfrac{1}{6}\Delta_0^2\right).$$

What can we learn from this figure? If the error behaves like ch^p, where $c \neq 0$, then its logarithm is approximately $p\ln h + \ln|c|$. Therefore, for small h, one can expect each ℓth curve in the top row to behave like the first curve, scaled by ℓ (since $p = \ell$ for the ℓth curve). This is not the case in the first two columns, since $h > 0$ is not small enough, but the pattern becomes more visible when h decreases; the reader could try $h = \tfrac{1}{1000}$ to confirm that this asymptotic behaviour indeed takes place. On the other hand, replacing h by $\tfrac{1}{2}h$ should lower each curve by an amount roughly equal to $\ln 2 \approx 0.6931$ and this can

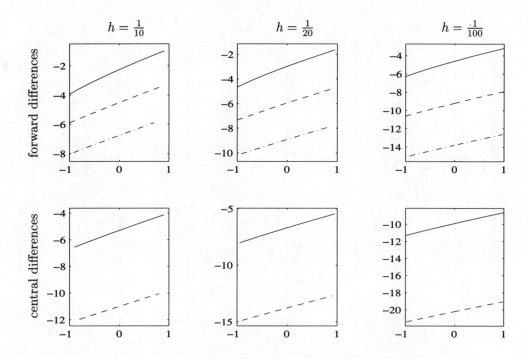

Figure 7.1 The error (on a logarithmic scale) in the approximation of z', where $z(x) = xe^x$, $-1 \leq x \leq 1$. Forward differences of size $\mathcal{O}(h)$ (solid line), $\mathcal{O}(h^2)$ (broken line) and $\mathcal{O}(h^3)$ (broken-and-dotted line) feature in the first row, while the second row presents central differences of size $\mathcal{O}(h^2)$ (solid line) and $\mathcal{O}(h^4)$ (broken line).

be observed by comparing the first two columns. Likewise, the curves in the third column are each lowered by about $\ln 5 \approx 1.6094$ in comparison with the second column.

Similar information is displayed in Figure 7.2, namely the logarithm of the error in approximating z'', where $z(x) = 1/(1 + x)$, by forward differences (in the top row) and central differences (in the second row). The specific approximants can be easily derived from (7.6) and (7.9) respectively. The pattern is similar, except that the singularity at $x = -1$ means that the quality of approximation deteriorates at the left end of the scale; it is always important to bear in mind that estimates based on Taylor expansions break down near singularities. ◇

Needless to say, there is no bar on using several finite difference operators in a single formula (cf. Exercise 7.5). However, other things being equal, we usually prefer to employ central differences.

There are two important exceptions. Firstly, realistic grids do not in fact extend from $-\infty$ to ∞; this was just a convenient assumption, which has simplified the

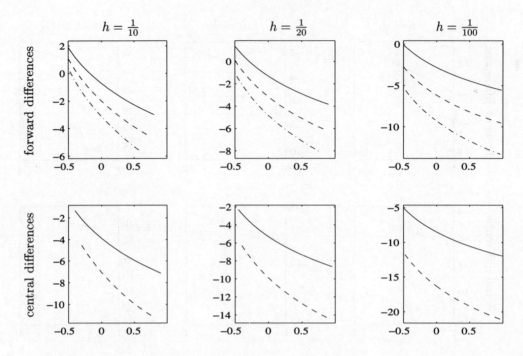

Figure 7.2 The error (on a logarithmic scale) in the approximation of z'', where $z(x) = 1/(1+x)$, $-\frac{1}{2} \le x \le 4$. The meaning of the curves is identical to that in Figure 7.1.

notation a great deal. Of course, we can employ finite differences on finite grids, except that the procedure might break down near the boundary. 'One-sided' finite differences possess obvious advantages in such situations. Secondly, for some PDEs the exact solution of the equation displays an innate 'preference' toward one spatial direction over the other, and in this case it is a good policy to let the approximation to the derivative reflect this fact. This behaviour is displayed by certain hyperbolic equations and we will encounter it in Chapter 14.

Finally, it is perfectly possible to approximate derivatives on non-equidistant grids. This, however, is by and large outside the scope of this book, except for a brief discussion in the next section of approximation near curved boundaries.

7.2   The five-point formula for $\nabla^2 u = f$

Perhaps the most important and ubiquitous PDE is the *Poisson* equation

$$\nabla^2 u = f, \qquad (x,y) \in \Omega, \tag{7.13}$$

where

$$\nabla^2 = \frac{\partial^2}{\partial x^2} + \frac{\partial^2}{\partial y^2},$$

Figure 7.3   An example of a computational grid for a two-dimensional domain Ω.
•, internal points; ○, near-boundary points; ×, a boundary point.

$f = f(x, y)$ is a known, continuous function and the domain $\Omega \subset \mathbb{R}^2$ is bounded, open, connected and has a piecewise-smooth boundary. We hasten to add that this is not the most general form of the Poisson equation – in fact we are allowed any number of space dimensions, not just two, Ω need not be bounded and its boundary, as well as the function f, can satisfy far less demanding smoothness requirements. However, the present framework is sufficient for our purpose.

Like any partial differential equation, for its solution (7.13) must be accompanied by a boundary condition. We assume the *Dirichlet* condition, namely that

$$u(x, y) = \phi(x, y), \qquad (x, y) \in \partial\Omega. \tag{7.14}$$

An implementation of finite differences always commences by inscribing a grid into the domain of interest. In our case we impose on $\mathrm{cl}\,\Omega$ a square grid $\Omega_{\Delta x}$ parallel to the axes, with an equal spacing of Δx in both spatial directions (Figure 7.3). In other words, we choose $\Delta x > 0$, $(x_0, y_0) \in \Omega$ and let $\Omega_{\Delta x}$ be the set of all points of the form $(x_0 + k\Delta x, \ y_0 + \ell\Delta x)$ that reside in the closure of Ω. We denote

$$\mathbf{I}_{\Delta x} := \left\{ (k, \ell) \in \mathbb{Z}^2 : (x_0 + k\Delta x, y_0 + \ell\Delta x) \in \mathrm{cl}\,\Omega \right\},$$
$$\mathbf{I}^{\circ}_{\Delta x} := \left\{ (k, \ell) \in \mathbb{Z}^2 : (x_0 + k\Delta x, y_0 + \ell\Delta x) \in \Omega \right\},$$

and, for every $(k, \ell) \in \mathbf{I}^{\circ}_{\Delta x}$, let $u_{k,\ell}$ stand for the approximation to the solution $u(x_0 + k\Delta x, y_0 + \ell\Delta x)$ of the Poisson equation (7.13) at the relevant grid point. Note that, of course, there is no need to approximate grid points $(k, \ell) \in \mathbf{I}_{\Delta x} \setminus \mathbf{I}^{\circ}_{\Delta x}$, since they lie on $\partial\Omega$ and there exact values are given by (7.14).

Wishing to approximate ∇^2 by finite differences, our first observation is that we are no longer allowed the comfort of sequences that stretch all the way to $\pm\infty$; whether in the x- or the y-direction, $\partial\Omega$ acts as a barrier and we cannot use grid points outside $\mathrm{cl}\,\Omega$ to assist in our approximation.

Our first finite difference scheme approximates $\nabla^2 u$ at the (k,ℓ)th grid point as a linear combination of the five values $u_{k,\ell}, u_{k\pm1,\ell}, u_{k,\ell\pm1}$ and it is valid only if the immediate horizontal and vertical neighbours of (k,ℓ), namely $(k\pm1,\ell)$ and $(k,\ell\pm1)$ respectively, are in $\mathbf{I}_{\Delta x}$. We say, for the purposes of our present discussion, that such a point $(x_0+k\Delta x,\, y_0+\ell\Delta x)$ is an *internal point*. In general, the set $\Omega_{\Delta x}$ consists of three types of points: *boundary points*, which lie on $\partial\Omega$ and whose value is known by virtue of (7.14); internal points, which will be soon subjected to our scrutiny; and the remaining *near-boundary points*, where we can no longer employ finite differences on an equidistant grid so that a special approach is required. Needless to say, the definition of internal and near-boundary points changes if we employ a different configuration of points in our finite difference scheme (Section 7.3).

Let us suppose that $(k,\ell)\in\mathbf{I}^{\circ}_{\Delta x}$ corresponds to an internal point. Following our recommendation from Section 7.1, we use central differences. Of course, our grid is now two-dimensional and we can use differences in either dimension. This creates no difficulty whatsoever, as long as we distinguish clearly the space dimension with respect to which our operator acts. We do this by appending a subscript, e.g. $\Delta_{0,x}$.

Let $v=v(x,y)$, $(x,y)\in\mathrm{cl}\,\Omega$, be an arbitrary sufficiently smooth function. It follows at once from (7.9) that, for every internal grid point,

$$\frac{\partial^2 v}{\partial x^2}\bigg|_{\substack{x=x_0+k\Delta x,\\ y=y_0+\ell\Delta x}} = \frac{1}{(\Delta x)^2}\Delta_{0,x}^2 v_{k,\ell} + \mathcal{O}\big((\Delta x)^2\big),$$

$$\frac{\partial^2 v}{\partial y^2}\bigg|_{\substack{x=x_0+k\Delta x,\\ y=y_0+\ell\Delta x}} = \frac{1}{(\Delta x)^2}\Delta_{0,y}^2 v_{k,\ell} + \mathcal{O}\big((\Delta x)^2\big),$$

where $v_{k,\ell}$ is the value of v at the (k,ℓ)th grid point. Therefore,

$$\frac{1}{(\Delta x)^2}(\Delta_{0,x}^2 + \Delta_{0,y}^2)$$

approximates ∇^2 to order $\mathcal{O}\big((\Delta x)^2\big)$. This motivates the replacement of the Poisson equation (7.13) by the *five point* finite difference scheme

$$\frac{1}{(\Delta x)^2}(\Delta_{0,x}^2 + \Delta_{0,y}^2)u_{k,\ell} = f_{k,\ell} \tag{7.15}$$

at every pair (k,ℓ) that corresponds to an internal grid point. Of course, $f_{k,\ell} = f(x_0+k\Delta x, y_0+\ell\Delta x)$. More explicitly, (7.15) can be written in the form

$$u_{k-1,\ell} + u_{k+1,\ell} + u_{k,\ell-1} + u_{k,\ell+1} - 4u_{k,\ell} = (\Delta x)^2 f_{k,\ell}, \tag{7.16}$$

and this motivates its name, the five-point formula. In lieu of the Poisson equation, we have a linear combination of the values of u at an (internal) grid point and at the immediate horizontal and vertical neighbours of this point.

Another way of depicting (7.15) is via a *computational stencil* (also known as a *computational molecule*). This is a pictorial representation that is self-explanatory (and becomes indispensable for more complicated finite difference schemes, which involve a larger number of points), as follows:

$$u_{k,\ell} = (\Delta x)^2 f_{k,\ell}$$

Thus, the equation (7.16) links five values of u in a linear fashion. Unless they lie on the boundary, these values are unknown. The main idea of the finite difference method is to associate with every grid point having an index in $\mathbf{I}_{\Delta x}^{\circ}$ (that is, every internal and near-boundary point) a single linear equation, for example (7.16). This results in a system of linear equations whose solution is our approximation $\boldsymbol{u} := (u_{k,\ell})_{(k,\ell)\in \mathbf{I}_{\Delta x}^{\circ}}$.

Three questions are critical to the performance of finite differences:

- Is the linear system nonsingular, so that the finite difference solution $\boldsymbol{u}$ exists and is unique?

- Suppose that a unique $\boldsymbol{u} = \boldsymbol{u}_{\Delta x}$ exists for all sufficiently small Δx, and let $\Delta x \to 0$. Is it true that the numerical solution converges to the exact solution of (7.13)? What is the asymptotic magnitude of the error?

- Are there efficient and robust means to solve the linear system, which is likely to consist of a very large number of equations?

We defer the third question to Chapters 9–12, where the theme of the numerical solution of large, sparse algebraic linear systems will be debated at some length. Meantime, we address ourselves to the first two questions in the special case when Ω is a square. Without loss of generality, we let $\Omega = \{(x,y) : 0 < x, y < 1\}$. This leads to considerable simplification since, provided we choose $\Delta x = 1/(m+1)$, say, for an integer m, and let $x_0 = y_0 = 0$, all grid points are either internal or boundary (Figure 7.4).

$\diamond$ **The Laplace equation**   Prior to attempting to prove theorems on the behaviour of numerical methods, it is always a good practice to run a few simple programs and obtain a 'feel' for what we are, after all, trying to prove. The computer is the mathematical equivalent of a physicist's laboratory!

In this spirit we apply the five-point formula (7.15) to the *Laplace equation* $\nabla^2 u = 0$ in the unit square $(0,1)^2$, subject to the boundary conditions

$$u(x,0) \equiv 0, \qquad u(x,1) = \frac{1}{(1+x)^2 + 1}, \qquad 0 \le y \le 1,$$

$$u(0,y) = \frac{y}{1+y^2}, \qquad u(1,y) = \frac{y}{4+y^2}, \qquad 0 \le x \le 1.$$

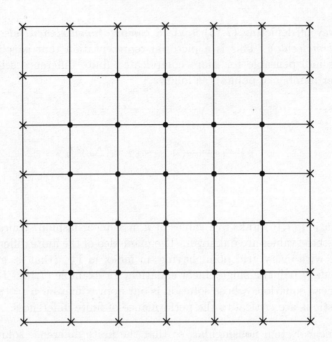

Figure 7.4 Computational grid for a unit square. As for Figure 7.3, internal and boundary points are denoted by solid circles and crosses, respectively.

Figure 7.5 displays the exact solution of this equation,

$$u(x,y) = \frac{y}{(1+x)^2 + y^2}, \qquad 0 \le x,y \le 1,$$

as well as its numerical solution by means of the five-point formula with $m = 5$, $m = 11$ and $m = 23$; this corresponds to $\Delta x = \frac{1}{6}$, $\Delta x = \frac{1}{12}$ and $\Delta x = \frac{1}{24}$ respectively. The size of the grid halves in each consecutive numerical trial and it is evident from the figure that the error decreases by a factor of four. This is consistent with an error decay of $\mathcal{O}\big((\Delta x)^{-2}\big)$, which is hinted at in our construction of (7.15) and will be proved in Theorem 7.2. $\diamond$

Recall that we wish to address ourselves to two questions. Firstly, is the linear system (7.16) nonsingular? Secondly, does its solution converge to the exact solution of the Poisson equation (7.13) as $\Delta x \to 0$? In the case of a square, both questions can be answered by employing a similar construction.

The function $u_{k,\ell}$ is defined on a two-dimensional grid and, to write the linear equations (7.16) formally in a matrix/vector notation, we need to rearrange $u_{k,\ell}$ into a one-dimensional column vector $\boldsymbol{u} \in \mathbb{R}^s$, where $s = m^2$. In other words, for any permutation $\{(k_i, \ell_i)\}_{i=1,2,\ldots,s}$ of the set $\{(k,\ell)\}_{k,\ell=1,2,\ldots,m}$ we can let

$$\boldsymbol{u} = \begin{bmatrix} u_{k_1,\ell_1} \\ u_{k_2,\ell_2} \\ \vdots \\ u_{k_s,\ell_s} \end{bmatrix}$$

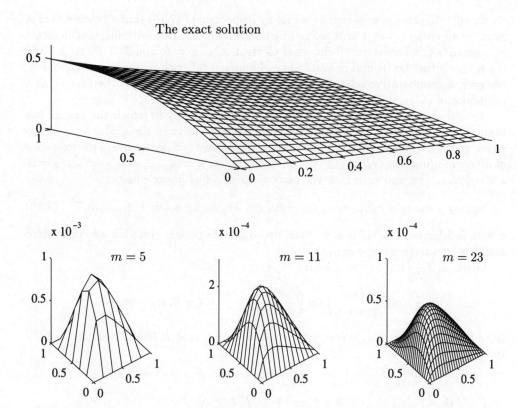

Figure 7.5   The exact solution of the Laplace equation discussed in the text and the errors of the five-point formula for $m = 5$, $m = 11$ and $m = 23$, with 25, 121 and 529 grid points respectively.

and write (7.16) in the form

$$Au = b, \qquad (7.17)$$

where A is an $s \times s$ matrix, while $b \in \mathbb{R}^s$ includes both the inhomogeneous part $(\Delta x)^2 f_{k,\ell}$, similarly ordered, and the contribution of the boundary values. Since any permutation of the s grid points provides for a different arrangement, there are $s! = (m^2)!$ distinct ways of deriving (7.17). Fortunately, none of the features that are important to our present analysis depends on the specific ordering of the $u_{k,\ell}$.

Lemma 7.1   *The matrix A in (7.17) is symmetric and the set of its eigenvalues is*

$$\sigma(A) = \{\lambda_{\alpha,\beta} \,:\, \alpha, \beta = 1, 2, \ldots, m\},$$

where

$$\lambda_{\alpha,\beta} = -4\left\{\sin^2\left[\frac{\alpha\pi}{2(m+1)}\right] + \sin^2\left[\frac{\beta\pi}{2(m+1)}\right]\right\}, \qquad \alpha, \beta = 1, 2, \ldots, m. \qquad (7.18)$$

Proof   To prove symmetry, we notice by inspection of (7.16) that all elements of A must be all either -4 or 1 or 0, according to the following rule. All diagonal elements $a_{\gamma,\gamma}$ equal -4, whereas an off-diagonal element $a_{\gamma,\delta}$, $\gamma \neq \delta$, equals 1 if (i_γ, j_γ) and (i_δ, j_δ) are either horizontal or vertical neighbours, 0 otherwise. Being a neighbour is, however, a commutative relation: if (i_γ, j_γ) is a neighbour of (i_δ, j_δ) then (i_δ, j_δ) is a neighbour of (i_γ, j_γ). Therefore $a_{\gamma,\delta} = a_{\delta,\gamma}$ for all $\gamma, \delta = 1, 2, \ldots, s$.

To find the eigenvalues of A we disregard the exact way in which the matrix has been composed – after all, symmetric permutations conserve eigenvalues – and, instead, go back to the equations (7.16). Suppose that we can demonstrate the existence of a nonzero function $(v_{k,\ell})_{k,\ell=0,1,\ldots,m+1}$ such that $v_{k,0} = v_{k,m+1} = v_{0,\ell} = v_{m+1,\ell} = 0$, $k, \ell = 1, 2, \ldots, m$, and such that the homogeneous set of linear equations

$$v_{k-1,\ell} + v_{k+1,\ell} + v_{k,\ell-1} + v_{k,\ell+1} - 4v_{k,\ell} = \lambda v_{k,\ell}, \qquad k, \ell = 1, 2, \ldots, m, \qquad (7.19)$$

is satisfied for some λ. It follows that, up to rearrangement, $(v_{k,\ell})$ is an eigenvector and λ is a corresponding eigenvalue of A.

Given $\alpha, \beta \in \{1, 2, \ldots, m\}$, we let

$$v_{k,\ell} = \sin\left(\frac{k\alpha\pi}{m+1}\right)\sin\left(\frac{\ell\beta\pi}{m+1}\right), \qquad k, \ell = 0, 1, \ldots, m+1.$$

Note that, as required, $v_{k,0} = v_{k,m+1} = v_{0,\ell} = v_{m+1,\ell} = 0$, $k, \ell = 1, 2, \ldots, m$. Substituting into (7.19),

$$v_{k-1,\ell} + v_{k+1,\ell} + v_{k,\ell-1} + v_{k,\ell+1} - 4v_{k,\ell}$$

$$= \left\{\sin\left[\frac{(k-1)\alpha\pi}{m+1}\right] + \sin\left[\frac{(k+1)\alpha\pi}{m+1}\right]\right\}\sin\left(\frac{\ell\beta\pi}{m+1}\right) \qquad (7.20)$$

$$+ \sin\left(\frac{k\alpha\pi}{m+1}\right)\left\{\sin\left[\frac{(\ell-1)\beta\pi}{m+1}\right] + \sin\left[\frac{(\ell+1)\beta\pi}{m+1}\right]\right\} - 4\sin\left(\frac{k\alpha\pi}{m+1}\right)\sin\left(\frac{\ell\beta\pi}{m+1}\right).$$

We exploit the trigonometric identity

$$\sin(\theta - \psi) + \sin(\theta + \psi) = 2\sin\theta\cos\psi$$

to simplify (7.20), obtaining for the right-hand side

$$= 2\sin\left(\frac{k\alpha\pi}{m+1}\right)\cos\left(\frac{\alpha\pi}{m+1}\right)\sin\left(\frac{\ell\beta\pi}{m+1}\right) + 2\sin\left(\frac{k\alpha\pi}{m+1}\right)\sin\left(\frac{\ell\beta\pi}{m+1}\right)\cos\left(\frac{\beta\pi}{m+1}\right)$$

$$- 4\sin\left(\frac{k\alpha\pi}{m+1}\right)\sin\left(\frac{\ell\beta\pi}{m+1}\right)$$

$$= -2\left[2 - \cos\left(\frac{\alpha\pi}{m+1}\right) - \cos\left(\frac{\beta\pi}{m+1}\right)\right]\sin\left(\frac{k\alpha\pi}{m+1}\right)\sin\left(\frac{\ell\beta\pi}{m+1}\right)$$

$$= -4\left\{\sin^2\left[\frac{\alpha\pi}{2(m+1)}\right] + \sin^2\left[\frac{\beta\pi}{2(m+1)}\right]\right\}v_{k,\ell}, \qquad k, \ell = 1, 2, \ldots, m.$$

Note that we have used in the last line the trigonometric identity

$$1 - \cos\theta = 2\sin^2\tfrac{1}{2}\theta.$$

We have thus demonstrated that (7.19) is satisfied by $\lambda = \lambda_{\alpha,\beta}$, and this completes the proof of the lemma.   ■

Corollary   *The matrix A is negative definite and, a fortiori, nonsingular.*

Proof   We have just shown that A is symmetric, and it follows from (7.18) that all its eigenvalues are negative. Therefore (see A.1.5.1) it is negative definite and nonsingular.   ■

◇ **Eigenvalues of the Laplace operator**   Before we continue with the orderly flow of our exposition, it is instructive to comment on how the eigenvalues and eigenvectors of the matrix A are related to the eigenvalues and eigenfunctions of the Laplace operator ∇^2 in the unit square.

The function v is said to be an *eigenfunction* of ∇^2 in a domain Ω and λ is the corresponding *eigenvalue* if v vanishes along $\partial\Omega$ and satisfies within Ω the equation $\nabla^2 v = \lambda v$. The linear system (7.19) is nothing other than a five-point discretization of this equation for $\Omega = (0,1)^2$.

The eigenfunctions and eigenvalues of ∇^2 can be evaluated easily and explicitly in the unit square. Given any two positive integers α, β, we let $v(x,y) = \sin(\alpha\pi x)\sin(\beta\pi y)$, $x, y \in [0,1]$. Note that, as required, v obeys zero Dirichlet boundary conditions. It is trivial to verify that $\nabla^2 v = -(\alpha^2 + \beta^2)\pi^2 v$, hence v is indeed an eigenfunction and the corresponding eigenvalue is $-(\alpha^2 + \beta^2)\pi^2$. It is possible to prove that *all* eigenfunctions of ∇^2 in $(0,1)^2$ have this form.

The vector $v_{k,\ell}$ from the proof of Lemma 7.1 can be obtained by sampling of the eigenfunction v at the grid points $\left\{\left(\frac{k}{m+1}, \frac{\ell}{k+1}\right)\right\}_{k,\ell=0,1,\ldots,m+1}$ (for $\alpha, \beta = 1, 2, \ldots, m$ only; the matrix A, unlike ∇^2, is finite-dimensional!), whereas $(\Delta x)^{-2}\lambda_{\alpha,\beta}$ is a good approximation to $-(\alpha^2 + \beta^2)\pi^2$ provided α and β are small in comparison with m. Expanding $\sin^2\theta$ in a power series and bearing in mind that $(m+1)\Delta x = 1$, we readily obtain

$$
\begin{aligned}
\frac{\lambda_{\alpha,\beta}}{(\Delta x)^2} &= -4\left(\left\{\left[\frac{\alpha\pi}{2(m+1)}\right]^2 - \frac{1}{3}\left[\frac{\alpha\pi}{2(m+1)}\right]^4 + \cdots\right\}\right. \\
&\quad \left. + \left\{\left[\frac{\beta\pi}{2(m+1)}\right]^2 - \frac{1}{3}\left[\frac{\beta\pi}{2(m+1)}\right]^4 + \cdots\right\}\right) \\
&= -(\alpha^2 + \beta^2)\pi^2 + \tfrac{1}{12}(\alpha^4 + \beta^4)\pi^4(\Delta x)^2 + \mathcal{O}((\Delta x)^4).
\end{aligned}
$$

Hence, $\lambda_{\alpha,\beta}$ approximate the exact eigenvalues of the Laplace operator.   ◇

Let u be the exact solution of (7.13) in a unit square. We set $\tilde{u}_{k,\ell} = u(k\Delta x, \ell\Delta x)$ and denote by $e_{k,\ell}$ the error of the five-point formula (7.15) at the (k,ℓ)th grid point, $e_{k,\ell} = u_{k,\ell} - \tilde{u}_{k,\ell}$, $k, \ell = 0, 1, \ldots, m+1$. The five-point equations (7.15) are represented in the matrix form (7.17) and e denotes an arrangement of $\{e_{k,\ell}\}$ into a vector in $\mathbb{R}^s$, $s = m^2$, whose ordering is identical to that of u. We are measuring the magnitude of e by the Euclidean norm $\|\cdot\|$ (A.1.3.3).

Theorem 7.2   *Subject to sufficient smoothness of the function f and the boundary conditions, there exists a number $c > 0$, independent of Δx, such that*

$$\|e\| \leq c(\Delta x)^2, \qquad \Delta x \to 0. \tag{7.21}$$

Proof   Since $(\Delta x)^{-2}(\Delta_{0,x}^2 + \Delta_{0,y}^2)$ approximates ∇^2 locally to order $\mathcal{O}((\Delta x)^2)$, it is true that

$$\tilde{u}_{k-1,\ell} + \tilde{u}_{k+1,\ell} + \tilde{u}_{k,\ell-1} + \tilde{u}_{k,\ell+1} - 4\tilde{u}_{k,\ell} = (\Delta x)^2 f_{k,\ell} + \mathcal{O}((\Delta x)^4) \tag{7.22}$$

for $\Delta x \to 0$. We subtract (7.22) from (7.16) and the outcome is

$$e_{k-1,\ell} + e_{k+1,\ell} + e_{k,\ell-1} + e_{k,\ell+1} - 4e_{k,\ell} = \mathcal{O}((\Delta x)^4), \qquad \Delta x \to 0,$$

or, in a vector notation (and paying due heed to the fact that $u_{k,\ell}$ and $\tilde{u}_{k,\ell}$ coincide along the boundary)

$$A e = \boldsymbol{\delta}_{\Delta x}, \tag{7.23}$$

where $\boldsymbol{\delta}_{\Delta x} \in \mathbb{R}^{m^2}$ is such that $\|\boldsymbol{\delta}_{\Delta x}\| = \mathcal{O}((\Delta x)^4)$. It follows from (7.23) that

$$e = A^{-1}\boldsymbol{\delta}_{\Delta x}. \tag{7.24}$$

Recall from Lemma 7.1 that A is symmetric. Hence so is A^{-1} and its Euclidean norm $\|A^{-1}\|$ is the same as its spectral radius $\rho(A^{-1})$ (A.1.5.2). The latter can be computed at once from (7.18), since $\lambda \in \sigma(B)$ is the same as $\lambda^{-1} \in \sigma(B^{-1})$ for any nonsingular matrix B. Thus, bearing in mind that $(m+1)\Delta x = 1$,

$$\rho(A^{-1}) = \max_{\alpha,\beta=1,2,\ldots,m} \frac{1}{4}\left\{\sin^2\left[\frac{\alpha\pi}{2(m+1)}\right] + \sin^2\left[\frac{\beta\pi}{2(m+1)}\right]\right\}^{-1} = \frac{1}{8\sin^2\frac{1}{2}\Delta x\pi}.$$

Since

$$\lim_{\Delta x \to 0}\left[\frac{(\Delta x)^2}{8\sin^2\frac{1}{2}\Delta x\pi}\right] = \frac{1}{2\pi^2},$$

it follows that for any constant $c_1 > (2\pi^2)^{-1}$ it is true that

$$\|A^{-1}\| = \rho(A^{-1}) \leq c_1(\Delta x)^{-2}, \qquad \Delta x \to 0. \tag{7.25}$$

Provided that f and the boundary conditions are sufficiently smooth,[2] u is itself sufficiently differentiable and there exists a constant $c_2 > 0$ such that $\|\boldsymbol{\delta}(\Delta x)\| \leq c_2(\Delta x)^4$ (recall that $\boldsymbol{\delta}$ depends solely on the exact solution). Substituting this and (7.25) into the inequality (7.24) yields (7.21) with $c = c_1 c_2$.   ∎

Our analysis can be generalized to rectangles, L-shaped domains etc., provided the ratios of all sides are rational numbers (cf. Exercise 7.7). In general, unfortunately, the grid contains near-boundary points, at which the five-point formula (7.15) cannot be implemented. To see this, it is enough to look at a single spatial dimension; without loss of generality let us suppose that we are seeking to approximate ∇^2 at the point P in Figure 7.6.

---

[2] We prefer not to be very specific here, since issues of the smoothness and differentiability of Poisson equations are notoriously difficult. However, these requirements are satisfied in most cases of practical interest.

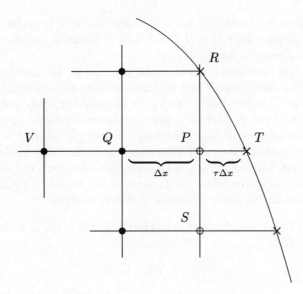

Figure 7.6   Computational grid near a curved boundary.

Given $z(x)$, we first approximate z'' at $P \sim x_0$ (we disregard the variable y, which plays no part in this process) as a linear combination of the values of z at P, $Q \sim x_0 - \Delta x$ and $T \sim x_0 + \tau \Delta x$. Expanding z in a Taylor series about x_0, we can easily show that

$$\frac{1}{(\Delta x)^2}\left[\frac{2}{\tau+1}z(x_0 - \Delta x) - \frac{2}{\tau}z(x_0) + \frac{2}{\tau(\tau+1)}z(x_0 + \tau\Delta x)\right]$$

$$= z''(x_0) + \tfrac{1}{2}(\tau - 1)z'''(x_0)\Delta x + \mathcal{O}((\Delta x)^2).$$

Unless $\tau = 1$, when everything reduces to the central difference approximation, the error is just $\mathcal{O}(\Delta x)$. To recover order $\mathcal{O}((\Delta x)^2)$, consistently with the five-point formula at internal points, we add the function value at $V \sim x_0 - 2\Delta x$ to the linear combination, whereby expansion in a Taylor series about x_0 yields

$$z''(x_0) = \frac{1}{(\Delta x)^2}\left[\frac{\tau-1}{\tau+2}z(x_0 - 2\Delta x) + \frac{2(2-\tau)}{\tau+1}z(x_0 - \Delta x) - \frac{3-\tau}{\tau}z(x_0)\right.$$

$$\left. + \frac{6}{\tau(\tau+1)(\tau+2)}z(x_0 + \Delta x)\right] + \mathcal{O}((\Delta x)^2).$$

A good approximation to $\nabla^2 u$ at P should involve, therefore, six points, P, Q, R, S, T and V. Assuming that P corresponds to the grid point (k^0, ℓ^0), say, we obtain the linear equation

$$\frac{\tau-1}{\tau+2}u_{k^0-2,\ell^0} + \frac{2(2-\tau)}{\tau+1}u_{k^0-1,\ell^0} + \frac{6}{\tau(\tau+1)(\tau+2)}u_{k^0+\tau,\ell^0} + u_{k^0,\ell^0-1}$$

$$+ u_{k^0,\ell^0+1} - \frac{3+\tau}{\tau}u_{k^0,\ell^0} = (\Delta x)^2 f_{k^0,\ell^0} \qquad (7.26)$$

where $u_{k^0+\tau, \ell^0}$ is the boundary point T, whose value is provided from the Dirichlet boundary condition. Note that if $\tau = 1$ and P becomes an internal point then this reduces to the five-point formula (7.16).

A similar treatment can be used in the y-direction. Of course, regardless of direction, we need Δx small enough that we have sufficient information to implement (7.26) or other $\mathcal{O}((\Delta x)^2)$ approximants to ∇^2 at all near-boundary points. The outcome, in tandem with (7.16) at internal points, is a linear algebraic equation for every grid point – whether internal or near-boundary – where the solution is unknown.

We will not extend Theorem 7.2 here and prove that the rate of decay of the error is $\mathcal{O}((\Delta x)^2)$ but will set ourselves a less ambitious goal: to prove that the linear algebraic system is nonsingular. First, however, we require a technical lemma that is of great applicability in many branches of matrix analysis.

Lemma 7.3 (The Gerschgorin criterion)   *Let $B = (b_{k,\ell})$ be an arbitrary complex $d \times d$ matrix. Then*

$$\sigma(B) \subset \bigcap_{i=1}^{d} \mathbb{S}_i,$$

where

$$\mathbb{S}_i = \left\{ z \in \mathbb{C} : |z - b_{i,i}| \leq \sum_{j=1,\, j\neq i}^{d} |b_{i,j}| \right\}$$

and $\sigma(B)$ is the set of the eigenvalues of B. Moreover, $\lambda \in \sigma(B)$ may lie on $\partial \mathbb{S}_{i^0}$ for some $i^0 \in \{1, 2, \ldots, d\}$ only if $\lambda \in \partial \mathbb{S}_i$ for all $i = 1, 2, \ldots, d$. The $\mathbb{S}_i$ are known as Gerschgorin discs.

Proof   This is relegated to Exercise 7.8, where it is broken down into a number of easily manageable chunks. ∎

There is another part of the Gerschgorin criterion that plays no role whatsoever in our discussion but we mention it as a matter of independent mathematical interest. Thus, suppose that

$$\{1, 2, \ldots, d\} = \{i_1, i_2, \ldots, i_r\} \cup \{j_1, j_2, \ldots, j_{d-r}\}$$

such that $\mathbb{S}_{i_\alpha} \cap \mathbb{S}_{i_\beta} \neq \emptyset$, $\alpha, \beta = 1, 2, \ldots, r$ and $\mathbb{S}_{i_\alpha} \cap \mathbb{S}_{j_\beta} = \emptyset$, $i = 1, 2, \ldots, r$, $j = 1, 2, \ldots, d - r$. Let $\mathcal{S} := \cup_{\alpha=1}^{r} \mathbb{S}_{i_\alpha}$. Then the set $\mathcal{S}$ includes *exactly* r eigenvalues of B.

Theorem 7.4   *Let $A\boldsymbol{u} = \boldsymbol{b}$ be the linear system obtained by employing the five-point formula (7.16) at internal points and the formula (7.26) or its reflections and extensions (catering for the case when one horizontal and one vertical neighbour are missing) at near-boundary points. Then A is nonsingular.*

Proof   No matter how we arrange the unknowns into a vector, thereby determining the ordering of the rows and the columns of A, each row of A corresponds to a single equation. Therefore all the elements along the ith row vanish, except for those that feature in the linear equation that is obeyed at the grid point corresponding to this row. It follows from an inspection of (7.16) and (7.26) that

$a_{i,i} < 0$ and that all the nonzero components $a_{i,j}$, $j \neq i$, are nonnegative. Moreover, it is easy to verify that the sum of the components in (7.16) and the sum of the components in (7.26) are both zero – but not all these components need feature along the ith row of A, since some may correspond to boundary points. Therefore $\sum_{j \neq i} |a_{i,j}| + a_{i,i} \leq 0$ and it follows that the origin may not lie in the interior of the Gerschgorin disc $\mathbb{S}_j$. Thus, by Lemma 7.3, $0 \in \sigma(A)$ only if $0 \in \partial \mathbb{S}_i$ for *all* rows i.

At least one equation has a neighbour on the boundary. Let this equation correspond to the i^0 row of A. Then $\sum_{j \neq i^0} |a_{i^0,j}| < |a_{i^0,i^0}|$, therefore $0 \notin \mathbb{S}_{i^0}$. We deduce that it is impossible for 0 to lie on the boundaries of all the discs $\mathbb{S}_i$, hence, by Lemma 7.3, it is not an eigenvalue. This means that A is nonsingular and the proof is complete.   ∎

7.3   Higher-order methods for $\nabla^2 u = f$

The Laplace operator ∇^2 has a key role in many important equations of mathematical physics; to mention just two, the parabolic *diffusion* equation

$$\frac{\partial u}{\partial t} = \nabla^2 u, \qquad u = u(x, y, t),$$

and the hyperbolic *wave* equation

$$\frac{\partial^2 u}{\partial t} = \nabla^2 u, \qquad u = u(x, y, t).$$

Therefore, the five-point approximation formula (7.15) is one of the workhorses of numerical analysis. This ubiquity, however, motivates a discussion of higher-order computational schemes.

Truncating (7.8) after two terms, we obtain in place of (7.15) the scheme

$$\frac{1}{(\Delta x)^2} \left[\Delta_{0,x}^2 + \Delta_{0,y}^2 - \tfrac{1}{12} \left(\Delta_{0,x}^4 + \Delta_{0,y}^4 \right) \right] u_{k,\ell} = f_{k,\ell}. \tag{7.27}$$

More economically, this can be written as the computational stencil

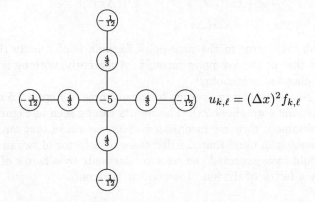

Although the error is $\mathcal{O}((\Delta x)^4)$, (7.27) is not a popular method. It renders too many points near-boundary, even in a square grid, which means that they require laborious special treatment. Worse, it gives linear systems that are considerably more expensive to solve than, for example, those generated by the five-point scheme. In particular, the fast solvers from Sections 12.1 and 12.2 cannot be implemented in this setting.

A more popular alternative is to approximate $\nabla^2 u$ at the (k, ℓ)th grid point by means of all its eight nearest neighbours: horizontal, vertical and diagonal. This results in the *nine-point formula*

$$\frac{1}{(\Delta x)^2} \left(\Delta_{0,x}^2 + \Delta_{0,y}^2 + \tfrac{1}{6} \Delta_{0,x}^2 \Delta_{0,y}^2 \right) u_{k,\ell} = f_{k,\ell}, \tag{7.28}$$

more familiarly known in the computational stencil notation

$$u_{k,\ell} = (\Delta x)^2 f_{k,\ell}$$

To analyse the error in (7.28) we recall from Section 7.1 that $\Delta_0 = \mathcal{E}^{1/2} - \mathcal{E}^{-1/2}$ and $\mathcal{E} = \mathrm{e}^{\Delta x \mathcal{D}}$. Therefore, expanding in a Taylor series in Δx,

$$\begin{aligned}
\Delta_0^2 &= \mathcal{E} - 2\mathcal{I} + \mathcal{E}^{-1} = \mathrm{e}^{\Delta x \mathcal{D}} - 2\mathcal{I} + \mathrm{e}^{-\Delta x \mathcal{D}} \\
&= h^2 \mathcal{D}^2 + \tfrac{1}{12}(\Delta x)^4 \mathcal{D}^4 + \mathcal{O}((\Delta x)^6).
\end{aligned}$$

Since this is valid for both spatial variables and $\mathcal{D}_x, \mathcal{D}_y$ commute, , substitution in (7.28) yields

$$\begin{aligned}
&\frac{1}{(\Delta x)^2} \left(\Delta_{0,x}^2 + \Delta_{0,y}^2 + \tfrac{1}{6} \Delta_{0,x}^2 \Delta_{0,y}^2 \right) \\
&= \frac{1}{(\Delta x)^2} \left\{ \left[(\Delta x)^2 \mathcal{D}_x^2 + \tfrac{1}{12}(\Delta x)^4 \mathcal{D}_x^4 + \mathcal{O}((\Delta x)^6) \right] + \left[(\Delta x)^2 \mathcal{D}_y^2 + \tfrac{1}{12}(\Delta x)^4 \mathcal{D}_y^4 \right. \right. \\
&\qquad\qquad \left. \left. + \mathcal{O}((\Delta x)^6) \right] + \tfrac{1}{6} \left[(\Delta x)^2 \mathcal{D}_x^2 + \mathcal{O}((\Delta x)^4) \right] \left[(\Delta x)^2 \mathcal{D}_y^2 + \mathcal{O}((\Delta x)^4) \right] \right\} \\
&= (\mathcal{D}_x^2 + \mathcal{D}_y^2) + \tfrac{1}{12}(\Delta x)^2 (\mathcal{D}_x^2 + \mathcal{D}_y^2)^2 + \mathcal{O}((\Delta x)^4) \\
&= \nabla^2 + \tfrac{1}{12}(\Delta x)^2 \nabla^4 + \mathcal{O}((\Delta x)^2). \tag{7.29}
\end{aligned}$$

In other words, the error in the nine-point formula is of *exactly* the same order of magnitude as that of the five-point formula. Apparently, nothing is gained by incorporating the diagonal neighbours!

Not giving in to despair, we return to the example of Figure 7.5 and recalculate it with the nine-point formula (7.28). The results can be seen in Figure 7.7 and, as can be easily ascertained, *they are inconsistent with our claim that the error decays like* $\mathcal{O}((\Delta x)^2)$! Bearing in mind that Δx decreases by a factor of two in each consecutive graph, we would have expected the error to attenuate by a factor of four – instead it attenuates by a factor of sixteen. Too good to be true?

Figure 7.7   The errors in the solution of the Laplace equation from Figure 7.5 by the nine-point formula, for $m = 5$, $m = 11$ and $m = 23$.

The reason for this spectacular behaviour is, to put it bluntly, our laziness. Had we attempted to solve a Poisson equation with a nontrivial inhomogeneous term the decrease in the error would have been consistent with our error estimate (cf. Figure 7.8). Instead, we applied the nine-point scheme to the Laplace equation, which, as far as (7.28) is concerned, is a special case. We now give a hand-waving explanation of this phenomenon.

It follows from (7.29) that the nine-point scheme is an approximation of order $\mathcal{O}\big((\Delta x)^4\big)$ to the 'equation'

$$\left[\nabla^2 + \tfrac{1}{12}(\Delta x)^2 \nabla^4\right] u = f. \tag{7.30}$$

Setting aside the dependence of (7.30) on Δx and the whole matter of boundary conditions, the operator $\mathcal{M}_{\Delta x} := \mathcal{I} + \tfrac{1}{12}(\Delta x)^2 \nabla^2$ is invertible for sufficiently small Δx. Multiplying (7.30) by its inverse while letting $f \equiv 0$ indicates that the nine-point scheme bears an error of $\mathcal{O}\big((\Delta x)^4\big)$ when applied to the Laplace equation. This explains the swift decay of the error in Figure 7.7.

The Laplace equation has many applications and this superior behaviour of the nine-point formula is a matter of interest. Remarkably, the logic that has led to an explanation of this phenomenon can be extended to cater for Poisson equations as well. Suppose thus that $\Delta x > 0$ is small enough, so that $\mathcal{M}_{\Delta x}^{-1}$ exists, and act with this operator on both sides of (7.30). Therefore we have

$$\nabla^2 u = \mathcal{M}_{\Delta x}^{-1} f, \tag{7.31}$$

a new Poisson equation for which the nine-point formula produces an error of order $\mathcal{O}\big((\Delta x)^4\big)$. The only snag is that the right-hand side differs from that in (7.13), but this is easy to put right. We replace f in (7.31) by a function $\tilde{f}$ such that

$$\tilde{f}(x,y) = f(x,y) + \tfrac{1}{12}\nabla^2 f(x,y) + \mathcal{O}\big((\Delta x)^4\big), \qquad (x,y) \in \Omega.$$

Since $\mathcal{M}_{\Delta x}^{-1} \tilde{f} = f + \mathcal{O}\big((\Delta x)^4\big)$, the new equation (7.31) differs from (7.13) only in its $\mathcal{O}\big((\Delta x)^4\big)$ terms. In other words, the nine-point formula, when applied to $\nabla^2 u = \tilde{f}$,

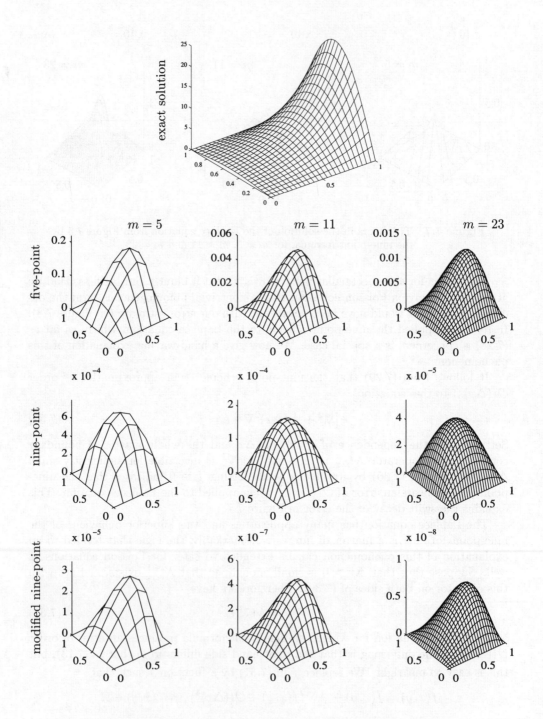

Figure 7.8   The exact solution of the Poisson equation (7.33) and the errors with the five-point, the nine-point and the modified nine-point schemes.

yields an $\mathcal{O}\big((\Delta x)^4\big)$ approximation to the original Poisson equation (7.13) with the same boundary conditions.

Although it is sometimes easy to derive $\tilde{f}$ by symbolic differentiation, perhaps computer-assisted, for simple functions f, it is easier to produce it by finite differences. Since

$$\frac{1}{(\Delta x)^2}(\Delta_{0,x}^2 + \Delta_{0,y}^2) = \nabla^2 + \mathcal{O}\big((\Delta x)^2\big),$$

it follows that

$$\tilde{f} := \big[\mathcal{I} + \tfrac{1}{12}(\Delta_{0,x}^2 + \Delta_{0,y}^2)\big] f$$

is of just the right form. Therefore, the scheme

$$\frac{1}{(\Delta x)^2}\left(\Delta_{0,x}^2 + \Delta_{0,y}^2 + \tfrac{1}{6}\Delta_{0,x}^2\Delta_{0,y}^2\right) u_{k,\ell} = \big[\mathcal{I} + \tfrac{1}{12}\left(\Delta_{0,x}^2 + \Delta_{0,y}^2\right)\big] f_{k,\ell}, \qquad (7.32)$$

which we can also write as

is $\mathcal{O}\big((\Delta x)^4\big)$ as an approximation of the Poisson equation (7.13). We call it the *modified nine-point scheme*.

The extra cost incurred in the modification of the nine-point formula is minimal, since the function f is known and so the cost of forming $\tilde{f}$ is extremely low. The rewards, however, are very rich indeed.

$\diamond$ **The modified nine-point scheme in action**...   Figure 7.8 displays the solution of the Poisson equation

$$\nabla^2 u = x^2 + y^2, \qquad 0 < x, y < 1, \qquad (7.33)$$

$$u(x,0) \equiv 0, \qquad u(x,1) = \tfrac{1}{2}x^2, \qquad 0 \le y \le 1,$$
$$u(0,y) = \sin \pi y, \qquad u(1,y) = \mathrm{e}^\pi \sin \pi y + \tfrac{1}{2}y^2, \qquad 0 \le x \le 1.$$

The first row displays the solution of (7.33) using the five-point formula (7.15). The error is quite large, but the exact solution,

$$u(x,y) = \mathrm{e}^{\pi x} \sin \pi y + \tfrac{1}{2}(xy)^2, \qquad 0 \le x, y \le 1,$$

can be as much as $\mathrm{e}^\pi + \tfrac{1}{2} \approx 23.6407$, and even the numerical solution for $m = 5$ comes within 1% of it. The most important observation, however, is that the error is attenuated roughly by a factor of four each time Δx is halved.

The outcome of the calculation with the nine-point formula (7.28) is displayed in the second row and, without any doubt, it is much better – about 200 times smaller than the corresponding values for the five-point scheme. There is a good reason: the error expansion of the five-point formula is

$$\frac{1}{(\Delta x)^2}(\Delta_{0,x}^2 + \Delta_{0,y}^2) - \nabla^2 = \tfrac{1}{12}(\Delta x)^2(\nabla^4 - \mathcal{D}_x^2\mathcal{D}_y^2) + \mathcal{O}\big((\Delta x)^4\big), \qquad \Delta x \to 0,$$

(the verification is left to the reader in Exercise 7.9) while the error expansion of the nine-point formula can be deduced at once from (7.29),

$$\frac{1}{(\Delta x)^2}\big(\Delta_{0,x}^2 + \Delta_{0,y}^2 + \tfrac{1}{6}\Delta_{0,x}^2\Delta_{0,y}^2\big) - \nabla^2 = \tfrac{1}{12}(\Delta x)^2\nabla^4 + \mathcal{O}\big((\Delta x)^4\big), \quad \Delta x \to 0.$$

As far as the Poisson equation (7.33) is concerned, we have

$$\begin{aligned}
(\nabla^4 - \mathcal{D}_x^2\mathcal{D}_y^2)u &= 2\pi^4 \mathrm{e}^{\pi x}\sin\pi y, \\
\nabla^4 u &\equiv 4,
\end{aligned} \qquad 0 \le x, y \le 1.$$

Hence the principal error term for the five-point formula can be as much as 1127 times larger than the corresponding term for the nine-point scheme in $(0,1)^2$. This is not a general feature of the underlying numerical methods!

Perhaps less striking, but nonetheless an important observation, is that the error associated with the nine-point scheme decays in Figure 7.8 by roughly four with each halving of Δx, consistently with the general theory and similarly to the behaviour of the five-point formula.

The bottom row in Figure 7.8 displays the outcome of the calculation with the modified nine-point formula (7.32). It is evident not just that the absolute magnitude of the error is vastly smaller (we do better with 25 grid points than the 'plain' nine-point formula with 529 grid points!) but also that its decay, roughly by a factor of 64 whenever Δx is halved, is consistent with the expected error $\mathcal{O}\big((\Delta x)^4\big)$. ⬦

Comments and bibliography

The numerical solution of the Poisson equation by means of finite differences features in many texts, e.g. Ames (1977). Diligent readers who wish first to acquaint themselves with the analytic theory of the Poisson equation (and more general elliptic equations) might consult the classical text of Agmon (1965). The best reference, however, is probably Hackbusch (1992), since it combines both analytic and numerical aspects of the subject.

The modified nine-point formula (7.32) is perhaps not as well known as it ought to be and part of the blame lies in the original name, *Mehrstellenverfahren* (Collatz, 1966). We prefer to avoid this sobriquet in the text, to deter over-enthusiastic instructors from making its spelling an issue in examinations.

Anticipating the discussion of Fourier transforms in Chapters 11–14, it is worthwhile to remark briefly on the connection of the latter with finite difference operators. Denoting the Fourier transform of a function f by $\hat{f}$,[3] it is easy to prove that

$$\widehat{\mathcal{E}f}(\xi) = \mathrm{e}^{\mathrm{i}\xi h}\hat{f}(\xi), \qquad \xi \in \mathbb{C},$$

---

[3]If you do not know the definition of the Fourier transform, skip this section, possibly returning to it later.

therefore

$$\widehat{\Delta_+ f}(\xi) = (e^{i\xi h} - 1)\hat{f}(\xi),$$
$$\widehat{\Delta_- f}(\xi) = (1 - e^{-i\xi h})\hat{f}(\xi),$$
$$\widehat{\Delta_0 f}(\xi) = (e^{i\xi h/2} - e^{-i\xi h/2})\hat{f}(\xi) = 2i\left(\sin\frac{\xi h}{2}\right)\hat{f}(\xi), \qquad \xi \in \mathbb{C}.$$
$$\widehat{\Upsilon_0 f}(\xi) = \tfrac{1}{2}(e^{i\xi h/2} + e^{i\xi h/2})\hat{f}(\xi) = \left(\cos\frac{\xi h}{2}\right)\hat{f}(\xi),$$

Likewise,

$$\widehat{\mathcal{D}f}(\xi) = i\xi\hat{f}(\xi), \qquad \xi \in \mathbb{C}.$$

The calculus of finite differences can be relocated to Fourier space. For example, the Fourier transform of $h^{-2}[f(x+h) - 2f(x) + f(x-h)]$ is

$$\frac{1}{h^2}\widehat{\Delta_0^2 f}(\xi) = \frac{e^{i\xi h} - 2 + e^{-i\xi h}}{2h^2}\hat{f}(\xi) = \left(-\tfrac{1}{2}\xi^2 + \tfrac{1}{24}h^2\xi^4 - \ldots\right)\hat{f}(\xi),$$

which we recognize as the transform of $f'' - \tfrac{1}{12}h^2 f^{(4)} + \cdots$.

The subject matter of this chapter can be generalized in numerous directions. Much of it is straightforward intellectually, although it might require lengthy manipulation and unwieldy algebra. An example is the generalization of methods for the Poisson equation to more variables. An engineer recognizes three spatial variables, while the agreed number of spatial dimensions in theoretical physics changes by the month (or so it seems), but numerical analysis can cater for all these situations (cf. Exercises 7.10 and 7.11). Of course, the cost of linear algebra is bound to increase with dimension, but such considerations are easier to deal within the framework of Chapters 9–12.

Finite differences extend to non-rectangular meshes. Although, for example, a honeycomb mesh such as the following

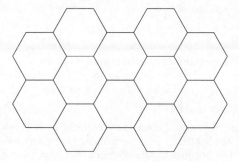

is considerably more difficult to use in practical programs, it occasionally confers advantages in dealing with curved boundaries. An example of a finite difference scheme in a honeycomb is provided by

$$u_{k,\ell} = (\Delta x)^2 f_{k,\ell}.$$

A slightly more complicated grid features in Exercise 7.13. There are, however, limits to the practicability of fancy grids – Escher's tessalations and Penrose tiles are definitely out of the question!

A more wide-ranging approach to curved and complicated geometries is to use nonuniform meshes. This allows a snug fit to difficult boundaries and also opens up the possibility of 'zooming in' on portions of the set Ω where, for some reason, the solution is more problematic or error-prone, and employing a finer grid there. This is relatively easy in one dimension (see Fornberg & Sloan (1994) for a very general approach) but, as far as finite differences are concerned, virtually hopeless in several spatial dimensions. Fortunately, the finite element method, which is reviewed in Chapter 8, provides a relatively accessible means of working with multivariate nonuniform meshes.

Yet another possibility, increasingly popular because of its efficiency in parallel computing architectures, is *domain decomposition* (Chan & Mathew, 1994; Le Tallec, 1994). The main idea is to tear the set Ω into smaller subsets, where the Poisson (or another) equation can be solved on a distinct grid (and possibly even on a different computer), subsequently 'glueing' different results together by solving smaller problems along the interfaces.

The scope of this chapter has been restricted to Dirichlet boundary conditions, but Poisson equations can be computed just as well with Neumann or mixed conditions. Moreover, finite differences and tricks like the *Mehrstellenverfahren* can be applied to other linear elliptic equations. Most notably, the computational stencil

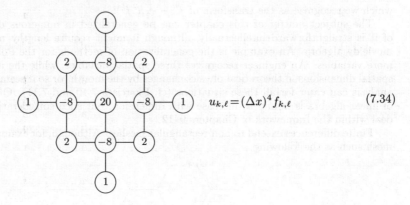

$$u_{k,\ell} = (\Delta x)^4 f_{k,\ell} \qquad (7.34)$$

represents an $\mathcal{O}\big((\Delta x)^2\big)$ method for the *biharmonic equation*

$$\nabla^4 u = f, \qquad (x,y) \in \Omega$$

(see Exercise 7.12).

However, no partial differential equation competes with Poisson in regard to its importance and ubiquity in applications. It is no wonder that there exist so many computational algorithms for (7.13), and not just finite difference, but also finite element (Chapter 8), spectral, and boundary element methods. The fastest to date is the *multipole* method, recently introduced by Carrier, Greengard and Rokhlin (1988).

This chapter would not be complete without mentioning a probabilistic method for solving the five-point system (7.15) in a special case, the Laplace equation (i.e., $f \equiv 0$). Although solution methods for linear algebraic systems belong in Chapters 9–12, we prefer to describe this method here, so that nobody mistakes it for a *viable* algorithm. Assume for example a computational grid in the shape of Manhattan Island, New York City, with streets and avenues forming rows and columns, respectively. (We need to disregard the Broadway and few other thoroughfares that fail to conform with the grid structure, but it is the principle that matters!) Suppose that a drunk emerges from a pub at the intersection of 6th Avenue and 20th Street. Being drunk, our friend turns at random in one of four available directions and staggers ahead. Having reached the next intersection, the drunk again turns at random, and so on and so forth. Disregarding the obvious perils of muggers, New York City drivers, pollution etc., the person in question can terminate this random walk (for this is what it is in mathematical terminology) only by reaching the harbour and falling into the water – an event that is bound to take place in finite time with probability one. Depending on the exact spot where our imbibing friend takes a dip, a fine is paid to the New York Harbor Authority. In other words, we have a 'fine function' $\phi(x, y)$, defined for all (x, y) along Manhattan's boundary, and the Harbor Authority is paid $\$\phi(x^0, y^0)$ if the drunk falls into the water at the point (x^0, y^0).

Suppose next that n drunks emerge from the same pub and each performs an independent meander through the city grid. Let $\tilde{u}_n$ be the average fine paid by the n drunks. It is then possible to prove that

$$\tilde{u} := \lim_{n \to \infty} \tilde{u}_n$$

is the solution of the five-point formula for the Laplace equation (with the Dirichlet boundary condition ϕ) at the grid point (20th Street, 6th Avenue).

Before trying this algorithm (hopefully, playing the part of the Harbor Authority), the reader had better be informed that the speed of convergence of $\tilde{u}_n$ to $\tilde{u}$ is $\mathcal{O}(1/\sqrt{n})$. In other words, to obtain four significant digits we need 10^8 drunks.

This is an example of a *Monte Carlo* method (Kalos & Whitlock, 1986) and we hasten to add that such algorithms can be very valuable in other areas of scientific computing.

Agmon, S. (1965), *Lectures on Elliptic Boundary Value Problems,* Van Nostrand, Princeton, NJ.

Ames, W.F. (1977), *Numerical Methods for Partial Differential Equations* (2nd ed.), Academic Press, New York.

Carrier, J., Greengard, L. and Rokhlin, V. (1988), A fast adaptive multipole algorithm for particle simulations, *SIAM Journal of Scientific and Statistical Computing* **9**, 669–686.

Chan, T.F. and Mathew, T.P. (1994), Domain decomposition algorithms, *Acta Numerica* **3**, 61–144.

Collatz, L. (1966), *The Numerical Treatment of Differential Equations* (3rd ed.), Springer-Verlag, Berlin.

Fornberg, B. and Sloan, D.M. (1994), A review of pseudospectral methods for solving partial differential equations, *Acta Numerica* **3**, 203–268.

Hackbusch, W. (1992), *Elliptic Differential Equations: Theory and Numerical Treatment,* Springer-Verlag, Berlin.

Kalos, M.H. and Whitlock, P.A. (1986), *Monte Carlo Methods,* Wiley, New York.

Le Tallec, P. (1994), Domain decomposition methods in computational mechanics, *Computational Mechanics Advances* **1**, 121–220.

Exercises

7.1   Prove the identities

$$\Delta_- + \Delta_+ = 2\Upsilon_0 \Delta_0, \qquad \Delta_- \Delta_+ = \Delta_0^2.$$

7.2   Show that formally

$$\mathcal{E} = \sum_{j=0}^{\infty} \Delta_-^j.$$

7.3   Demonstrate that for every $s \geq 1$ there exists a constant $c_s \neq 0$ such that

$$\frac{\mathrm{d}^s}{\mathrm{d}x^s} z(x) - \frac{1}{h^s} \left(\Delta_+^s - \tfrac{1}{2} s \Delta_+^{s+1} \right) z(x) = c_s \frac{\mathrm{d}^{s+2}}{\mathrm{d}x^{s+2}} z(x) h^2 + \mathcal{O}(h^3), \qquad h \to 0,$$

for every sufficiently smooth function z. Evaluate c_s explicitly for $s = 1, 2$.

7.4   For every $s \geq 1$ find a constant $d_s \neq 0$ such that

$$\frac{\mathrm{d}^{2s}}{\mathrm{d}x^{2s}} z(x) - \frac{1}{h^{2s}} \Delta_0^{2s} z(x) = d_s \frac{\mathrm{d}^{2s+2}}{\mathrm{d}x^{2s+2}} z(x) h^2 + \mathcal{O}(h^4), \qquad h \to 0,$$

for every sufficiently smooth function z. Compare the sizes of d_1 and c_2; what does this tell you about the error in the forward difference and central difference approximants?

7.5   In this exercise we consider finite difference approximants to the derivative that use one point to the left and $s \geq 1$ points to the right of x.

a Determine constants α_j, $j = 1, 2, \ldots$, such that

$$\mathcal{D} = \frac{1}{h} \left(\beta \Delta_- + \sum_{j=1}^{\infty} \alpha_j \Delta_+^j \right),$$

where $\beta \in \mathbb{R}$ is given.

b Given an integer $s \geq 1$, show how to choose the parameter β so that

$$\mathcal{D} = \frac{1}{h} \left(\beta \Delta_- + \sum_{j=1}^{s} a_j \Delta_+^j \right) + \mathcal{O}(h^{s+1}), \qquad h \to 0.$$

7.6   Determine the order (in the form $\mathcal{O}((\Delta x)^p)$) of the finite difference approximation to $\partial^2 / \partial x \partial y$ given by the computational stencil

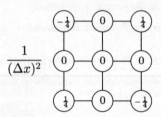

7.7   The five-point formula (7.15) is applied in an L-shaped domain of the form

and we assume that all grid points are either internal or boundary (this is possible if the ratios of the sides are rational). Prove *without relying on Theorem 7.4 or on its method of proof* that the underlying matrix is nonsingular. [*Hint: Nothing prevents you from relying on the method of proof of Lemma 7.1.*]

7.8 In this exercise we prove, step by step, the Gerschgorin criterion, which has been stated in Lemma 7.3.

a Let $C = (c_{i,j})$ be an arbitrary $d \times d$ singular complex matrix. Then there exists $\boldsymbol{x} \in \mathbb{C}^d \setminus \{\boldsymbol{0}\}$ such that $C\boldsymbol{x} = \boldsymbol{0}$. Choose $\ell \in \{1, 2, \ldots, d\}$ such that

$$|x_\ell| = \max_{j=1,2,\ldots,d} |x_j| > 0.$$

By considering the ℓth row of $C\boldsymbol{x}$, prove that

$$|c_{\ell,\ell}| \leq \sum_{j=1,\, j\neq\ell}^d |c_{\ell,j}|. \tag{7.35}$$

b Let B be a $d \times d$ matrix and choose $\lambda \in \sigma(B)$. Substituting $C = B - \lambda I$ in (7.35) prove that $\lambda \in \mathbb{S}_\ell$ (the Gerschgorin discs $\mathbb{S}_i$ have been defined in Lemma 7.3). Hence deduce that

$$\sigma(B) \subset \bigcap_{i=1}^d \mathbb{S}_i.$$

c Suppose that the inequality (7.35) holds as an *equality* for some $\ell \in \{1, \ldots, d\}$ and show that this equality implies that

$$|c_{k,k}| = \sum_{j=1,\, j\neq k}^d |c_{k,j}|, \qquad k = 1, 2, \ldots.$$

Deduce that if $\lambda \in \sigma(B)$ lies on $\partial\mathbb{S}_\ell$ for one ℓ then $\lambda \in \partial\mathbb{S}_k$ for all $k = 1, 2, \ldots, d$.

7.9 Prove that

$$\frac{1}{(\Delta x)^2}(\Delta_{0,x}^2 + \Delta_{0,y}^2) - \nabla^2 = \tfrac{1}{12}(\Delta x)^2(\nabla^4 - \mathcal{D}_x^2\mathcal{D}_y^2) + \mathcal{O}\big((\Delta x)^4\big), \qquad \Delta x \to 0.$$

7.10 We consider the d-dimensional Laplace operator

$$\nabla^2 = \sum_{j=1}^d \frac{\partial^2}{\partial x_j^2}.$$

Prove a d-dimensional generalization of (7.15),

$$\frac{1}{(\Delta x)^2} \sum_{j=1}^d \Delta_{0,x_j}^2 = \nabla^2 + \mathcal{O}\big((\Delta x)^2\big).$$

7.11⋆   Let ∇^2 again denote the d-dimensional Laplace operator and set

$$\mathcal{L}_{\Delta x} = -\tfrac{2}{3}\mathcal{I} + \tfrac{2}{3}\sum_{j=1}^{d}\Delta_{0,x_j}^2 + \tfrac{2}{3}\prod_{j=1}^{d}\left(\mathcal{I} + \tfrac{1}{2}\Delta_{0,x_j}^2\right),$$

$$\mathcal{M}_{\Delta x} = \mathcal{I} + \tfrac{1}{12}\sum_{j=1}^{d}\Delta_{0,x_j}^2.$$

a Prove that

$$\frac{1}{(\Delta x)^2}\mathcal{L}_{\Delta x} = \nabla^2 + \tfrac{1}{12}(\Delta x)^2\nabla^4 + \mathcal{O}\big((\Delta x)^4\big), \qquad \Delta x \to 0$$

and

$$\mathcal{M}_{\Delta x} = \mathcal{I} + \tfrac{1}{12}(\Delta x)^2\nabla^2 + \mathcal{O}\big((\Delta x)^4\big), \qquad \Delta x \to 0.$$

b Deduce that the method

$$\mathcal{L}_{\Delta x}u_{k_1,k_2,\ldots,k_d} = (\Delta x)^2\mathcal{M}_{\Delta x}f_{k_1,k_2,\ldots,k_d}$$

solves the d-dimensional Poisson equation $\nabla^2 u = f$ with an order-$\mathcal{O}\big((\Delta x)^4\big)$ error. [*This is the multivariate generalization of the modified nine-point formula* (7.32).]

7.12   Prove that the computational stencil (7.34) for the solution of the biharmonic equation $\nabla^4 u = f$ is equivalent to the finite difference representation

$$(\Delta_{0,x}^2 + \Delta_{0,y}^2)^2 u_{k,\ell} = (\Delta x)^4 f_{k,\ell}$$

thereby deducing the order of the error.

7.13⋆   Find the equivalent of the five-point formula in a computational grid of equilateral triangles,

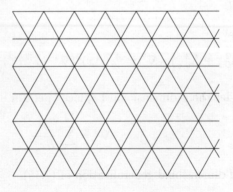

Your scheme should couple each internal grid point with its six nearest neighbours. What is the order of the error?

8

The finite element method

8.1 Two-point boundary value problems

The *finite element method* (FEM) presents to all those weaned on finite differences an entirely novel approach to the computation of a numerical solution for differential equations. Although it is often encapsulated in a few buzzwords – 'weak solution', 'Galerkin', 'finite element functions' – an understanding of the FEM calls not just for a different frame of mind but also for the comprehension of several principles. Each principle is important but it is their combination that makes the FEM into such an effective computational tool.

Instead of commencing our exposition from the deep end, let us first examine in detail a simple example of the Poisson equation in just *one* space variable. In principle, such an equation is $u'' = f$, but this is clearly too trivial for our purposes, since it can be readily solved by integration. Instead, we adopt a more ambitious goal and examine linear *two-point boundary value problems*

$$-\frac{\mathrm{d}}{\mathrm{d}x}\left[a(x)\frac{\mathrm{d}u}{\mathrm{d}x}\right] + b(x)u = f, \qquad 0 \le x \le 1, \tag{8.1}$$

where a, b and f are given functions, a is differentiable and $a(x) > 0$, $b(x) \ge 0$, $0 < x < 1$. Any equation of the form (8.1) must be specified in tandem with proper initial or boundary data. For the time being, we assume Dirichlet boundary conditions

$$u(0) = \alpha, \qquad u(1) = \beta. \tag{8.2}$$

Two-point boundary problems (8.1) abound in applications, e.g. in mechanics, and their numerical solution is of independent interest. However, in the context of this section, our main motivation is to use them as a paradigm for a general linear boundary value problem and as a vehicle for the description of the FEM.

Throughout this chapter we extensively employ the terminology of linear spaces, inner products and norms. The reader might wish to consult Appendix section A.2.1 for relevant concepts and definitions.

Instead of estimating the solution of (8.1) on a grid, a practice that has underlain the discourse of previous chapters, we wish to approximate u in a finite-dimensional space. We choose a function φ_0 that obeys the boundary conditions (8.2) and a set of m linearly independent functions $\varphi_1, \varphi_2, \ldots, \varphi_m$ that satisfy the zero boundary conditions $\varphi_\ell(0) = \varphi_\ell(1) = 0$, $\ell = 1, 2, \ldots, m$. All these functions need to satisfy certain smoothness conditions, but it is best to leave this question in abeyance for a while. Our goal is to represent u approximately in the form

$$u_m(x) = \varphi_0(x) + \sum_{\ell=1}^{m} \gamma_\ell \varphi_\ell(x), \qquad 0 \le x \le 1, \tag{8.3}$$

where $\gamma_1, \gamma_2, \ldots, \gamma_m$ are real constants. In other words, let

$$\overset{\circ}{\mathbb{H}}_m := \mathrm{Sp}\,\{\varphi_1, \varphi_2, \ldots, \varphi_m\},$$

the *span* of $\varphi_1, \varphi_2, \ldots, \varphi_m$, be the set of all linear combinations of these functions. Since the φ_ℓ, $\ell = 1, 2, \ldots, m$, are linearly independent, $\overset{\circ}{\mathbb{H}}_m$ is an m-dimensional linear space. An alternative phrasing of (8.3) is that we seek

$$u_m - \varphi_0 \in \overset{\circ}{\mathbb{H}}_m$$

so that, in some sense, u_m approximates the solution of (8.1). Note that every member of $\overset{\circ}{\mathbb{H}}_m$ obeys zero boundary conditions. Hence, by design, u_m is bound to satisfy the boundary conditions (8.2). For future reference, we record this first principle of the FEM.

- *Approximate the solution in a finite-dimensional space.*

What might we mean, however, by the phrase 'approximates the solution of (8.1)'? One possibility, that we have already seen in Chapter 3, is *collocation*: we choose $\gamma_1, \gamma_2, \ldots, \gamma_m$ so as to satisfy the differential equation (8.1) at m distinct points in $[0, 1]$. Here, though, we shall apply a different and more general line of reasoning. For any choice of $\gamma_1, \gamma_2, \ldots, \gamma_m$ consider the *defect*

$$d_m(x) := -\frac{\mathrm{d}}{\mathrm{d}x}\left[a(x)\frac{\mathrm{d}u_m(x)}{\mathrm{d}x}\right] + b(x)u_m(x) - f(x), \qquad 0 < x < 1.$$

Were u_m the solution of (8.1), the defect would be identically zero. Hence, the nearer d_m is to the zero function, the better we can expect our candidate solution to be.

In the fortunate case when d_m itself lies in the space $\overset{\circ}{\mathbb{H}}_m$ for all $\gamma_1, \gamma_2, \ldots, \gamma_m$, the problem is simple. The zero function in an inner product space is identified by being orthogonal to all members of that space. Hence, we equip $\overset{\circ}{\mathbb{H}}_m$ with an inner product $\langle \cdot, \cdot \rangle$ and seek parameters $\gamma_0, \gamma_1, \ldots, \gamma_m$ such that

$$\langle d_m, \varphi_k \rangle = 0, \qquad k = 1, 2, \ldots, m. \tag{8.4}$$

Since $\{\varphi_1, \varphi_2, \ldots, \varphi_m\}$ is, by design, a basis of $\overset{\circ}{\mathbb{H}}_m$, it follows from (8.4) that d_m is orthogonal to all members of $\overset{\circ}{\mathbb{H}}_m$, hence that it is the zero function. The orthogonality conditions (8.4) are called the *Galerkin equations*.

In general, however, $d_m \notin \overset{\circ}{\mathbb{H}}_m$ and we cannot expect to solve (8.1) exactly from within the finite-dimensional space. Nonetheless, the principle of the last paragraph still holds good: we wish d_m to obey the orthogonality conditions (8.4). In other words, the goal that underlies our discussion is to render the defect orthogonal to the finite-dimensional space $\overset{\circ}{\mathbb{H}}_m$.[1] This, of course, represents valid reasoning only if $\overset{\circ}{\mathbb{H}}_m$ approximates in some sense the infinite-dimensional linear space of all the candidate solutions of (8.1); more about this later. The second principle of the FEM is thus as follows.

- *Choose the approximant so that the defect is orthogonal to the space $\overset{\circ}{\mathbb{H}}_m$.*

Using the representation (8.3) and the linearity of the differentiation operator, the defect becomes

$$d_m = - \left[(a\varphi_0')' + \sum_{\ell=1}^m \gamma_\ell (a\varphi_\ell')' \right] + b \left[\varphi_0 + \sum_{\ell=1}^m \gamma_\ell \varphi_\ell \right] - f$$

and substitution in (8.4) results, after an elementary rearrangement of terms, in

$$\sum_{\ell=1}^m \gamma_\ell \left[\langle -(a\varphi_\ell')', \varphi_k \rangle + \langle b\varphi_\ell, \varphi_k \rangle \right] = \langle f, \varphi_k \rangle - [-(a\varphi_0')', \varphi_k \rangle + \langle b\varphi_0, \varphi_k \rangle],$$

$$k = 1, 2, \ldots, m. \qquad (8.5)$$

On the face of it, (8.4) is a linear system of m equations for the m unknowns $\gamma_1, \gamma_2, \ldots, \gamma_m$. However, before we rush to solve it, we must first perform a crucial operation, integration by parts.

Let us suppose that $\langle \cdot, \cdot \rangle$ is the standard Euclidean inner product over functions (the L_2 inner product; see A.2.1.4),

$$\langle v, w \rangle = \int_0^1 v(\tau) w(\tau) \, d\tau.$$

It is defined over all functions v and w such that

$$\int_0^1 |v(\tau)|^2 \, d\tau, \quad \int_0^1 |w(\tau)|^2 \, d\tau < \infty. \qquad (8.6)$$

Hence, (8.5) assumes the form

$$\sum_{\ell=1}^m \gamma_\ell \left(-\int_0^1 \{-[a(\tau)\varphi_\ell'(\tau)]'\varphi_k(\tau)\} \, d\tau + \int_0^1 b(\tau)\varphi_\ell(\tau)\varphi_k(\tau) \, d\tau \right)$$

$$= \int_0^1 f(\tau)\varphi_k(\tau) \, d\tau - \left(-\int_0^1 \{-[a(\tau)\varphi_0'(\tau)]'\varphi_k(\tau)\} \, d\tau + \int_0^1 b(\tau)\varphi_0(\tau)\varphi_k(\tau) \, d\tau \right),$$

$$k = 1, 2, \ldots, m.$$

Since the function φ_k vanishes at the endpoints, integration by parts may be carried out:

$$\int_0^1 \{-[a(\tau)\varphi_\ell'(\tau)]'\} \varphi_k(\tau) \, d\tau = -a(\tau)\varphi_\ell(\tau)\varphi_k(\tau) \Big|_0^1 + \int_0^1 a(\tau)\varphi_\ell'(\tau)\varphi_k'(\tau) \, d\tau$$

$$= \int_0^1 a(\tau)\varphi_\ell'(\tau)\varphi_k'(\tau) \, d\tau, \qquad \ell = 0, 1, \ldots, m.$$

---

[1]Collocation fits easily into this formulation, provided the inner product is properly defined (cf. Exercise 8.1).

The outcome,

$$\sum_{\ell=1}^{m} a_{k,\ell}\gamma_\ell = \int_0^1 f(\tau)\varphi_k(\tau)\,\mathrm{d}\tau - a_{k,0}, \qquad k = 1, 2, \ldots, m, \tag{8.7}$$

where

$$a_{k,\ell} := \int_0^1 [a(\tau)\varphi'_\ell(\tau)\varphi'_k(\tau) + b(\tau)\varphi_\ell(\tau)\varphi_k(\tau)]\,\mathrm{d}\tau, \quad k = 1, 2, \ldots, m, \ \ell = 0, 1, \ldots, m,$$

is a form of the Galerkin equations suitable for numerical work.

The reason why (8.7) is preferable to (8.5) lies in the choice of the functions φ_ℓ, which will be discussed soon. The whole point is that good choices of the basis functions (within the FEM framework) possess quite poor smoothness properties – in fact, the more we can lower the differentiability requirements, the wider the class of desirable basis functions that we might consider. Integration by parts takes away one derivative, hence we need no longer insist that the φ_ℓ are twice-differentiable in order to satisfy (8.6). Even once-differentiability is, in fact, too strong. Since the value of an integral is independent of the values that the integrand assumes on a *finite* set of points, it is sufficient to choose basis functions that are *piecewise differentiable* in $[0, 1]$. We will soon see that it is this lowering of the smoothness requirements through the agency of an integration by parts that confers important advantages on the finite element method. This is therefore our third principle.

- *Integrate by parts to depress to the maximal extent possible the differentiability requirements of the space* $\overset{\circ}{\mathbb{H}}_m$.

The importance of lowered smoothness and integration by parts ranges well beyond the FEM and numerical analysis.

$\diamond$ **Weak solutions**   Let us pause and ponder for a while the meaning of the term 'exact solution of a differential equation', which we are using so freely and with such apparent abandon on these pages. Thus, suppose that we have an equation of the form $\mathcal{L}\boldsymbol{u} = \boldsymbol{f}$, where $\mathcal{L}$ is a differential operator – ordinary or partial, in one or several dimensions, linear or nonlinear – provided with the right boundary and/or initial conditions. One candidate for the term 'exact solution' is a function $\boldsymbol{u}$ that obeys the equation and the 'side' conditions – the *classical* solution. However, there is an alternative. Suppose that we are given an infinite-dimensional linear space $\overset{\circ}{\mathbb{H}}$ that is rich enough to include in its closure all functions of interest. Given a 'candidate function' $\boldsymbol{v} \in \overset{\circ}{\mathbb{H}}$, we define the defect as $\boldsymbol{d}(\boldsymbol{v}) := \mathcal{L}\boldsymbol{v} - \boldsymbol{f}$ and say that $\boldsymbol{v}$ is the *weak solution* of the differential equation if $\langle \boldsymbol{d}(\boldsymbol{v}), \boldsymbol{w} \rangle = 0$ for every $\boldsymbol{w} \in \overset{\circ}{\mathbb{H}}$.

On the face of it, we have not changed much – the naive point of view is that if the defect is orthogonal to the whole space then it is the zero function, hence $\boldsymbol{v}$ obeys the differential equation in the classical sense. This is a fallacy, inherited from a finite-dimensional intuition! The whole point is that, astutely integrating by parts, we are usually able to lower the differentiability requirements of $\overset{\circ}{\mathbb{H}}$ to roughly half those in the original equation. In other words, it is entirely possible for an equation to possess a weak solution that is neither a classical solution nor, indeed, can even be subjected to the action of the differential operator $\mathcal{L}$ in a naive way.

The distinction between classical and weak solutions makes little sense in the context of initial value problems for ODEs, since there the Lipschitz condition ensures the existence of a unique classical solution (which, of course, is also a weak solution). This is not the case with boundary value problems and much of modern PDE analysis hinges upon the concept of a weak solution.   ◇

Suppose that the coefficients $a_{k,\ell}$ in (8.7) have been evaluated, whether explicitly or by means of quadrature (see Section 3.1). The system (8.7) comprises m linear equation in m unknowns and our next step is to employ a computer to solve it, thereby recovering the coefficients $\gamma_1, \gamma_2, \ldots, \gamma_m$ that render the best (in the sense of the underlying inner product) linear combination (8.3).

To introduce the FEM, we need another crucial ingredient, namely a specific choice of the set $\overset{\circ}{\mathbb{H}}_m$. There are, in principle, two objectives that we might seek in this choice. On the one hand, we might wish to choose the $\varphi_1, \varphi_2, \ldots, \varphi_m$ that, in some sense, are the most 'dense' in the infinite-dimensional space inhabited by the exact (weak) solution of (8.1). In other words, we might wish

$$\|u_m - u\| = \langle u_m - u, u_m - u \rangle^{1/2}$$

to be the smallest possible in $(0,1)$. This is a perfectly sensible choice that results in *spectral methods*. On the other hand there is, however, an alternative goal. The dimension of the finite-dimensional space from which we are seeking the solution will be often very large, perhaps not in the particular case of (8.7), when $m \approx 100$ is at the upper end of what is reasonable, but definitely so in a multivariate case. To implement (8.7) (or its multivariate brethren) we need to calculate approximately m^2 integrals and solve a linear system of m equations. This might be a very expensive process and thus a second reasonable goal is to choose $\varphi_1, \varphi_2, \ldots, \varphi_m$ so as to make this task much easier.

The penultimate and crucial principle of the FEM is thus designed to save computational cost.

- *Choose each function φ_k so that it vanishes along most of $(0,1)$, thereby ensuring that $\varphi_k \varphi_\ell \equiv 0$ for most choices of $k, \ell = 1, 2, \ldots, m$.*

In other words, each function φ_k is supported on a relatively small set $\mathbb{E}_k \subset (0,1)$, say, and $\mathbb{E}_k \cap \mathbb{E}_\ell = \emptyset$ for as many $k, \ell = 1, 2, \ldots, m$ as possible.

Recall that, by virtue of integration by parts, we have narrowed the differentiability requirements so much that piecewise linear functions are perfectly acceptable in $\overset{\circ}{\mathbb{H}}_m$. Settting $h = 1/(m+1)$, we choose

$$\varphi_k(x) = \begin{cases} 1 - k + \dfrac{x}{h}, & (k-1)h \leq x \leq kh, \\ 1 + k - \dfrac{x}{h}, & kh \leq x \leq (k+1)h, \\ 0, & |x - kh| \geq h, \end{cases} \qquad k = 1, 2, \ldots, m.$$

In other words, $\varphi_k = \psi(x/h - k)$, $k = 1, 2, \ldots, m$, where ψ is the *chapeau function* (also known as the *hat function*), represented as follows:

It can also be written in the form

$$\psi(x) = (x+1)_+ - 2(x)_+ + (x-1)_+, \qquad x \in \mathbb{R}, \tag{8.8}$$

where

$$(t)_+ := \begin{cases} t, & t \geq 0, \\ 0, & t < 0, \end{cases} \qquad t \in \mathbb{R}.$$

The advantages of this cryptic notation will become clear later. At present we draw attention to the fact that $\mathbb{E}_k = ((k-1)h, (k+1)h)$, therefore

$$\mathbb{E}_k \cap \mathbb{E}_\ell = \begin{cases} ((k-1)h, (k+1)h), & k = \ell, \\ ((k-1)h, kh), & k = \ell + 1, \\ (kh, (k+1)h), & k = \ell - 1, \\ \emptyset, & |k - \ell| \geq 2, \end{cases} \qquad k, \ell = 1, 2, \ldots, m.$$

Therefore the matrix in (8.7) becomes tridiagonal. This means firstly that we need to evaluate just $\mathcal{O}(m)$, rather than $\mathcal{O}(m^2)$, integrals and secondly that the solution of triangular linear systems is very easy indeed (see Section 10.1).

◇ **Spectral methods *vs* the FEM**   We attempt to solve the equation

$$-\frac{\mathrm{d}}{\mathrm{d}x}\left[(1+x)\frac{\mathrm{d}u}{\mathrm{d}x}\right] + \frac{1}{1+x}u = \frac{2}{1+x}, \tag{8.9}$$

accompanied by the Dirichlet boundary conditions

$$u(0) = 0, \qquad u(1) = 1,$$

using two distinct choices of the functions $\varphi_1, \varphi_2, \ldots, \varphi_m$. Firstly, we let

$$\varphi_k(x) = \sin k\pi x, \qquad 0 \leq x \leq 1, \quad k = 1, 2, \ldots, m \tag{8.10}$$

and force the boundary conditions by setting

$$\varphi_0(x) = \sin \frac{\pi x}{2}, \qquad 0 \leq x \leq 1.$$

This is an example of a spectral method.

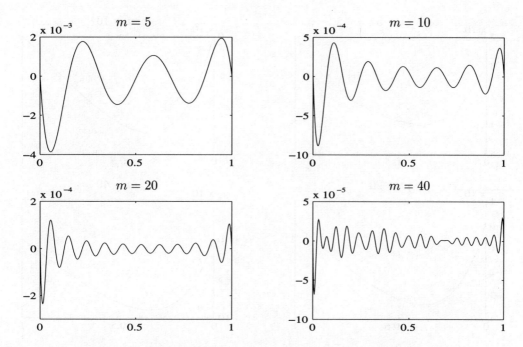

Figure 8.1   The error in (8.3) when the equation (8.9) is solved using the spectral method (8.10).

In fairness (and with understatement), we confess that (8.10) does not represent a particularly appropriate choice of $\overset{\circ}{\mathbb{H}}_m$ for the equation in question. Methods that employ trigonometric functions are in fact considerably better suited to equations with periodic boundary conditions.

The sprinter in FEM colours is the piecewise linear approximation that we have just described, i.e.,

$$\varphi_k(x) = \psi((m+1)x - k), \qquad 0 \le x \le 1 \qquad k = 1, 2, \ldots, m. \qquad (8.11)$$

Boundary conditions are recovered by choosing

$$\varphi_0(x) = \psi((m+1)(x-1)), \qquad 0 \le x \le 1;$$

note that $\varphi_0(0) = 0$, $\varphi_0(1) = 1$, as required. Of course, the support of φ_0 extends beyond $(0, 1)$, but this is of no consequence, since its integration is restricted to the interval $(m/(m+1), 1)$. The errors for (8.10) and (8.11) are displayed in Figures 8.1 and 8.2 respectively. At a first glance, both methods are performing well but the FEM has a slight edge, because of the large wiggles in Figure 8.1 near the endpoints. This *Gibbs effect,* sometimes also known as *aliasing,* is the penalty that we endure for attempting to approximate the non-periodic solution

$$u(x) = \frac{2x}{1+x}, \qquad 0 \le x \le 1,$$

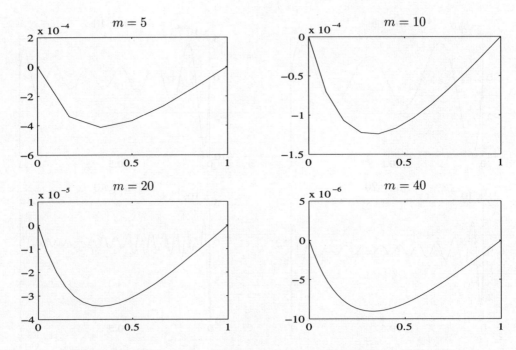

Figure 8.2 The error in (8.3) when the equation (8.9) is solved using the FEM (8.11).

with trigonometric functions. If we disregard the vicinity of the endpoints, the 'spectral' method performs slightly better.

The error decays roughly quadratically in both figures, the latter consistently with the estimate $\|u_m - u\| = \mathcal{O}(m^{-2})$ (which, as we will see in Section 8.2, happens to be the correct order of magnitude). Had (8.10) been a genuine spectral method and had the boundary conditions been periodic, the error would have decayed at an *exponential* speed. Leaping to the defence of the FEM, we remark that it took more than a hundredfold in terms of computer time to produce Figure 8.1, in comparison with Figure 8.2.   ◇

All the essential ingredients that, together, make the FEM are in place, except for one. To clarify it, we describe an alternative methodology leading to the Galerkin equations (8.7).

Many differential equations of practical interest start their life as *variational problems* and only subsequently are converted to a more familiar form by using the *Euler–Lagrange* equations. This gives little surprise to physicists, since the primary truth about physical models is not that derivatives (velocity, momentum, acceleration etc.) are somehow linked to the state of a system but that they arrange themselves according to the familiar principle of least action and least expenditure of energy. It is only mathematical ingenuity that renders this in the terminology of differential equations!

In a general variational problem we are given a *functional* $\mathcal{J} : \mathbb{H} \to \mathbb{R}$, where $\mathbb{H}$ is some function space, and we wish to find a function $u \in \mathbb{H}$ such that

$$J(u) = \min_{v \in \mathbb{H}} J(v).$$

Let us consider the following variational problem. Three functions, a, b and f are given in the interval $(0,1)$ in which we stipulate $a(x) > 0$ and $b(x) \geq 0$. The space $\mathbb{H}$ consists of all functions v that obey the boundary conditions (8.2) and

$$\int_0^1 v^2(\tau)\,\mathrm{d}\tau, \ \int_0^1 [v'(\tau)]^2\,\mathrm{d}\tau \ < \ \infty,$$

and we let

$$J(v) := \int_0^1 \left\{ a(\tau)[v'(\tau)]^2 + b(\tau)[v(\tau)]^2 - 2f(\tau)v(\tau) \right\}\mathrm{d}\tau, \qquad v \in \mathbb{H}. \tag{8.12}$$

It is possible to prove that $\inf_{v \in \mathbb{H}} J(v) > -\infty$ (cf. Exercise 8.7). Moreover, since the space $\mathbb{H}$ is complete,[2] every infimum is attainable within it and the operations inf and min become equal. Hence our variational problem always possesses a solution.

The space $\mathbb{H}$ is not a *linear* function space since (unless $\alpha = \beta = 0$) it is not closed under addition or multiplication by a scalar. However, choose an arbitrary $u \in \mathbb{H}$ and let $\overset{\circ}{\mathbb{H}}$ be the subset of $\mathbb{H}$ that includes all functions obeying *zero boundary conditions*. Unlike $\mathbb{H}$, the set $\overset{\circ}{\mathbb{H}}$ is a linear space and it is trivial to prove that each function $v \in \mathbb{H}$ can be written in a unique way as $v = u + w$, where $w \in \overset{\circ}{\mathbb{H}}$. We denote this by $\mathbb{H} = u + \overset{\circ}{\mathbb{H}}$ and say that $\mathbb{H}$ is an *affine* space.

Let us suppose that $u \in \mathbb{H}$ minimizes J. In other words, and bearing in mind that $\mathbb{H} = u + \overset{\circ}{\mathbb{H}}$,

$$J(u) \leq J(u+v), \qquad v \in \overset{\circ}{\mathbb{H}}. \tag{8.13}$$

We choose $v \in \overset{\circ}{\mathbb{H}} \setminus \{0\}$ and a real number $\varepsilon \neq 0$. Then

$$\begin{aligned}
J(u + \varepsilon v) &= \int_0^1 \left[a(u' + \varepsilon v')^2 + b(u + \varepsilon v) - 2f(u + \varepsilon v) \right]\mathrm{d}\tau \\
&= \int_0^1 \left\{ [(u')^2 + 2\varepsilon u'v' + \varepsilon^2 (v')^2] + b\left(u^2 + 2\varepsilon uv + \varepsilon^2 v^2 \right) - 2f(u + \varepsilon v) \right\}\mathrm{d}\tau \\
&= \int_0^1 \left[a(u')^2 + bu^2 - 2fu \right]\mathrm{d}\tau + 2\varepsilon \int_0^1 (au'v' + buv - fv)\,\mathrm{d}\tau \\
&\quad + \varepsilon^2 \int_0^1 \left[a(v')^2 + bv^2 \right]\mathrm{d}\tau \\
&= J(u) + 2\varepsilon \int_0^1 (au'v' + buv - fv)\,\mathrm{d}\tau + \varepsilon^2 \int_0^1 \left[a(v')^2 + bv^2 \right]\mathrm{d}\tau. \tag{8.14}
\end{aligned}$$

---

[2]Unless you know functional analysis, do not try to prove this – accept it as an act of faith.... This is perhaps the place to mention that $\mathbb{H}$ is rich enough to contain all piecewise differentiable functions in $(0,1)$ but, in order to be complete, it must contain many other functions as well.

To be consistent with (8.13) (v being replaced by εv) we require

$$2\varepsilon \int_0^1 (au'v' + buv)\, \mathrm{d}\tau + \varepsilon^2 \int_0^1 \left[a(v')^2 + bv^2\right] \mathrm{d}\tau \geq 0.$$

Note that the integral is nonnegative. As $|\varepsilon| > 0$ can be made arbitrarily small, we can make the second term negligible, thereby deducing that

$$\varepsilon \int_0^1 (au'v' + buv - fv)\, \mathrm{d}\tau \geq 0.$$

Recall that no assumptions have been made with regard to ε, except that it is nonzero and that its magnitude is adequately small. In particular, the inequality is valid when we replace ε with $-\varepsilon$, and we therefore deduce that

$$\int_0^1 \left[a(\tau)u'(\tau)v'(\tau) + b(\tau)u(\tau)v(\tau) - f(\tau)v(\tau)\right] \mathrm{d}\tau = 0. \tag{8.15}$$

We have just proved that (8.15) is necessary for u to be the solution of the variational problem, and it is easy to demonstrate that it is also sufficient. Thus, assuming that the identity is true, (8.14) (with $\varepsilon = 1$) gives

$$\mathcal{J}(u + v) = \mathcal{J}(u) + \int_0^1 \left[a(v')^2 + bv^2\right] \mathrm{d}\tau \geq \mathcal{J}(u), \qquad v \in \overset{\circ}{\mathbb{H}}.$$

Since $\mathbb{H} = u + \overset{\circ}{\mathbb{H}}$, it follows that u indeed minimizes $\mathcal{J}$ in $\mathbb{H}$.

Identity (8.15) possesses a further remarkable property: it is the weak form (in the Euclidean norm) of the two-point boundary value problem (8.1). This is easy to ascertain using integration by parts in the first term. Since $v \in \overset{\circ}{\mathbb{H}}$, it vanishes at the endpoints and (8.15) becomes

$$\int_0^1 \left\{-[a(\tau)u'(\tau)]' + b(\tau)u(\tau) - f(\tau)\right\} v(\tau)\, \mathrm{d}\tau = 0, \qquad v \in \overset{\circ}{\mathbb{H}}.$$

In other words, the function u is a solution of the variational problem (8.12) *if and only if it is the weak solution of the differential equation* (8.1).[3] We thus say that the two-point boundary value problem (8.1) is the *Euler–Lagrange* equation of (8.12).

Traditionally, variational problems have been converted into their Euler–Lagrange counterparts but, as far as numerical solution is concerned, we may attempt to approximate (8.12) rather than (8.1). The outcome is the *Ritz method*.

Let $\varphi_1, \varphi_2, \ldots, \varphi_m$ be linearly independent functions in $\overset{\circ}{\mathbb{H}}$ and choose an arbitrary $\varphi_0 \in \mathbb{H}$. As before, $\overset{\circ}{\mathbb{H}}_m$ is the m-dimensional linear space which is spanned by φ_k, $k = 1, 2, \ldots, m$. We seek a minimum of $\mathcal{J}$ in the m-dimensional affine space $\varphi_0 + \overset{\circ}{\mathbb{H}}_m$. In other words, we seek a vector $\boldsymbol{\gamma} = [\begin{array}{cccc} \gamma_1 & \gamma_2 & \cdots & \gamma_m \end{array}]^T \in \mathbb{R}^m$ that minimizes

$$\mathcal{J}_m(\boldsymbol{\delta}) := \mathcal{J}\left(\varphi_0 + \sum_{\ell=1}^m \delta_\ell \varphi_\ell\right), \qquad \boldsymbol{\delta} \in \mathbb{R}^m.$$

---

[3]This proves, incidentally, that the solution of (8.12) is unique, but you may try to prove uniqueness directly from (8.14).

The functional $\mathcal{J}_m$ acts on just m variables and its minimization can be accomplished, by well-known rules of calculus, by letting the gradient equal zero. Since $\mathcal{J}_m$ is quadratic in its variables,

$$\mathcal{J}_m(\boldsymbol{\delta}) = \int_0^1 \left[a\left(\varphi_0' + \sum_{\ell=1}^m \delta_\ell \varphi_\ell' \right)^2 + b\left(\varphi_0 + \sum_{\ell=1}^m \delta_\ell \varphi_\ell \right)^2 - 2f\left(\varphi_0 + \sum_{\ell=1}^m \delta_\ell \varphi_\ell \right) \right] d\tau,$$

the gradient is easy to calculate. Thus,

$$\frac{1}{2} \frac{\partial \mathcal{J}_m(\boldsymbol{\delta})}{\partial \delta_k} = \sum_{\ell=1}^m \int_0^1 \left(a\varphi_\ell' \varphi_k' + b\varphi_\ell \varphi_\ell \right) d\tau + \int_0^1 \left(a\varphi_0' \varphi_k' + b\varphi_0 \varphi_k \right) d\tau - \int_0^1 f\varphi_k \, d\tau.$$

Letting $\partial \mathcal{J}_m / \partial \delta_k = 0$ for $k = 1, 2, \ldots, m$ recovers exactly the form (8.7) of the Galerkin equations.

A careful reader will observe that setting the gradient to zero is merely a necessary condition for a minimum. For sufficiency we require in addition that the Hessian matrix $\left(\partial^2 \mathcal{J}_m / \partial \delta_k \partial \delta_j \right)_{k,j=1}^m$ is nonnegative definite. This is easy to prove (see Exercise 8.6).

What have we gained from the Ritz method? On the face of it, not much except for some additional insight, since it results in the same linear equations as the Galerkin method. This, however, ceases to be true for many other equations; in these cases Ritz and Galerkin result in genuinely different computational schemes. Moreover, the variational formulation provides us with an important clue about how to deal with more complicated boundary conditions.

There is an important mismatch between the boundary conditions for variational problems and those for differential equations. Each differential equation requires the right amount of boundary data. For example, (8.1) requires two conditions, of the form

$$\alpha_{0,i} u(0) + \alpha_{1,i} u'(0) + \beta_{0,i} u(1) + \beta_{1,i} u'(1) = \gamma_i, \qquad i = 1, 2,$$

such that

$$\text{rank} \begin{bmatrix} \alpha_{0,1} & \alpha_{1,1} & \beta_{0,1} & \beta_{1,1} \\ \alpha_{0,2} & \alpha_{1,2} & \beta_{0,2} & \beta_{1,2} \end{bmatrix} = 2.$$

Observe that (8.2) is a simple special case. However, a variational problem happily survives with less than a full complement of boundary data. For example, (8.12) can be defined with just a single boundary value, $u(0) = \alpha$, say. The rule is to replace in the Euler–Lagrange equations each 'missing' boundary condition by a *natural* boundary condition. For example, we complement $u(0) = \alpha$ with $u'(1) = 0$. (The proof is virtually identical to the reasoning that has led us from (8.12) to the corresponding Euler–Lagrange equation (8.1), except that we need to use the natural boundary condition when integrating by parts.)

In the Ritz method we traverse the path connecting variational problems and differential equations in the opposite direction, from (8.1) to (8.15), say. This means that, whenever the two-point boundary value problem is provided with a natural boundary condition, we disregard it in the formation of the space $\mathbb{H}$. In other words, the function φ_0 need obey only the *essential* boundary conditions that survive in the variational problem. The *quid pro quo* for the disappearance of, say, $u'(1) = 0$ is that

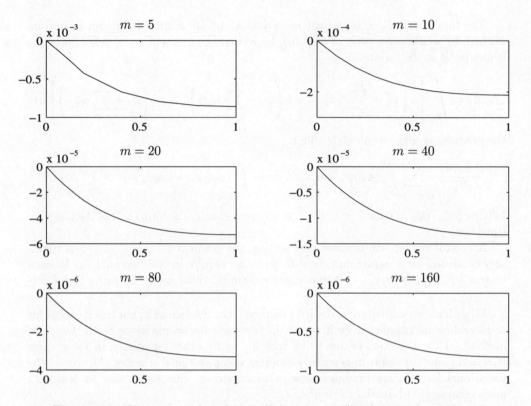

Figure 8.3   The error in the solution of the equation (8.16) with boundary data
$u(0) = 0$, $u'(1) = 0$, by the Ritz–Galerkin method with chapeau functions.

we need to add φ_{m+1} (defined consistently with (8.11)) to our space and an extra
equation, for $k = m + 1$, to the linear system (8.7); otherwise, by default, we are
imposing $u(1) = 0$, which is wrong.

◇ **A natural boundary condition**   Consider the equation

$$-u'' + u = 2\mathrm{e}^{-x}, \qquad 0 \leq x \leq 1, \tag{8.16}$$

given in tandem with the boundary conditions

$$u(0) = 0, \qquad u'(1) = 0.$$

The exact solution is easy to find: $u(x) = x\mathrm{e}^{-x}$, $0 \leq x \leq 1$.

Figure 8.3 displays the numerical solution of (8.16) using the piecewise linear
chapeau functions (8.8). Note that there is no need to provide the 'boundary
function' φ_0 at all. It is evident from the figure that the algorithm, as expected, is clever enough to recover the correct natural boundary condition at
$x = 1$. Another observation, which the figure shares with Figure 8.2, is that
the decay of the error is consistent with $\mathcal{O}(h)$.

Why not, one may ask, impose the natural boundary condition at $x = 1$? The
obvious reason is that we cannot employ a chapeau function for that purpose,

since its derivative will be discontinuous at the end-point. Of course, we might instead use a more complicated function but, unsurprisingly, such functions complicate matters – needlessly.   ◇

A natural boundary condition is just one of several kinds of boundary data that undergo change when differential equations are solved with the FEM. We do not wish to delve further into this issue, which is more than adequately covered in specialized texts. However, and to remind the reader of the need for proper respect towards boundary data, we hereby formulate our last principle of the FEM.

- *Retain only essential boundary conditions.*

Throughout this section we have identified several principles that together combine into the finite element method. Let us repeat them with some reformulation and reordering.

- *Approximate the solution in a finite-dimensional space $\varphi_0 + \overset{\circ}{\mathbb{H}}_m \subset \mathbb{H}$.*

- *Retain only essential boundary conditions.*

- *Choose the approximant so that the defect is orthogonal to $\overset{\circ}{\mathbb{H}}_m$ or, alternatively, so that a variational problem is minimized in $\overset{\circ}{\mathbb{H}}_m$.*

- *Integrate by parts to depress to the maximal extent possible the differentiability requirements of the space $\overset{\circ}{\mathbb{H}}_m$.*

- *Choose each function in a basis of $\overset{\circ}{\mathbb{H}}_m$ so that it vanishes along much of the spatial domain of interest, thereby ensuring that the intersection between the supports of most of the basis functions is empty.*

Needless to say, there is much more to the FEM than these five principles. In particular, we wish to specify $\overset{\circ}{\mathbb{H}}_m$ so that for sufficiently large m the numerical solution converges to the exact (weak) solution of the underlying equation – and, preferably, converges fairly fast. This is a subject that has attracted enough research to fill many a library shelf. The next section presents a brief review of the FEM in a more general setting, with an emphasis on the choice of $\overset{\circ}{\mathbb{H}}_m$ that ensures convergence to the exact solution.

8.2 A synopsis of FEM theory

In this section we present an outline of the finite element theory. We mostly dispense with proofs. The reason is that an honest exposition of the FEM is bound to be based on the theory of Sobolev spaces and relatively advanced functional-analytic concepts. There are many excellent texts on the FEM which we list at the end of this chapter, to which we refer the more daring and inquisitive reader.

The object of our attention is the boundary value problem

$$\mathcal{L}u = f, \qquad \boldsymbol{x} \in \Omega, \tag{8.17}$$

where $u = u(\boldsymbol{x})$, the function $f = f(\boldsymbol{x})$ is bounded and $\Omega \subset \mathbb{R}^d$ is a a open, bounded, connected set with sufficiently smooth boundary. $\mathcal{L}$ is a linear differential operator,

$$\mathcal{L} = \sum_{k=0}^{2\nu} \sum_{\substack{i_1+i_2+\cdots+i_d=k \\ i_1,i_2,\ldots,i_d \geq 0}} c_{i_1,i_2,\ldots,i_d}(\boldsymbol{x}) \frac{\partial^k}{\partial x_1^{i_1} \partial x_2^{i_2} \cdots \partial x_d^{i_d}}.$$

The equation (8.17) is accompanied by ν boundary conditions along $\partial\Omega$ – some might be essential, other natural, but we will not delve further into this issue.

Let $\mathbb{H}$ be the affine space of all functions that act in Ω, whose νth derivative is integrable[4] and which obey all *essential* boundary conditions along $\partial\Omega$. We denote by $\overset{\circ}{\mathbb{H}} \subseteq \mathbb{H}$ the linear space of all functions in $\mathbb{H}$ that satisfy zero boundary conditions.

We equip ourselves with the Euclidean inner product

$$\langle v, w \rangle = \int_\Omega v(\boldsymbol{\tau}) w(\boldsymbol{\tau}) \, \mathrm{d}\boldsymbol{\tau}, \qquad v, w \in \mathbb{H},$$

and the inherited Eulidean norm

$$\|v\| = \{\langle v, v \rangle\}^{1/2}, \qquad v \in \mathbb{H}.$$

Note that we have designed $\overset{\circ}{\mathbb{H}}$ so that terms of the form $\langle \mathcal{L}v, w \rangle$ make sense for every $v, w \in \mathbb{H}$, but this is true only subject to integration by parts ν times, to depress the degree of derivatives inside the integral from 2ν down to ν. If $d \geq 2$ we need to use various multivariate counterparts of integration, of which perhaps the most useful are the *divergence theorem*

$$\int_\Omega \nabla[a(\boldsymbol{x})\nabla v(\boldsymbol{x})] w(\boldsymbol{x}) \, \mathrm{d}\boldsymbol{x} = \int_{\partial\Omega} a(\boldsymbol{s}) w(\boldsymbol{s}) \frac{\partial v(\boldsymbol{s})}{\partial n} \, \mathrm{d}\boldsymbol{s} - \int_\Omega a(\boldsymbol{x})[\nabla v(\boldsymbol{x})] \cdot [\nabla w(\boldsymbol{x})] \, \mathrm{d}\boldsymbol{x},$$

and *Green's formula*

$$\int_\Omega [\nabla^2 v(\boldsymbol{x})] w(\boldsymbol{x}) \, \mathrm{d}\boldsymbol{x} + \int_\Omega [\nabla v(\boldsymbol{x})] \cdot [\nabla w(\boldsymbol{x})] \, \mathrm{d}\boldsymbol{x} = \int_{\partial\Omega} \frac{\partial v(\boldsymbol{s})}{\partial n} w(\boldsymbol{s}) \, \mathrm{d}\boldsymbol{s}.$$

Here $\nabla = \begin{bmatrix} \partial/\partial x_1 & \partial/\partial x_2 & \cdots & \partial/\partial x_d \end{bmatrix}^T$, while $\partial/\partial n$ is the derivative in the direction of the outward normal to the boundary $\partial\Omega$.[5]

The linear differential operator $\mathcal{L}$, (8.17), is said to be

self-adjoint	if $\langle \mathcal{L}v, w \rangle = \langle v, \mathcal{L}w \rangle$ for all $v, w \in \overset{\circ}{\mathbb{H}}$;
elliptic	if $\langle \mathcal{L}v, v \rangle > 0$ for all $v \in \overset{\circ}{\mathbb{H}}$; and
positive definite	if it is both self-adjoint and elliptic.

---

[4]As we have already seen in Section 8.1, this does not mean that the νth derivative exists everywhere in Ω.

[5]By rights, this means that the Laplace operator should be denoted by $\nabla^T \nabla$, rather than ∇^2, and that, faithful to our convention, we should have used boldface to remind ourselves that ∇ is a vector. Regretfully, and with a heavy sigh, we yield to the wider convention.

An important example of a positive-definite operator is

$$\mathcal{L} = -\sum_{i=1}^{d} \frac{\partial}{\partial x_i} \sum_{j=1}^{d} b_{i,j}(\boldsymbol{x}) \frac{\partial}{\partial x_j}, \tag{8.18}$$

where the matrix $B(\boldsymbol{x}) = (b_{i,j}(\boldsymbol{x}))$, $i,j = 1, 2, \ldots, d$, is symmetric and positive definite for every $\boldsymbol{x} \in \Omega$. To prove this we use a variant of the divergence theorem. Since $w \in \mathring{\mathbb{H}}$, it vanishes along the boundary $\partial \Omega$ and it is easy to verify that

$$\langle \mathcal{L}v, w \rangle = -\int_{\Omega} \left\{ \sum_{i=1}^{d} \frac{\partial}{\partial x_i} \left[\sum_{j=1}^{d} b_{i,j}(\boldsymbol{x}) \frac{\partial v(\boldsymbol{x})}{\partial x_j} \right] \right\} w(\boldsymbol{x}) \, \mathrm{d}\boldsymbol{x}$$

$$= \int_{\Omega} \sum_{i=1}^{d} \sum_{j=1}^{d} \left[\frac{\partial v(\boldsymbol{x})}{\partial x_i} \right] b_{i,j}(\boldsymbol{x}) \left[\frac{\partial w(\boldsymbol{x})}{\partial x_j} \right] \, \mathrm{d}\boldsymbol{x}. \tag{8.19}$$

Since $b_{i,j} \equiv b_{j,i}$, $i,j = 1, 2, \ldots, d$, we deduce that the last expression is symmetric in v and w. Therefore $\langle \mathcal{L}v, w \rangle = \langle v, \mathcal{L}w \rangle$ and $\mathcal{L}$ is self-adjoint. To prove ellipticity we let $w = v \not\equiv 0$ in (8.19); then

$$\langle \mathcal{L}v, v \rangle = \int_{\Omega} \sum_{i=1}^{d} \sum_{j=1}^{d} \left[\frac{\partial v(\boldsymbol{x})}{\partial x_i} \right] b_{i,j}(\boldsymbol{x}) \left[\frac{\partial v(\boldsymbol{x})}{\partial x_j} \right] \, \mathrm{d}\boldsymbol{x} > 0$$

by definition of the positive definiteness of matrices (A.1.3.5).

Note that both the negative of the Laplace operator, $-\nabla^2$, and the one-dimensional operator

$$-\frac{\mathrm{d}}{\mathrm{d}x} \left[a(x) \frac{\mathrm{d}}{\mathrm{d}x} \right] + b(x), \tag{8.20}$$

where $a(x) > 0$ and $b(x) \geq 0$ in the interval of interest, are special cases of (8.18); therefore they are positive definite.

Whenever a differential operator $\mathcal{L}$ is positive definite, we can identify the differential equation (8.17) with a variational problem, thereby setting the stage for the Ritz method.

Theorem 8.1 *Provided that the operator $\mathcal{L}$ is positive definite, (8.17) is the Euler–Lagrange equation of the variational problem*

$$\mathcal{J}(v) := \langle \mathcal{L}v, v \rangle - 2\langle f, v \rangle, \qquad v \in \mathbb{H}. \tag{8.21}$$

The weak solution of $\mathcal{L}u = f$ is therefore the unique minimum of $\mathcal{J}$ in $\mathbb{H}$.[6]

Proof We generalize an argument that has already been set out in Section 8.1 for the special case of the two-point boundary value problem (8.20).

---

[6]An unexpected (and very valuable) consequence of this theorem is the existence and uniqueness of the solution of (8.17) in $\mathbb{H}$. Therefore Theorem 8.1 – like much of the material in this section – is relevant to both the analytic and numerical aspects of elliptic PDEs.

Because of ellipticity, the variational functional $\mathcal{J}$ possesses a minimum (see Exercise 8.7). Let us denote a local minimum by $u \in \mathbb{H}$. Therefore, for any given $v \in \overset{\circ}{\mathbb{H}}$ and sufficiently small $|\varepsilon|$ we have

$$\mathcal{J}(u) \leq \mathcal{J}(u + \varepsilon v) = \langle \mathcal{L}(u + \varepsilon v), u + \varepsilon v \rangle - 2\langle f, u + \varepsilon v \rangle.$$

The operator $\mathcal{L}$ being linear, this results in

$$\mathcal{J}(u) \leq [\langle \mathcal{L}u, u \rangle - 2\langle f, u \rangle] + \varepsilon [\langle \mathcal{L}v, u \rangle + \langle \mathcal{L}u, v \rangle - 2\langle f, v \rangle] + \varepsilon^2 \langle \mathcal{L}v, v \rangle$$

and self-adjointness together with linearity yield

$$\mathcal{J}(u) \leq \mathcal{J}(u) + 2\varepsilon\langle \mathcal{L}u - f, v \rangle + \varepsilon^2 \langle \mathcal{L}v, v \rangle.$$

In other words,

$$2\varepsilon\langle \mathcal{L}u - f, v \rangle + \varepsilon^2 \langle \mathcal{L}v, v \rangle \geq 0 \tag{8.22}$$

for all $v \in \overset{\circ}{\mathbb{H}}$ and sufficiently small $|\varepsilon|$.

Suppose that u is not a weak solution of (8.17). Then there exists $v \in \overset{\circ}{\mathbb{H}}$, $v \not\equiv 0$, such that $\langle \mathcal{L}u - f, v \rangle \neq 0$. We may assume without loss of generality that this inner product is negative, otherwise we replace v with $-v$. It follows that, choosing sufficiently small $\varepsilon > 0$, we may render the expression on the left of (8.22) negative. Since this is forbidden by the inequality, we deduce that no such $v \in \overset{\circ}{\mathbb{H}}$ exists and u is indeed a weak solution of (8.17).

Assume, though, that $\mathcal{J}$ has several local minima in $\overset{\circ}{\mathbb{H}}$ and denote two such distinct functions by u_1 and u_2. Repeating our analysis with $\varepsilon = 1$ whilst replacing v by $u_2 - u_1 \in \overset{\circ}{\mathbb{H}}$ results in

$$\begin{aligned} \mathcal{J}(u_2) &= \mathcal{J}(u_1 + (u_2 - u_1)) \\ &= \mathcal{J}(u_1) + 2\langle \mathcal{L}u_1 - f, u_2 - u_1 \rangle + \langle \mathcal{L}(u_2 - u_1), u_2 - u_1 \rangle. \end{aligned} \tag{8.23}$$

We have just proved that $\langle \mathcal{L}u_1 - f, u_2 - u_1 \rangle = 0$, since u_1 locally minimizes $\mathcal{J}$. Moreover, $\mathcal{L}$ is elliptic and $u_2 \neq u_1$, therefore $\langle \mathcal{L}(u_2 - u_1), u_2 - u_1 \rangle > 0$. Substitution into (8.23) yields the contradictory inequality $\mathcal{J}(u_1) < \mathcal{J}(u_1)$, thereby leading us toward the conclusion that $\mathcal{J}$ possesses a single minimum in $\overset{\circ}{\mathbb{H}}$. ∎

◇ **When is a zero really a zero?**   An important yet subtle point in the theory of function spaces is the identity of the zero function. In other words, when are u_1 and u_2 really different? Suppose for example that u_2 is the same as u_1, except that it has a different value at just one point. This, clearly, will pass unnoticed by our inner product, which consists of integrals. In other words, if u_1 is a minimum of $\mathcal{J}$ (and a weak solution of (8.17)), then so is u_2; in this sense there is no uniqueness. In order to be distinct *in the sense of the function space* $\overset{\circ}{\mathbb{H}}$, u_1 and u_2 need to satisfy $\|u_2 - u_1\| > 0$. In the language of measure theory, they must differ by a function of positive Lebesgue measure. The truth, seldom spelt out in elementary texts, is that a normed function space (i.e., a linear function space equipped with a norm) consists not of functions but of *equivalence classes* of functions: u_1 and u_2 are in the same equivalence class if $\|u_2 - u_1\| = 0$ (that is, if $u_2 - u_1$ is of measure zero). This is an artefact of function spaces defined on continua that has no counterpart in the more familiar vector spaces such as $\mathbb{R}^d$. Fortunately, as soon as this point is comprehended, we can, like everybody else, go back to our habit of referring to the members of $\overset{\circ}{\mathbb{H}}$ as 'functions'. $\diamond$

The *Ritz method* for (8.17) (where $\mathcal{L}$ is presumed positive definite) is a straightforward generalization of the corresponding algorithm from the last section. Again, we choose $\varphi_0 \in \mathbb{H}$, let m linearly independent vectors $\varphi_1, \varphi_2, \ldots, \varphi_m \in \overset{\circ}{\mathbb{H}}$ span a finite-dimensional linear space $\overset{\circ}{\mathbb{H}}$ and seek $\boldsymbol{\gamma} = \begin{bmatrix} \gamma_1 & \gamma_2 & \cdots & \gamma_m \end{bmatrix}^T \in \mathbb{R}^m$ so as to minimize

$$\mathcal{J}_m(\boldsymbol{\delta}) := \mathcal{J}\left(\varphi_0 + \sum_{\ell=1}^m \delta_\ell \varphi_\ell\right), \qquad \boldsymbol{\delta} \in \mathbb{R}^m$$

We set the gradient of $\mathcal{J}_m$ to $\mathbf{0}$, and this results in the m linear equations (8.7), where

$$a_{k,\ell} = \langle \mathcal{L}\varphi_k, \varphi_\ell \rangle, \qquad k = 1, 2, \ldots, m, \quad \ell = 0, 1, \ldots, m. \tag{8.24}$$

Incidentally, the self-adjointness of $\mathcal{L}$ means that $a_{k,\ell} = a_{\ell,k}$, $k, \ell = 1, 2, \ldots, m$. This saves roughly half the work of evaluating integrals. Moreover, the symmetry of a matrix often simplifies the task of its numerical solution.

The general *Galerkin method* is also an easy generalization of the algorithm presented in Section 8.1 for the ODE (8.1). Again, we seek $\boldsymbol{\gamma}$ such that

$$\left\langle \mathcal{L}\left(\varphi_0 + \sum_{\ell=1}^m \gamma_\ell \varphi_\ell\right) - f, \varphi_k \right\rangle = 0, \qquad k = 1, 2, \ldots, m. \tag{8.25}$$

In other words, we endeavour to approximate a weak solution from a finite-dimensional space.

We have stipulated that $\mathcal{L}$ is linear, and this means that (8.25) is, again, nothing other than the linear system (8.7) with coefficients defined by (8.24). However, (8.25) makes sense even for nonlinear operators.

The existence and uniqueness of the solution of the Ritz–Galerkin equations (8.7) has already been addressed in Theorem 8.1. Another important statement is the *Lax–Milgram theorem*, which requires more than ellipticity but considerably less than self-adjointness. Moreover, it also provides a most valuable error estimate.

Given any $v \in \mathbb{H}$, we let

$$\|v\|_H := \left(\|v\|^2 + \langle \mathcal{L}v, v \rangle\right)^{1/2}.$$

It is possible to prove that $\| \cdot \|_H$ is a norm – in fact, this is a special case of the famed *Sobolev norm* and it is the correct way of measuring distances in $\mathbb{H}$. We say that the linear differential operator $\mathcal{L}$ is

bounded   if there exists $\delta > 0$ such that $|\langle \mathcal{L}v, w \rangle| \leq \delta \|v\|_H \times \|w\|_H$ for every $v, w \in \mathbb{H}$;   and

coercive   if there exists $\kappa > 0$ such that $\langle \mathcal{L}v, v \rangle \geq \kappa \|v\|_H^2$ for every $v \in \mathbb{H}$.

To be precise, it is not $\mathcal{L}$ but the *bilinear form* $\bar{a}(v, w) := \langle \mathcal{L}v, w \rangle$ which might be bounded or coercive. However, and given that we will not have the pleasure of the company of $\bar{a}$ again in this book, we prefer to employ a non-standard terminology.

Theorem 8.2 (The Lax–Milgram theorem)   *Let $\mathcal{L}$ be linear, bounded and coercive and let $\mathbb{V}$ be a linear subspace of $\mathring{\mathbb{H}}$. There exists a unique $\tilde{u} \in \mathbb{V}$ such that*

$$\langle \mathcal{L}\tilde{u} - f, v \rangle = 0, \qquad v \in \mathbb{V}$$

and

$$\|\tilde{u} - u\|_H \leq \frac{\delta}{\kappa} \inf \{ \|v - u\|_H : v \in \varphi_0 + \mathbb{V} \}, \tag{8.26}$$

where $\varphi_0 \in \mathbb{H}$ is arbitrary and u is a weak solution of (8.17) in $\mathbb{H}$.   ∎

The inequality (8.26) is sometimes called the *Céa Lemma*.

The space $\mathbb{V}$ need not be finite dimensional. In fact, it could be the space $\mathring{\mathbb{H}}$ itself, in which case we would deduce from the first part of the theorem that the weak solution of (8.17) exists and is unique. Thus, exactly like Theorem 8.1, the Lax–Milgram theorem can be used for analytic, as well as for numerical ends.

A proof of the coercivity and boundedness of $\mathcal{L}$ is typically much more difficult than a proof of positive definiteness. It suffices to say here that, for most domains of interest, it is possible to prove that the operator $-\nabla^2$ satisfies the conditions of the Lax–Milgram theorem. An essential step in this proof is the *Poincaré inequality*: there exists a constant c, dependent only on Ω, such that

$$\|v\| \leq c \left\| \sum_{i=1}^{d} \frac{\partial v}{\partial x_i} \right\|, \qquad v \in \mathring{\mathbb{H}}.$$

As far as the FEM is concerned, however, the error estimate (8.26) is the most valuable consequence of the theorem. On the right-hand side we have a constant, δ/κ, which is independent of the choice of $\mathring{\mathbb{H}}_m = \mathbb{V}$ and of the norm of the *distance* of the exact solution u from the affine space $\varphi_0 + \mathring{\mathbb{H}}_m$. Of course, $\inf_{v \in \varphi_0 + \mathring{\mathbb{H}}_m} \|v - u\|_H$ is unknown, since we do not know u. The one piece of information, however, that is definitely true about u is that it lies in $\mathbb{H} = \varphi_0 + \mathring{\mathbb{H}}$. Therefore the distance from u to $\varphi_0 + \mathring{\mathbb{H}}_m$ can be bounded in terms of the distance of an *arbitrary* member $w \in \varphi_0 + \mathring{\mathbb{H}}$ from $\varphi_0 + \mathring{\mathbb{H}}_m$. The final observation is that φ_0 makes no difference to our estimates and we hence deduce that, subject to linearity, boundedness and coercivity, the estimation of the error in the Galerkin method can be replaced by an approximation-theoretical problem: given a function $w \in \mathring{\mathbb{H}}$ find the distance $\inf_{v \in \mathring{\mathbb{H}}_m} \|w - v\|_H$.

In particular, the question of the *convergence* of the FEM reduces, subject to the conditions of Theorem 8.2, to the following question in approximation theory. Suppose that we have an infinite sequence of linear spaces $\mathring{\mathbb{H}}_{m_1}, \mathring{\mathbb{H}}_{m_2}, \mathring{\mathbb{H}}_{m_3}, \ldots \subset \mathring{\mathbb{H}}$, where $\dim \mathring{\mathbb{H}}_{m_i} = m_i$ and the sequence $\{m_i\}_{i=1}^{\infty}$ ascends monotonically to infinity. Is it true that

$$\lim_{i \to \infty} \|u_{m_i} - u\|_H = 0,$$

where u_{m_i} is the Galerkin solution in the space $\varphi_0 + \mathring{\mathbb{H}}_{m_i}$? In the light of the inequality (8.26) and of our discussion, a sufficient condition for convergence is that for every $v \in \mathring{\mathbb{H}}$ it is true that

$$\lim_{i \to \infty} \inf_{w \in \mathring{\mathbb{H}}_{m_i}} \|v - w\|_H = 0. \tag{8.27}$$

It now pays to recall, when talking of the FEM, that the spaces $\mathring{\mathbb{H}}_{m_i}$ are spanned by functions with small support. In other words, each $\mathring{\mathbb{H}}_{m_i}$ possesses a basis

$$\varphi_1^{[i]}, \varphi_2^{[i]}, \ldots, \varphi_{m_i}^{[i]} \in \mathring{\mathbb{H}}_{m_i}$$

such that each $\varphi_j^{[i]}$ is supported on the open set $\mathbb{E}_j^{[i]} \subset \mathring{\mathbb{H}}_{m_i}$, and $\mathbb{E}_k^{[i]} \cap \mathbb{E}_\ell^{[i]} = \emptyset$ for most choices of $k, \ell = 1, 2, \ldots, m_i$. In practical terms, this means that the d-dimensional set Ω needs to be partitioned as follows:

$$\mathrm{cl}\, \Omega = \bigcup_{\alpha=1}^{n_i} \mathrm{cl}\, \Omega_\alpha^{[i]}, \qquad \text{where} \qquad \Omega_\alpha^{[i]} \cap \Omega_\beta^{[i]} = \emptyset, \quad \alpha \neq \beta.$$

Each $\Omega_\alpha^{[i]}$ is called an *element,* hence the name 'finite element method'. We allow each support $\mathbb{E}_j^{[i]}$ to extend across a small number of elements. Hence, $\mathbb{E}_k^{[i]} \cap \mathbb{E}_\ell^{[i]}$ consists exactly of the sets $\Omega_\alpha^{[i]}$ (and possibly their boundaries) shared by both supports. This implies that an overwhelming majority of intersections is empty.

Recall the solution of (8.7) using chapeau functions. In that case $m_i = i$, $n_i = i+1$,

$$\varphi_k^{[i]} = \psi\left(x - \frac{k}{i+1}\right), \qquad k = 1, 2, \ldots, i$$

(ψ having been defined in (8.8)),

$$\Omega_\alpha^{[i]} = \left(\frac{\alpha-1}{i+1}, \frac{\alpha}{i+1}\right), \qquad \alpha = 1, 2, \ldots, i+1,$$

and

$$\mathbb{E}_j^{[i]} = \Omega_{j-1}^{[i]} \cup \Omega_j^{[i]}, \qquad j = 1, 2, \ldots, i.$$

Further examples, in two spatial dimensions, feature in Section 8.3.

What are reasonable requirements for a 'finite element space' $\mathring{\mathbb{H}}_{m_i}$? Firstly, of course, $\mathring{\mathbb{H}}_{m_i} \subset \mathring{\mathbb{H}}$, and this means that all members of the set must be sufficiently smooth to be subjected to the weak form (i.e., after integration by parts) of action by $\mathcal{L}$. Secondly, each set $\Omega_\alpha^{[i]}$ must contain functions $\varphi_j^{[i]}$ of sufficient number and variety to be able to approximate well arbitrary functions; recall (8.27). Thirdly, as i increases and the partition is being refined, we wish to ensure that the diameters of all elements ultimately tend to zero. It is usual to express this as the requirement that $\lim_{i \to \infty} h_i = 0$, where

$$h_i = \max_{\alpha=1,2,\ldots,n_i} \mathrm{diam}\, \Omega_\alpha^{[i]}$$

is the *diameter* of the *i*th partition.[7] This does not mean that we need to refine all elements at an equal speed – an important feature of the FEM is that it lends itself to local refinement, and this confers important practical advantages. Our fourth and final requirement is that, as $i \to \infty$, the geometry of the elements does not become too 'difficult': in practical terms, the elements are likely to be polytopes (for example, polygons in $\mathbb{R}^2$) and we wish all their angles to be bounded away from zero as $i \to \infty$.

The latter two conditions are relatively simple to formulate and enforce, but the first two require further attention and elaboration. As far as the smoothness of $\varphi_j^{[i]}$, $j = 1, 2, \ldots, m_i$, is concerned, the obvious difficulty is likely to be smoothness *across element boundaries*, since it is in general easy to specify arbitrarily smooth functions within each $\Omega_\alpha^{[i]}$. On the other hand, 'approximability' of the finite element space $\overset{\circ}{\mathbb{H}}_{m_i}$ is all about what happens *inside* each element.

Our policy in the remainder of this chapter is to use elements $\Omega_\alpha^{[i]}$ that are all linear translates of the same 'master element', similarly to the chapeau function (8.8) being defined in the interval $[-1, 1]$ and then translated to arbitrary intervals. Specifically, for $d = 2$, our interest centres on translates of triangular elements (not necessarily all with identical angles) and quadrilateral elements. We choose functions $\varphi_j^{[i]}$ that are *polynomial* within each element – obviously the question of smoothness is relevant only across element boundaries. Needless to say, our refinement condition $\lim_{i \to \infty} h_i = 0$ and our ban on arbitrarily acute angles are strictly enforced.

We say that the space $\overset{\circ}{\mathbb{H}}_{m_i}$ is of *smoothness q* if each function $\varphi_j^{[i]}$, $j = 1, 2, \ldots, m_i$, is $q - 1$ times smoothly differentiable in all of Ω and q times differentiable inside each element $\mathbb{E}_\alpha^{[i]}$, $\alpha = 1, 2, \ldots, n_i$. (The latter requirement is not necessary within our framework, since we have already required all functions $\varphi_j^{[i]}$ to be polynomials. It is stated for the sake of conformity with more general finite element spaces.) Furthermore, the space $\overset{\circ}{\mathbb{H}}_{m_i}$ is of *accuracy p* if, within each element $\Omega_\alpha^{[i]}$ the functions $\varphi_j^{[i]}$ span the set $\mathbb{P}_p^d[\boldsymbol{x}]$ of all *d*-dimensional polynomials of *total degree p*. The latter encompasses all functions of the form

$$\sum_{\substack{\ell_1 + \cdots + \ell_d \leq p \\ \ell_1, \ldots, \ell_d \geq 0}} c_{\ell_1, \ell_2, \ldots, \ell_d} x_1^{\ell_1} x_2^{\ell_2} \cdots x_d^{\ell_d}$$

where $c_{\ell_1, \ell_2, \ldots, \ell_d} \in \mathbb{R}$ for all $\ell_1, \ell_2, \ldots, \ell_d$.

Let us illustrate the above concepts for the case of (8.1) and chapeau functions. Firstly, each translate of (8.8) is continuous throughout $\Omega = (0, 1)$ but not differentiable throughout the whole interval, hence $q - 1 = 0$ and we deduce a smoothness $q = 1$. Secondly, each element $\Omega_\alpha^{[i]}$ is the support of both $\varphi_{\alpha-1}^{[i]}$ and $\varphi_\alpha^{[i]}$ (with obvious modification for $\alpha = 1$ and $\alpha = i + 1$). Both are linear functions, the first increasing, with slope $+h^{-1}$, and the second decreasing with slope $-h^{-1}$. Hence linear independence allows the conclusion that every linear function can be expressed inside $\Omega_\alpha^{[i]}$ as a linear combination of $\varphi_{\alpha-1}^{[i]}$ and $\varphi_\alpha^{[i]}$. Since $\mathbb{P}_1^1[\boldsymbol{x}] = \mathbb{P}_1[x]$, the set of all univariate linear functions, it follows that $\overset{\circ}{\mathbb{H}}_{m_1}$ is of accuracy $p = 1$.

---

[7]The quantity diam $\mathbb{U}$, where $\mathbb{U}$ is a bounded set, is defined as the least radius of a ball into which this set can be inscribed. It is called the *diameter* of $\mathbb{U}$.

Much effort has been spent in the last few pages to argue that there is an intimate connection between smoothness, accuracy and the error estimate (8.26). Unfortunately, this is as far as we can go without venturing into much deeper waters of functional analysis – except for stating, without any proof, a theorem that quantifies this connection in explicit terms.

Theorem 8.3 *Let $\mathcal{L}$ obey the conditions of Theorem 8.2 and suppose that we solve the equation (8.17) by the FEM, subject to the aforementioned restrictions (the shape of the elements, $\lim_{m\to\infty} h_i = 0$ etc.), with smoothness and accuracy $q = p \geq \nu$. Then there exists a constant $c > 0$, independent of i, such that*

$$\|u_m - u\|_H \leq c h_i^{p+1-\nu} \|u^{(p+1)}\|, \qquad i = 1, 2, \ldots. \tag{8.28}$$

■

Returning to the chapeau functions and their solution of the two-point boundary value equation (8.1), we use the inequality (8.28) to confirm our impression from Figures 8.2 and 8.3, namely that the error is $\mathcal{O}(h_i)$.

Theorem 8.3 is just a sample of the very rich theory of the FEM. Error bounds are available in a variety of norms (often with more profound significance to the underlying problem than the Euclidean norm) and subject to different conditions. However, inequality (8.28) is sufficient for the applications to the Poisson equation in the next section.

8.3 The Poisson equation

As we have already seen in the last section, the operator $\mathcal{L} = -\nabla^2$ is positive definite, being a special case of (8.18). Moreover, we have claimed that, for most realistic domains Ω, it is coercive and bounded. The coefficients (8.24) of the Ritz–Galerkin equations are simply given by

$$a_{k,\ell} = \int_\Omega (\nabla\varphi_k) \cdot (\nabla\varphi_\ell) \, d\boldsymbol{x}, \qquad k, \ell = 1, 2, \ldots, m. \tag{8.29}$$

Letting $d = 2$, we assume that the boundary $\partial\Omega$ is composed from a finite number of straight segments and partition Ω into triangles. The only restriction is that no vertex of one triangle may lie on the edge of another; vertices must be shared. In other words, a configuration like

(where the position of a vertex is emphasized by '•') is not allowed. Figures 8.5 and 8.6 display a variety of triangulations that conform with this rule.

In light of (8.28), we require for convergence that $p, q \geq 1$, where p and q are accuracy and smoothness respectively. This is similar to the situation that we have already encountered in Section 8.1 and we propose to address it with a similar remedy, namely by choosing $\varphi_1, \varphi_2, \ldots, \varphi_m$ as *piecewise linear* functions. Each function in $\mathbb{P}_1^2$ can be represented in the form

$$g(x, y) = \alpha + \beta x + \gamma y \tag{8.30}$$

for some $\alpha, \beta, \gamma \in \mathbb{R}$. Each function φ_k supported by an element Ω_j, $j = 1, 2, \ldots, n$, is consequently of the form (8.30). Thus, to be accurate to order $p = 1$, each element must support at least three linearly independent functions. Recall that smoothness $q = 1$ means that every linear combination of the functions $\varphi_1, \varphi_2, \ldots, \varphi_m$ is continuous in Ω and this, obviously, need be checked only across element boundaries.

We have already seen in Section 8.1 one construction that provides both for accuracy $p = 1$ and for continuity with piecewise linear functions. The idea is to choose a basis of piecewise linear functions that vanish at all the vertices, except that at each vertex one of the functions equals $+1$. Chapeau functions are one example of such *cardinal functions* and they have counterparts in $\mathbb{R}^2$. Figure 8.4 displays three examples of *pyramid functions,* the planar cardinal functions, within their support (the set of all values of the argument where they are nonzero). Unfortunately, it also demonstrates that using cardinal functions in $\mathbb{R}^2$ is, in general, a poor idea. The number of elements in each support may change from vertex to vertex and the description of each cardinal function, although easy in principle, is quite messy and inconvenient for practical work.

Figure 8.4   Pyramid functions for different configurations of vertices.

The correct procedure is to represent the approximation inside each Ω_j by data at its vertices. As long as we adopt this approach, it is of no importance how many different triangles meet at each vertex and we can apply the same algorithm to all elements.

Let the triangle in question be

We determine the piecewise linear approximation s by interpolating at the three vertices. According to (8.30), this results in the linear system

$$\alpha + x_\ell \beta + y_\ell \gamma = g_\ell, \qquad \ell = 1, 2, 3,$$

where g_ℓ is the interpolated value at (x_ℓ, y_ℓ). Since the three vertices are not collinear, the system is nonsingular and can be solved with ease. This procedure (which, formally, is completely equivalent to the use of cardinal functions) ensures accuracy of order $p = 1$.

We need to prove that the above approach produces a function that is continuous throughout Ω, since this is equivalent to $q = 1$, the required degree of smoothness. This, however, follows from our construction. Recall that we need to prove continuity only across element boundaries. Suppose, without loss of generality, that the line segment joining (x_1, y_1) and (x_2, y_2) is not part of $\partial\Omega$ (otherwise there would be nothing to prove). The function s reduces along a straight line to a *linear function in one variable,* hence it is determined uniquely by the two interpolated values g_1 and g_2 at the endpoints. Since these endpoints are shared by the triangle that adjoins along this edge, it follows that s is continuous there. A similar argument extends to all internal edges of the triangulation. We conclude that $p = q = 1$, hence the error (in a correct norm) decays like $\mathcal{O}(h)$, where h is the diameter of the triangulation.

A practical solution using the FEM requires us to assemble the *stiffness matrix*

$$A = (a_{k,\ell})_{k,\ell=1}^m.$$

The dimension being $d = 2$, (8.29) formally becomes

$$
\begin{aligned}
a_{k,\ell} &= \iint_\Omega \left(\frac{\partial \varphi_k}{\partial x} \frac{\partial \varphi_\ell}{\partial x} + \frac{\partial \varphi_k}{\partial y} \frac{\partial \varphi_\ell}{\partial y} \right) \mathrm{d}x\,\mathrm{d}y \\
&= \sum_{j=1}^n \iint_{\Omega_j} \left(\frac{\partial \varphi_k}{\partial x} \frac{\partial \varphi_\ell}{\partial x} + \frac{\partial \varphi_k}{\partial y} \frac{\partial \varphi_\ell}{\partial y} \right) \mathrm{d}x\,\mathrm{d}y \\
&= \sum_{j=1}^n a_{k,\ell,j} \qquad k, \ell = 1, 2, \ldots, m.
\end{aligned}
$$

Inside the jth element the quantity $a_{k,\ell,j}$ vanishes, unless both φ_k and φ_ℓ are supported there. In the latter case, each is a linear function and, at least in principle, all integrals can be calculated (probably using quadrature). This, however, fails to take account of a subtle change of basis that we have just introduced in our characterization of the approximant inside each element in terms of its values on the vertices. Of course, except for vertices that happen to lie on $\partial\Omega$, these values are unknown and their computation is the whole purpose of the exercise. *The values at the vertices are our new unknowns* and we thereby rephrase the Ritz problem as follows: out of all possible piecewise linear functions that are consistent with our partition (that is, are linear inside each element), find the one that minimizes the functional

$$
\mathcal{J}(v) = \iint_\Omega \left[\left(\frac{\partial v}{\partial x} \right)^2 + \left(\frac{\partial v}{\partial y} \right)^2 \right] \, \mathrm{d}x \, \mathrm{d}y - 2 \iint_\Omega f v \, \mathrm{d}x \, \mathrm{d}y
$$
$$
= \sum_{j=1}^n \iint_{\Omega_j} \left[\left(\frac{\partial v}{\partial x} \right)^2 + \left(\frac{\partial v}{\partial y} \right)^2 \right] \, \mathrm{d}x \, \mathrm{d}y - 2 \sum_{j=1}^n \iint_{\Omega_j} f v \, \mathrm{d}x \, \mathrm{d}y.
$$

Inside each Ω_j the function v is linear, $v(x,y) = \alpha_j + \beta_j x + \gamma_j y$, and explicitly

$$
\iint_{\Omega_j} \left[\left(\frac{\partial v}{\partial x} \right)^2 + \left(\frac{\partial v}{\partial y} \right)^2 \right] \, \mathrm{d}x \, \mathrm{d}y = (\beta_j^2 + \gamma_j^2) \, \mathrm{area}\,(\Omega_j).
$$

As far as the second integral is concerned, we usually discretize it by quadrature and this, again, results in a function of α_j, β_j and γ_j.

With a little help from elementary analytic geometry, this can be expressed in terms of the values of v at the vertices. Let these be v_1, v_2, v_3, say, and assume that the corresponding (inner) angles of the triangle are $\theta_1, \theta_2, \theta_3$ respectively, where $\theta_1 + \theta_2 + \theta_3 = \pi$. Letting $\sigma_k = 1/(2\tan\theta_k)$, $k = 1, 2, 3$, we obtain

$$
\iint_{\Omega_j} \left[\left(\frac{\partial v}{\partial x} \right)^2 + \left(\frac{\partial v}{\partial y} \right)^2 \right] \mathrm{d}x \, \mathrm{d}y = \begin{bmatrix} v_1 & v_2 & v_3 \end{bmatrix} \begin{bmatrix} \sigma_2 + \sigma_3 & -\sigma_3 & -\sigma_2 \\ -\sigma_3 & \sigma_1 + \sigma_3 & -\sigma_1 \\ -\sigma_2 & -\sigma_1 & \sigma_1 + \sigma_2 \end{bmatrix} \begin{bmatrix} v_1 \\ v_2 \\ v_3 \end{bmatrix}.
$$

Meting out a similar treatment to the second integral and repeating this procedure for all elements in the triangulation, we finally represent the variational functional, acting on piecewise linear functions, in the form

$$
\mathcal{J}(v) = \tfrac{1}{2} v^T \tilde{A} v - f^T v, \tag{8.31}
$$

where v is the vector of the values of the function v at the m internal vertices (the number of such vertices is the same as the dimension of the space – why?). The $m \times m$ *stiffness matrix* $\tilde{A} = (\tilde{a}_{k,\ell})_{k,\ell=1}^m$ is assembled from the contributions of individual vertices. Obviously, $\tilde{a}_{k,\ell} = 0$ unless k and ℓ are indices of neighbouring vertices. The vector $f \in \mathbb{R}^m$ is constructed similarly, except that it also contains the contributions of boundary vertices.

Setting the gradient of (8.31) to zero results in the linear algebraic system

$$
\tilde{A} v = f
$$

which we need to solve, e.g. by the methods of Chapters 9–12.

Our extensive elaboration of the construction of (8.31) illustrates the point that it is substantially more difficult to work with the FEM than with finite differences. The extra effort, however, is a price that we pay for extra flexibility.

Figure 8.5 displays the solution of the Poisson equation (7.33) on three meshes. These meshes are *hierarchical* – each is constructed by refining the previous one – and of increasing fineness. The graph to the left displays the mesh, while the shape on the right is the numerical solution as constructed from linear pieces (compare with the exact solution at the top of Figure 7.8).

The advantages of the FEM are apparent if we are faced with difficult geometries and, even more profoundly, when it is known *a priori* that the solution is likely to be more problematic in part of the domain of interest and we wish to 'zoom in' on the triangulation there. For example, suppose that a Poisson equation is given in a domain with a *re-entrant corner* (for example, an L-shaped domain). We can expect the solution to be more difficult near such a corner and it is a good policy to refine the triangulation there.

As an example, let us consider the equation

$$\nabla^2 u + 2\pi^2 \sin \pi x \sin \pi y = 0, \qquad (x,y) \in \Omega = [-1,1]^2 \setminus [0,1]^2, \qquad (8.32)$$

with zero Dirichlet boundary conditions along $\partial\Omega$. The exact solution is simply $u(x,y) = \sin \pi x \sin \pi y$.

Figure 8.6 displays the triangulations and underlying numerical solutions for three meshes, that are increasingly refined. The triangulation is substantially finer near the re-entrant corner, as it should be, but perhaps the most important observation is that this does not require more effort than, say, the uniform tessalation of Figure 8.5. In fact, both figures were produced by an identical program, but with different input! Although writing such a program is more of a challenge than coding finite differences, the rewards are very rich indeed. . . .

The error in Figures 8.5 and 8.6 is consistent with the bound $\|u_m - u\|_H \leq ch\|u''\|$ (where h is the diameter of the triangulation), and this is consistent with (8.28). In particular, its rate of decay (as a function of h) in Figure 8.5 is similar to those of the five-point formula and the (unmodified) nine-point formula in Figure 7.8. At first glance, this might perhaps seem contradictory; did we not state in Chapter 7 that the error of the five-point formula (7.15) is $\mathcal{O}(h^2)$? True enough, except that we have been using different criteria to measure the error. Suppose, thus, that $u_{k,\ell} \approx u(k\Delta x, \ell\Delta x) + c_{k,\ell}h^2$ at all the grid points. Provided that $h = \mathcal{O}(m^{-2})$ (note that the number m means here the *total* number of variables in the whole grid), that there are $\mathcal{O}(m^2)$ grid points and that the error coefficients $c_{k,\ell}$ are of roughly similar order of magnitude, it is easy to verify that

$$\left\{ \sum_{(k,\ell) \text{ in the grid}} [u_{k,\ell} - u(k\Delta x, \ell\Delta x)]^2 \right\}^{1/2} = \mathcal{O}(h).$$

This corresponds to the Euclidean norm in the finite element space. Although the latter is distinct from the Sobolev norm $\| \cdot \|_H$ of inequality (8.28), our argument indicates why the two error estimates are similar.

As was the case with finite difference schemes, the aforementioned accuracy sometimes falls short of that desired. This motivates the discussion of function bases having superior smoothness and accuracy properties. In one dimension this is straightforward, at least on the conceptual level: we need to replace piecewise linear functions with

Figure 8.5　The solution of the Poisson equation (7.33) in a square domain with various triangulations.

Figure 8.6 The solution of the Poisson equation (8.32) in an L-shaped domain with various triangulations.

splines, functions that are kth degree polynomials, say, in each element and possess $k-1$ smooth derivatives in the whole interval of interest. A convenient basis of kth degree splines is provided by *B-splines,* which are distinguished by having the least possible support, extending across $k+1$ consecutive elements. In a general partition $\xi_0 < \xi_1 < \cdots < \xi_n$, say, a kth degree B-spline is defined explicitly by the formula

$$B_j^{[k]}(x) = \sum_{\ell=j}^{k+j+1} \left(\sum_{i=j,\, i\neq\ell}^{k+j+1} \frac{1}{\xi_i - \xi_\ell} \right) (x - \xi_\ell)_+^k .$$

Comparison with (8.8) ascertains that chapeau functions are nothing else but linear B-splines.

The task in hand is more complicated in the case of two-dimensional triangulation, because of our dictum that everything needs to be formulated in terms of function values in an individual element and across its boundary. As a matter of fact, we have used only the values at the boundary – specifically, at the vertices – but this is about to change.

A general quadratic in $\mathbb{R}^2$ is of the form

$$s(x,y) = \alpha + \beta x + \gamma y + \delta x^2 + \eta xy + \zeta y^2;$$

we note that it has six parameters. Likewise, a general cubic has ten parameters (verify!). We need to specify the correct number of interpolation points in the (closed) triangle. Two choices that give orders of accuracy $p=2$ and $p=3$ are

and

respectively. Unfortunately, their smoothness is just $q=1$ since, although a unique univariate quadratic or cubic, respectively, can be fitted along each edge (hence continuity), a tangental derivative might well be discontinuous. A superior interpolation pattern is

$$(8.33)$$

where '◉' means that we interpolate both function values and both spatial derivatives. We require altogether ten data items, and this is exactly the number of degrees of freedom in a bivariate cubic. Moreover, it is possible to show that the *Hermite* interpolation of both function values and (directional) derivatives along each edge results in both function and derivative smoothness there, hence $q=2$.

Interpolation patterns like (8.33) are indispensable when, instead of the Laplace operator we consider the *biharmonic operator* ∇^4, since then $\nu = 2$ and we need $q \geq 2$ (se Exercise 8.5).

We conclude this chapter with few words on piecewise linear interpolation with *quadrilateral* elements. The main problem in this context is that the bivariate linear function has three parameters – exactly right for a triangle but problematic in a quadrilateral. Recall that we must place interpolation points to attain continuity in the whole domain, and this means that at least two such points must reside along each edge. The standard solution of this conundrum is to restrict attention to rectangles (aligned with the axes) and interpolate with functions of the form

$$s(x, y) = s_1(x)s_2(y), \qquad \text{where} \qquad s_1(x) := \alpha + \beta x, \quad s_2(y) := \gamma + \delta y.$$

Obviously, piecewise linears are a special case, but now we have four parameters, just right to interpolate at the four corners,

Along both horizontal edges s_2 is constant and s_1 is uniquely specified by the values at the corners. Therefore, the function s along each horizontal edge is independent of the interpolated values elsewhere in the rectangle. Since an identical statement is true for the vertical edges, we deduce that the interpolant is continuous and that $q = 1$.

Comments and bibliography

Weak solutions and Sobolev spaces are two inseparable themes that permeate the modern theory of linear elliptic differential equations (Agmon, 1965; John, 1982). The capacity of the FEM to fit so snugly into this framework is not just a matter of æsthetics. It also, as we have had a chance to observe in this chapter, provides for truly powerful error estimates and for a computational tool that can cater for a wide range of difficult situations – curved geometries, problems with internal interfaces, solutions with singularities.... Yet, the FEM is considerably less popular in applications than the finite difference method. The two reasons are the considerably more demanding theoretical framework and the more substantial effort required to program the FEM. If all you need is to solve the Poisson equation in a square, say, with nice boundary conditions, then probably there is absolutely no need to bother with the FEM (unless off-the-shelf software is available), since finite differences will do perfectly well. More difficult problems, e.g. the equations of elasticity theory, the Navier–Stokes equations etc. justify the additional effort involved in mastering and using finite elements.

It is legitimate, however, to query how genuine are weak solutions? Anybody familiar with the capacity of mathematicians to generalize from the useful yet mundane to the beautiful yet useless has every right to feel sceptical. The simple answer is that they occur in many application areas, in linear as well as nonlinear PDEs and in variational problems. Moreover, seemingly 'nice' problems often have weak solutions. For a simple example, borrowed from Gelfand & Fomin (1963), we turn to the calculus of variations. Let

$$\mathcal{J}(v) := \int_{-1}^{1} v^2(\tau)[2\tau - v'(\tau)]^2 \, d\tau, \qquad v(-1) = 0, \quad v(1) = 1.$$

A nice cosy problem which, needless to say, should have a nice cosy solution. And it does!
The exact solution can be written down explicitly,

$$u(x) = \begin{cases} x^2, & 0 \le x \le 1 \\ 0, & -1 \le x \le 0 \end{cases}$$

However, the underlying Euler–Lagrange equation is

$$y\left(4x^2 + 2y - y'^2 - yy''\right) = 0 \qquad (8.34)$$

and includes a second derivative, while the function u fails to be twice-differentiable at the
origin. The solution of (8.34) exists only in a weak sense!

Lest the last example sounds a mite artificial (and it is – artificiality is the price of
simplicity!), let us add that many equations of profound interest in applications can be
investigated only in the context of weak solutions and Sobolev spaces. A thoroughly modern
applied mathematician must know a great deal of mathematical analysis.

An unexpected luxury for students of the FEM is the abundance of excellent books in the
subject, e.g. Axelsson & Barker (1984); Brenner & Scott (1994); Hackbusch (1992); Johnson
(1987); Mitchell & Wait (1977). Arguably, the most readable introductory text is Strang &
Fix (1973) – and it is rare for a book in a fast-moving subject to stay at the top of the hit
parade for more than twenty years! The most comprehensive exposition of the subject, short
of research papers and specialized monographs, is Ciarlet (1976). The reader is referred to
this FEM feast for a thorough and extensive exposition of themes upon which we have touched
briefly – error bounds, design of finite elements in multivariate spaces – and many themes that
have not been mentioned in this chapter. In particular, we encourage interested readers to
consult more advanced monographs on the generalization of finite element functions to $d \ge 3$,
on the attainment of higher smoothness conditions and on elements with curved boundaries.
Things are often not what they seem to be in Sobolev spaces and it is always worthwhile,
when charging the computational ramparts, to ensure adequate pure-mathematical covering
fire.

These remarks will not be complete without mentioning recent work that blends concepts from finite element, finite difference and spectral methods. A whole new menagerie of
concepts has emerged in the last decade: boundary element methods, h-p formulation of the
FEM, hierarchical bases.... Only the future will tell how much will survive and find its way
into textbooks, but these are exciting times at the frontiers of the FEM.

Agmon, S. (1965), *Lectures on Elliptic Boundary Value Problems,* Van Nostrand, Princeton,
    NJ.

Axelsson, O. and Barker, V.A. (1984), *Finite Element Solution of Boundary Value Problems:
    Theory and Computation,* Academic Press, Orlando, FL.

Brenner, S.C. and Scott, L.R. (1994), *The Mathematical Theory of Finite Element Methods,*
    Springer-Verlag, New York.

Ciarlet, P.G. (1976), *Numerical Analysis of the Finite Element Method,* North-Holland, Am-
    sterdam.

Gelfand, I.M. and Fomin, S.V. (1963), *Calculus of Variations,* Prentice–Hall, Englewood
    Cliffs, NJ.

Hackbusch, W. (1992), *Elliptic Differential Equations: Theory and Numerical Treatment,*
    Springer-Verlag, Berlin.

John, F. (1982), *Partial Differential Equations* (4th ed.), Springer-Verlag, New York.

Johnson, C. (1987), *Numerical Solution of Partial Differential Equations by the Finite Element Method,* Cambridge University Press, Cambridge.

Mitchell, A.R. and Wait, R. (1977), *The Finite Element Method in Partial Differential Equations,* Wiley, London.

Strang, G. and Fix, G.J. (1973), *An Analysis of the Finite Element Method,* Prentice–Hall, Englewood Cliffs, NJ.

Exercises

8.1 Demonstrate that the collocation method (3.12) finds in the interval $[t_n, t_{n+1}]$ an approximation to the weak solution of the ordinary differential system $\boldsymbol{y}' = \boldsymbol{f}(t, \boldsymbol{y})$, $\boldsymbol{y}(t_0) = \boldsymbol{y}_0$, from the space $\mathbb{P}_\nu$ of ν-degree polynomials, provided that we employ the inner product

$$\langle \boldsymbol{v}, \boldsymbol{w} \rangle = \sum_{j=1}^{\nu} \boldsymbol{v}(t_n + c_j h)^T \boldsymbol{w}(t_n + c_j h),$$

where $h = t_{n+1} - t_n$. [*Strictly speaking,* $\langle \cdot, \cdot \rangle$ *is a* semi-inner product, *since it is not true that* $\langle \boldsymbol{v}, \boldsymbol{v} \rangle = 0$ *implies* $\boldsymbol{v} \equiv \boldsymbol{0}$.]

8.2 Find explicitly the coefficients $a_{k,\ell}$, $k, \ell = 1, 2, \ldots, m$, for the equation $-y'' + y = f$, assuming that the space $\overset{\circ}{\mathbb{H}}_m$ is spanned by chapeau functions on an equidistant grid.

8.3 Suppose that the equation (8.1) is solved by the Galerkin method with chapeau functions on a non-equidistant grid. In other words, we are given $0 = t_0 < t_1 < t_2 < \cdots < t_m < t_{m+1} = 1$ so that each φ_j is supported in (t_{j-1}, t_{j+1}), $j = 1, 2, \ldots, m$. Prove that the linear system (8.7) is nonsingular. [*Hint: Use the Gerschgorin criterion (Lemma 7.3).*]

8.4 Let a be a given positive univariate function and

$$\mathcal{L} := \frac{\partial^2}{\partial x^2} \left[a(x) \frac{\partial^2}{\partial x^2} \right].$$

Assuming zero Dirichlet boundary conditions, prove that $\mathcal{L}$ is positive definite in the Euclidean norm.

8.5 Prove that the biharmonic operator ∇^4, acting in a parallelepiped in $\mathbb{R}^d$, is positive definite in the Euclidean norm.

8.6 Let $\mathcal{J}$ be given by (8.21), suppose that the operator $\mathcal{L}$ is positive definite and let

$$\mathcal{J}_m(\boldsymbol{\delta}) := \mathcal{J}\left(\varphi_0 + \sum_{\ell=1}^{m} \delta_\ell \varphi_\ell \right), \qquad \boldsymbol{\delta} \in \mathbb{R}^m.$$

Prove that the matrix

$$\left(\frac{\partial^2 \mathcal{J}_m(\boldsymbol{\delta})}{\partial \delta_k \partial \delta_\ell}\right)_{k,\ell=1,2,\ldots,m}$$

is positive definite, thereby deducing that the solution of the Ritz equations is indeed the global minimum of $\mathcal{J}_m$.

8.7   Let $\mathcal{L}$ be an elliptic differential operator and f a given bounded function.

a Prove that the numbers

$$c_1 := \min_{\substack{v \in \overset{\circ}{\mathbb{H}} \\ \|v\|=1}} \langle \mathcal{L}v, v \rangle \qquad \text{and} \qquad c_2 := \max_{\substack{v \in \overset{\circ}{\mathbb{H}} \\ \|v\|=1}} \langle f, v \rangle$$

are bounded and that $c_1 > 0$.

b Given $w \in \mathbb{H}$, prove that

$$\langle \mathcal{L}w, w \rangle - 2\langle f, w \rangle \geq \|w\|^2 c_1 - 2\|w\| c_2.$$

[*Hint: Write $w = \kappa v$, where $\|v\| = 1$ and $|\kappa| = \|w\|$.*]

c Deduce that

$$\langle \mathcal{L}w, w \rangle - 2\langle f, w \rangle \geq -\frac{c_2^2}{c_1}, \qquad w \in \mathbb{H},$$

thereby proving that the functional $\mathcal{J}$ from (8.21) has a bounded minimum.

8.8   Find explicitly a cardinal piecewise linear function (a pyramid function) in a domain partitioned into equilateral triangles (cf. the graph in Exercise 7.13).

8.9   Nine interpolation points are specified in a rectangle:

Prove that they can be interpolated by a function of the form $s(x,y) = s_1(x)s_2(y)$, where both s_1 and s_2 are quadratics. Find the orders of the accuracy and of the smoothness of this procedure.

8.10   The Poisson equation is solved in a square partition by the FEM in a manner that has been described in Section 8.3. In each square element the approximant is the function $s(x,y) = s_1(x)s_2(y)$, where s_1 and s_2 are linear, and it is being interpolated at the vertices. Derive explicitly the entries $\tilde{a}_{k,\ell}$ of the stiffness matrix $\tilde{A}$ from (8.31).

8.11 Prove that the four interpolatory conditions specified at the vertices of the three-dimensional *tetrahedral* element

can be satisfied by a piecewise linear function.

Gaussian elimination for sparse linear equations

9.1 Banded systems

Whether the objective is to solve the Poisson equation using finite differences or with finite elements, the outcome of discretization is a set of linear algebraic equations, e.g. (7.16) or (8.7). The solution of such equations ultimately constitutes the lion's share of computational expenses. This is true not just with regard to the Poisson equation or even elliptic PDEs since, as will become apparent in Chapter 13, the practical computation of parabolic PDEs also requires the solution of linear algebraic systems.

The systems (7.16) and (8.7) share two important characteristics. Our first observation is that in practical situations such systems are likely to be very large. Thus, five-point equations in an 81×81 grid result in 6400 equations. Even this might sound large to the uninitiated but it is, actually, relatively modest compared to what is encountered on a daily basis in real-life situations. Consider equations of motion (whether of fluids or solids), for example. The universe is three-dimensional and typical GFD (geophysical fluid dynamics) codes employ fourteen variables – three each for position and velocity, one each for density, pressure, temperature and, say, concentrations of five chemical elements. (If you think that fourteen variables is excessive, you might be interested to learn that in combustion theory, say, even this is regarded as rather modest.) Altogether, and unless some convenient symmetries allow us to simplify the task in hand, we are solving equations in a three-dimensional parallelepiped. Requiring 81 grid points in each spatial dimension spells $14 \times 80^3 = 7\,168\,000$ coupled linear equations!

The cost of computation using the familiar *Gaussian elimination* is $\mathcal{O}(d^3)$ for a $d \times d$ system, and this renders it useless for systems of size such as the above.[1] Even were we able to design a computer that can perform $(64\,000\,000)^3 \approx 2.6 \times 10^{23}$ operations, say, in a reasonable time, the outcome is bound to be useless because of an accumulation of roundoff error.[2]

---

[1] A brief remark about the $\mathcal{O}()$ notation. Often, the meaning of '$f(x) = \mathcal{O}(x^\alpha)$ as $x \to x_0$' is that $\lim_{x \to x_0} x^{-\alpha} f(x)$ exists and is bounded. The $\mathcal{O}()$ notation in this section can be formally defined in a similar manner, but it is perhaps more helpful to interpret $\mathcal{O}(d^3)$, say, in a more intuitive fashion: it means that a quantity equals roughly a constant times d^3.

[2] Of course, everybody knows that there are no such computers. Are they possible, however? Assuming serial computer architecture and considering that signals travel (at most) at the speed of light, the distance between the central processing unit and each random access memory cell should be at an atomic level.

Specifically, a back-of-the-envelope computation indicates that, were *all* the expense just in communication (at the speed of light!) and were the whole calculation to be completed in less than 24

Fortunately, linear systems originating in finite differences or finite elements have one redeeming grace: they are *sparse*. In other words, each variable is coupled to just a small number of other variables (typically, neighbouring grid points or neighbouring vertices) and an overwhelming majority of elements in the matrix vanish. For example, in each row and column of a matrix originating in the five-point formula (7.16) at most four off-diagonal elements are nonzero. This abundance of zeros and the special structure of the matrix allow us to implement Gaussian elimination in a manner that brings systems with 80^2 equations into the realm of microcomputers and allows the sufficiently rapid solution of $7\,168\,000$-variable systems on (admittedly, parallel) supercomputers.

The subject of our attention is the linear system

$$A\boldsymbol{x} = \boldsymbol{b}, \tag{9.1}$$

where the $d \times d$ real matrix A and the vector $\boldsymbol{b} \in \mathbb{R}^d$ are given. We assume that A is nonsingular and *well conditioned*; the latter means, roughly, that A is sufficiently far from being singular that its numerical solution by Gaussian elimination or its variants is always viable and does not require any special techniques such as pivoting (A.1.4.4). Elements of A, $\boldsymbol{x}$ and $\boldsymbol{b}$ will be denoted by $a_{k,\ell}$, x_k and b_ℓ respectively, $k, \ell = 1, 2, \ldots, d$. The size of d will play no further direct role, but it is always important to bear in mind that it motivates the whole discussion.

We say that A is a *banded* matrix of *bandwidth* s if $a_{k,\ell} = 0$ for every $k, \ell \in \{1, 2, \ldots, d\}$ such that $|k - \ell| > s$. Familiar examples are *tridiagonal* $(s = 1)$ and *quindiagonal* $(s = 2)$ matrices.

Recall that, subject to mild restrictions, a $d \times d$ matrix A can be factorized into the form

$$A = LU \tag{9.2}$$

where

$$
L = \begin{bmatrix}
1 & 0 & \cdots & 0 \\
\ell_{2,1} & 1 & \ddots & \vdots \\
\vdots & \ddots & \ddots & 0 \\
\ell_{d,1} & \cdots & \ell_{d,d-1} & 1
\end{bmatrix}
\quad \text{and} \quad
U = \begin{bmatrix}
u_{1,1} & u_{1,2} & \cdots & u_{1,d} \\
0 & u_{2,2} & \ddots & \vdots \\
\vdots & \ddots & \ddots & u_{d-1,d} \\
0 & \cdots & 0 & u_{d,d}
\end{bmatrix}
$$

are lower triangular and upper triangular respectively (Appendix A.1.4.5). In general, it costs $\mathcal{O}(d^3)$ operations to calculate L and U. However, if A has bandwidth s then this can be significantly reduced.

---

hours on a serial computer – a reasonable requirement, e.g. in calculations originating in weather prediction – the average distance between the CPU and every memory cell should be roughly 10^{-7} millimetres, barely twice the radius of a hydrogen atom.

This, needless to say, is in the realm of fantasy. Even the bravest souls in the miniaturisation business dare not contemplate realistic computers of this size and, anyway, quantum effects are bound to make an atomic-sized computer an uncertain (in Heisenberg's sense) proposition.

To demonstrate that this is indeed the case (and, incidentally, to measure exactly the extent of the savings) we assume that $a_{1,1} \neq 0$, and we let

$$\ell_1 := \begin{bmatrix} 1 \\ a_{2,1}/a_{1,1} \\ a_{3,1}/a_{1,1} \\ \vdots \\ a_{d,1}/a_{1,1} \end{bmatrix}, \qquad \boldsymbol{u}^T = \begin{bmatrix} a_{1,1} & a_{1,2} & \cdots & a_{1,d} \end{bmatrix}$$

and set $\tilde{A} := A - \boldsymbol{\ell u}^T$. Regardless of the bandwidth of A, the matrix $\tilde{A}$ has zeros along its first row and column and we find

$$L = \begin{bmatrix} \boldsymbol{\ell} & \begin{matrix} \boldsymbol{0}^T \\ \hat{L} \end{matrix} \end{bmatrix}, \qquad U = \begin{bmatrix} \begin{matrix} \boldsymbol{u}^T \\ \boldsymbol{0} & \hat{U} \end{matrix} \end{bmatrix},$$

where $\hat{A} = \hat{L}\hat{U}$, the matrix $\hat{A}$ having been obtained from $\tilde{A}$ by deleting the first row and column. Setting the first column of L and the first row of U to $\boldsymbol{\ell}$ and $\boldsymbol{u}^T$ respectively, we therefore reduce the problem of LU-factorizing the $d \times d$ matrix A to that of an LU factorization of the $(d-1) \times (d-1)$ matrix $\hat{A}$. For a general matrix A it costs $\mathcal{O}(d^2)$ operations to evaluate $\hat{A}$, but the operation count is smaller if A is of bandwidth s. Since just $(s+1)$ top components of $\boldsymbol{\ell}$ and $(s+1)$ leftward components of $\boldsymbol{u}^T$ are nonzero, we need to form just the top $(s+1) \times (s+1)$ minor of $\boldsymbol{\ell u}^T$. In other words, $\mathcal{O}(d^2)$ is replaced by $\mathcal{O}((s+1)^2)$.

Continuing by induction we obtain progressively smaller matrices and, after $d-1$ such steps, derive an *operation count* $\mathcal{O}((s+1)^2 d)$ for the LU factorization of a banded matrix. We assume, of course, that the *pivots* $a_{1,1}, \hat{a}_{1,1}, \ldots$ never vanish, otherwise the above procedure could not be completed, but mention in passing that substantial savings accrue even when there is a need for pivoting (cf. Exercise 9.1).

The matrices L and U share the bandwidth of A. This is a very important observation, since a common mistake is to regard the difficulty of solving (9.1) with very large A as being associated solely with the number of operations. *Storage* plays a crucial role as well! In place of d^2 storage 'units' required for a dense $d \times d$ matrix, a banded matrix requires just $\approx (2s+1)d$. Provided that $s \ll d$, this often makes as much difference as the reduction in the operation count. Since L and U also have bandwidth s and we obviously have no need to store known zeros, or for that matter known ones along the diagonal of L, we can reuse computer memory that has been devoted to the storing of A (in a sparse representation!) to store L and U instead.

Having obtained the LU factorization, we can solve (9.1) with relative ease by solving first $L\boldsymbol{y} = \boldsymbol{b}$ and then $U\boldsymbol{x} = \boldsymbol{y}$. On the face of it, this sounds like yet another mathematical nonsense – instead of solving one $d \times d$ linear system, we solve two! However, since both L and U are banded, this can be done considerably more cheaply. In general, for a dense matrix A, the operation count is $\mathcal{O}(d^2)$, but this can be substantially reduced for banded matrices. Writing $L\boldsymbol{y} = \boldsymbol{b}$ in a form that pays heed to sparsity, we have

$$\begin{aligned} y_1 &= b_1, \\ \ell_{2,1} y_1 + y_2 &= b_2, \end{aligned}$$

$$\ell_{3,1}y_1 + \ell_{3,2}y_2 + y_3 = b_3,$$

$$\vdots$$

$$\ell_{1,s}y_1 + \cdots + \ell_{s-1,s}y_{s-1} + y_s = b_s$$

$$\ell_{k-s,k}y_{k-s} + \cdots + \ell_{k-1,k}y_{k-1} + y_k = b_k, \qquad k = s+1, s+2, \ldots, d,$$

and hence $\mathcal{O}(sd)$ operations. A similar argument applies to $U\boldsymbol{x} = \boldsymbol{y}$.

Let us count the blessings of bandedness. Firstly, LU factorization 'costs' $\mathcal{O}(s^2 d)$ operations, rather than $\mathcal{O}(d^3)$. Secondly, the storage requirement is $\mathcal{O}((2s+1)d)$, compared with d^2 for a dense matrix. Finally, provided that we have already derived the factorization, the solution of (9.1) entails just $\mathcal{O}(sd)$ operations in place of $\mathcal{O}(d^2)$.

◇ **A few examples of banded matrices**   The savings due to the exploitation of bandedness are at their most striking in the case of tridiagonal matrices. Then we need just $\mathcal{O}(d)$ operations for the factorization and a similar number for the solution of triangular systems, and just $4d - 2$ real numbers (inclusive of the vector $\boldsymbol{x}$) need be stored at any one time. The implementation of banded LU factorization in the case $s = 1$ is sometimes known as the *Thomas algorithm*.

A more interesting case is presented by the five-point equations (7.16) in a square. To present them in the form (9.1) we need to arrange, at least formally, the two-dimensional $m \times m$ grid from Figure 7.4 into a vector in $\mathbb{R}^d$, $d = m^2$. Although there are many distinct ways of doing it (to be exact, $d!$; can you prove this?), the most obvious is simply to append the columns of an array to each other, starting from the leftmost. Appropriately, this is known as *natural ordering*,

$$\boldsymbol{x} = \begin{bmatrix} u_{1,1} \\ u_{2,1} \\ \vdots \\ u_{m,1} \\ u_{1,2} \\ \vdots \\ u_{m,2} \\ \vdots \end{bmatrix}, \qquad \tilde{\boldsymbol{b}} = (\Delta x)^2 \begin{bmatrix} f_{1,1} \\ f_{2,1} \\ \vdots \\ f_{m,1} \\ f_{1,2} \\ \vdots \\ f_{m,2} \\ \vdots \end{bmatrix},$$

and the vector $\boldsymbol{b}$ is composed of $\tilde{\boldsymbol{b}}$ and the boundary points' contribution:

$$b_1 = \tilde{b}_1 - [u(\Delta x, 0) + u(0, \Delta x)],$$
$$b_2 = \tilde{b}_2 - u(2\Delta x, 0),$$

$$\vdots$$

$$b_{m+1} = \tilde{b}_{m+1} - u(0, 2\Delta x),$$
$$b_{m+2} = \tilde{b}_{m+2},$$

$$\vdots$$

It is convenient to represent the matrix A as composed of m blocks, each of size $m \times m$. For example, $m = 4$ results in the 16×16 matrix

$$
\left[\begin{array}{cccc|cccc|cccc|cccc}
-4 & 1 & 0 & 0 & 1 & 0 & 0 & 0 & 0 & 0 & 0 & 0 & 0 & 0 & 0 & 0 \\
1 & -4 & 1 & 0 & 0 & 1 & 0 & 0 & 0 & 0 & 0 & 0 & 0 & 0 & 0 & 0 \\
0 & 1 & -4 & 1 & 0 & 0 & 1 & 0 & 0 & 0 & 0 & 0 & 0 & 0 & 0 & 0 \\
0 & 0 & 1 & -4 & 0 & 0 & 0 & 1 & 0 & 0 & 0 & 0 & 0 & 0 & 0 & 0 \\
\hline
1 & 0 & 0 & 0 & -4 & 1 & 0 & 0 & 1 & 0 & 0 & 0 & 0 & 0 & 0 & 0 \\
0 & 1 & 0 & 0 & 1 & -4 & 1 & 0 & 0 & 1 & 0 & 0 & 0 & 0 & 0 & 0 \\
0 & 0 & 1 & 0 & 0 & 1 & -4 & 1 & 0 & 0 & 1 & 0 & 0 & 0 & 0 & 0 \\
0 & 0 & 0 & 1 & 0 & 0 & 1 & -4 & 0 & 0 & 0 & 1 & 0 & 0 & 0 & 0 \\
\hline
0 & 0 & 0 & 0 & 1 & 0 & 0 & 0 & -4 & 1 & 0 & 0 & 1 & 0 & 0 & 0 \\
0 & 0 & 0 & 0 & 0 & 1 & 0 & 0 & 1 & -4 & 1 & 0 & 0 & 1 & 0 & 0 \\
0 & 0 & 0 & 0 & 0 & 0 & 1 & 0 & 0 & 1 & -4 & 1 & 0 & 0 & 1 & 0 \\
0 & 0 & 0 & 0 & 0 & 0 & 0 & 1 & 0 & 0 & 1 & -4 & 0 & 0 & 0 & 1 \\
\hline
0 & 0 & 0 & 0 & 0 & 0 & 0 & 0 & 1 & 0 & 0 & 0 & -4 & 1 & 0 & 0 \\
0 & 0 & 0 & 0 & 0 & 0 & 0 & 0 & 0 & 1 & 0 & 0 & 1 & -4 & 1 & 0 \\
0 & 0 & 0 & 0 & 0 & 0 & 0 & 0 & 0 & 0 & 1 & 0 & 0 & 1 & -4 & 1 \\
0 & 0 & 0 & 0 & 0 & 0 & 0 & 0 & 0 & 0 & 0 & 1 & 0 & 0 & 1 & -4
\end{array}\right].
$$
$$(9.3)$$

In general, it is easy to see that A has a bandwidth $s = m$. In other words, we need just $\mathcal{O}(m^4)$ operations to LU-factorize it – a large number, yet significantly smaller than $\mathcal{O}(m^6)$, the operation count for 'dense' LU factorization.

Note that a matrix might possess a large number of zeros inside the band; (9.3) is a case in point. These zeros will, in all likelihood, be destroyed (or 'filled in') throughout LU factorization. The banded algorithm guarantees only that zeros *outside* the band are retained! $\diamond$

Whenever a matrix originates in a one-dimensional arrangement of a multivariate grid, the exact nature of the ordering is likely to have a bearing on the bandwidth. Indeed, the secret of efficient implementation of banded LU factorization for matrices that originate in a planar (or higher-dimensional) grid is in finding a good arrangement of grid points. An example of a bad arrangement is provided by *red–black* ordering, which, as far as the matrix (9.3) is concerned, is

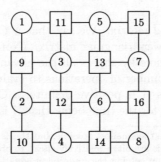

In other words, the grid is viewed as a chequerboard and black squares are selected

before the red ones. The outcome,

$$
\begin{bmatrix}
\times & \circ & \circ & \circ & \circ & \circ & \circ & \circ & \times & \circ & \times & \circ & \circ & \circ & \circ & \circ \\
\circ & \times & \circ & \circ & \circ & \circ & \circ & \circ & \circ & \times & \times & \circ & \times & \circ & \circ & \circ \\
\circ & \circ & \times & \circ & \circ & \circ & \circ & \circ & \times & \circ & \times & \times & \times & \circ & \circ & \circ \\
\circ & \circ & \circ & \times & \circ & \circ & \circ & \circ & \circ & \times & \circ & \times & \circ & \times & \circ & \circ \\
\circ & \circ & \circ & \circ & \times & \circ & \circ & \circ & \circ & \circ & \times & \circ & \times & \circ & \times & \circ \\
\circ & \circ & \circ & \circ & \circ & \times & \circ & \circ & \circ & \circ & \circ & \times & \times & \times & \circ & \times \\
\circ & \circ & \circ & \circ & \circ & \circ & \times & \circ & \circ & \circ & \circ & \circ & \times & \circ & \times & \times \\
\circ & \circ & \circ & \circ & \circ & \circ & \circ & \times & \circ & \circ & \circ & \circ & \circ & \times & \circ & \times \\
\times & \times & \times & \circ & \circ & \circ & \circ & \circ & \times & \circ & \circ & \circ & \circ & \circ & \circ & \circ \\
\circ & \times & \circ & \times & \circ & \circ & \circ & \circ & \circ & \times & \circ & \circ & \circ & \circ & \circ & \circ \\
\times & \circ & \times & \circ & \times & \circ & \circ & \circ & \circ & \circ & \times & \circ & \circ & \circ & \circ & \circ \\
\circ & \times & \times & \times & \circ & \times & \circ & \circ & \circ & \circ & \circ & \times & \circ & \circ & \circ & \circ \\
\circ & \circ & \times & \circ & \times & \times & \times & \circ & \circ & \circ & \circ & \circ & \times & \circ & \circ & \circ \\
\circ & \circ & \circ & \times & \circ & \times & \circ & \times & \circ & \circ & \circ & \circ & \circ & \times & \circ & \circ \\
\circ & \circ & \circ & \circ & \times & \circ & \times & \circ & \circ & \circ & \circ & \circ & \circ & \circ & \times & \circ \\
\circ & \circ & \circ & \circ & \circ & \times & \times & \times & \circ & \circ & \circ & \circ & \circ & \circ & \circ & \times
\end{bmatrix}
\qquad (9.4)
$$

is of bandwidth $s = 10$ and is quite useless as far as sparse LU factorization is concerned. (Here and in the sequel we adopt the notation whereby '$\times$' and '$\circ$' stand for nonzero and zero components respectively; after all, the exact numerical values of these components have no bearing on the underlying problem.) We mention in passing that, although the red–black ordering is an exceedingly poor idea if the goal is to minimize the bandwith, it has certain virtues in the context of iterative methods (cf. Chapter 10).

In many situations it is relatively easy to find a configuration that results in a small bandwidth, but occasionally this might present quite a formidable problem. This is in particular the case with linear equations that originate from tessalations of two-dimensional sets into irregular triangles. There exist combinatorial algorithms that help to arrange arbitrary sets of equations into matrices having a 'good' bandwidth,[3] but they are outside the scope of this exposition.

9.2 Graphs of matrices and perfect Cholesky factorization

Let A be a symmetric $d \times d$ matrix. We say that a *Cholesky factorization* of A is LL^T, where L is a $d \times d$ lower-triangular matrix. (Note that the diagonal elements of L need not be ones.) Cholesky shares all the advantages of an LU factorization but requires only half the number of operations to evaluate and half the memory to store. Moreover, as long as A is positive definite, a Cholesky factorization always exists (A.1.4.6). We assume for the time being that A is indeed positive definite.

---

[3] A 'good' bandwidth is seldom the smallest possible, but this is frequently the case with combinatorial algorithms. 'Good' solutions are often relatively cheap to obtain, but finding the best solution is often much more expensive than solving the underlying equations with even the most inefficient ordering.

It follows at once from the proof of Theorem 7.4 that every matrix A obtained from the five-point formula and reasonable boundary schemes is negative definite. A similar statement is true in regard to matrices that arise when the finite element method is applied to the Poisson equation, provided that piecewise linear basis functions are used and that the geometry is sufficiently simple. Since we can solve $-A\boldsymbol{x} = -\boldsymbol{b}$ in place of $A\boldsymbol{x} = \boldsymbol{b}$, the stipulation of positive definiteness is not as contrived as it might perhaps seem at first glance.

We have already seen in Section 9.1 that the secret of good LU (and, for that matter, Cholesky) factorization is in the ordering of the equations and variables. In the case of a grid, this corresponds to ordering the grid points, but remember from the discussion in Chapter 7 that each such point corresponds to an equation and to a variable! Given a matrix A, we wish to find an ordering of equations and an ordering of variables so that the outcome is amenable to efficient factorization. Any rearrangement of equations (hence, of the rows of A) is equivalent to the product PA, where P is a $d \times d$ *permutation matrix* (A.1.2.5). Likewise, relabelling the variables is tantamount to rearranging the columns of A, hence to the product AQ, where Q is a permutation matrix. (Bearing in mind the purpose of the whole exercise, namely the solution of the linear system (9.1), we need to replace $\boldsymbol{b}$ by $P\boldsymbol{b}$ and $\boldsymbol{x}$ by $Q\boldsymbol{x}$; cf. Exercise 9.5.) The outcome, PAQ, retains symmetry and positive definiteness if $Q = P^T$, hence we herewith assume that rows and columns are always reordered in unison. In practical terms, it means that if the (k,ℓ)th grid point corresponds to the jth equation, the variable $u_{k,\ell}$ becomes the jth unknown.

The matrix A is sparse and the purpose of a good Cholesky factorization is to retain as many zeros as possible. More formally, in any particular (symmetric) ordering of equations, we say that the *fill-in* is the number of pairs (i,j), $1 \le j < i \le d$, such that $a_{i,j} = 0$ and $\ell_{i,j} \neq 0$. Our goal is to devise an ordering that minimizes fill-in. In particular, we say that a Cholesky factorization of a specific matrix A is *perfect* if there exists an ordering that yields zero fill-in: every zero is retained.

An example of a perfect factorization is provided by a banded symmetric matrix, where we assume that no components vanish within the band. This is clear from the discussion in Section 9.1, which generalizes at once to a Cholesky factorization. Another example, at the other extreme, is a completely dense matrix (which, of course, is banded with a bandwidth of $s = d - 1$, but this abuses the spirit, if not the letter, of the definition of bandedness).

A convenient way of analysing the sparsity patterns of symmetric matrices is afforded by *graph theory*. We have already mentioned graphs in a less formal setting, in the discussion at the end of Chapter 3.

Formally, a *graph* is the set $\mathbb{G} = \{\mathbb{V}, \mathbb{E}\}$, where $\mathbb{V} = \{1, 2, \ldots, d\}$ and $\mathbb{E} \subseteq \mathbb{V}^2$ consists of pairs of the form (i,j), $i < j$. The elements of $\mathbb{V}$ and $\mathbb{E}$ are said to be the *vertices* and *edges* of $\mathbb{G}$, respectively.

We say that $\mathbb{G}$ is the graph of a symmetric $d \times d$ matrix A if $(i,j) \in \mathbb{E}$ if and only if $a_{i,j} \neq 0$, $1 \le i < j \le d$. In other words, $\mathbb{G}$ displays the *sparsity pattern* of A, which we have presented in (9.4) using the symbols 'o' and '×'.

Although a graph can be represented by listing all the edges one by one, it is considerably more convenient to illustrate it pictorially in a self-explanatory manner.

We now give a few examples of matrices (represented by their sparsity pattern) and their graphs.

tridiagonal:
$$
\begin{bmatrix}
\times & \times & \circ & \circ & \circ & \circ \\
\times & \times & \times & \circ & \circ & \circ \\
\circ & \times & \times & \times & \circ & \circ \\
\circ & \circ & \times & \times & \times & \circ \\
\circ & \circ & \circ & \times & \times & \times \\
\circ & \circ & \circ & \circ & \times & \times
\end{bmatrix}
\Rightarrow
$$

(1)—(2)—(3)—(4)—(5)—(6)

quindiagonal:
$$
\begin{bmatrix}
\times & \times & \times & \circ & \circ & \circ \\
\times & \times & \times & \times & \circ & \circ \\
\times & \times & \times & \times & \times & \circ \\
\circ & \times & \times & \times & \times & \times \\
\circ & \circ & \times & \times & \times & \times \\
\circ & \circ & \circ & \times & \times & \times
\end{bmatrix}
\Rightarrow
$$

(1)—(2)—(3)—(4)—(5)—(6)

arrowhead:
$$
\begin{bmatrix}
\times & \times & \times & \times & \times & \times \\
\times & \times & \circ & \circ & \circ & \circ \\
\times & \circ & \times & \circ & \circ & \circ \\
\times & \circ & \circ & \times & \circ & \circ \\
\times & \circ & \circ & \circ & \times & \circ \\
\times & \circ & \circ & \circ & \circ & \times
\end{bmatrix}
\Rightarrow
$$

(2) (3) (4) (5) (6) all connected to (1)

cyclic:
$$
\begin{bmatrix}
\times & \times & \circ & \circ & \circ & \times \\
\times & \times & \times & \circ & \circ & \circ \\
\circ & \times & \times & \times & \circ & \circ \\
\circ & \circ & \times & \times & \times & \circ \\
\circ & \circ & \circ & \times & \times & \times \\
\times & \circ & \circ & \circ & \times & \times
\end{bmatrix}
\Rightarrow
$$

(1)—(2)—(3)—(4)—(5)—(6)—(1)

The graph of a matrix often reveals at a single glance its structure, which might not be evident from the sparsity pattern. Thus, consider the matrix

$$
\begin{bmatrix}
\times & \circ & \times & \circ & \times & \circ \\
\circ & \times & \circ & \times & \times & \circ \\
\times & \circ & \times & \circ & \circ & \times \\
\circ & \times & \circ & \times & \circ & \times \\
\times & \times & \circ & \circ & \times & \circ \\
\circ & \circ & \times & \times & \circ & \times
\end{bmatrix}.
$$

At a first glance, there is nothing to link it to any of the four matrices that we have just displayed, but its graph, tells a different story – it is nothing other than the cyclic matrix in disguise! To notice this, just relabel vertices as follows

$$1 \to 1, \quad 2 \to 5, \quad 3 \to 2, \quad 4 \to 4, \quad 5 \to 6, \quad 6 \to 3.$$

This, of course, is equivalent to reordering (simultaneously) equations and variables.

An ordered set of edges $\{(i_k, j_k)\}_{k=1}^{\nu} \subseteq \mathbb{E}$ is called a *path* joining the vertices α and β if $\alpha \in \{i_1, j_1\}$, $\beta \in \{i_\nu, j_\nu\}$ and for every $k = 1, 2, \ldots, \nu - 1$ the set $\{i_k, j_k\} \cap \{i_{k+1}, j_{k+1}\}$ contains exactly one member. It is a *simple path* if it does not visit any vertex more than once. We say that $\mathbb{G}$ is a *tree* if each two members of $\mathbb{V}$ are joined by a unique simple path. Both tridiagonal and arrowhead matrices correspond to trees, but this is not the case with either quindiagonal or cyclic matrices when $\nu \geq 3$.

Given a tree $\mathbb{G}$ and an arbitrary vertex $r \in \mathbb{V}$, the pair $\mathbb{T} = \langle \mathbb{G}, r \rangle$ is called a *rooted tree*, while r is said to be the *root*. Unlike an ordinary graph, $\mathbb{T}$ admits a natural partial ordering, which can best be explained by an analogy with a family tree. Thus, the root r is the *predecessor* of all the vertices in $\mathbb{V} \setminus \{r\}$ and these vertices are *successors* of r. Moreover, every $\alpha \in \mathbb{V} \setminus \{r\}$ is joined to v by a simple path and we designate each vertex along this path, except for r and α, as a *predecessor* of α and a *successor* of r. We say that the rooted tree $\mathbb{T}$ is *monotonically ordered* if each vertex is labelled before all its predecessors; in other words, we label the vertices from the top of the tree to the root. (As we have already said it, relabelling a graph is tantamount to permuting the rows and the columns of the underlying matrix.)

Every rooted tree can be monotonically ordered and, in general, such an ordering is not unique. We now give three monotone orderings of the same rooted tree:

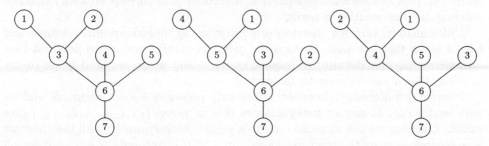

Theorem 9.1   *Let A be a symmetric matrix whose graph $\mathbb{G}$ is a tree. Choose a root $r \in \{1, 2, \ldots, d\}$ and assume that the rows and columns of A have been arranged so that $\mathbb{T} = \langle \mathbb{G}, r \rangle$ is monotonically ordered. Given that $A = LL^T$ is a Cholesky factorization, it is true that*

$$\ell_{k,j} = \frac{a_{k,j}}{\ell_{j,j}}, \qquad k = j+1, j+2, \ldots, d, \quad j = 1, 2, \ldots, d-1. \tag{9.5}$$

Therefore $\ell_{k,j} = 0$ whenever $a_{k,j} = 0$ and the matrix A can be Cholesky-factorized perfectly.

Proof   The coefficients of L can be written down explicitly (A.1.4.5). In particular,

$$\ell_{k,j} = \frac{1}{\ell_{j,j}} \left(a_{k,j} - \sum_{i=1}^{j-1} \ell_{k,i} \ell_{j,i} \right), \qquad k = j+1, j+2, \ldots, d, \quad j = 1, 2, \ldots, d-1. \tag{9.6}$$

It follows at once that the statement of the theorem is true with regard to the first column, since (9.6) yields $\ell_{k,1} = a_{k,1}/\ell_{1,1}$, $k = 2, 3, \ldots, d$. We continue by induction on j. Suppose thus that the theorem is true for $j = 1, 2, \ldots, q-1$, where $q \in \{2, 3, \ldots, d-1\}$. The rooted tree $\mathbb{T}$ is monotonically ordered, and this means that for every $i = 1, 2, \ldots, d-1$ there exists a *unique* vertex $\gamma_i \in \{i+1, i+2, \ldots, d\}$ such that $(i, \gamma_i) \in \mathbb{E}$.

Now, choose any $k \in \{q+1, q+2, \ldots, d\}$. If $\ell_{k,i} \neq 0$ for some $i \in \{1, 2, \ldots, q-1\}$ then, by the induction assumption, $a_{k,i} \neq 0$ also. This implies $(i, k) \in \mathbb{E}$, hence $k = \gamma_i$. We deduce that $q \neq \gamma_i$, therefore $(i, q) \notin \mathbb{E}$. Consequently $a_{i,q} = 0$ and, exploiting again the induction assumption, $\ell_{i,q} = 0$.

By an identical argument, if $\ell_{q,i} \neq 0$ for some $i \in \{1, 2, \ldots, q-1\}$ then $q = \gamma_i$, hence $k \neq \gamma_i$ for all $k = q+1, q+2, \ldots, d$, and this implies in turn that $\ell_{k,i} = 0$.

We let $j = q$ in (9.6). Since, as we have just proved, $\ell_{k,i}\ell_{q,i} = 0$ for all $i = 1, 2, \ldots, q-1$ and $k = i+1, i+2, \ldots, d$, the sum in (9.6) vanishes and we deduce that $\ell_{k,q} = a_{k,q}/\ell_{q,q}$, $k = q+1, q+2, \ldots, d$. An inductive proof of the theorem is thus complete. $\blacksquare$

An important observation is that the expense of Cholesky factorization of a matrix consistently with the conditions of Theorem 9.1 is proportional to the number of nonzero elements under the main diagonal. This is certainly true as far as $\ell_{k,j}$, $1 \leq j < k \leq d$, is concerned and it is possible to verify (see Exercise 9.8) that this is also the case for the calculation of $\ell_{1,1}, \ell_{2,2}, \ldots, \ell_{d,d}$.

Monotone ordering can lead to spectacular savings and a striking example is the arrowhead matrix. If we factorized it in a naive fashion we could easily cause total fill-in, but, provided we rearrange the rows and columns to correspond with monotone ordering, the factorization is perfect.

Unfortunately, very few matrices of interest are symmetric, positive definite and have a graph that is a tree. Perhaps the only truly useful example is provided by a (symmetric, positive definite) tridiagonal matrix – and we do not need graph theory to tell us that it can be perfectly factorized!

Positive definiteness is, however, not strictly necessary for our argument and we have used it only as an cast-iron guarantee that no pivots $\ell_{1,1}, \ell_{2,2}, \ldots, \ell_{d-1,d-1}$ ever vanish. Likewise, we can dispense – up to a point – with symmetry. All that matters is a symmetric sparsity structure, namely $a_{k,j} \neq 0$ if and only if $a_{j,k} \neq 0$ for all $d = 1, 2, \ldots, d$. We have LU factorization in place of Cholesky factorization, but a generalization of Theorem 9.1 presents no insurmountable difficulties.

The one truly restrictive assumption is that the graph is a tree and this renders Theorem 9.1 of little immediate interest in applications. There are three reasons why we nevertheless attend to it. Firstly, it provides the flavour of considerably more substantive results on matrices and graphs. Secondly, it is easy to generalize the theorem to *partitioned trees,* graphs where we have a tree-like structure, provided that instead of vertices we allow subsets of $\mathbb{V}$. An example illustrating this concept and its application to perfect factorization is presented in the comments at the end of the chapter. Finally, the idea of using graphs to investigate the sparsity structure and factorization of matrices is such a beautiful example of lateral thinking in mathematics that its presentation can surely be justified on purely æsthetic grounds.

We complete this brief review of graph theory and the factorization of sparse matrices by remarking that, of course, there are many matrices whose graphs are not trees, yet which can be perfectly factorized. We have already seen that this is the case with a quindiagonal matrix and a less trivial example will be presented in the comments below. It is possible to characterize all graphs that correspond to matrices with a perfect (Cholesky or LU) factorization, but that requires considerably deeper graph theory.

Comments and bibliography

The solution of sparse algebraic systems is one of the main themes of modern scientific computing. Factorization that exploits sparsity is just one, and not necessarily the preferred, option and we shall examine alternative approaches – specifically, iterative methods and fast Poisson solvers – in Chapters 10–12.

References on sparse factorization (sometimes dubbed *direct solution*, to distinguish it from iterative methods) abound and we refer the reader to Duff *et al.* (1986); George & Liu (1981); Tewarson (1973). Likewise, many textbooks present graph theory and we single out Harary (1969) and Golumbic (1980). The latter includes an advanced treatment of graph-theoretical methods in sparse matrix factorization, including the characterization of all matrices that can be factorized perfectly.

It is natural to improve upon the concept of a banded matrix by allowing the bandwidth to vary. For example, consider the 8×8 matrix

$$
\begin{bmatrix}
\times & \times & \circ & \circ & \circ & \circ & \circ & \circ \\
\times & \times & \times & \times & \circ & \circ & \circ & \circ \\
\circ & \times & \times & \times & \circ & \circ & \circ & \circ \\
\circ & \times & \times & \times & \times & \times & \circ & \circ \\
\circ & \circ & \circ & \times & \times & \times & \circ & \circ \\
\circ & \circ & \circ & \times & \times & \times & \times & \circ \\
\circ & \circ & \circ & \circ & \circ & \times & \times & \times \\
\circ & \circ & \circ & \circ & \circ & \circ & \times & \times
\end{bmatrix}
$$

The portion of the matrix enclosed between the solid lines is called the *envelope*. It is easy to demonstrate that LU factorization can be performed in such a manner that fill-in cannot occur outside the envelope; cf. Exercise 9.2. It is evident even in this simple example that 'envelope factorization' might result in considerable savings over 'banded factorization'.

Sometimes it is easy to find a good banded structure or a good envelope of a given matrix by inspection – a procedure that might involve the rearrangement of equations and variables. In general, this is a substantially more formidable task, which needs to be accomplished by a combinatorial algorithm. An effective, yet relatively simple, such method is the *reverse Cuthill–McKee* algorithm. We do not dwell further on this theme, which is explained well in George & Liu (1981).

Throughout our discussion of banded and envelope algorithms we have tacitly assumed that the underlying matrix A is symmetric or, at the very least, has a symmetric sparsity structure. As we have already commented, this is eminently sensible whenever we consider, for example, the equations that occur when the Poisson equation is approximated by the five-point formula. However, it is possible to extend much of the theory to arbitrary matrices. A good deal of the extension is trivial. For example, if there are nonzero elements in the set $\{a_{k,j} : k-2 \leq j \leq k+1\}$ then (assuming, as always, that the underlying factorization is well conditioned) $A = LU$, where $\ell_{k,j} = 0$ for $j \leq k - 3$ and $u_{k,j} = 0$ for $j \geq k + 2$. However, the question of how to arrange elements of a nonsymmetric matrix so that it is amenable to this kind of treatment is substantially more formidable. Note that the correspondence between graphs and matrices assumes symmetry. Where the more advanced concepts of graph theory – specifically, directed graphs – have been applied to nonsymmetric sparsity structures, the results so far have met with only modest success.

To appreciate the power of graph theory in revealing the 'connectedness' of a *symmetric* sparse matrix, thereby permitting the intelligent exploitation of sparsity, consider the following example. At first glance, the matrix

$$
\begin{bmatrix}
\times & \circ & \circ & \circ & \circ & \circ & \times & \circ & \circ & \circ & \circ & \circ & \circ \\
\circ & \times & \circ & \circ & \times & \circ & \circ & \circ & \times & \times & \circ & \circ & \times \\
\circ & \circ & \times & \circ & \circ & \times & \times & \circ & \circ & \circ & \times & \circ & \circ \\
\circ & \circ & \circ & \times & \circ & \circ & \circ & \circ & \circ & \times & \circ & \circ & \circ \\
\circ & \times & \circ & \circ & \times & \circ & \circ & \circ & \circ & \times & \circ & \times & \times \\
\circ & \circ & \times & \circ & \circ & \times & \circ & \circ & \circ & \circ & \circ & \circ & \times \\
\times & \circ & \times & \circ & \circ & \circ & \times & \circ & \circ & \circ & \times & \circ & \circ \\
\circ & \circ & \circ & \circ & \circ & \circ & \circ & \times & \circ & \circ & \times & \circ & \circ \\
\circ & \times & \circ & \circ & \circ & \circ & \circ & \circ & \times & \circ & \circ & \circ & \circ \\
\circ & \times & \circ & \times & \circ & \circ & \circ & \circ & \circ & \times & \circ & \circ & \times \\
\circ & \circ & \times & \circ & \circ & \circ & \times & \times & \circ & \circ & \times & \circ & \circ \\
\circ & \circ & \circ & \circ & \times & \circ & \circ & \circ & \circ & \circ & \circ & \times & \circ \\
\circ & \times & \circ & \circ & \times & \times & \circ & \circ & \circ & \times & \circ & \circ & \times
\end{bmatrix}.
$$

might appear as a completely unstructured mishmash of noughts and crosses, but we claim nonetheless that it can be perfectly factorized. This becomes more apparent upon an examination of its graph:

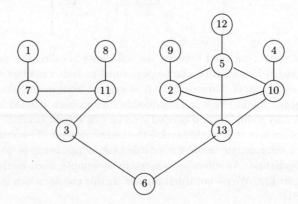

Although this is not a tree, a tree-like structure is apparent. In fact, it is a *partitioned tree*, a 'super-graph' that can be represented as a tree of of 'super-vertices' that are themselves graphs.

The equations and unknowns are ordered so that the graph is traversed from top to bottom. This can be done in a variety of different ways and we herewith choose an ordering that keeps all vertices in each 'super-vertex' together; this is not really necessary but makes the exposition simpler. The outcome of this permutation is displayed in a block form, corresponding to the structure of the partitioned tree, as follows:

	1	4	8	9	12	7	11	3	2	5	10	13	6
1	x	o	o	o	o	x	o	o	o	o	o	o	o
4	o	x	o	o	o	o	o	o	o	o	x	o	o
8	o	o	x	o	o	o	x	o	o	o	o	o	o
9	o	o	o	x	o	o	o	o	x	o	o	o	o
12	o	o	o	o	x	o	o	o	o	x	o	o	o
7	x	o	o	o	o	x	x	x	o	o	o	o	o
11	o	o	x	o	o	x	x	x	o	o	o	o	o
3	o	o	o	o	o	x	x	x	o	o	o	o	x
2	o	o	o	x	o	o	o	o	x	x	x	x	o
5	o	o	o	o	x	o	o	o	x	x	x	x	o
10	o	x	o	o	o	o	o	o	x	x	x	x	o
13	o	o	o	o	o	o	o	o	x	x	x	x	x
6	o	o	o	o	o	o	o	x	o	o	o	x	x

It is now clear how to proceed. Firstly, factorize the vertices $1, 4, 8, 9$ and 12. This obviously causes no fill-in. If you cannot see it at once, consider the equivalent problem of Gaussian elimination. In the first column we need to eliminate just a single nonzero component and we do this by subtracting a multiple of row 1 from row 7. Likewise, we eliminate one component in the first row (remember that we need to maintain symmetry!). Neither operation causes fill-in and we proceed similarly with $4, 8, 9, 12$.

Having factorized the aforementioned five rows and columns, we are left with an 8×8 problem. We next factorize $7, 11, 3$, in this order; again, there is no fill-in. Finally, we factorize $2, 5, 10$ and, then, 13. The outcome is a perfect Cholesky factorization.

The last example, contrived as it might be, provides some insight into one of the most powerful techniques in the factorization of sparse matrices, the method of *partitioned trees*. Given a sparse matrix A with a graph $\mathbb{G} = \{\mathbb{V}, \mathbb{E}\}$, we partition the set of vertices

$$\mathbb{V} = \bigcup_{i=1}^{s} \mathbb{V}_i \quad \text{such that} \quad \mathbb{V}_i \cap \mathbb{V}_j = \emptyset \quad \text{for every} \quad i, j = 1, 2, \ldots, s, \quad i \neq j.$$

Letting $\tilde{\mathbb{V}} := \{1, 2, \ldots, s\}$, we construct the set $\tilde{\mathbb{E}} \subseteq \tilde{\mathbb{V}} \times \tilde{\mathbb{V}}$ by assigning to it every pair (i, j) for which $i < j$ and there exist $\alpha \in \mathbb{V}_i$ and $\beta \in \mathbb{V}_j$ such that either $(\alpha, \beta) \in \mathbb{E}$ or $(\beta, \alpha) \in \mathbb{E}$. The set $\tilde{\mathbb{G}} := \{\tilde{\mathbb{V}}, \tilde{\mathbb{E}}\}$ is itself a graph. Suppose that we have partitioned $\mathbb{V}$ so that $\tilde{\mathbb{G}}$ is a tree. In that case we know from Theorem 9.1 that, by selecting a root in $\tilde{\mathbb{V}}$ and imposing monotone ordering on the partitioned tree, we can factorize the matrix without any fill-in taking place *between* partitioned vertices. There might well be fill-in inside each set $\mathbb{V}_i$, $i = 1, 2, \ldots, s$. However, provided that these sets are either small (as in the extreme case of $s = d$, when each $\mathbb{V}_i$ is a singleton) or fairly dense (in our example, all the sets $\mathbb{V}_i$ are completely dense), the fill-in is likely to be modest.

Sometimes it is possible to find a good partitioned tree structure for a graph just by inspecting it, but there exist algorithms that produce good partitions automatically. Such methods are extremely unlikely to produce the best possible partition (in the sense of minimizing the fill-in), but it should be clear by now that, when it comes to combinatorial algorithms, the best is often the mortal enemy of the good.

We conclude these remarks with few sobering thoughts. Most numerical analysts regard direct factorization methods as *passé* and inferior to iterative algorithms. Bearing in mind the power of modern iterative schemes for linear equations – multigrid, preconditioned conjugate gradients, generalized minimal residuals (GMRes)... – this is probably a sensible approach. However, direct factorization has its place, not just in the obvious instances such as banded matrices; it can also, as we will note in Chapter 10, join forces with the (iterative) method of conjugate gradients to produce one of the most effective solvers of linear algebraic systems.

Duff, I.S., Erisman, A.M. and Reid, J.K. (1986), *Direct Methods for Sparse Matrices*, Oxford University Press, Oxford.

George, A. and Liu, J.W.-H. (1981), *Computer Solution of Large Sparse Positive Definite Systems*, Prentice–Hall, Englewood Cliffs, NJ.

Golumbic, M.C. (1980), *Algorithmic Graph Theory and Perfect Graphs*, Academic Press, New York.

Harary, F. (1969), *Graph Theory*, Addison–Wesley, Reading, MA.

Tewarson, R.P. (1973), *Sparse Matrices*, Academic Press, New York.

Exercises

9.1   Let A be a $d \times d$ nonsingular matrix with bandwidth $s \geq 1$ and suppose that Gaussian elimination with *column pivoting* is used to solve the system $A\boldsymbol{x} = \boldsymbol{b}$. This means that, before eliminating all nonzero compenents under the main diagonal in the jth column, where $j \in \{1, 2, \ldots, d-1\}$, we first find $|a_{k_j,j}| := \max_{i=j,j+1,\ldots,d} |a_{i,j}|$ and next exchange the jth and the k_jth rows of A, as well as the corresponding components of $\boldsymbol{b}$ (cf. A.1.4.4).

  a Identify all the coefficients of the intermediate equations that might be filled in during this procedure.

  b Prove that the operation count of LU factorization with column pivoting is $\mathcal{O}(s^2 d)$.

9.2   Let A be a $d \times d$ symmetric positive definite matrix and for every $j = 1, 2, \ldots, d-1$ define $k_j \in \{1, 2, \ldots, j\}$ as the least integer such that $a_{k_j,j} \neq 0$. We may assume without loss of generality that $k_1 \leq k_2 \leq \cdots \leq k_{d-1}$, since otherwise A can be brought into this form by row and column exchanges.

  a Prove that the number of operations required to find a Cholesky factorization of A is

$$\mathcal{O}\left(\sum_{j=1}^{d-1} (j - k_j + 1) \right).$$

  b Demonstrate that the result of an operation count for a Cholesky factorization of banded matrices is a special case of this formula.

9.3   Find the bandwidth of the $m^2 \times m^2$ matrix that is obtained from the nine-point equations (7.28) with natural ordering.

9.4 Suppose that a square is triangulated in the following fashion.

The Poisson equation in a square is discretized using the FEM and piecewise linear functions with this triangulation. Observe that every vertex can be identified with an unknown in the linear system (8.7).

a Arranging equations in an $m \times m$ grid in natural ordering, find the bandwidth of the $m^2 \times m^2$ matrix.

b What is the graph of this matrix?

9.5 Let P and Q be two $d \times d$ *permutation matrices* (A.1.2.5) and let $\tilde{A} := PAQ$, where the $d \times d$ matrix A is nonsingular. Prove that if $\tilde{x} \in \mathbb{R}^d$ is the solution of $\tilde{A}\tilde{x} = Pb$ then $x = Q\tilde{x}$ solves (9.1).

9.6 Construct the graphs of the matrices with the following sparsity patterns:

a
$$
\begin{bmatrix}
\times & \circ & \circ & \times & \circ & \circ & \circ & \circ \\
\circ & \times & \times & \circ & \times & \times & \times & \circ \\
\circ & \circ & \times & \circ & \times & \circ & \circ & \times \\
\times & \circ & \circ & \times & \circ & \circ & \times & \circ \\
\circ & \times & \times & \circ & \times & \circ & \circ & \circ \\
\circ & \times & \circ & \circ & \circ & \times & \circ & \times \\
\circ & \times & \circ & \times & \circ & \circ & \times & \circ \\
\circ & \circ & \times & \circ & \circ & \times & \circ & \times
\end{bmatrix} ;
$$

b
$$
\begin{bmatrix}
\times & \circ & \circ & \circ & \times & \times & \circ & \circ \\
\circ & \times & \circ & \circ & \circ & \circ & \circ & \times \\
\circ & \circ & \times & \times & \times & \circ & \times & \circ \\
\circ & \circ & \times & \times & \circ & \circ & \circ & \circ \\
\times & \circ & \times & \circ & \times & \circ & \circ & \times \\
\times & \circ & \circ & \circ & \circ & \times & \circ & \circ \\
\circ & \circ & \times & \circ & \circ & \circ & \times & \circ \\
\circ & \times & \circ & \circ & \times & \circ & \circ & \times
\end{bmatrix} ;
$$

c
$$
\begin{bmatrix}
\times & \circ & \circ & \circ & \times & \circ & \circ & \circ \\
\circ & \times & \circ & \times & \times & \times & \circ & \times \\
\circ & \circ & \times & \circ & \times & \circ & \circ & \circ \\
\circ & \times & \circ & \times & \circ & \circ & \circ & \circ \\
\times & \times & \times & \circ & \times & \circ & \times & \circ \\
\circ & \times & \circ & \circ & \circ & \times & \circ & \circ \\
\circ & \circ & \circ & \circ & \times & \circ & \times & \circ \\
\circ & \times & \circ & \circ & \circ & \circ & \circ & \times
\end{bmatrix} ;
$$

d
$$
\begin{bmatrix}
\times & \circ & \times & \times & \circ & \circ & \circ & \circ \\
\circ & \times & \circ & \circ & \times & \times & \times & \times \\
\times & \circ & \times & \circ & \circ & \circ & \times & \circ \\
\times & \circ & \circ & \times & \circ & \circ & \circ & \times \\
\circ & \times & \circ & \times & \times & \circ & \circ & \circ \\
\circ & \times & \circ & \circ & \circ & \times & \circ & \circ \\
\circ & \times & \times & \circ & \circ & \circ & \times & \circ \\
\circ & \times & \circ & \times & \circ & \circ & \circ & \times
\end{bmatrix} .
$$

Identify all trees. Suggest a monotone ordering for each tree.

9.7 We say that a graph $\mathbb{G} = \{\mathbb{V}, \mathbb{E}\}$ is *connected* if any two vertices in $\mathbb{V}$ can be joined by a path of edges from $\mathbb{E}$; otherwise it is *disconnected*. Let A be a

$d \times d$ symmetric matrix with graph $\mathbb{G}$. Prove that if $\mathbb{G}$ is disconnected then, after rearrangement of rows and columns, A can be written in the form

$$A = \begin{bmatrix} A_1 & O \\ O & A_2 \end{bmatrix},$$

where A_j is a matrix of size $d_j \times d_j$, $j = 1, 2$, and $d_1 + d_2 = d$.

9.8⋆   Theorem 9.1 states that the number of operations needed to form $\ell_{k,j}$, $1 \leq j < k \leq d$, is proportional to the number of sparse components underneath the diagonal of A. The purpose of this question is to prove a similar statement with regard to the diagonal terms $\ell_{k,k}$, $k = 1, 2, \ldots, d$. To this end one might use the explicit formula

$$\ell_{k,k}^2 = a_{k,k} - \sum_{i=1}^{k-1} \ell_{k,i}^2, \qquad k = 1, 2, \ldots, d.$$

and count the total number of nonzero terms $\ell_{k,i}^2$ in all the sums.

9.9⋆   Prove that a symmetric, positive definite matrix with the sparsity pattern

$$
\begin{bmatrix}
\times & \circ & \circ & \circ & \circ & \times & \circ & \circ & \circ & \circ & \times & \circ & \circ & \circ & \circ & \circ \\
\circ & \times & \circ & \circ & \circ & \circ & \times & \circ & \circ & \circ & \circ & \times & \circ & \circ & \circ & \circ \\
\circ & \circ & \times & \circ & \circ & \circ & \circ & \times & \circ & \circ & \circ & \circ & \times & \circ & \circ & \circ \\
\circ & \circ & \circ & \times & \circ & \circ & \circ & \circ & \times & \circ & \circ & \circ & \circ & \times & \circ & \circ \\
\circ & \circ & \circ & \circ & \times & \circ & \circ & \circ & \circ & \times & \circ & \circ & \circ & \circ & \times & \circ \\
\times & \circ & \circ & \circ & \circ & \times & \circ & \circ & \circ & \circ & \times & \circ & \circ & \circ & \circ & \circ \\
\circ & \times & \circ & \circ & \circ & \circ & \times & \circ & \circ & \circ & \circ & \times & \circ & \circ & \circ & \circ \\
\circ & \circ & \times & \circ & \circ & \circ & \circ & \times & \circ & \circ & \circ & \circ & \times & \circ & \circ & \circ \\
\circ & \circ & \circ & \times & \circ & \circ & \circ & \circ & \times & \circ & \circ & \circ & \circ & \times & \circ & \circ \\
\circ & \circ & \circ & \circ & \times & \circ & \circ & \circ & \circ & \times & \circ & \circ & \circ & \circ & \times & \circ \\
\times & \circ & \circ & \circ & \circ & \times & \circ & \circ & \circ & \circ & \times & \circ & \circ & \circ & \circ & \times \\
\circ & \times & \circ & \circ & \circ & \circ & \times & \circ & \circ & \circ & \circ & \times & \circ & \circ & \circ & \times \\
\circ & \circ & \times & \circ & \circ & \circ & \circ & \times & \circ & \circ & \circ & \circ & \times & \circ & \circ & \times \\
\circ & \circ & \circ & \times & \circ & \circ & \circ & \circ & \times & \circ & \circ & \circ & \circ & \times & \circ & \times \\
\circ & \circ & \circ & \circ & \times & \circ & \circ & \circ & \circ & \times & \circ & \circ & \circ & \circ & \times & \times \\
\circ & \circ & \circ & \circ & \circ & \circ & \circ & \circ & \circ & \circ & \times & \times & \times & \times & \times & \times
\end{bmatrix}
$$

possesses a perfect Cholesky factorization.

10

Iterative methods for sparse linear equations

stationary schemes

solution of the linear system

$$Ax = b, \qquad (10.1)$$

matrix and $b \in \mathbb{R}^d$. The most general iterative $0, 1, \ldots$ and $x^{[0]}, x^{[1]}, \ldots, x^{[k]} \in \mathbb{R}^d$ generates a ls, it is a family of functions $\{h_k\}_{k=0}^{\infty}$ such that

$$\overbrace{\mathbb{R}^d} \to \mathbb{R}^d, \qquad k = 0, 1, \ldots,$$

$$], \ldots, x^{[k]}), \qquad k = 0, 1, \ldots. \qquad (10.2)$$

ith regard to the scheme (10.2) is about its for every *starting value* $x^{[0]} \in \mathbb{R}^d$?[1] Secondly, it bound to be the true solution of the linear

he true solution, the scheme is, obviously, un-terative methods are equally good. Our main omputational cost, we must consider how fast e expense of each iteration.

to be *linear* if each h_k is linear in all its ar-s solely on $x^{[k-m+1]}, x^{[k-m+2]}, \ldots, x^{[k]}$, $k = m-1, m, \ldots$. Finally, an m-step method is *stationary* if the function h_k does not vary with k for $k \geq m-1$. Each of these three concepts represents a considerable simplification and it makes good sense to focus our effort on the most elementary model possible: a *linear, one-step stationary* scheme. In that case (10.2) becomes

$$x^{[k+1]} = Hx^{[k]} + v, \qquad (10.3)$$

---

[1]We are content when methods for *nonlinear* systems, e.g. functional iteration or the Newton–Raphson algorithm (cf. Chapter 6), converge for a suitably large set of starting values. When it comes to linear systems, however, we are more greedy!

where the $d \times d$ *iteration matrix* H and $v \in \mathbb{R}^d$ are independent of k. (Of course, both H and v must depend on A and b, otherwise convergence is impossible.)

Lemma 10.1   *Given an arbitrary linear system* (10.1), *a linear, one-step stationary scheme* (10.3) *converges to a unique bounded limit* $\hat{x} \in \mathbb{R}^d$, *regardless of the choice of the starting value* $x^{[0]}$, *if and only if* $\rho(H) < 1$, *where* $\rho(\cdot)$ *denotes the spectral radius* (A.1.5.2). *Provided that* $\rho(H) < 1$, $\hat{x}$ *is the correct solution of the linear system* (10.1) *if and only if*

$$v = (I - H)A^{-1}b. \tag{10.4}$$

Proof   Let us commence by assuming $\rho(H) < 1$. In this case we claim that

$$\lim_{k \to \infty} H^k = O. \tag{10.5}$$

To prove this statement, we make the simplifying assumption that H has a complete set of eigenvectors, hence that there exist a nonsingular $d \times d$ matrix V and a diagonal $d \times d$ matrix D such that $H = VDV^{-1}$ (A.1.5.3). Hence $H^2 = \left(VDV^{-1}\right) \times \left(VDV^{-1}\right) = VD^2V^{-1}$, $H^3 = VD^3V^{-1}$ and, in general, it is trivial to prove by induction that $H^k = VD^kV^{-1}$, $k = 0, 1, 2, \ldots$. Therefore, passing to the limit,

$$\lim_{k \to \infty} H^k = V \left(\lim_{k \to \infty} D^k \right) V^{-1}.$$

The elements along the diagonal of D are the eigenvalues of H, hence $\rho(H) < 1$ implies $D^k \overset{k \to \infty}{\longrightarrow} O$ and we deduce (10.5).

Unless the set of eigenvectors is complete, (10.5) can be proved just as easily by using a Jordan factorization (see A.1.5.6 and Exercise 10.1).

Our next assertion is that

$$x^{[k]} = H^k x^{[0]} + (I - H)^{-1}(I - H^k)v, \qquad k = 0, 1, 2, \ldots; \tag{10.6}$$

note that $\rho(H) < 1$ implies $1 \notin \sigma(H)$, where $\sigma(H)$ is the set of all eigenvectors (the *spectrum*) of H, therefore the inverse of $I - H$ exists.

The proof is by induction. It is obvious that (10.6) is true for $k = 0$. Hence, let us assume it for $k \geq 0$ and attempt its verification for $k + 1$. Using the definition (10.3) of the iterative scheme in tandem with the induction assumption (10.6), we readily obtain

$$\begin{aligned}
x^{[k+1]} &= Hx^{[k]} + v = H\left[H^k x^{[0]} + (I - H)^{-1}(I - H^k)v\right] + v \\
&= H^{k+1}x^{[0]} + \left[(I - H)^{-1}(H - H^{k+1}) + (I - H)^{-1}(I - H)\right]v \\
&= H^{k+1}x^{[0]} + (I - H)^{-1}(I - H^{k+1})v
\end{aligned}$$

and the proof of (10.6) is complete.

Letting $k \to \infty$ in (10.6), (10.5) implies at once that the iterative process converges,

$$\lim_{k \to \infty} x^{[k]} = \hat{x} := (I - H)^{-1}v. \tag{10.7}$$

We next consider the case $\rho(H) \geq 1$. Provided that $1 \notin \sigma(H)$, the matrix $I - H$ is invertible and $\hat{x} = (I - H)^{-1} v$ is the only possible bounded limit of the iterative scheme. For, suppose the existence of a bounded limit $\hat{y}$. Then

$$\hat{y} = \lim_{k \to \infty} x^{[k+1]} = H \lim_{k \to \infty} x^{[k]} + v = H\hat{y} + v, \tag{10.8}$$

therefore $\hat{y} = \hat{x}$.

Even if $1 \in \sigma(H)$ it remains true that every possible limit $\hat{y}$ must obey (10.8). To see this, let w be an eigenvector corresponding to the eigenvalue 1. Substitution into (10.8) verifies that $\hat{y} + w$ is also a solution. Hence either there is no limit or the limit is not unique and depends on the starting value; both cases are categorized under 'absence of convergence'. We thus assume that $1 \notin \sigma(H)$.

Choose $\lambda \in \sigma(H)$ such that $|\lambda| = \rho(H)$ and let w be a unit-length eigenvector corresponding to λ: $Hw = \lambda w$ and $\|w\| = 1$. ($\| \cdot \|$ denotes here – and elsewhere in this chapter – the Euclidean norm.)

We need to show that there always exists a starting value $x^{[0]} \in \mathbb{R}^d$ for which the scheme (10.3) fails to converge.

Case 1   Let $\lambda \in \mathbb{R}$. Note that, λ being real, so is w. We choose $x^{[0]} = w + \hat{x}$ and claim that

$$x^{[k]} = \lambda^k w + \hat{x}, \qquad k = 0, 1, \ldots. \tag{10.9}$$

As we have already seen, (10.9) is true when $k = 0$. By induction,

$$\begin{aligned} x^{[k+1]} &= H\left(\lambda^k w + \hat{x}\right) + v = \lambda^k Hw + (I - H)^{-1}[H + (I - H)]v \\ &= \lambda^{k+1} w + \hat{x}, \qquad k = 0, 1, \ldots, \end{aligned}$$

and we deduce (10.9).

Because $|\lambda| \geq 1$, (10.9) implies that

$$\|x^{[k]} - \hat{x}\| = |\lambda|^k \geq 1, \qquad k = 0, 1, \ldots.$$

Therefore it is impossible for the sequence $\{x^{[k]}\}_{k=0}^{\infty}$ to converge to $\hat{x}$.

Case 2   Suppose that λ is complex. Therefore $\bar{\lambda}$ is also an eigenvalue of H (the bar denotes complex conjugation). Since $Hw = \lambda w$, conjugation implies $H\bar{w} = \bar{\lambda}\bar{w}$, hence $\bar{w}$ is a unit-length eigenvector corresponding to the eigenvalue $\bar{\lambda}$. Furthermore, $\bar{\lambda} \neq \lambda$, hence w and $\bar{w}$ are linearly independent, otherwise they would have corresponded to the same eigenvalue. We define a function

$$g(z) := \|zw + \bar{z}\bar{w}\|, \qquad z \in \mathbb{C}.$$

It is trivial to verify that $g : \mathbb{C} \to \mathbb{R}$ is continuous, hence it attains its minimum in every closed, bounded subset of $\mathbb{C}$, in particular, in the unit circle. Therefore,

$$\inf_{-\pi \leq \theta \leq \pi} \|e^{i\theta} w + e^{-i\theta} \bar{w}\| = \min_{-\pi \leq \theta \leq \pi} \|e^{i\theta} w + e^{-i\theta} \bar{w}\| = \nu \geq 0,$$

say. Suppose that $\nu = 0$. Then there exists $\theta_0 \in [-\pi, \pi]$ such that

$$\|e^{i\theta_0} w + e^{-i\theta_0} \bar{w}\| = 0,$$

therefore $\bar{w} = e^{2i\theta_0}w$, in contradiction to the linear independence of w and $\bar{w}$. Consequently $\nu > 0$.

The function g is homogeneous,

$$g(re^{i\theta}) = r\|e^{i\theta}w + e^{-i\theta}\bar{w}\| = rg(e^{i\theta}), \qquad r > 0, \quad |\theta| \leq \pi,$$

hence

$$g(z) = |z|\, g\left(\frac{z}{|z|}\right) \geq \nu|z|, \qquad z \in \mathbb{C} \setminus \{0\}. \tag{10.10}$$

We let $x^{[0]} = w + \bar{w} + \hat{x} \in \mathbb{R}^d$. An inductive argument identical to the proof of (10.9) affirms that

$$x^{[k]} = \lambda^k w + \bar{\lambda}^k \bar{w} + \hat{x}, \qquad k = 0, 1, \ldots.$$

Therefore, substituting in the inequality (10.10),

$$\|x^{[k]} - \hat{x}\| = g(\lambda^k) \geq |\lambda|^k \nu \geq \nu, \qquad k = 0, 1, \ldots.$$

As in case 1, we obtain a sequence that is bounded away from its only possible limit, $\hat{x}$; therefore it cannot converge.

To complete the proof, we need to demonstrate that (10.4) is true, but this is trivial. The exact solution of (10.1) being $x = A^{-1}b$, (10.4) follows by substitution into (10.8). ∎

⬦ **Incomplete LU factorization**   Suppose that we can write the matrix A in the form $A = \tilde{A} - E$, the underlying assumption being that the LU factorization of the nonsingular matrix $\tilde{A}$ can be evaluated with ease. For example, $\tilde{A}$ might be banded or (in the case of symmetric sparsity structure) have a graph that is a tree; cf. Chapter 9. Moreover, we assume that E is small in comparison with $\tilde{A}$. Writing (10.1) in the form

$$\tilde{A}x = Ex + b$$

suggests the iterative scheme

$$\tilde{A}x^{[k+1]} = Ex^{[k]} + b, \qquad k = 0, 1, \ldots, \tag{10.11}$$

incomplete LU factorization (ILU). Its implementation requires just a single LU (or Cholesky, if $\tilde{A}$ is symmetric) factorization, which can be reused in each iteration.

To write (10.11) in the form (10.3), we let $H = -\tilde{A}^{-1}E$ and $v = \tilde{A}^{-1}b$. Therefore

$$(I - H)A^{-1}b = (I - \tilde{A}^{-1}E)(\tilde{A} - E)^{-1}b = \tilde{A}^{-1}(\tilde{A} - E)(\tilde{A} - E)^{-1}b$$

$$= \tilde{A}^{-1}b = v,$$

consistently with (10.4). Note that this definition of H and v is purely formal. In reality we never compute them explicitly; we use (10.11) instead.   ⬦

The ILU iteration (10.11) is an example of a *regular splitting*. With greater generality, we make the splitting $A = P - N$, where P is a nonsingular matrix, and consider the iterative scheme

$$P\boldsymbol{x}^{[k+1]} = N\boldsymbol{x}^{[k]} + \boldsymbol{b}, \qquad k = 0, 1, \ldots . \tag{10.12}$$

The underlying assumption is that a system having matrix P can be solved with ease, whether by LU factorization or by other means. Note that, formally, $H = P^{-1}N = P^{-1}(P - A) = I - P^{-1}A$ and $\boldsymbol{v} = P^{-1}\boldsymbol{b}$.

Theorem 10.2   *Suppose that both A and $P + P^T - A$ are symmetric and positive definite. Then the method (10.12) converges.*

Proof   Let $\lambda \in \mathbb{C}$ be an arbitrary eigenvector of the iteration matrix H and suppose that $\boldsymbol{w}$ is a corresponding eigenvector. Recall that $H = I - P^{-1}A$, therefore

$$(I - P^{-1}A)\boldsymbol{w} = \lambda\boldsymbol{w}.$$

We multiply both sides by the matrix P, and this results in

$$(1 - \lambda)P\boldsymbol{w} = A\boldsymbol{w}. \tag{10.13}$$

Our first conclusion is that $\lambda \neq 1$, otherwise (10.13) implies $A\boldsymbol{w} = \boldsymbol{0}$, which contradicts our assumption that A, being positive definite, is nonsingular.

We deduce further from (10.13) that

$$\bar{\boldsymbol{w}}^T A\boldsymbol{w} = (1 - \lambda)\,\bar{\boldsymbol{w}}^T P\boldsymbol{w}. \tag{10.14}$$

However, A is symmetric, therefore $\bar{\boldsymbol{y}}^T A\boldsymbol{y}$ is real for every $\boldsymbol{y} \in \mathbb{C}^d$. Therefore, taking conjugates in (10.14),

$$\bar{\boldsymbol{w}}^T A\boldsymbol{w} = \overline{(1 - \lambda)\,\bar{\boldsymbol{w}}^T P\boldsymbol{w}} = (1 - \bar{\lambda})\,\bar{\boldsymbol{w}}^T P^T\boldsymbol{w}.$$

This, together with (10.14) and $\lambda \neq 1$, implies the identity

$$\left(\frac{1}{1 - \lambda} + \frac{1}{1 - \bar{\lambda}} - 1\right)\bar{\boldsymbol{w}}^T A\boldsymbol{w} = \bar{\boldsymbol{w}}^T(P + P^T - A)\boldsymbol{w}. \tag{10.15}$$

We note first that

$$\frac{1}{1 - \lambda} + \frac{1}{1 - \bar{\lambda}} - 1 = \frac{2 - 2\,\mathrm{Re}\,\lambda - |1 - \lambda|^2}{|1 - \lambda|^2} = \frac{1 - |\lambda|^2}{|1 - \lambda|^2} \in \mathbb{R}.$$

Next, we let $\boldsymbol{w} = \boldsymbol{w}_{\mathrm{R}} + i\boldsymbol{w}_{\mathrm{I}}$, where both $\boldsymbol{w}_{\mathrm{R}}$ and $\boldsymbol{w}_{\mathrm{I}}$ are real vectors, and take real parts in (10.15). On the left-hand side we obtain

$$\mathrm{Re}\left[(\boldsymbol{w}_{\mathrm{R}} - i\boldsymbol{w}_{\mathrm{I}})^T A(\boldsymbol{w}_{\mathrm{R}} + i\boldsymbol{w}_{\mathrm{I}})\right] = \boldsymbol{w}_{\mathrm{R}}^T A\boldsymbol{w}_{\mathrm{R}} + \boldsymbol{w}_{\mathrm{I}}^T A\boldsymbol{w}_{\mathrm{I}}$$

and a similar identity is true on the right with A replaced by $P + P^T - A$. Therefore

$$\frac{1 - |\lambda|^2}{|1 - \lambda|^2}\left(\boldsymbol{w}_{\mathrm{R}}^T A\boldsymbol{w}_{\mathrm{R}} + \boldsymbol{w}_{\mathrm{I}}^T A\boldsymbol{w}_{\mathrm{I}}\right) = \boldsymbol{w}_{\mathrm{R}}^T(P + P^T - A)\boldsymbol{w}_{\mathrm{R}} + \boldsymbol{w}_{\mathrm{I}}^T(P + P^T - A)\boldsymbol{w}_{\mathrm{I}}. \tag{10.16}$$

Recall that both A and $P + P^T - A$ are positive definite. It is impossible for both $\boldsymbol{w}_\mathrm{R}$ and $\boldsymbol{w}_\mathrm{I}$ to vanish (since this would imply that $\boldsymbol{w} = \boldsymbol{0}$), therefore

$$\boldsymbol{w}_\mathrm{R}^T A \boldsymbol{w}_\mathrm{R} + \boldsymbol{w}_\mathrm{I}^T A \boldsymbol{w}_\mathrm{I}, \ \boldsymbol{w}_\mathrm{R}^T (P + P^T - A) \boldsymbol{w}_\mathrm{R} + \boldsymbol{w}_\mathrm{I}^T (P + P^T - A) \boldsymbol{w}_\mathrm{I} > 0.$$

We therefore conclude from (10.16) that

$$\frac{1 - |\lambda|^2}{|1 - \lambda|^2} > 0,$$

hence $|\lambda| < 1$. This is true for every $\lambda \in \sigma(H)$, consequently $\rho(H) < 1$ and we use Lemma 10.1 to argue that the iterative scheme (10.12) converges. ∎

◇ **Tridiagonal matrices**   A relatively simple demonstration of the power of Theorem 10.2 is provided by *tridiagonal* matrices. Thus, let us suppose that the $d \times d$ symmetric matrix

$$A = \begin{bmatrix} \alpha_1 & \beta_1 & 0 & \cdots & & 0 \\ \beta_1 & \alpha_2 & \beta_2 & & \ddots & \vdots \\ 0 & \ddots & \ddots & & \ddots & 0 \\ \vdots & \ddots & & \beta_{d-2} & \alpha_{d-1} & \beta_{d-1} \\ 0 & \cdots & & 0 & \beta_{d-1} & \alpha_d \end{bmatrix}$$

is positive definite. Our claim is that the regular splittings

$$P = \begin{bmatrix} \alpha_1 & 0 & \cdots & 0 \\ 0 & \alpha_2 & \ddots & \vdots \\ \vdots & \ddots & \ddots & 0 \\ 0 & \cdots & 0 & \alpha_d \end{bmatrix}, \quad N = - \begin{bmatrix} 0 & \beta_1 & 0 & \cdots & & 0 \\ \beta_1 & 0 & \beta_2 & & \ddots & \vdots \\ 0 & \ddots & \ddots & & \ddots & 0 \\ \vdots & \ddots & & \beta_{d-2} & 0 & \beta_{d-1} \\ 0 & \cdots & & 0 & \beta_{d-1} & 0 \end{bmatrix}$$

$$(10.17)$$

and

$$P = \begin{bmatrix} \alpha_1 & 0 & \cdots & & \cdots & 0 \\ \beta_1 & \alpha_2 & 0 & & & \vdots \\ 0 & \ddots & \ddots & & \ddots & \vdots \\ \vdots & & \ddots & \beta_{d-2} & \alpha_{d-1} & 0 \\ 0 & \cdots & & 0 & \beta_{d-1} & \alpha_d \end{bmatrix}, \quad N = - \begin{bmatrix} 0 & \beta_1 & \cdots & 0 \\ 0 & 0 & \ddots & \vdots \\ \vdots & \ddots & \ddots & \beta_{d-1} \\ 0 & \cdots & 0 & 0 \end{bmatrix}$$

$$(10.18)$$

– the *Jacobi splitting* and the *Gauss–Seidel splitting* respectively – result in convergent schemes (10.12).

Since A is positive definite, Theorem 10.2 and the positive definiteness of the matrix $Q := P + P^T - A$ imply convergence. For the splitting (10.17) we readily obtain

$$
Q = \begin{bmatrix}
\alpha_1 & -\beta_1 & 0 & \cdots & & 0 \\
-\beta_1 & \alpha_2 & -\beta_2 & \ddots & & \vdots \\
0 & \ddots & \ddots & \ddots & & 0 \\
\vdots & \ddots & -\beta_{d-2} & \alpha_{d-1} & -\beta_{d-1} \\
0 & \cdots & 0 & -\beta_{d-1} & \alpha_d
\end{bmatrix}.
$$

Our claim is that Q is indeed positive definite, and this follows from the positive definiteness of A. Specifically, A is positive definite if and only if $x^T A x > 0$ for all $x \in \mathbb{R}^d$, $x \neq 0$ (A.1.3.5). But

$$
x^T A x = \sum_{j=1}^{d} \alpha_j^2 x_j^2 + 2 \sum_{j=1}^{d-1} \beta_j x_j x_{j+1} = \sum_{j=1}^{d} \alpha_j^2 y_j^2 - 2 \sum_{j=1}^{d-1} \beta_j y_j y_{j+1} = y^T Q y,
$$

where $y_j = (-1)^j x_j$, $j = 1, 2, \ldots, d$. Therefore $y^T Q y > 0$ for every $y \in \mathbb{R}^d \setminus \{0\}$ and we deduce that the matrix Q is indeed positive definite.

The proof for (10.18) is, if anything, even easier, since Q is simply the diagonal matrix

$$
Q = \begin{bmatrix}
\alpha_1 & 0 & \cdots & 0 \\
0 & \alpha_2 & \ddots & \vdots \\
\vdots & \ddots & \ddots & 0 \\
0 & \cdots & 0 & \alpha_d
\end{bmatrix}.
$$

Since A is positive definite and $\alpha_j = e_j^T A e_j > 0$, $j = 1, 2, \ldots$, where $e_j \in \mathbb{R}^d$ is the jth unit vector, $j = 1, 2, \ldots, d$, it follows at once that Q also is positive definite.

Figure 10.1 displays the error in the solution of a $d \times d$ tridiagonal system with

$$
\begin{aligned}
\alpha_1 &= d, & \alpha_j &= 2j(d - j - 1) + d, & j &= 2, 3, \ldots, d - 1, & \alpha_d &= d, \\
\beta_j &= -j(d - j), & & & j &= 1, 2, \ldots, d - 1, \\
b_j &\equiv 1, & x_j^{[0]} &\equiv 0, & j &= 1, 2, \ldots, d. & & \text{(10.19)}
\end{aligned}
$$

It is trivial to use the Gerschgorin criterion (Lemma 7.3) to prove that the underlying matrix A is positive definite.

The system has been solved with both the Jacobi splitting (10.17) (top row in the figure) and the Gauss–Seidel splitting (10.18) (bottom row) for $d = 10, 20, 30$. Even superficial examination of the figure reveals a number of interesting features.

- Both Jacobi and Gauss–Seidel converge. This should come as no surprise since, as we have just proved, provided A is tridiagonal its positive definiteness is sufficient for both methods to converge.

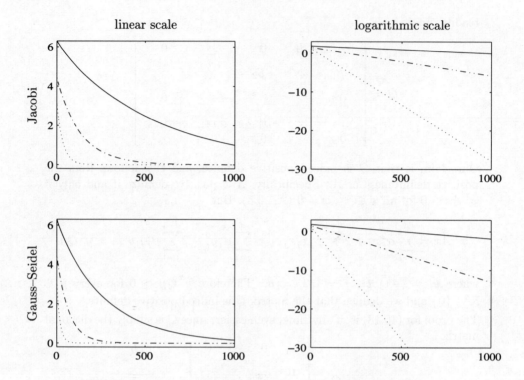

Figure 10.1   The error in the Jacobi and Gauss–Seidel splittings for the
system (10.19) with $d = 10$ (dotted line), $d = 20$ (broken-and-dotted line)
and $d = 40$ (solid line). The first column displays the error (in the Euclidean
norm) on a linear scale; the second column shows its logarithm.

- Convergence proceeds at a geometric speed; this is obvious from the
  second column, since the logarithm of the error is remarkably close to a
  linear function. This is not very surprising either since it is implicit in
  the proof of Lemma 10.1 (and made explicit in Exercise 10.2) that, at
  least asymptotically, the error decays like $[\rho(H)]^k$.

- The rate of convergence is slow and deteriorates markedly as d increases.
  This is a worrying feature since in practical computation we are interested in equations of considerably larger size than $d = 40$.

- Gauss–Seidel is better than Jacobi. Actually, careful examination of the
  rate of decay (which, obviously, is more transparent in the logarithmic
  scale) reveals that the error of Gauss–Seidel decays at twice the speed of
  Jacobi! In other words, we need just half the steps to attain the specified
  accuracy.

In the next section we will observe that the disappointing rate of decay of
both methods, as well as the better performance of Gauss–Seidel, are a fairly
general phenomenon, not just a feature of the system (10.19).   ◇

10.2 Classical iterative methods

Let A be a real $d \times d$ matrix. We split it in the form

Here the $d \times d$ matrices D, L_0 and U_0 are the diagonal, minus the strictly lower-triangular and minus the strictly upper-triangular portions of A, respectively. We assume that $a_{j,j} \neq 0$, $j = 1, 2, \ldots, d$. Therefore D is nonsingular and we let

$$L := D^{-1}L_0, \qquad U := D^{-1}U_0.$$

The *Jacobi* iteration is defined by setting in (10.3)

$$H = B := L + U, \qquad v := D^{-1}b \tag{10.20}$$

or, equivalently, considering a regular splitting (10.12) with $P = D$, $N = L_0 + U_0$. Likewise, we define the *Gauss–Seidel* iteration by specifying

$$H = \mathcal{L} := (I - L)^{-1}U, \qquad v := (I - L)^{-1}D^{-1}b, \tag{10.21}$$

and this is the same as the regular splitting $P = D - L_0$, $N = U_0$. Observe that (10.17) and (10.18) are nothing other than the Jacobi and Gauss–Seidel splittings, respectively, as applied to tridiagonal matrices.

The list of classical iterative schemes would be incomplete without mentioning the *successive over-relaxation* (SOR) scheme, which is defined by setting

$$H = \mathcal{L}_\omega := (I - \omega L)^{-1}[(1 - \omega)I + \omega U], \qquad v = \omega(I - \omega L)^{-1}D^{-1}b, \tag{10.22}$$

where $\omega \in [1, 2)$ is a parameter. Although this might not be obvious at a glance, the SOR scheme can be represented alternatively as a regular splitting with

$$P = \frac{1}{\omega}D - L_0, \qquad N = \left(\frac{1}{\omega} - 1\right)D + U_0 \tag{10.23}$$

(Exercise 10.4).

Note that Gauss–Seidel is simply a special case of SOR, with $\omega = 1$. However, it makes good sense to single it out for special treatment.

All three methods (10.20)–(10.22) are consistent with (10.4), therefore Lemma 10.1 implies that if they converge, the limit is necessarily the true solution of the linear system.

The 'H-v' notation is helpful within the framework of Lemma 10.1 but on the whole it is somewhat opaque. The three methods can be presented in a much more user-friendly manner. Thus, writing the system (10.1) in the form

$$\sum_{j=1}^{d} a_{\ell,j}x_j = b_\ell, \qquad \ell = 1, 2, \ldots,$$

the Jacobi iteration reads

$$\sum_{j=1}^{\ell-1} a_{\ell,j} x_j^{[k]} + a_{\ell,\ell} x_\ell^{[k+1]} + \sum_{j=\ell+1}^{d} a_{\ell,j} x_j^{[k]} = b_\ell, \qquad \ell = 1, 2, \ldots, d, \quad k = 0, 1, \ldots.$$

while the Gauss–Seidel scheme becomes

$$\sum_{j=1}^{\ell} a_{\ell,j} x_j^{[k+1]} + \sum_{j=\ell+1}^{d} a_{\ell,j} x_j^{[k]} = b_\ell, \qquad \ell = 1, 2, \ldots, d, \quad k = 0, 1, \ldots.$$

In other words, the main difference between Jacobi and Gauss–Seidel is that in the first we always express each new component of $x^{[k+1]}$ solely in terms of $x^{[k]}$, while in the latter we use the elements of $x^{[k+1]}$ whenever they are available. This is an important distinction as far as implementation is concerned. In each iteration (10.20) we need to store both $x^{[k]}$ and $x^{[k+1]}$ and this represents a major outlay in terms of computer storage – recall that x is a 'stretched' computational grid. (Of course, we do not store or even generate the matrix A, which originates in highly sparse finite difference or finite element equations. Instead, we need to know the 'rule' for constructing each linear equation, e.g. the five-point formula.) Clever programming and exploitation of the sparsity pattern can reduce the required amount of storage but this cannot ever compete with (10.21): in Gauss–Seidel we can throw away any ℓth component of $x^{[k]}$ as soon as $x_\ell^{[k+1]}$ has been generated, since both quantites may share the same storage.

The SOR iteration (10.22) can be also written in a similar form. Multiplying (10.23) by ω results in

$$\omega \sum_{j=1}^{\ell-1} a_{\ell,j} x_j^{[k+1]} + a_{\ell,\ell} x_\ell^{[k+1]} + (\omega - 1) a_{\ell,\ell} x_\ell^{[k]} + \omega \sum_{j=\ell+1}^{d} a_{\ell,j} x_j^{[k]} = \omega b_j,$$

$$\ell = 1, 2, \ldots, d, \quad k = 0, 1, \ldots.$$

Since precise estimates depend on the sparsity pattern of A, it is apparent that the cost of a single SOR iteration is not substantially larger than its counterpart for either Jacobi or Gauss–Seidel. Moreover, SOR shares with Gauss–Seidel the important virtue of requiring just a single copy of x to be stored at any one time.

The SOR iteration and its special case, the Gauss–Seidel method, share another feature. Their precise definition depends upon the ordering of the equations and the unknowns. As we have already seen in Chapter 9, the rearrangement of equations and unknowns is tantamount to acting on A with permutation matrices on the left and on the right respectively, and these two operations, in general, result in different iterative schemes. It is entirely possible that one of these arrangements converges, while the other fails to do so!

We have already observed in Section 10.1 that both Jacobi and Gauss–Seidel converge whenever A is a tridiagonal, symmetric, positive definite matrix and it is not difficult to verify that this is also the case with SOR for every $1 \leq \omega < 2$ (cf. Exercise 10.5).

As far as convergence is concerned, Jacobi and Gauss–Seidel share similar behaviour for a wide range of linear systems. Thus, let A be *strictly diagonally dominant*.

This means that

$$|a_{\ell,\ell}| \geq \sum_{\substack{j=1 \\ j \neq \ell}}^{d} |a_{\ell,j}|, \qquad \ell = 1, 2, \ldots, d \tag{10.24}$$

and the inequality is sharp for at least one $\ell \in \{1, 2, \ldots, d\}$. (Some definitions require sharp inequality for *all* ℓ, but the present, weaker, definition is just perfect for our purposes.)

Theorem 10.3   *If the matrix A is strictly diagonally dominant then both the Jacobi and Gauss–Seidel methods converge.*

Proof   According to (10.20) and (10.24),

$$\sum_{j=1}^{d} |b_{\ell,j}| = \sum_{\substack{j=1 \\ j \neq \ell}}^{d} |b_{\ell,j}| = \frac{1}{|a_{\ell,\ell}|} \sum_{\substack{j=1 \\ j \neq \ell}}^{d} |a_{\ell,j}| \leq 1, \qquad \ell = 1, 2, \ldots, d$$

and the inequality is sharp for at least one ℓ. Therefore $\rho(B) < 1$ by the Gerschgorin criterion (Lemma 7.3) and the Jacobi iteration converges.

The proof for Gauss–Seidel is slightly more complicated; essentially, we need to revisit the proof of Lemma 7.3 (i.e., Exercise 7.8) in a different framework. Choose $\lambda \in \sigma(\mathcal{L}_1)$ with an eigenvector $\boldsymbol{w}$. Therefore, multiplying $\mathcal{L}$ and $\boldsymbol{v}$ from (10.21) first by $I - L$ and then by D,

$$U_0 \boldsymbol{w} = \lambda(D - L_0)\boldsymbol{w}.$$

It is convenient to rewrite this in the form

$$\lambda a_{\ell,\ell} w_\ell = \sum_{j=\ell+1}^{d} a_{\ell,j} w_j - \lambda \sum_{j=1}^{\ell-1} a_{\ell,j} w_j, \qquad \ell = 1, 2, \ldots, d.$$

Therefore, by the triangle inequality,

$$|\lambda| \, |a_{\ell,\ell}| \, |w_\ell| \leq \sum_{j=\ell+1}^{d} |a_{\ell,j}| \, |w_j| + |\lambda| \sum_{j=1}^{\ell-1} |a_{\ell,j}| \, |w_j| \tag{10.25}$$

$$\leq \left(\sum_{j=\ell+1}^{d} |a_{\ell,j}| + |\lambda| \sum_{j=1}^{\ell-1} |a_{\ell,j}| \right) \max_{j=1,2,\ldots,d} |w_j|, \quad j = 1, \ldots, d. \tag{10.26}$$

Let $\alpha \in \{1, 2, \ldots, d\}$ be such that

$$|w_\alpha| = \max_{j=1,2,\ldots,d} |w_j| > 0.$$

Substituting into (10.26), we have

$$|\lambda| \, |a_{\alpha,\alpha}| \leq \sum_{j=\alpha+1}^{d} |a_{\alpha,j}| + |\lambda| \sum_{j=1}^{\alpha-1} |a_{\alpha,j}|.$$

Let us assume that $|\lambda| \geq 1$. We deduce that

$$|a_{\alpha,\alpha}| \leq \sum_{\substack{j=1 \\ j \neq \alpha}}^{d} |a_{\alpha,j}|,$$

and this can be consistent with (10.24) only if the weak inequality holds as an equality. Substitution in (10.25), in tandem with $|\lambda| \geq 1$, results in

$$|\lambda| \sum_{\substack{j=1 \\ j \neq \alpha}}^{d} |a_{\alpha,j}| \, |w_\alpha| \leq \sum_{j=\alpha+1}^{d} |a_{\alpha,j}| \, |w_j| + |\lambda| \sum_{j=1}^{\alpha-1} |a_{\alpha,j}| \, |w_j| \leq |\lambda| \sum_{\substack{j=1 \\ j \neq \alpha}}^{d} |a_{\alpha,j}| \, |w_j|.$$

This, however, can be consistent with $|w_\alpha| = \max |w_j|$ only if $|w_\ell| = |w_\alpha|$, $\ell = 1, 2, \ldots, d$. Therefore, every $\ell \in \{1, 2, \ldots, d\}$ can play the role of α and

$$|a_{\ell,\ell}| = \sum_{\substack{j=1 \\ j \neq \ell}}^{d} |a_{\ell,j}|, \qquad \ell = 1, 2, \ldots, d,$$

in defiance of the definition of strict diagonal dominance. Therefore, having led to a contradiction, our assumption that $|\lambda| \geq 1$ is wrong. We deduce that $\rho(\mathcal{L}_1) < 1$, hence Lemma 10.1 implies convergence. ∎

Another, less trivial, example where Jacobi and Gauss–Seidel converge in tandem is provided in the following theorem, which we state without proof.

Theorem 10.4 (The Stein–Rosenberg theorem) *Suppose that $a_{\ell,\ell} \neq 0$, $\ell = 1, 2, \ldots, d$, and that all the components of B are nonnegative. Then*

either	$\rho(\mathcal{L}_1) = \rho(B) = 0$	*or* $\rho(\mathcal{L}_1) < \rho(B) < 1$
or	$\rho(\mathcal{L}_1) = \rho(B) = 1$	*or* $\rho(\mathcal{L}_1) > \rho(B) > 1.$

Hence, the Jacobi and Gauss–Seidel methods are either simultaneously convergent or simultaneously divergent. ∎

◇ **An example of divergence** Lest there is an impression that iterative methods are bound to converge or that they always share similar behaviour with regard to convergence, we give here a trivial counterexample,

$$A = \begin{bmatrix} 3 & 2 & 1 \\ 2 & 3 & 2 \\ 1 & 2 & 3 \end{bmatrix}.$$

The matrix is symmetric and positive definite; its eigenvalues are 2 and $\frac{1}{2}(7 \pm \sqrt{33}) > 0$. It is easy to verify, either directly or from Theorem 10.2 or Exercise 10.3, that $\rho(\mathcal{L}_1) < 1$ and the Gauss–Seidel method converges. However, $\rho(B) = \frac{1}{6}(1 + \sqrt{33}) > 1$ and the Jacobi method diverges.

An interesting variation on the last example is provided by the matrix

$$A = \begin{bmatrix} 3 & 2 & 1 \\ -2 & 3 & 2 \\ -1 & -2 & 3 \end{bmatrix}.$$

The spectrum of B is $\{0, \pm i\}$, therefore the Jacobi method diverges marginally. Gauss–Seidel, however, proudly converges, since $\sigma(\mathcal{L}_1) = \{0, \frac{1}{54}(23 \pm \sqrt{97})\} \in (0, 1)$.

Let us exchange the second and third rows and the second and third columns of A. The outcome is

$$\begin{bmatrix} 3 & 1 & 2 \\ -1 & 3 & -2 \\ -2 & 2 & 3 \end{bmatrix}.$$

The spectral radius of Jacobi remains intact, since it does not depend upon ordering. However, the eigenvalues of the new $\mathcal{L}_1$ are 0 and $\frac{1}{54}(31 \pm \sqrt{1393})$, spectral radius exceeds unity and the iteration diverges.

This demonstrates not just the sensitivity of (10.21) to ordering but also that the Jacobi iteration need not be the underachieving sibling of Gauss–Seidel; by replacing 3 with $3 + \varepsilon$, where $0 < \varepsilon \ll 1$, along the diagonal, we render Jacobi convergent (this is an immediate consequence of the Gerschgorin criterion), while continuity of the eigenvalues of $\mathcal{L}_1$ as a function of ε means that Gauss–Seidel (in the second ordering) still diverges. $\diamond$

To gain intuition with respect to the behaviour of classical iterative methods, let us first address ourselves in some detail to the matrix

$$A = \begin{bmatrix} -2 & 1 & 0 & \cdots & 0 \\ 1 & -2 & 1 & \ddots & \vdots \\ 0 & \ddots & \ddots & \ddots & 0 \\ \vdots & \ddots & 1 & -2 & 1 \\ 0 & \cdots & 0 & 1 & -2 \end{bmatrix}. \tag{10.27}$$

It is clear why such an A is relevant to our discussion, since it is obtained from a central difference approximation to a second derivative.

A $d \times d$ matrix $T = (t_{k,\ell})_{k,\ell=1}^d$ is said to be *Toeplitz* if it is constant along all its diagonals, in other words, if there exist numbers $\tau_{-d+1}, \tau_{-d+2}, \ldots, \tau_0, \ldots, \tau_{d-1}$ such that

$$t_{k,\ell} = \tau_{k-\ell}, \qquad k, \ell = 1, 2, \ldots, d.$$

The matrix A, (10.27), is a Toeplitz matrix with

$$\tau_{-d+1} = \cdots = \tau_{-2} = 0, \quad \tau_{-1} = 1, \quad \tau_0 = -2, \quad \tau_1 = 1, \quad \tau_2 = \cdots = \tau_{d-1} = 0.$$

We say that a matrix is TST if it is Toeplitz, symmetric and tridiagonal. Therefore, T is TST if

$$\tau_j = 0, \quad |j| = 2, 3, \ldots, d - 1, \qquad \tau_{-1} = \tau_1.$$

TST matrices are important both for fast solution of the Poisson equation (cf. Chapter 12) and to stability analysis of discretized PDEs of evolution (cf. Chapter 13) but, in the present context, we note that the matrix A, (10.27), is TST.

Lemma 10.5   *Let T be a $d \times d$ TST matrix and $\alpha := t_0$, $\beta := t_{-1} = t_1$. Then the eigenvalues of T are*

$$\lambda_j = \alpha + 2\beta \cos\left(\frac{\pi j}{d+1}\right), \qquad j = 1, 2, \ldots, d, \tag{10.28}$$

each with corresponding orthogonal eigenvector $\boldsymbol{q}_j$, where

$$q_{j,\ell} = \sqrt{\frac{2}{d+1}} \sin\left(\frac{\pi j \ell}{d+1}\right), \qquad j, \ell = 1, 2, \ldots, d. \tag{10.29}$$

Proof   Although it is an easy matter to verify (10.28) and (10.29) directly from the definition of a TST matrix, we adopt a more roundabout approach, since this will pay dividends later in this section.

We assume that $\beta \neq 0$, otherwise T reduces to a multiple of the identity matrix and the lemma is trivial. Let us suppose that λ is an eigenvalue of T with corresponding eigenvector $\boldsymbol{q}$. Letting $q_0 = q_{d+1} = 0$, we can write $A\boldsymbol{q} = \lambda\boldsymbol{q}$ in the form

$$\beta q_{\ell-1} + \alpha q_\ell + \beta q_{\ell+1} = \lambda q_\ell, \qquad \ell = 1, 2, \ldots, d$$

or, after a minor rearrangement,

$$\beta q_{\ell+1} + (\alpha - \lambda)q_\ell + \beta q_{\ell-1} = 0, \qquad \ell = 1, 2, \ldots, d.$$

This is a special case of a difference equation (4.19) and its general solution is

$$q_\ell = a\eta_+^\ell + b\eta_-^\ell, \qquad \ell = 0, 1, \ldots, d+1,$$

where $\eta_\pm$ are zeros of the characteristic polynomial

$$\beta\eta^2 + (\alpha - \lambda)\eta + \beta = 0.$$

In other words,

$$\eta_\pm = \frac{1}{2\beta}\left\{\lambda - \alpha \pm \sqrt{(\lambda - \alpha)^2 - 4\beta^2}\right\}. \tag{10.30}$$

The constants a and b are determined by requiring $q_0 = q_{d+1} = 0$. The first condition yields $a + b = 0$, therefore $q_\ell = a(\eta_+^\ell - \eta_-^\ell)$, where $a \neq 0$ is arbitrary. To fulfil the second condition we need

$$\eta_+^{d+1} = \eta_-^{d+1}.$$

There are $d + 1$ roots to this equation, namely

$$\eta_+ = \eta_- \exp\left(\frac{2\pi i j}{d+1}\right), \qquad j = 0, 1, \ldots, d, \tag{10.31}$$

but we can discard at once the case $j = 0$, since it corresponds to $\eta_- = \eta_+$, hence to $q_\ell \equiv 0$.

We multiply (10.31) by $\exp[-\pi i j/(d+1)]$ and substitute the values of $\eta_\pm$ from (10.30). Therefore

$$\left(\lambda - \alpha + \sqrt{(\lambda - \alpha)^2 - 4\beta^2}\right) \exp\left(\frac{-\pi i j}{d+1}\right) = \left(\lambda - \alpha - \sqrt{(\lambda - \alpha)^2 - 4\beta^2}\right) \exp\left(\frac{\pi i j}{d+1}\right)$$

for some $j \in \{1, 2, \ldots, d\}$. Rearrangement yields

$$\sqrt{(\lambda - \alpha)^2 - 4\beta^2} \cos\left(\frac{\pi j}{d+1}\right) = (\lambda - \alpha) i \sin\left(\frac{\pi j}{d+1}\right)$$

and, squaring, we deduce

$$\left[(\lambda - \alpha)^2 - 4\beta^2\right] \cos^2\left(\frac{\pi j}{d+1}\right) = -(\lambda - \alpha)^2 \sin^2\left(\frac{\pi j}{d+1}\right).$$

Therefore

$$(\lambda - \alpha)^2 = 4\beta^2 \cos^2\left(\frac{\pi j}{d+1}\right)$$

and, taking the square roots, we obtain

$$\lambda = \alpha \pm 2\beta \cos\left(\frac{\pi j}{d+1}\right).$$

Taking the plus sign we recover (10.28) with $\lambda = \lambda_j$, while the minus repeats $\lambda = \lambda_{d+1-j}$ and can be discarded. This concurs with the stipulated form of the eigenvalues.

Substituting (10.28) into (10.30), we readily obtain

$$\eta_\pm = \cos\left(\frac{\pi j}{d+1}\right) \pm i \sin\left(\frac{\pi j}{d+1}\right) = \exp\left(\frac{\pm\pi i j}{d+1}\right),$$

therefore

$$q_{j,\ell} = a(\eta_+^\ell - \eta_-^\ell) = 2ai \sin\left(\frac{\pi j \ell}{d+1}\right), \qquad j, \ell = 1, 2, \ldots, d.$$

This will demonstrate that (10.29) is true, if we can determine a value of a such that

$$\sum_{\ell=1}^{d} q_{j,\ell}^2 = 1$$

(note that symmetry implies that eigenvectors are orthogonal, A.1.3.2.) It is an easy exercise to show that

$$\sum_{\ell=1}^{d} \sin^2\left(\frac{\pi j \ell}{d+1}\right) = \tfrac{1}{2}(d+1), \qquad j = 1, 2, \ldots, d,$$

thereby providing a value of a that is consistent with (10.29). ∎

Corollary *All $d \times d$ TST matrices commute with each other.*

Proof   According to (10.29), all such matrices share the same set of eigenvectors, hence they commute (A.1.5.4).   ∎

Let us now return to the matrix A and to our discussion of classical iterative methods. It follows at once from (10.20) that the iteration matrix B is also a TST matrix, with $\alpha = 0$ and $\beta = \frac{1}{2}$. Therefore

$$\rho(B) = \cos\left(\frac{\pi}{d+1}\right) \approx 1 - \frac{\pi^2}{2d^2} < 1. \tag{10.32}$$

In other words, the Jacobi method converges; but we already know this from Section 10.1. However, (10.32) gives us an extra morsel of information, namely the speed of convergence. The news is not very good, unfortunately: the error is attenuated by $\mathcal{O}(d^{-2})$ in each iteration. In other words, if d is large, convergence up to any reasonable tolerance is very slow indeed.

Instead of debating Gauss–Seidel, we next leap all the way to the SOR scheme – after all, Gauss–Seidel is nothing other than SOR with $\omega = 1$. Although the matrix $\mathcal{L}_\omega$ is no longer Toeplitz, symmetric or tridiagonal, the method of proof of Lemma 10.5 is equally effective. Thus, let $\lambda \in \sigma(\mathcal{L}_\omega)$ and denote by $\boldsymbol{q}$ a corresponding eigenvector. It follows from (10.22) that

$$[(1 - \omega)I + \omega U]\,\boldsymbol{q} = \lambda(I - \omega L)\boldsymbol{q}.$$

Therefore, letting $q_0 = q_{d+1} = 0$, we obtain

$$-2(1 - \omega)q_\ell - \omega q_{\ell+1} = \lambda(\omega q_{\ell-1} - 2q_\ell), \qquad \ell = 1, 2, \dots, d,$$

which we again rewrite as a difference equation

$$\omega q_{\ell+1} - 2(\lambda - 1 + \omega)q_\ell + \omega \lambda q_{\ell-1} = 0, \qquad \ell = 1, 2, \dots, d.$$

The solution is again

$$q_\ell = a(\eta_+^\ell - \eta_-^\ell), \qquad \ell = 0, 1, \dots, d+1,$$

except that (10.30) needs to be replaced by

$$\eta_\pm = \lambda - 1 + \omega \pm \sqrt{(\lambda - 1 + \omega)^2 - \omega^2 \lambda}. \tag{10.33}$$

We set $\eta_+^{d+1} = \eta_-^{d+1}$ and proceed as in the proof of Lemma 10.5. Substitution of the values of $\eta_\pm$ from (10.33) results in

$$(\lambda - 1 + \omega)^2 = \omega^2 \chi \lambda, \tag{10.34}$$

where

$$\chi = \cos^2\left(\frac{\pi\ell}{d+1}\right)$$

for some $\ell \in \{1, 2, \dots, d\}$.

In the special case of Gauss–Seidel, (10.34) yields

$$\lambda^2 = \omega^2 \chi \lambda$$

and we deduce that

$$0, \cos^2\left(\frac{\pi}{d+1}\right), \cos^2\left(\frac{2\pi}{d+1}\right), \ldots, \cos^2\left(\frac{d\pi}{d+1}\right) \subseteq \sigma(\mathcal{L}_1)$$

$$\subseteq \{0\} \cup \left\{ \cos^2\left(\frac{\pi\ell}{d+1}\right) : \ell = 1, 2, \ldots, d \right\}.$$

In particular,

$$\rho(\mathcal{L}_1) = \cos^2\left(\frac{\pi}{d+1}\right) \approx 1 - \frac{\pi^2}{d^2}. \tag{10.35}$$

Comparison with (10.32) demonstrates that, as far as the specific matrix A is concerned, Gauss–Seidel converges at exactly *twice* the rate of Jacobi. In other words, each iteration of Gauss–Seidel is, at least asymptotically, as effective as two iterations of Jacobi! Recall that Gauss–Seidel has important advantages over Jacobi in terms of storage, while the number of operations in each iteration is identical in both schemes. Thus, remarkably, it appears that Gauss–Seidel wins on every score. There is, however, an important reason why the Jacobi iteration is of interest and we address ourselves to this theme later in this section. At present, we wish to debate the convergence of SOR for different values of $\omega \in [1, 2)$. Note that our goal is not merely to check the convergence and assess its speed. The whole point of using SOR, rather than Gauss–Seidel, rests in the exploitation of the parameter to accelerate convergence. We already know from (10.35) that a particular choice of ω is associated with convergence; now we seek to improve upon this result by identifying ω_{opt}, the optimal value of ω.

We distinguish between the following cases.

Case 1 $\chi\omega^2 \leq 4(\omega - 1)$, hence the roots of (10.34) form a complex conjugate pair. It is easy, substituting the explicit value of χ, to verify that this is indeed the case when

$$2\frac{1 - |\sin[\pi\ell/(d+1)]|}{\cos^2[\pi\ell/(d+1)]} \leq \omega \leq 2\frac{1 + |\sin[\pi\ell/(d+1)]|}{\cos^2[\pi\ell/(d+1)]}.$$

Moreover,

$$\frac{1 + |\sin[\pi\ell/(d+1)]|}{\cos^2[\pi\ell/(d+1)]} \geq 1$$

and we restrict our attention to $\omega \leq 2$. Therefore case 1 corresponds to

$$\tilde{\omega} := 2\frac{1 - |\sin[\pi\ell/(d+1)]|}{\cos^2[\pi\ell/(d+1)]} \leq \omega < 2. \tag{10.36}$$

The two solutions of (10.34) are

$$\lambda = 1 - \omega + \tfrac{1}{2}\chi\omega^2 \pm \sqrt{\left(1 - \omega + \tfrac{1}{2}\chi\omega^2\right)^2 - (1 - \omega)^2}; \tag{10.37}$$

consequently

$$|\lambda|^2 = \left(1 - \omega + \tfrac{1}{2}\chi\omega^2\right)^2 + \left[(1 - \omega)^2 - \left(1 - \omega + \tfrac{1}{2}\chi\omega^2\right)^2\right] = (\omega - 1)^2$$

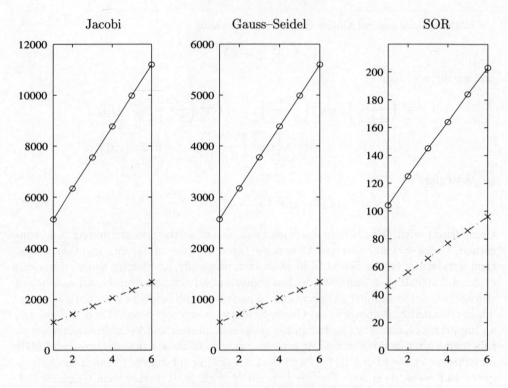

Figure 10.2 The number of iterations in the solution of $Ax = b$, where A is given by (10.27), $x^{[0]} = 0$ and $b = 1$, required to reach accuracy up to given tolerance. The x-axis displays the number of correct significant digits, while the number of iterations can be read from the y-axis. The broken-and-dotted line corresponds to $d = 25$ and the solid line to $d = 50$.

and we obtain

$$|\lambda| = \omega - 1. \tag{10.38}$$

Case 2 $\chi\omega^2 \geq 4(\omega - 1)$ and both zeros of (10.34) are real. Differentiating (10.34) with respect to ω yields

$$2(\lambda - 1 + \omega)(\lambda_\omega + 1) = 2\chi\omega\lambda + \chi\omega^2\lambda_\omega,$$

where $\lambda_\omega = \mathrm{d}\lambda/\mathrm{d}\omega$. Therefore λ_ω may vanish only for

$$\lambda = \frac{1 - \omega}{1 - \chi\omega}.$$

Substitution into (10.34) results in $\chi(1 - \omega) = 1 - \chi\omega$, hence in $\chi = 1$ – but this is impossible, because $\chi = \cos^2[\pi\ell/(d + 1)] \in (0, 1)$. Therefore $\lambda_\omega \neq 0$ in

$$1 < \omega \leq \tilde{\omega}$$

(compare with (10.36)) and the zeros of (10.34) are monotone in this interval. Since they are continuous functions of ω, it follows that, to track the zero of largest magnitude, it is enough to restrict attention to the endpoints 1 and $\tilde{\omega}$.

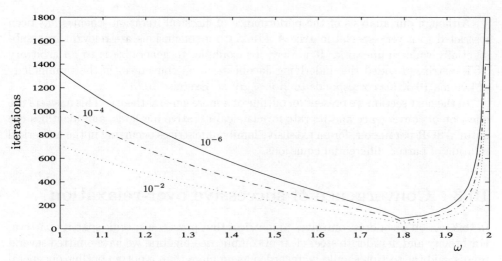

Figure 10.3   The number of iterations required to attain a given accuracy with SOR for different values of ω ($d = 25$).

At $\omega = 1$ we are back to the Gauss–Seidel case and the spectral radius is given by (10.35). At the other endpoint (which is the meeting point of cases 1 and 2) we have $\chi\omega^2 - 4\omega + 4 = 0$, hence

$$\omega = \frac{2\left(1 - \sqrt{1 - \chi}\right)}{\chi}.$$

We have taken the minus in front of the square root, otherwise $\omega \notin [1, 2)$. Therefore, by (10.38),

$$|\lambda| = \rho(\chi) := 2\frac{1 - \sqrt{1 - \chi}}{\chi} - 1.$$

The above is true for every $\chi = \cos^2[\pi\ell/(d+1)]$, $\ell = 1, 2, \ldots, d$. It is, however, elementary to verify that ρ increases strictly monotonically as a function of χ. Thus, the maximum is attained for the largest value of χ, i.e. when $\ell = 1$, and we deduce that

$$\omega_{\mathrm{opt}} = 2\frac{1 - \sin[\pi/(d+1)]}{\cos^2[\pi/(d+1)]} \tag{10.39}$$

and

$$\rho(\mathcal{L}_{\omega_{\mathrm{opt}}}) = \omega_{\mathrm{opt}} - 1 = \left\{\frac{1 - \sin[\pi/(d+1)]}{\cos[\pi/(d+1)]}\right\}^2 \approx 1 - \frac{2\pi}{d}. \tag{10.40}$$

Casting our eyes at the expressions (10.32), (10.35) and (10.40), for the spectral radii of Jacobi, Gauss–Seidel and SOR (with ω_{opt}) respectively, it is difficult not to notice the drastic improvement inherent in the SOR scheme. This observation is vividly demonstrated in Figure 10.2, where the three methods are employed to solve $A\boldsymbol{x} = \boldsymbol{b}$, where A is given by (10.27), $\boldsymbol{b} = \boldsymbol{1}$ and $\boldsymbol{x}^{[0]} = \boldsymbol{0}$. Gauss–Seidel is, predictably, twice as efficient as Jacobi, but SOR leaves both far behind!

Figure 10.3 displays the number of iterations required to approach the solution of $A\boldsymbol{x} = \boldsymbol{b}$ (for $d = 25$, with the same $\boldsymbol{b}$ and $\boldsymbol{x}^{[0]}$ as in Figure 10.2) to within a given accuracy. The sensitivity of the rate of convergence to the value of ω is striking.

Although our analysis of the performance of classical iteration schemes has been restricted to a very special matrix A, (10.27), its conclusions are relevant in a substantially wider framework. It is easy, for example, to generalize it to an arbitrary TST matrix, provided the underlying Jacobi iteration converges (in the terminology of Lemma 10.5 this corresponds to $2|\beta| \leq |\alpha|$; cf. Exercise 10.6).

In the next section we present an outline of a more general theory. This answers the question of convergence and its rate for classical iterative methods, and identifies the optimal SOR parameter, for an extensive family of matrices occurring in the numerical solution of partial differential equations.

10.3   Convergence of successive over-relaxation

In this section we address ourselves to matrices that possess a specific sparsity pattern. For brevity and in order to steer clear of complicated algebra, we have omitted several proofs, and the section should be regarded as no more than a potted outline of general SOR theory.

We have already had an opportunity to observe in Chapter 9 that the terminology of graph theory confers a useful means to express sparsity patterns. As we wish at present to discuss matrices that are neither symmetric nor conform with symmetric sparsity patterns, we say that $\mathbb{G} = \{\mathbb{V}, \mathbb{E}\}$ is the *digraph* (an abbreviation for *directed graph*) of a $d \times d$ matrix A if $\mathbb{V} = \{1, 2, \ldots, d\}$ and $(m, \ell) \in \mathbb{E}$ for $m, \ell \in \{1, 2, \ldots, d\}$, $m \neq \ell$, if and only if $a_{m,\ell} \neq 0$.

To distinguish between $(\ell, m) \in \mathbb{E}$ and $(m, \ell) \in \mathbb{E}$, a pictorial representation of a digraph uses arrows to denote the 'direction'. Thus, (ℓ, m) is represented by a curve joining the ℓth and the mth vertex, with an arrow pointing at the latter. The restriction $\ell \neq m$ is often dropped in a definition of a digraph. We insist on it in the present context because it simplifies the notation to some extent.

Given a $d \times d$ matrix A, we say that $\boldsymbol{j} \in \mathbb{Z}^d$ is an *ordering vector* if $|j_\ell - j_m| = 1$ for every $(\ell, m) \in \mathbb{E}$. Moreover, $\boldsymbol{j}$ is a *compatible ordering vector* if, in addition,

$$
\begin{aligned}
\ell \geq m + 1 &\implies j_\ell - j_m = +1, \\
\ell \leq m - 1 &\implies j_\ell - j_m = -1.
\end{aligned}
$$

Lemma 10.6   *If the matrix A has an ordering vector then there exists a permutation matrix P such that the matrix $\tilde{A} := PAP^{-1}$ has a compatible ordering vector.*

Proof   Each similarity transformation by a permutation matrix is merely a relabeling of variables; this theme underlies the discussion in Section 9.2.[2] Therefore, the graph of $\tilde{A}$ is simply $\tilde{\mathbb{G}} = \{\boldsymbol{\pi}(\mathbb{V}), \boldsymbol{\pi}(\mathbb{E})\}$, where $\boldsymbol{\pi}$ is the corresponding permutation of $\{1, 2, \ldots, d\}$; in other words, $\boldsymbol{\pi}(\mathbb{V}) = \{\pi(1), \pi(2), \ldots, \pi(d)\}$ and $(\pi(\ell), \pi(m)) \in \boldsymbol{\pi}(\mathbb{E})$ if and only if $(\ell, m) \in \mathbb{E}$.

Let $\boldsymbol{j}$ be the ordering vector whose existence has been stipulated in the statement of the lemma and set $i_\ell := j_{\pi^{-1}(\ell)}$, $\ell = 1, 2, \ldots, d$, where $\boldsymbol{\pi}^{-1}$ is the inverse permutation

---

[2]It is trivial to prove that P^{-1} is also a permutation matrix which merely reverses the action of P.

of π. It is easy to verify that i is an ordering vector of $\tilde{A}$ since $(\ell, m) \in \pi(\mathbb{E})$ implies $(\pi^{-1}(\ell), \pi^{-1}(m)) \in \mathbb{E}$. Therefore, by the definition of an ordering vector, $|j_{\pi^{-1}(\ell)} - j_{\pi^{-1}(m)}| = 1$. Recalling the way we have constructed the vector i, it follows that $|i_\ell - i_m| = 1$ and that i is indeed an ordering vector of $\tilde{A}$.

Let us choose a permutation π such that $i_1 \leq i_2 \leq \cdots \leq i_d$; this, of course, corresponds to $j_{\pi^{-1}(1)} \leq j_{\pi^{-1}(2)} \leq \cdots \leq j_{\pi^{-1}(d)}$ and can always be done. Given $(\ell, m) \in \pi(\mathbb{E})$, $\ell \geq m + 1$, we then obtain $i_\ell - i_m \geq 0$; therefore, i being an ordering vector, $i_\ell - i_m = +1$. Likewise $(\ell, m) \in \pi(\mathbb{E})$, $\ell \leq m - 1$, implies that $i_\ell - i_m = -1$. We thus deduce that i is a compatible ordering vector of $\tilde{A}$. ∎

◇ **Ordering vectors**   Consider a matrix A with the following sparsity pattern:

$$
\begin{bmatrix}
\times & \circ & \times & \circ & \times \\
\circ & \times & \times & \circ & \circ \\
\times & \times & \times & \circ & \circ \\
\circ & \circ & \circ & \times & \times \\
\times & \circ & \circ & \times & \times
\end{bmatrix}.
$$

Here $\mathbb{E} = \{(1,3), (3,1), (1,5), (5,1), (2,3), (3,2), (4,5), (5,4)\}$ and it is easy to verify that

$$
j = \begin{bmatrix} 2 \\ 2 \\ 1 \\ 2 \\ 3 \end{bmatrix}
$$

is an ordering vector. However, it is not a compatible ordering vector since, for example, $(3,1) \in \mathbb{E}$ and $j_3 - j_1 = -1$.

To construct a compatible ordering vector i for a permutation π, we require $j_{\pi^{-1}(1)} \leq j_{\pi^{-1}(2)} \leq j_{\pi^{-1}(3)} \leq j_{\pi^{-1}(4)} \leq j_{\pi^{-1}(5)}$. This will be the case if we let, for example, $\pi^{-1}(1) = 3$, $\pi^{-1}(2) = 1$, $\pi^{-1}(3) = 2$, $\pi^{-1}(4) = 4$ and $\pi^{-1}(5) = 5$. In other words,

$$
\pi = \begin{bmatrix} 2 \\ 3 \\ 1 \\ 4 \\ 5 \end{bmatrix}, \quad
P = \begin{bmatrix}
0 & 0 & 1 & 0 & 0 \\
1 & 0 & 0 & 0 & 0 \\
0 & 1 & 0 & 0 & 0 \\
0 & 0 & 0 & 1 & 0 \\
0 & 0 & 0 & 0 & 1
\end{bmatrix} \quad \text{and} \quad
\tilde{A} = \begin{bmatrix}
\times & \times & \times & \circ & \circ \\
\times & \times & \circ & \circ & \times \\
\times & \circ & \times & \circ & \circ \\
\circ & \circ & \circ & \times & \times \\
\circ & \times & \circ & \times & \times
\end{bmatrix}.
$$

Incidentally, i is not the only possible compatible ordering vector. To demonstrate this and, incidentally, that a compatible ordering vector need not be unique, we render the digraph of A pictorially,

and observe that A is itself a permutation of a tridiagonal matrix. It is easy to identify a compatible ordering vector for any tridiagonal matrix with nonvanishing off-diagonal elements – a task that is relegated to Exercise 10.8 – hence producing yet another compatible ordering vector for a permutation of A.   ◇

The existence of a compatible ordering vector confers on a matrix several interesting properties that are of great relevance to the behaviour of classical iterative methods.

Lemma 10.7   *If the matrix A has a compatible ordering vector then the function*

$$g(s,t) := \det\left(tL_0 + \frac{1}{t}U_0 - sD\right), \qquad s \in \mathbb{R}, \quad t \in \mathbb{R} \setminus \{0\}$$

is independent of t.

Proof   Since A and $H(s,t) := tL_0 + (1/t)U_0 - sD$ share the same sparsity pattern for all $t \neq 0$, it follows from the definition that every compatible ordering vector of A is also a compatible ordering vector of $H(s,t)$. In particular, we deduce that the matrix $H(s,t)$ possesses a compatible ordering vector.

By the definition of a determinant,

$$g(s,t) = \sum_{\pi \in \Pi_d} (-1)^{|\pi|} \prod_{i=1}^{d} h_{i,\pi(i)}(s,t),$$

where $H(s,t) = (h_{j,\ell}(s,t))_{j,\ell=1}^{d}$, Π_d is the set of all permutations of $\{1,2,\ldots,d\}$ and $|\pi|$ is the sign of $\pi \in \Pi_d$. Since

$$h_{j,\ell}(s,t) = -t^{\sigma_{j-\ell}} s^{1-|\sigma_{j-\ell}|} a_{j,\ell}, \qquad j,\ell = 1,2,\ldots,d,$$

where

$$\sigma_m = \begin{cases} +1, & m > 0, \\ -1, & m < 0, \\ 0, & m = 0, \end{cases}$$

we deduce that

$$g(s,t) = (-1)^d \sum_{\pi \in \Pi_d} (-1)^{|\pi|} t^{d_{\mathrm{L}}(\pi) - d_{\mathrm{U}}(\pi)} s^{d - d_{\mathrm{L}}(\pi) - d_{\mathrm{U}}(\pi)} \prod_{i=1}^{d} a_{i,\pi(i)}, \qquad (10.41)$$

where $d_{\mathrm{L}}(\pi)$ and $d_{\mathrm{U}}(\pi)$ denote the number of elements $\ell \in \{1,2,\ldots,d\}$ such that $\ell > \pi(\ell)$ and $\ell < \pi(\ell)$ respectively.

Let j be the compatible ordering vector of $H(s,t)$, whose existence we have already deduced from the statement of the lemma, and choose an arbitrary $\pi \in \Pi_d$ such that

$$a_{1,\pi(1)}, a_{2,\pi(2)}, \ldots, a_{d,\pi(d)} \neq 0.$$

It follows from the definition that

$$d_{\mathrm{L}}(\pi) = \sum_{\substack{\ell=1 \\ \pi(\ell)<\ell}}^{d} [j_\ell - j_{\pi(\ell)}], \qquad d_{\mathrm{U}}(\pi) = \sum_{\substack{\ell=1 \\ \pi(\ell)>\ell}}^{d} [j_{\pi(\ell)} - j_\ell],$$

hence

$$d_{\mathrm{L}}(\boldsymbol{\pi}) - d_{\mathrm{U}}(\boldsymbol{\pi}) = \sum_{\substack{\ell=1 \\ \pi(\ell)\neq\ell}}^{d} [j_\ell - j_{\pi(\ell)}] = \sum_{\ell=1}^{d} [j_\ell - j_{\pi(\ell)}] = \sum_{\ell=1}^{d} j_\ell - \sum_{\ell=1}^{d} j_{\pi(\ell)}.$$

Recall, however, that $\boldsymbol{\pi}$ is a permutation of $\{1, 2, \ldots, d\}$; therefore

$$\sum_{\ell=1}^{d} j_{\pi(\ell)} = \sum_{\ell=1}^{d} j_\ell$$

and $d_{\mathrm{L}}(\boldsymbol{\pi}) - d_{\mathrm{U}}(\boldsymbol{\pi}) = 0$ for every $\boldsymbol{\pi} \in \Pi_d$ such that $a_{\ell,\pi(\ell)} \neq 0$, $\ell = 1, 2, \ldots, d$. Therefore

$$t^{d_{\mathrm{L}}(\boldsymbol{\pi})-d_{\mathrm{U}}(\boldsymbol{\pi})} \prod_{i=1}^{d} a_{i,\pi(i)} = \prod_{i=1}^{d} a_{i,\pi(i)}, \qquad \boldsymbol{\pi} \in \Pi_d,$$

and it follows from (10.41) that $g(s,t)$ is indeed independent of $t \in \mathbb{R} \setminus \{0\}$. ∎

Theorem 10.8 *Suppose that the matrix A has a compatible ordering vector and let $\mu \in \mathbb{C}$ be an eigenvalue of the matrix B, the iteration matrix of the Jacobi method. Then also*

(i) *$-\mu \in \sigma(B)$ and the multiplicities of $+\mu$ and $-\mu$ (as eigenvalues of B) are identical;*

(ii) *Given any $\omega \in (0, 2)$, every $\lambda \in \mathbb{C}$ that obeys the equation*

$$\lambda + \omega - 1 = \omega\mu\lambda^{1/2} \tag{10.42}$$

belongs to $\sigma(\mathcal{L}_\omega)$;[3]

(iii) *For every $\lambda \in \sigma(\mathcal{L}_\omega)$, $\omega \in (0, 2)$, there exists $\mu \in \sigma(B)$ so that the equation (10.42) holds.*

Proof According to Lemma 10.7, the presence of a compatible ordering vector of A implies that

$$\det(L_0 + U_0 - \mu D) = g(\mu, 1) = g(\mu, -1) = \det(-L_0 - U_0 - \mu D).$$

Moreover, $\det(-C) = (-1)^d \det C$ for any $d \times d$ matrix C and we thus deduce that

$$\det(L_0 + U_0 - \mu D) = (-1)^d \det(L_0 + U_0 + \mu D). \tag{10.43}$$

By the definition of an eigenvalue, $\mu \in \sigma(B)$ if and only if $\det(B - \mu I) = 0$ (see A.1.5.1). But, according to (10.20) and (10.43),

$$\det(B - \mu I) = \det\left[D^{-1}(L_0 + U_0) - \mu I\right] = \det[D^{-1}(L_0 + U_0 - \mu D)]$$

$$= \frac{1}{\det D} \det(L_0 + U_0 - \mu D) = \frac{(-1)^d}{\det D} \det(L_0 + U_0 + \mu D)$$

$$= (-1)^d \det(B + \mu I).$$

---

[3] The SOR iteration has been defined in (10.22) for $\omega \in [1, 2)$, while now we allow $\omega \in (0, 2)$. This should cause no difficulty whatsoever.

This proves (i).

The matrix $I - \omega L$ is lower triangular with ones across the diagonal, therefore $\det(I - \omega L) \equiv 1$. Hence, it follows from the definition (10.22) of $\mathcal{L}_\omega$ that

$$\det(\mathcal{L}_\omega - \lambda I) = \det\left\{(I - \omega L)^{-1}[\omega U + (1 - \omega)I] - \lambda I\right\}$$

$$= \frac{1}{\det(I - \omega L)}\det[\omega U + \omega\lambda L - (\lambda + \omega - 1)I]$$

$$= \det[\omega U + \omega\lambda L - (\lambda + \omega - 1)I]. \tag{10.44}$$

Suppose that $\lambda = 0$ lies in $\sigma(\mathcal{L}_\omega)$. Then (10.44) implies that $\det[\omega U - (\omega - 1)I] = 0$. Recall that U is strictly upper triangular, therefore

$$\det[\omega U - (\omega - 1)I] = (1 - \omega)^d$$

and we deduce $\omega = 1$. It is trivial to check that $(\lambda, \omega) = (0, 1)$ obeys the equation (10.42). Conversely, if $\lambda = 0$ satisfies (10.42) then we immediately deduce that $\omega = 1$ and, by (10.44), $0 \in \sigma(\mathcal{L}_1)$. Therefore, (ii) and (iii) are true in the special case $\lambda = 0$.

To complete the proof of the theorem, we need to discuss the case $\lambda \neq 0$. According to (10.44),

$$\frac{1}{\omega^d \lambda^{d/2}}\det(\mathcal{L}_\omega - \lambda I) = \det\left(\lambda^{1/2}L + \lambda^{-1/2}U - \frac{\lambda + \omega - 1}{\omega\lambda^{1/2}}I\right)$$

and, again using Lemma 10.7,

$$\frac{1}{\omega^d \lambda^{d/2}}\det(\mathcal{L}_\omega - \lambda I) = \det\left(L + U - \frac{\lambda + \omega - 1}{\omega\lambda^{1/2}}I\right) = \det\left(B - \frac{\lambda + \omega - 1}{\omega\lambda^{1/2}}I\right). \tag{10.45}$$

Let $\mu \in \sigma(B)$ and suppose that λ obeys the equation (10.42). Then

$$\mu = \frac{\lambda + \omega - 1}{\omega\lambda^{1/2}}$$

and substitution in (10.45) proves that $\det(\mathcal{L}_\omega - \lambda I) = 0$, hence $\lambda \in \sigma(\mathcal{L}_\omega)$. On the other hand, if $\lambda \in \sigma(\mathcal{L}_\omega)$, $\lambda \neq 0$, then, according to (10.45), $(\lambda + \omega - 1)/(\omega\lambda^{1/2}) \in \sigma(B)$. Consequently, there exists $\mu \in \sigma(B)$ such that (10.42) holds; thus the proof of the theorem is complete. ∎

Corollary   *Let A be a tridiagonal matrix and suppose that $a_{j,\ell} \neq 0$, $|j - \ell| \leq 1$, $j, \ell = 1, 2, \ldots$. Then $\rho(\mathcal{L}_1) = [\rho(B)]^2$.*

Proof   We have already mentioned that every tridiagonal matrix with a nonvanishing off-diagonal has a compatible ordering vector, a statement whose proof has been consigned to Exercise 10.8. Therefore (10.42) holds and, ω being unity, reduces to

$$\lambda = \mu\lambda^{1/2}.$$

In other words, either $\lambda = 0$ or $\lambda = \mu^2$. If $\lambda = 0$ for all $\lambda \in \sigma(\mathcal{L}_1)$ then part (iii) of the theorem implies that all the eigenvalues of B vanish as well. Hence $\rho(\mathcal{L}_1) = [\rho(B)]^2 = 0$. Otherwise, there exists $\lambda \neq 0$ in $\sigma(\mathcal{L}_1)$ and, since $\lambda = \mu^2$, parts (ii) and

(iii) of the theorem imply that $\rho(\mathcal{L}_1) \geq [\rho(B)]^2$ and $\rho(\mathcal{L}_1) \leq [\rho(B)]^2$ respectively. This proves the corollary. ∎

The statement of the corollary should not come as a surprise, since we have already observed behaviour consistent with $\rho(\mathcal{L}_1) = [\rho(B)]^2$ in Figure 10.1 and proved it in Section 10.2 for a specific TST matrix.

The importance of Theorem 10.8 ranges well beyond a comparison of the Jacobi and Gauss–Seidel schemes. It comes into its own when applied to SOR and its convergence.

Theorem 10.9 *Let A be a $d \times d$ matrix. If $\rho(\mathcal{L}_\omega) < 1$ and the SOR iteration (10.22) converges then necessarily $\omega \in (0, 2)$. Moreover, if A has a compatible ordering vector and all the eigenvalues of B are real then the iteration converges for every $\omega \in (0, 2)$ if and only if $\rho(B) < 1$ and the Jacobi method converges for the same matrix.*

Proof Let $\sigma(\mathcal{L}_\omega) = \{\lambda_1, \lambda_2, \ldots, \lambda_d\}$, therefore

$$\det \mathcal{L}_\omega = \prod_{\ell=1}^{d} \lambda_\ell. \tag{10.46}$$

Similarly to our derivation of (10.44), we have

$$\det \mathcal{L}_\omega = \det[\omega U - (\omega - 1)I] = (1 - \omega)^d$$

and substitution in (10.46) leads to the inequality

$$\rho(\mathcal{L}_\omega) = \max_{\ell=1,2,\ldots,d} |\lambda_\ell| \geq \left| \prod_{\ell=1}^{d} \lambda_\ell \right|^{1/d} = |1 - \omega|.$$

Therefore $\rho(\mathcal{L}_\omega) < 1$ is inconsistent with either $\omega \leq 0$ or $\omega \geq 2$ and the first statement of the theorem is true.

We next suppose that A possesses a compatible ordering vector. Thus, according to Theorem 10.8, for every $\lambda \in \sigma(\mathcal{L}_\omega)$ there exists $\mu \in \sigma(B)$ such that $p(\lambda^{1/2}) = 0$, where $p(z) := z^2 - \omega\mu z + (\omega - 1)$ (equivalently, λ is a solution of (10.42)).

Recall the *Cohn–Schur criterion* (Lemma 4.9): Both zeros of the quadratic $\alpha w^2 + \beta w + \gamma$, $\alpha \neq 0$, reside in the closed complex unit disc if and only if $|\alpha|^2 \geq |\gamma|^2$ and $(|\alpha|^2 - |\gamma|^2)^2 \geq |\alpha\bar{\beta} - \beta\bar{\gamma}|^2$. Similarly, it is possible to prove that both zeros of this quadratic reside in the *open* unit disc if and only if

$$|\alpha|^2 > |\gamma|^2 \quad \text{and} \quad (|\alpha|^2 - |\gamma|^2)^2 > |\alpha\bar{\beta} - \beta\bar{\gamma}|^2.$$

Letting $\alpha = 1$, $\beta = -\omega\mu$, $\gamma = \omega - 1$ and bearing in mind that $\mu \in \mathbb{R}$, these two conditions become $(\omega - 1)^2 < 1$ (which is the same as $\omega \in (0, 2)$) and $\mu^2 < 1$. Therefore, provided $\rho(B) < 1$, it is true that $|\mu| < 1$ for all $\mu \in \sigma(B)$, therefore $|\lambda|^{1/2} < 1$ for all $\lambda \in \sigma(\mathcal{L}_\omega)$ and $\rho(\mathcal{L}_\omega) < 1$ for all $\omega \in (0, 2)$. Likewise, if $\rho(\mathcal{L}_\omega) < 1$ then part (iii) of Theorem 10.8 implies that all the eigenvalues of B reside in the open unit disc. ∎

The condition that all the zeros of B are real is satisfied in the important special case where B is symmetric. If it fails, the second statement of the theorem need not be true; cf. Exercise 10.10, where it is proved that $\rho(B) < 1$ and $\rho(\mathcal{L}_\omega) > 1$ for $\omega \in (1,2)$ for the matrix

$$
A = \begin{bmatrix}
2 & 1 & 0 & \cdots & 0 \\
-1 & 2 & 1 & \ddots & \vdots \\
0 & \ddots & \ddots & \ddots & 0 \\
\vdots & \ddots & -1 & 2 & 1 \\
0 & \cdots & 0 & -1 & 2
\end{bmatrix},
$$

provided that d is sufficiently large.

Within the conditions of Theorem 10.9 there is not much to choose between our three iterative procedures regarding convergence: either they all converge or they all fail on that score. The picture changes when we take the speed of convergence into account. Thus, the corollary to Theorem 10.8 affirms that Gauss–Seidel is asymptotically *twice as good* as Jacobi. Bearing in mind that Gauss–Seidel is but a special case of SOR, we thus expect to improve the rate of convergence further by choosing a superior value of ω.

Theorem 10.10 *Suppose that A possesses a compatible ordering vector, that $\sigma(B) \subset \mathbb{R}$ and that $\tilde{\mu} := \rho(B) < 1$. Then*

$$
\rho(\mathcal{L}_{\omega_{\mathrm{opt}}}) < \rho(\mathcal{L}_\omega), \qquad \omega \in (0,2) \setminus \{\omega_{\mathrm{opt}}\},
$$

where

$$
\omega_{\mathrm{opt}} := \frac{2}{1 + \sqrt{1 - \tilde{\mu}^2}} = 1 + \left(\frac{\tilde{\mu}}{1 + \sqrt{1 - \tilde{\mu}^2}} \right)^2 \in (1,2). \tag{10.47}
$$

Proof Although it might not be immediately obvious, the proof is but an elaboration of the detailed example from Section 10.2. Having already done all the hard work, we can allow ourselves to proceed at an accelerated pace.

Solving the quadratic (10.42) yields

$$
\lambda = \tfrac{1}{4} \left[\omega\mu \pm \sqrt{(\omega\mu)^2 - 4(\omega - 1)} \right]^2.
$$

According to Theorem 10.8, both roots reside in $\sigma(\mathcal{L}_\omega)$.

Since μ is real, the term inside the square root is nonpositive when

$$
\tilde{\omega} := \frac{2(1 - \sqrt{1 - \mu^2})}{\mu^2} \leq \omega < 2.
$$

In this case $\lambda \in \mathbb{C} \setminus \mathbb{R}$ and it is trivial to verify that $|\lambda| = \omega - 1$.

In the remaining portion of the range of ω both roots λ are positive and the larger one equals $\tfrac{1}{4}[f(\omega, |\mu|)]^2$, where

$$
f(\omega, t) := \omega t + \sqrt{(\omega t)^2 - 4(\omega - 1)}, \qquad \omega \in (0, \tilde{\omega}], \quad t \in [0,1).
$$

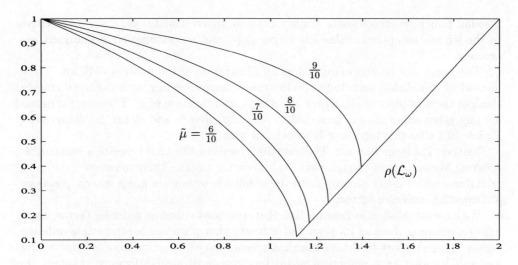

Figure 10.4   The graph of $\rho(\mathcal{L}_\omega)$ for different values of $\tilde{\mu}$.

It is an easy matter to ascertain that for any fixed $\omega \in (0, \tilde{\omega}]$ the function $f(\omega, \cdot)$ increases strictly monotonically for $t \in [0, 1)$. Likewise, $\tilde{\omega}$ increases strictly monotonically as a function of $\mu \in [0, 1)$. Therefore the spectral radius of $\mathcal{L}_\omega$ in the range $\omega \in \left(0, 2(1 - \sqrt{1 - \tilde{\mu}^2})\right)$ is $\frac{1}{4}f(\omega, \tilde{\mu})$. Note that the endpoint of the interval is

$$2(1 - \sqrt{1 - \tilde{\mu}^2}) = \frac{2}{1 + \sqrt{1 - \tilde{\mu}^2}} = \omega_{\mathrm{opt}},$$

as given in (10.47). The function $f(\cdot, t)$ decreases strictly monotonically in $\omega \in (0, \omega_{\mathrm{opt}})$ for any fixed $t \in [0, 1)$, thereby reaching its minimum at $\omega = \omega_{\mathrm{opt}}$.

As far as the interval $[\omega_{\mathrm{opt}}, 2)$ is concerned, the modulus of each λ 'corresponding' to $\mu \in \sigma(B)$, $|\mu| = \tilde{\mu}$, equals $\omega - 1$ and is at least as large as the magnitude of any other eigenvalues of $\mathcal{L}_\omega$. We thus deduce that

$$\rho(\mathcal{L}_\omega) = \begin{cases} \frac{1}{4}\left[\omega\tilde{\mu} + \sqrt{(\omega\tilde{\mu})^2 - 4(\omega - 1)}\right]^2, & \omega \in (0, \omega_{\mathrm{opt}}], \\ \omega - 1, & \omega \in [\omega_{\mathrm{opt}}, 2) \end{cases} \qquad (10.48)$$

and, for all $\omega \in (0, 2)$, $\omega \neq \omega_{\mathrm{opt}}$, it is true that

$$\rho(\mathcal{L}_\omega) > \rho(\mathcal{L}_{\omega_{\mathrm{opt}}}) = \left(\frac{\tilde{\mu}}{1 + \sqrt{1 - \tilde{\mu}^2}}\right)^2.$$

∎

Figure 10.4 displays $\rho(\mathcal{L}_\omega)$ for different values of $\tilde{\mu}$ for matrices that are consistent with the conditions of the theorem. As apparent from (10.48), each curve is composed of two smooth portions, joining at ω_{opt}.

In practical computation the value of $\tilde{\mu}$ is frequently estimated rather than derived in an explicit form. An important observation from Figure 10.4 is that it is always a sound policy to overestimate (rather than to underestimate) ω_{opt}, since the curve to the left of the optimal value has larger slope and overestimation is punished less severely.

The figure can be also employed as an illustration of the method of proof. Thus, instead of visualizing each individual curve as corresponding to a different matrix, think of them as plots of $|\lambda|$, where $\lambda \in \sigma(\mathcal{L}_\omega)$, as a function of ω. The spectral radius for any given value of ω is provided by the top curve – and it can be observed in Figure 10.4 that this top curve is associated with $\tilde{\mu}$.

Neither Theorem 10.8 nor Theorem 10.9 requires the knowledge of a compatible ordering vector – it is enough that such a vector exists. Unfortunately, it is not a trivial matter to verify directly from the definition whether a given matrix possesses a compatible ordering vector.

We have established in Lemma 10.6 that, provided A has an ordering vector, there exists a rearrangement of its rows and columns that possesses a compatible ordering vector. There is more to the lemma than meets the eye, since its proof is constructive and can be used as a numerical algorithm in a most straightforward manner. In other words, it is enough to find an ordering vector (provided that it exists!) and the algorithm from the proof of Lemma 10.6 takes care of compatibility!

A $d \times d$ matrix with a digraph $\mathbb{G} = \{\mathbb{V}, \mathbb{E}\}$ is said to possess *property* A if there exists a partition $\mathbb{V} = \mathbb{S}_1 \cup \mathbb{S}_2$, where $\mathbb{S}_1 \cap \mathbb{S}_2 = \emptyset$, such that for every $(j, \ell) \in \mathbb{E}$ either $j \in \mathbb{S}_1$, $\ell \in \mathbb{S}_2$ or $\ell \in \mathbb{S}_1$, $j \in \mathbb{S}_2$.

◇ **Property A**   As often in matters involving sparsity patterns, pictorial representation conveys more information than many a formal definition.

Let us consider, for example, a 6×6 matrix with a symmetric sparsity structure, as follows

$$\begin{bmatrix} \times & \times & \circ & \times & \circ & \circ \\ \times & \times & \times & \circ & \times & \circ \\ \circ & \times & \times & \times & \circ & \times \\ \times & \circ & \times & \times & \circ & \circ \\ \circ & \times & \circ & \circ & \times & \times \\ \circ & \circ & \times & \circ & \times & \times \end{bmatrix}$$

We claim that $\mathbb{S}_1 = \{1, 3, 5\}$, $\mathbb{S}_2 = \{2, 4, 6\}$ is a partition consistent with property A. To confirm this, write the digraph $\mathbb{G}$ in the following fashion.

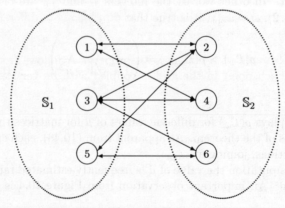

(In the interests of clarity, we have replaced each pair of arrows pointing in opposite directions by a single, 'two-sided' arrow.) Evidently, no edges join vertices in the same set, be it $\mathbb{S}_1$ or $\mathbb{S}_2$, and this is precisely the meaning of property A.

An alternative interpretation of property A comes to light when we rearrange the matrix in such a way that rows and columns corresponding to $\mathbb{S}_1$ precede those of $\mathbb{S}_2$. In our case, we permute rows and columns in the order $1, 3, 5, 2, 4, 6$ and the resultant sparsity pattern is then

$$
\begin{bmatrix}
\times & \circ & \circ & \times & \times & \circ \\
\circ & \times & \circ & \times & \times & \times \\
\circ & \circ & \times & \times & \circ & \times \\
\hline
\times & \times & \times & \times & \circ & \circ \\
\times & \times & \circ & \circ & \times & \circ \\
\circ & \times & \times & \circ & \circ & \times
\end{bmatrix} .
$$

In other words, the partitioned sparsity pattern consists of two diagonal blocks along the main diagonal. ◇

The importance of property A is encapsulated in the following result.

Lemma 10.11 *A matrix possesses property A if and only if it has an ordering vector.*

Proof Suppose first that a $d \times d$ matrix has property A and set

$$
j_\ell = \begin{cases} 1, & \ell \in \mathbb{S}_1, \\ 2, & \ell \in \mathbb{S}_2, \end{cases} \qquad \ell = 1, 2, \ldots, d.
$$

For any $(\ell, m) \in \mathbb{E}$ it is true that ℓ and m belong to different sets, therefore $j_\ell - j_m = \pm 1$ and we deduce that j is an ordering vector.

To establish the proof in the opposite direction assume that the matrix has an ordering vector j and let

$$
\mathbb{S}_1 := \{\ell \in \mathbb{V} : j_\ell \text{ is odd}\}, \qquad \mathbb{S}_2 := \{\ell \in \mathbb{V} : j_\ell \text{ is even}\}.
$$

Clearly, $\mathbb{S}_1 \cup \mathbb{S}_2 = \mathbb{V}$ and $\mathbb{S}_1 \cap \mathbb{S}_2 = \emptyset$, therefore $\{\mathbb{S}_1, \mathbb{S}_2\}$ is indeed a partition of $\mathbb{V}$. For any $\ell, m \in \mathbb{V}$ such that $(\ell, m) \in \mathbb{E}$ it follows from the definition of an ordering vector that $j_\ell - j_m = \pm 1$. In other words, the integers j_ℓ and j_m are of different parity, hence it follows from our construction that ℓ and m belong to different partition sets. Consequently, the matrix has property A. ∎

An important example of a matrix with property A follows from the five-point equations (7.16). Each point in the grid is coupled with its vertical and horizontal neighbours, hence we need to partition the grid points in such a way that $\mathbb{S}_1$ and $\mathbb{S}_2$ separate neighbours. This can be performed most easily in terms of *red–black ordering*, which we have already mentioned in Chapter 9. Thus, we traverse the grid as in natural ordering except that all grid points (ℓ, m) such that $\ell + m$ is odd, say, are consigned to $\mathbb{S}_1$ and all other points to $\mathbb{S}_2$.

An example, corresponding to a five-point formula in a 4×4 square, is presented in (9.4). Of course, the real purpose of the exercise is not simply to verify property

A or, equivalently, to prove that an ordering vector exists. Our goal is to identify a permutation that yields a compatible ordering vector. As we have already mentioned, this can be performed by the method of proof of Lemma 10.6. However, in the present circumstances we can single out such a vector directly *for the natural ordering*. For example, as far as (9.4) is concerned, we associate with every grid point (which, of course, corresponds to an equation and a variable in the linear system) an integer as follows:

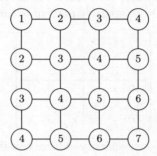

As can be easily verified, natural ordering results in a compatible ordering vector. All this can be easily generalized to rectangular grids of arbitrary size (see Exercise 10.11).

The exploitation of red–black ordering in the search for property A is not restricted to rectangular grids. Thus, consider the L-shaped grid

$$(10.49)$$

where '□' and '○' denote vertices in $\mathbb{S}_1$ and $\mathbb{S}_2$, respectively. Note that (10.49) serves a dual purpose: it is both the depiction of the computational grid and the graph of a matrix. It is quite clear that here this underlying matrix has property A. The task of finding explicitly a compatible ordering vector is relegated to Exercise 10.12.

10.4   The Poisson equation

Figure 10.5 displays the error attained by four different iterative methods, when applied to the Poisson equation (7.33) on a 16×16 grid. The first row depicts the line relaxation method (a variant of the incomplete LU factorization (10.11) – read on for details), the second corresponds to the Jacobi iteration (10.20), next comes the Gauss–Seidel method (10.21) and, finally, the bottom row displays the error in the successive over-relaxation method (10.22) with optimal choice of the parameter ω. Each column corresponds to a different number of iterations, specifically 50, 100 and 150, except that there is little point in displaying the error for ≥ 100 iterations for SOR since, remarkably, the error after 50 iterations is already close to machine accuracy![4]

Our first observation is that, evidently, all four methods converge. This is hardly a surprise in the case of Jacobi, Gauss–Seidel and SOR since we have already noted in the last section that the underlying matrix possesses property A. The latter feature explains also the very different rate of convergence: Gauss–Seidel converges at twice the pace of Jacobi while the speed of convergence of SOR is of a different order of magnitude altogether.

Another interesting observation pertains to the line relaxation method, a version of ILU from Section 10.1 where $\tilde{A}$ is the tridiagonal portion of A. Figure 10.5 suggests that line relaxation and Gauss–Seidel deliver very similar performances and we will prove later that this is indeed the case. We commence our discussion, however, with classical iterative methods.

Because, as we have noted in Section 10.3, the underlying matrix has a compatible ordering vector, we need to determine $\tilde{\mu} = \rho(B)$; and, by virtue of Theorems 10.8 and 10.10, $\tilde{\mu}$ determines completely both ω_{opt} and the rates of convergence of Gauss–Seidel and SOR.

Let $V = (v_{j,\ell})_{j,\ell=1}^{m}$ be an eigenvector of the matrix B from (10.20) and λ be the corresponding eigenvalue. We assume that the matrix A originates in the five-point formula (7.16) in a $m \times m$ square. Formally, V is a matrix; to obtain a genuine vector $\boldsymbol{v} \in \mathbb{R}^{m^2}$ we would need to stretch the grid, but this in fact will not be necessary.

Setting $v_{0,\ell},\, v_{m+1,\ell},\, v_{k,0},\, v_{k,m+1} := 0$, where $k, \ell = 1, 2, \ldots, m$, we can express $A\boldsymbol{v} = \lambda\boldsymbol{v}$ in the form

$$v_{j-1,\ell} + v_{j+1,\ell} + v_{j,\ell-1} + v_{j,\ell+1} = 4\lambda v_{j,\ell}, \qquad j, \ell = 1, 2, \ldots, m. \tag{10.50}$$

Our claim is that we are allowed to take

$$v_{j,\ell} = \sin\left(\frac{\pi p j}{m+1}\right) \sin\left(\frac{\pi q \ell}{m+1}\right), \qquad j, \ell = 1, 2, \ldots, m,$$

where p and q are arbitrary integers in $\{1, 2, \ldots, m\}$. If so, then

$$v_{j-1,\ell} + v_{j+1,\ell} = \left\{\sin\left[\frac{\pi p(j-1)}{m+1}\right] + \sin\left[\frac{\pi p(j+1)}{m+1}\right]\right\} \sin\left(\frac{\pi q \ell}{m+1}\right)$$

$$= 2v_{j,\ell}\cos\left(\frac{\pi p}{m+1}\right),$$

$$v_{j,\ell-1} + v_{j,\ell+1} = \sin\left(\frac{\pi p j}{m+1}\right)\left\{\sin\left[\frac{\pi q(\ell-1)}{m+1}\right] + \sin\left[\frac{\pi q(\ell+1)}{m+1}\right]\right\}$$

$$= 2v_{k,\ell}\cos\left(\frac{\pi q}{m+1}\right),$$

---

[4]To avoid any misunderstanding, this is the point to emphasize that by 'error' we mean departure from the solution of the corresponding five-point equations (7.16) *not* from the exact solution of the Poisson equation.

Figure 10.5 The error in the line relaxation, Jacobi, Gauss–Seidel and SOR (with ω_{opt}) methods for the Poisson equation (7.33) for $m = 16$ after 50, 100 and 150 iterations.

and substitution into (10.50) confirms that

$$\lambda = \lambda_{p,q} = \tfrac{1}{2}\left[\cos\left(\frac{\pi p}{m+1}\right) + \cos\left(\frac{\pi q}{m+1}\right)\right]$$

is an eigenvalue of A for every $p, q = 1, 2, \ldots, m$. This procedure yields all m^2 eigenvalues of A and we therefore deduce that

$$\tilde{\mu} = \rho(B) = \cos\left(\frac{\pi}{m+1}\right) \approx 1 - \frac{\pi^2}{2m^2}.$$

We next employ Theorem 10.8 to argue that

$$\rho(\mathcal{L}_1) = \tilde{\mu}^2 = \cos^2\left(\frac{\pi}{m+1}\right) \approx 1 - \frac{\pi^2}{m^2}. \tag{10.51}$$

Finally, (10.47) produces the optimal SOR parameter,

$$\omega_{\mathrm{opt}} = \frac{2}{1 + \sin[\pi/(m+1)]} = \frac{2\{1 - \sin[\pi/(m+1)]\}}{\cos^2[\pi/(m+1)]}$$

and

$$\rho(\mathcal{L}_{\omega_{\mathrm{opt}}}) = \frac{1 - \sin[\pi/(m+1)]}{1 + \sin[\pi/(m+1)]} \approx 1 - \frac{2\pi}{m}. \tag{10.52}$$

Note, incidentally, that (replacing m by d) our results are identical to the corresponding quantites for the TST matrix from Section 10.2; cf. (10.32), (10.35), (10.39) and (10.40). This is not a coincidence, since the TST matrix corresponds to a one-dimensional equivalent of the five-point formula.

The difference between (10.51) and (10.52) amounts to just a single power of m but glancing at Figure 10.5 ascertains that this seemingly minor distinction causes a most striking improvement in the speed of convergence.

Finally, we return to the top row of Figure 10.5, to derive the rate of convergence of the line relaxation method and explain its remarkable similarity to that for the Gauss–Seidel method.

In our implementation of the incomplete LU method in Section 10.1 we have split the matrix A into a tridiagonal portion and a remainder – the iteration is carried out on the tridiagonal part and, for reasons that have been clarified in Chapter 9, is very low in cost. In the context of five-point equations this splitting is termed *line relaxation*.

Provided the matrix A has been derived from the five-point formula in a square, we can write it in a block form that has been already implied in (9.3), namely

$$A = \begin{bmatrix} C & I & O & \cdots & O \\ I & C & I & \ddots & \vdots \\ O & \ddots & \ddots & \ddots & O \\ \vdots & \ddots & I & C & I \\ O & \cdots & O & I & C \end{bmatrix}, \tag{10.53}$$

where I and O are the $m \times m$ identity and zero matrices respectively, and

$$C = \begin{bmatrix} -4 & 1 & 0 & \cdots & 0 \\ 1 & -4 & 1 & \ddots & \vdots \\ 0 & \ddots & \ddots & \ddots & 0 \\ \vdots & \ddots & 1 & -4 & 1 \\ 0 & \ddots & 0 & 1 & -4 \end{bmatrix}.$$

In other words, A is block-TST and each block is itself a TST matrix.

Line relaxation (10.11) splits A into a tridiagonal part $\tilde{A}$ and a remainder $-E$. The matrix C being itself tridiagonal, we deduce that $\tilde{A}$ is block-diagonal, with C's along the main diagonal, while E is composed from the contribution of the off-diagonal blocks. Let λ and v be an eigenvalue and a corresponding eigenvector of the iteration matrix $\tilde{A}^{-1}E$. Therefore

$$Ev = \lambda \tilde{A} v$$

and, rendering as before the vector $v \in \mathbb{R}^{m^2}$ as an $m \times m$ matrix V, we obtain

$$v_{j,\ell-1} + v_{j,\ell+1} + \lambda(v_{j-1,\ell} - 4v_{j,\ell} + v_{j+1,\ell}) = 0, \qquad j,\ell = 1,2,\ldots,d. \qquad (10.54)$$

As before, we have assumed zero 'boundary values': $v_{j,0}, v_{j,m+1}, v_{0,\ell}, v_{m+1,\ell} = 0$, $j,\ell = 1,2,\ldots,d$.

Our claim (which, with the benefit of experience, is hardly surprising) is that

$$v_{j,\ell} = \sin\left(\frac{\pi p j}{m+1}\right) \sin\left(\frac{\pi q \ell}{m+1}\right), \qquad j,\ell = 1,2,\ldots,m,$$

for some $p,q \in \{1,2,\ldots,m\}$. Since

$$\sin\left[\frac{\pi p(j-1)}{m+1}\right] - 4\sin\left(\frac{\pi p j}{m+1}\right) + \sin\left[\frac{\pi p(j+1)}{m+1}\right] = 2\left[\cos\left(\frac{\pi p}{m+1}\right) - 2\right]\sin\left(\frac{\pi p j}{m+1}\right),$$

$$\sin\left[\frac{\pi q(\ell-1)}{m+1}\right] + \sin\left[\frac{\pi q(\ell+1)}{m+1}\right] = 2\cos\left(\frac{\pi q}{m+1}\right)\sin\left(\frac{\pi q \ell}{m+1}\right),$$

substitution in (10.54) results in

$$\lambda = \lambda_{p,q} = \frac{-\cos[\pi q/(m+1)]}{2 - \cos[\pi p/(m+1)]}.$$

Letting p,q range across $\{1,2,\ldots,m\}$, we recover all m^2 eigenvalues and, in particular, determine the spectral radius of the iteration matrix,

$$\rho(\tilde{A}^{-1}E) = \frac{\cos[\pi/(m+1)]}{2 - \cos[\pi/(m+1)]} \approx 1 - \frac{\pi^2}{m^2}. \qquad (10.55)$$

Comparison of (10.55) with (10.51) verifies our observation from Figure 10.5, that line relaxation and Gauss–Seidel have very similar rates of convergence. As a matter

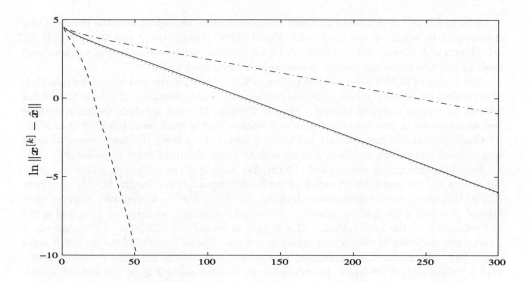

Figure 10.6 The logarithm of the error in Euclidean norm for Jacobi (broken-and-dotted line), Gauss–Seidel (dotted line), SOR (broken line) and line relaxation (solid line) for $m = 16$ in the first 300 iterations. The starting vector is $x^{[0]} = \mathbf{0}$.

of fact, it is easy to prove that Gauss–Seidel is marginally better, since

$$\cos^2 \varphi < \frac{\cos \varphi}{2 - \cos \varphi} < \cos^2 \varphi + \frac{\varphi^2}{2}, \qquad 0 < \varphi \le \pi.$$

Figure 10.6 displays (on a logarithmic scale) the decay of the Euclidean norm of the error after a given number of iterations – thus, the information in each three-dimensional surface from Figure 10.5 is reduced to a single number. Having analysed and understood the four methods in some detail, we can again observe and compare their features. There is however, an interesting new detail in Figure 10.6. In principle, the size of the spectral radius determines the speed of convergence only in an asymptotic sense and there is nothing in our analysis to tell how soon – or how late – the asymptotic regime occurs. However, we can observe in Figure 10.6 (and, for that matter, though for a different equation, in Figure 10.1) that the onset of asymptotic behaviour is pretty rapid for general starting vectors $x^{[0]}$.

Comments and bibliography

Numerical mathematics is an old art and its history is replete with the names of intellectual giants – Newton, Euler, Lagrange, Legendre, Gauss, Jacobi.... It is fair, however, to observe that the most significant milestone in its long journey has been the invention of the electronic computer. Numerical linear algebra, including iterative methods for sparse linear systems, is a case in point. A major research effort in the 1950s – the dawn of the computer era – led to an enhanced understanding of classical iterative methods and forms the cornerstone of our exposition.

A large number of textbooks and monographs concern themselves with the theme of this chapter and we single out the books of Axelsson (1994), Varga (1962) and Young (1971), cf. also Hageman & Young (1981). Readers who are at home with the terminology of functional analysis will also enjoy the concise monograph of Nevanlinna (1993).

The theory of SOR has been developed significantly beyond the material of Section 10.3. This involves much beautiful and intricate mathematics but is, arguably, of mainly theoretical interest for reasons that will become clearer in Chapter 11 – in a nutshell, we describe there how to accelerate Gauss–Seidel iteration in a manner that is much more powerful than SOR.

Classical iterative methods are neither the only nor, indeed, the best means to solve sparse linear systems by iteration, and we wish to single out three other approaches.

Recall the line relaxation method (10.11). We have split the matrix A, which has origi-nated from the five-point discretization of the Poisson equation in a square, into $\tilde{A} - E$, where $\tilde{A}$ is its tridiagonal part, subsequently iterating $\tilde{A}x^{[k+1]} = Ex^{[k]} + b$. Suppose, however, that instead of ordering the grid by columns (the natural ordering), we do so by rows and apply line relaxation to the new system. This is just as logical – or illogical – as employing a column-wise ordering but leads to a different iterative scheme (note that the matrix A stays intact but both x and b are permuted as a consequence of the row-wise rearrangement). Which variant should we adopt, by column or by row? A natural approach is to alternate. In other words, we write $A = A_x + A_y$, where A_x and A_y originate in central differencing in the $x-$ and y-directions respectively. Column-wise line relaxation (10.11) reads

$$(A_y - 2I)x^{[k+1]} = -(A_x + 2I)x^{[k]} + b, \qquad k = 0, 1, \ldots,$$

while a row-wise rearrangement results in

$$(A_x - 2I)x^{[k+1]} = -(A_y + 2I)x^{[k]} + b, \qquad k = 0, 1, \ldots.$$

For greater generality, we choose parameters $\alpha_0, \alpha_1, \ldots$ and iterate

$$\begin{aligned} (A_x - \alpha_{2k}I)x^{[2k+1]} &= -(A_y + \alpha_{2k})x^{[2k]} + b, \\ (A_y - \alpha_{2k+1}I)x^{[2k+2]} &= -(A_x + \alpha_{2k+1})x^{[2k+1]} + b, \end{aligned} \qquad k = 0, 1, \ldots.$$

This is the *alternate directions implicit* (ADI) method (Wachspress, 1966). As often in numerical analysis, the devil is in the parameter. However, it is known how to choose $\{\alpha_k\}$ so as to accelerate ADI a great deal. The outcome, at least in certain cases, e.g. the Poisson equation in a square, is an iterative method that clearly outperforms SOR. This, however, falls outside the scope of this volume.

Another example of an iterative nonstationary scheme is the method of *conjugate gradi-ents*. Let us suppose that a $d \times d$ matrix A is symmetric and positive definite. Wishing to solve the linear system (10.1), we choose a positive integer ν and let

$$f(x) := \tfrac{1}{2}x^T A^\nu x - b^T A^{\nu-1}x, \qquad x \in \mathbb{R}^d.$$

The function f is differentiable and its gradient is

$$\nabla f(x) = A^\nu x - A^{\nu-1}b = A^{\nu-1}(Ax - b).$$

Therefore, and since A is positive definite, $\nabla f(x) = 0$ if and only if $Ax = b$. Moreover, $\nabla^2 f(x) = A$ hence, exploiting again positive definiteness, we deduce that $\nabla f(x) = 0$ occurs at a unique minimum of f. Consequently,

$$f(\hat{x}) = \min_{x \in \mathbb{R}^d} f(x) \qquad \Longleftrightarrow \qquad \hat{x} \text{ is the solution of (10.1)}$$

and the underlying problem is equivalent to the minimization of f.

We next observe that

$$f(\boldsymbol{x}) = \tfrac{1}{2}(\boldsymbol{x} - \hat{\boldsymbol{x}})^T A^\nu (\boldsymbol{x} - \hat{\boldsymbol{x}}) - \tfrac{1}{2}\hat{\boldsymbol{x}}^T A^\nu \hat{\boldsymbol{x}}$$

and that its minimization is the same as that of the function

$$\tilde{f}(\boldsymbol{x}) := \tfrac{1}{2}(\boldsymbol{x} - \hat{\boldsymbol{x}})^T A^\nu (\boldsymbol{x} - \hat{\boldsymbol{x}}), \qquad \boldsymbol{x} \in \mathbb{R}^d;$$

after all, f and $\tilde{f}$ differ just by an constant that is independent of the variable $\boldsymbol{x}$. On the face of it, minimizing $\tilde{f}$ is nonsensical, since its definition involves $\hat{\boldsymbol{x}}$, the unknown goal of the whole calculation! Fortunately, although $\hat{\boldsymbol{x}}$ is unknown, we can easily obtain $A(\boldsymbol{x} - \hat{\boldsymbol{x}})$ for any $\boldsymbol{x} \in \mathbb{R}^d$ by observing that

$$A(\boldsymbol{x} - \hat{\boldsymbol{x}}) = (A\boldsymbol{x} - \boldsymbol{b}) - (A\hat{\boldsymbol{x}} - \boldsymbol{b}) = \boldsymbol{r}(\boldsymbol{x}),$$

where $\boldsymbol{r}(\boldsymbol{x}) := A\boldsymbol{x} - \boldsymbol{b}$ is the *residual* of $\boldsymbol{x} \in \mathbb{R}^d$ (note that $\boldsymbol{r}(\hat{\boldsymbol{x}}) = \boldsymbol{0}$, because $\hat{\boldsymbol{x}}$ is the solution of (10.1)). Consequently, we can write $\tilde{f}$ in the form

$$\tilde{f}(\boldsymbol{x}) = \tfrac{1}{2}\boldsymbol{r}(\boldsymbol{x})^T A^{\nu-2} \boldsymbol{r}(\boldsymbol{x}), \qquad \boldsymbol{x} \in \mathbb{R}^d. \tag{10.56}$$

We choose a starting vector $\boldsymbol{x}^{[0]} \in \mathbb{R}^d$ and let

$$\boldsymbol{x}^{[k+1]} = \boldsymbol{x}^{[k]} + \lambda_k \boldsymbol{d}^{[k]}, \qquad k = 0, 1, \ldots, \tag{10.57}$$

where the scalar λ_k and the *direction* $\boldsymbol{d}^{[k]} \in \mathbb{R}^d$ are yet to be defined. Letting $\boldsymbol{r}^{[k]} := \boldsymbol{r}(\boldsymbol{x}^{[k]})$ we deduce from (10.57) that the residual obeys the recursion

$$\boldsymbol{r}^{[k+1]} = \boldsymbol{r}^{[k]} + \lambda_k A \boldsymbol{d}^{[k]}, \qquad k = 0, 1, \ldots.$$

Therefore, substitution in (10.56) results in

$$\tilde{f}(\boldsymbol{x}^{[k+1]}) = \tfrac{1}{2}\boldsymbol{r}^{[k]^T} A^{\nu-2} \boldsymbol{r}^{[k]} + \lambda_k \boldsymbol{d}^{[k]^T} A^{\nu-1} \boldsymbol{d}^{[k]} + \tfrac{1}{2}\lambda_k^2 \boldsymbol{d}^{[k]^T} A^\nu \boldsymbol{d}^{[k]}, \qquad k = 0, 1, \ldots,$$

a quadratic equation in λ_k. Bearing in mind that our goal is to choose $\boldsymbol{x}^{[k+1]}$ so as to minimize $\tilde{f}$, we let λ_k be the minimum of the quadratic. Since A is positive definite, the minimum of $\tilde{f}$ is obtained by setting $\mathrm{d}\tilde{f}/\mathrm{d}\lambda_k$ to zero, and the outcome is

$$\lambda_k = -\frac{\boldsymbol{d}^{[k]^T} A^{\nu-1} \boldsymbol{d}^{[k]}}{\boldsymbol{d}^{[k]^T} A^\nu \boldsymbol{d}^{[k]}}.$$

A central idea behind the method of conjugate gradients is to choose the directions

$$\begin{aligned} \boldsymbol{d}^{[0]} &= -\boldsymbol{r}^{[0]}, \\ \boldsymbol{d}^{[k]} &= -\boldsymbol{r}^{[k]} + \beta_k \boldsymbol{d}^{[k-1]}, \qquad k = 1, 2, \ldots, \end{aligned}$$

Figure 10.7   The error in conjugate gradients and SOR (with ω_{opt}) for the Poisson
equation (7.33) with $m = 21$ after 25, 50, 75 and 100 iterations.

where

$$\beta_k := \frac{r^{[k]^T} A^\nu d^{[k-1]}}{d^{[k-1]^T} A^\nu d^{[k-1]}}.$$

It follows at once from our construction that $d^{[k]^T} A^\nu d^{[k-1]} = 0$ and, with somewhat greater
effort, it is possible to prove that $d^{[k]^T} A^\nu d^{[\ell]} = 0$ for all $\ell = 0, 1, \ldots, k - 1$. In other words,
the directions are orthogonal with respect to the weighted inner product $\langle u, v \rangle := u^T A^\nu v$,
where $u, v \in \mathbb{R}^d$ (this is a proper inner product since A is positive definite) (Axelsson, 1994;
Golub & van Loan, 1989).

Since orthogonal vectors are linearly independent, it follows that, in exact arithmetic,
the method of conjugate gradients converges to the exact solution of (10.1) in at most d
iterations; a result of doubtful practicality since d is likely to be large and computers operate
in inexact arithmetic. The reality, in fact, is considerably more interesting: the number of
iterations that suffices for the convergence of conjugate gradients (again, in exact arithmetic)
is not d but the number of different eigenvalues of A. This is a critical observation that
acts as a starting point in the development of the *preconditioned conjugate gradient* method
(Axelsson, 1994).

Figure 10.7 displays the error for the conjugate gradients method (with $\nu = 1$), as applied
to the five-point equations in a square with $m = 21$. For comparison, we also provide identical
information for the SOR method. There is little need to elaborate, since conjugate gradients clearly win handsomely – in fact, the error after 100 iterations consists of little more than

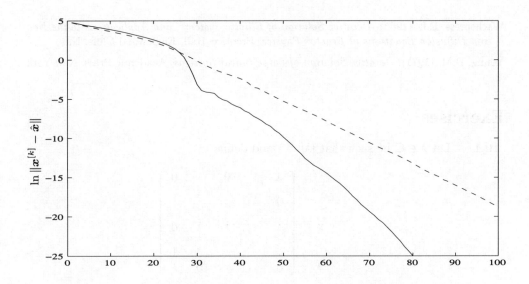

Figure 10.8 The logarithms of of the error (in the Euclidean norm) for conjugate gradients (solid line) and SOR (broken line) for $m = 21$ in the first 100 iterations. The starting vector is $x^{[0]} = 0$.

roundoff! And this is the outcome even without preconditioning, which has been mentioned in the last paragraph and which accelerates convergence a great deal further.

A further comparison between conjugate gradients and SOR is presented in Figure 10.8, where we plot the Euclidean norm of the error in the first 100 iterations for the problem from Figure 10.7. This is perhaps the place to comment that a general convergence theory for conjugate gradients is available and, had space allowed, we could have determined the exact decay of the error for the five-point equations in a square, in line with the theme of Section 10.4.

The method of conjugate gradients is the oldest representative of a large menagerie of so-called *Krylov space methods* (Freund *et al.*, 1992), which have proved themselves in the last decade to be a powerful tool in the numerical solution of linear (and even nonlinear) algebraic systems. Other interesting examples of Krylov space methods are *generalized minimal residuals (GMRes)* and *bi-conjugate gradients*.

Axelsson, O. (1994), *Iterative Solution Methods,* Cambridge University Press, Cambridge.

Freund, R.W., Golub, G.H. and Nachtigal, N.M. (1992), Iterative solution of linear systems, *Acta Numerica* **1**, 57–100.

Golub, G.H. and van Loan, C.F. (1989), *Matrix Computations* (2nd ed.), Johns Hopkins University Press, Baltimore.

Hageman, L.A. and Young, D.M. (1981), *Applied Iterative Methods,* Academic Press, New York.

Nevanlinna, O. (1993), *Convergence of Iterations for Linear Equations,* Birkhäuser, Basel.

Varga, R.S. (1962), *Matrix Iterative Analysis,* Prentice–Hall, Englewood Cliffs, NJ.

Wachspress, E.L. (1966), *Iterative Solution of Elliptic Systems, and Applications to the Neutron Diffusion Equations of Reactor Physics,* Prentice–Hall, Englewood Cliffs, NJ.

Young, D.M. (1971), *Iterative Solution of Large Linear Systems,* Academic Press, New York.

Exercises

10.1   Let $\lambda \in \mathbb{C}$ be such that $|\lambda| < 1$ and define

$$
J = \begin{bmatrix}
\lambda & 1 & 0 & \cdots & 0 \\
0 & \lambda & \ddots & \ddots & \vdots \\
\vdots & \ddots & \ddots & 1 & 0 \\
\vdots & & \ddots & \lambda & 1 \\
0 & \cdots & \cdots & 0 & \lambda
\end{bmatrix}.
$$

Prove that $\lim_{k\to\infty} J^k = O$.

10.2   Suppose that the $d \times d$ matrix H has a full set of eigenvectors, i.e., that there exist d linearly independent vectors $\boldsymbol{w}_\ell \in \mathbb{C}^d$ and numbers $\lambda_1, \lambda_2, \ldots, \lambda_d \in \mathbb{C}$ such that $H\boldsymbol{w}_\ell = \lambda_\ell \boldsymbol{w}_\ell$, $\ell = 1, 2, \ldots, d$. We consider the iterative scheme (10.3). Let $\boldsymbol{r}^{[k]} := (I - H)\boldsymbol{x}^{[k]} - \boldsymbol{v}$ be the residual in the kth iteration and suppose that

$$
\boldsymbol{r}^{[0]} = \sum_{\ell=1}^{d} \alpha_\ell \boldsymbol{w}_\ell
$$

(such $\alpha_1, \alpha_2, \ldots, \alpha_d$ always exist – why?). Prove that

$$
\boldsymbol{r}^{[k]} = \sum_{\ell=1}^{d} \alpha_\ell \lambda_\ell^k \boldsymbol{w}_\ell, \qquad k = 0, 1, \ldots.
$$

Outline an alternative proof of Lemma 10.1 using this representation.

10.3   Prove that the Gauss–Seidel iteration converges whenever the matrix A is symmetric and positive definite.

10.4   Show that the SOR method is a regular splitting (10.12) with

$$
P = \omega^{-1}D - L_0, \qquad N = (\omega^{-1} - 1)D + U_0.
$$

10.5   Let A be a symmetric, positive definite, tridiagonal matrix. Prove that the SOR method converges for this matrix and for $0 < \omega < 2$.

10.6   Let A be a TST matrix such that $a_{1,1} = \alpha$ and $a_{1,2} = \beta$. Show that the Jacobi iteration converges if $2|\beta| < |\alpha|$. Moreover, prove that if convergence is required for *all* $d \geq 1$ then this inequality is necessary as well as sufficient.

10.7 Demonstrate that

$$\sum_{\ell=1}^{d} \sin^2\left(\frac{\pi j\ell}{d+1}\right) = \tfrac{1}{2}(d+1), \qquad j = 1, 2, \ldots, d,$$

thereby verifying (10.29).

10.8 Let

$$A = \begin{bmatrix} \alpha_1 & \beta_1 & 0 & \cdots & & 0 \\ \gamma_1 & \alpha_2 & \beta_2 & \ddots & & \vdots \\ 0 & \ddots & \ddots & \ddots & & 0 \\ \vdots & \ddots & & \gamma_{d-2} & \alpha_{d-1} & \beta_{d-1} \\ 0 & \cdots & & 0 & \gamma_{d-1} & \alpha_d \end{bmatrix},$$

where $\beta_\ell, \gamma_\ell \neq 0$, $\ell = 1, 2, \ldots, d-1$. Prove that $\boldsymbol{j}$, where $j_\ell = \ell$, $\ell = 1, 2, \ldots, d$, is a compatible ordering vector of A.

10.9 Find an ordering vector for a matrix with the sparsity pattern

$$\begin{bmatrix} \times & \times & \circ & \times & \circ & \circ & \circ & \circ & \circ & \times \\ \times & \times & \times & \circ & \circ & \circ & \circ & \circ & \times & \circ \\ \circ & \times & \times & \times & \circ & \times & \circ & \circ & \circ & \circ \\ \times & \circ & \times & \times & \times & \circ & \circ & \circ & \circ & \circ \\ \circ & \circ & \circ & \times & \times & \times & \circ & \times & \circ & \circ \\ \circ & \circ & \times & \times & \times & \times & \times & \circ & \circ & \circ \\ \circ & \circ & \circ & \circ & \circ & \times & \times & \times & \circ & \times \\ \circ & \circ & \circ & \circ & \times & \circ & \times & \times & \times & \circ \\ \circ & \times & \circ & \circ & \circ & \circ & \circ & \times & \times & \times \\ \times & \circ & \circ & \circ & \circ & \circ & \times & \circ & \times & \times \end{bmatrix}.$$

10.10* We consider the $d \times d$ tridiagonal Toeplitz matrix

$$A = \begin{bmatrix} 2 & 1 & 0 & \cdots & 0 \\ -1 & 2 & 1 & \ddots & \vdots \\ 0 & \ddots & \ddots & \ddots & 0 \\ \vdots & \ddots & -1 & 2 & 1 \\ 0 & \cdots & 0 & -1 & 2 \end{bmatrix}.$$

a Prove that A possesses a compatible ordering vector.

b Find explicitly all the eigenvalues of the Jacobi iteration matrix B. [*Hint: solve explicitly the difference equation obeyed by the components of an eigenvector of B.*] Conclude that this iterative scheme diverges.

    c Using Theorem 10.8, or otherwise, show that $\rho(\mathcal{L}_\omega) > 1$ and that the SOR iteration diverges for all choices of $\omega \in (0,2)$.

10.11   Let A be an $m^2 \times m^2$ matrix, which originates in the implementation of the five-point formula in a square $m \times m$ grid. For every grid point (r,s) we let

$$j_{(r,s)} := m + s - r, \qquad r,s, = 1,2,\ldots,m,$$

and we construct a vector j by assembling the components in the same order as that used in the matrix A. Prove that j is an ordering vector of A and identify a permutation for which j is a compatible ordering vector.

10.12   Find a compatible ordering vector for the L-shaped grid (10.49).

11

Multigrid techniques

11.1 In lieu of a justification...

How good is the Gauss–Seidel iteration (10.21) at solving the five-point equations on an $m \times m$ grid? On the face of it, posing this question just after we have completed a whole chapter devoted to iterative methods is neither necessary nor appropriate. According to (10.51), the spectral radius of the iteration matrix is $\cos^2[\pi/(m+1)] \approx 1 - \pi^2 m^{-2}$ and inspection of the third row of Figure 10.5 will convince us that this presents a fair estimate of the behaviour of the scheme. Yet, by its very nature, the spectral radius displays the *asymptotic* attenuation rate of the error and it is entirely legitimate to query how well (or badly) Gauss–Seidel performs before the onset of its asymptotic regime.

Figure 11.1 displays the logarithm of the Euclidean norm of the residual for $m = 10, 20, 40, 80$; we remind the reader that, given the equation

$$Ax = b \tag{11.1}$$

and a sequence of iterants $\{x^{[k]}\}_{k=0}^{\infty}$, the residual is defined as $r^{[k]} = Ax^{[k]} - b$, $k \geq 0$.[1] The emerging picture is startling: the norm drops dramatically in the first few iterations! Only after a while does the rate of attenuation approach the linear curve predicted by the spectral radius of $\mathcal{L}_1$. Moreover, this phenomenon – unlike the asymptotic rate of decay of $\ln \|r^{[k]}\|$ – appears to be fairly independent of the magnitude of m.

A similar lesson can be drawn from Figure 11.2, where we have displayed in detail the residuals for the first six even numbers of iterations for $m = 20$. Evidently, a great deal of the error disappears very fast indeed, while after about ten iterations nothing much changes and each further iteration removes roughly $\cos^2(\pi/21) \approx 0.9778$ of the remaining residual.

The explanation of this phenomenon is quite interesting, as far as the understanding of Gauss–Seidel is concerned. More importantly, it provides a clue about how to accelerate iterative schemes for linear algebraic equations that originate in finite difference and finite element discretizations.

We hasten to confess that limitations of space and of the degree of mathematical sophistication that we allow ourselves in this book preclude us from providing a comprehensive explanation of the aforementioned phenomenon. Instead, we plan to indulge

---

[1]There exists an intimate connection between $r^{[k]}$ and the error $\varepsilon^{[k]} = x^{[k]} - \hat{x}$, where $\hat{x}$ is the solution of (11.1) – cf. Exercise 11.1.

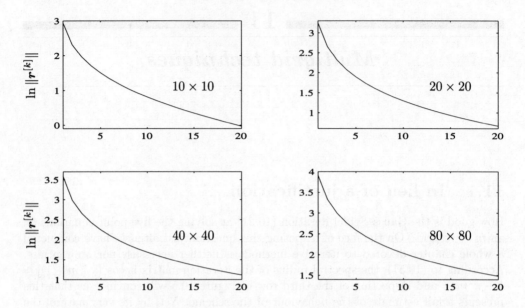

Figure 11.1 The logarithm of the norm of the residual in the first twenty Gauss–Seidel iterations for the five-point discretization of the Poisson equation (7.33).

Figure 11.2 The residual after various small numbers k of Gauss–Seidel iterations for the five-point discretization of the Poisson equation (7.33) with $m = 20$.

in a great deal of mathematical hand-waving, our excuse being that it conveys the spirit, if not the letter, of a complete analysis and sets us on the right path to exploit the phenomenon in presenting superior iterative schemes.

Let us subtract the exact five-point equations,

$$u_{j-1,\ell} + u_{j,\ell-1} + u_{j+1,\ell} + u_{j,\ell+1} - 4u_{j,\ell} = (\Delta x)^2 f_{j,\ell}, \qquad j,\ell = 1,2,\ldots,m,$$

from the Gauss–Seidel scheme

$$u_{j-1,\ell}^{[k+1]} + u_{j,\ell-1}^{[k+1]} + u_{j+1,\ell}^{[k]} + u_{j,\ell+1}^{[k]} - 4u_{j,\ell}^{[k+1]} = (\Delta x)^2 f_{j,\ell}, \qquad j,\ell = 1,2,\ldots,m.$$

The outcome is

$$\varepsilon_{j-1,\ell}^{[k+1]} + \varepsilon_{j,\ell-1}^{[k+1]} + \varepsilon_{j+1,\ell}^{[k]} + \varepsilon_{j,\ell+1}^{[k]} - 4\varepsilon_{j,\ell}^{[k+1]} = 0, \qquad j,\ell = 1,2,\ldots,m, \qquad (11.2)$$

where $\varepsilon_{j,\ell}^{[k]} := u_{j,\ell}^{[k]} - u_{j,\ell}$ is the error after k iterations at the (j,ℓ)th grid point. Since we assume Dirichlet boundary conditions, $u_{j,\ell}$ and $u_{j,\ell}^{[k]}$ are identical at all boundary grid points, therefore $\varepsilon_{j,\ell}^{[k]} = 0$ there.

Let

$$p^{[k]}(\theta,\psi) = \sum_{j=1}^{m} \sum_{\ell=1}^{m} \varepsilon_{j,\ell}^{[k]} \mathrm{e}^{\mathrm{i}(j\theta+\ell\psi)}, \qquad 0 \le \theta, \psi \le 2\pi,$$

be a *bivariate Fourier transform* of the sequence $\{\varepsilon_{j,\ell}^{[k]}\}_{j,\ell=1}^{m}$. We intend to consider Fourier transforms in a more formal setting in Section 12.2 and subsequently to employ them (to entirely different ends) in Chapters 13 and 14. In the present section our treatment of Fourier transfroms is therefore perfunctory and we will hint at proofs rather than providing them with any degree of detail.

We measure the magnitude of $p^{[k]}$ in the Euclidean norm

$$\|g\| = \left[\frac{1}{4\pi^2} \int_{-\pi}^{\pi} \int_{-\pi}^{\pi} |g(\theta,\psi)|^2 \, \mathrm{d}\theta \, \mathrm{d}\psi \right]^{1/2}.$$

Therefore

$$\begin{aligned}
\|p^{[k]}\|^2 &= \frac{1}{4\pi^2} \int_{-\pi}^{\pi} \int_{-\pi}^{\pi} \left| \sum_{j=1}^{m} \sum_{\ell=1}^{m} \varepsilon_{j,\ell}^{[k]} \mathrm{e}^{\mathrm{i}(j\theta+\ell\psi)} \right|^2 \mathrm{d}\theta \, \mathrm{d}\psi \\
&= \frac{1}{4\pi^2} \sum_{j_1=1}^{m} \sum_{j_2=1}^{m} \sum_{\ell_1=1}^{m} \sum_{\ell_2=1}^{m} \varepsilon_{j_1,\ell_1}^{[k]} \bar{\varepsilon}_{j_2,\ell_2}^{[k]} \int_{-\pi}^{\pi} \mathrm{e}^{\mathrm{i}(j_1-j_2)\theta} \, \mathrm{d}\theta \int_{-\pi}^{\pi} \mathrm{e}^{\mathrm{i}(\ell_1-\ell_2)\psi} \, \mathrm{d}\psi \\
&= \sum_{j=1}^{m} \sum_{\ell=1}^{m} |\varepsilon_{j,\ell}^{[k]}|^2 = \|\boldsymbol{\varepsilon}^{[k]}\|^2,
\end{aligned}$$

where

$$\|\boldsymbol{y}\| = \left(\sum_{j=1}^{m} \sum_{\ell=1}^{m} |y_{j,\ell}|^2 \right)^{1/2}$$

is the standard Euclidean norm on vectors in $\mathbb{C}^{m^2}$ (which, for convenience, we arrange in $m \times m$ arrays. Here $\| \cdot \|$ should not be mistaken for the Euclidean *matrix* norm; see A.1.3.4). We have outlined a proof of a remarkable identity, which is at the root of many applications of Fourier transforms: provided we measure both vectors and their transforms in the corresponding Euclidean norms, their magnitudes are the same.[2] Recalling our goal, to measure the rate of decay of the residuals, we deduce that monitoring $\|p^{[k]}\|$ or $\|\varepsilon^{[k]}\|$ is equivalent.

We multiply (11.2) by $\mathrm{e}^{\mathrm{i}(j\theta + \ell\psi)}$ and sum for $j, \ell = 1, 2, \ldots, d$. Since

$$\sum_{j=1}^{m} \sum_{\ell=1}^{m} \varepsilon_{j-1,\ell}^{[k+1]} \mathrm{e}^{\mathrm{i}(j\theta + \ell\psi)} = \sum_{j=0}^{m-1} \sum_{\ell=1}^{m} \varepsilon_{j,\ell}^{[k+1]} \mathrm{e}^{\mathrm{i}((j+1)\theta + \ell\psi)} = \mathrm{e}^{\mathrm{i}\theta} p^{[k]}(\theta, \psi) - \mathrm{e}^{\mathrm{i}(m+1)\theta} \sum_{\ell=1}^{m} \varepsilon_{m,\ell}^{\mathrm{i}\ell\psi},$$

applying similar algebra to the other terms in (11.2) we obtain the identity

$$(4 - \mathrm{e}^{\mathrm{i}\theta} - \mathrm{e}^{\mathrm{i}\psi}) p^{[k+1]}(\theta, \psi) = (\mathrm{e}^{-\mathrm{i}\theta} + \mathrm{e}^{-\mathrm{i}\psi}) p^{[k]}(\theta, \psi)$$

$$- \left\{ \mathrm{e}^{\mathrm{i}(m+1)\theta} \sum_{\ell=1}^{m} \varepsilon_{m,\ell}^{[k+1]} \mathrm{e}^{\mathrm{i}\ell\psi} + \mathrm{e}^{\mathrm{i}(m+1)\psi} \sum_{j=1}^{m} \varepsilon_{j,m}^{[k+1]} \mathrm{e}^{\mathrm{i}j\theta} \right.$$

$$\left. + \sum_{\ell=1}^{m} \varepsilon_{1,\ell}^{[k]} \mathrm{e}^{\mathrm{i}\ell\psi} + \sum_{j=1}^{m} \varepsilon_{j,1}^{[k]} \mathrm{e}^{\mathrm{i}j\theta} \right\}.$$

This is a moment when we commit a mathematical crime and assume that the term in the curly brackets is so small in comparison with $p^{[k]}$ and $p^{[k+1]}$ that it can be harmlessly neglected. Our half-hearted excuse is that this term sums over m components, whereas $p^{[k]}$, say, sums over m^2, and that if boundary conditions were periodic rather than Dirichlet, it would have disappeared altogether. However, the true justification, as for most other crimes, is that it pays.

The main idea now is to consider how fast the Gauss–Seidel iteration attentuates each individual *wavenumber* (θ, ψ). In other words, we are interested in the *local attenuation factor*

$$\rho^{[k]}(\theta, \psi) := \left| \frac{p^{[k+1]}(\theta, \psi)}{p^{[k]}(\theta, \psi)} \right|, \qquad |\theta|, |\psi| \leq \pi.$$

Having agreed that

$$(4 - \mathrm{e}^{\mathrm{i}\theta} - \mathrm{e}^{\mathrm{i}\psi}) p^{[k+1]}(\theta, \psi) \approx (\mathrm{e}^{-\mathrm{i}\theta} + \mathrm{e}^{-\mathrm{i}\psi}) p^{[k]}(\theta, \psi), \qquad |\theta|, |\psi| \leq \pi,$$

we can make the following estimate

$$\rho^{[k]}(\theta, \psi) \approx \tilde{\rho}(\theta, \psi) := \left| \frac{\mathrm{e}^{\mathrm{i}\theta} + \mathrm{e}^{\mathrm{i}\psi}}{4 - \mathrm{e}^{\mathrm{i}\theta} - \mathrm{e}^{\mathrm{i}\psi}} \right|, \qquad |\theta|, |\psi| \leq \pi. \tag{11.3}$$

---

[2]Lemma 13.9 provides a more formal statement of this important result, as well as a complete proof in a single dimension. A generalization to bivariate – indeed, multivariate – Fourier transforms is straightforward.

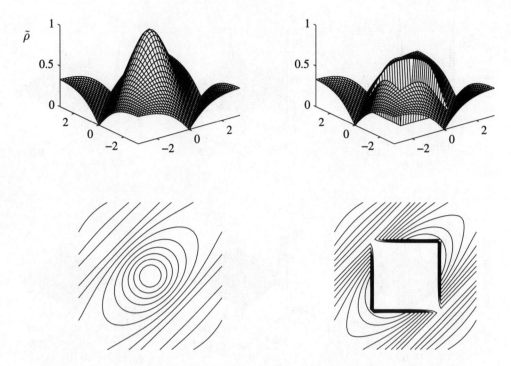

Figure 11.3 The function $\tilde{\rho}$ as a three-dimensional surface and as a contour plot. The left-hand column displays the whole square $[-\pi, \pi] \times [-\pi, \pi]$, while the right-hand column displays only the set $\mathbb{O}_0$.

Note that the function $\tilde{\rho}$ is independent of k. On the left in Figure 11.3 the whole square $[-\pi, \pi] \times [-\pi, \pi]$ is displayed and there are no surprises there; thus, $\tilde{\rho}(\theta, \psi) \leq 1$ for all $|\theta|, |\psi| \leq \pi$.

Considerably more interesting is the right-hand column of Figure 11.3, where the function $\tilde{\rho}$ is displayed just in the set

$$\mathbb{O}_0 := \left\{ (\theta, \psi) \: : \: \tfrac{1}{2}\pi \leq \max\{|\theta|, |\psi|\} \leq \pi \right\}$$

of *oscillatory* wavenumbers. A remarkable feature emerges: provided only such wavenumbers are considered, the function $\tilde{\rho}$ peaks at a value of $\frac{1}{2}$.

Forearmed with this observation, we formally evaluate the maximum of $\tilde{\rho}$ within the set $\mathbb{O}_0$. It is not difficult to verify that

$$\max_{(\theta,\psi)\in\mathbb{O}} \tilde{\rho}(\theta, \psi) = \tilde{\rho}\left(\tfrac{\pi}{2}, \tan^{-1} \tfrac{3}{4} \right) = \tfrac{1}{2}.$$

This confirms our observation: as soon as we disregard non-oscillatory wavenumbers, *the amplitude of the error is halved in each iteration!* This at last explains the phenomenon that we have observed in Figures 11.1 and 11.2. To start with, the error is typically a linear combination of many wavenumbers, oscillatory as well as non-oscillatory. A Gauss–Seidel iteration attenuates oscillatory components much faster, and this means that the contribution of the latter is, to all practical purposes,

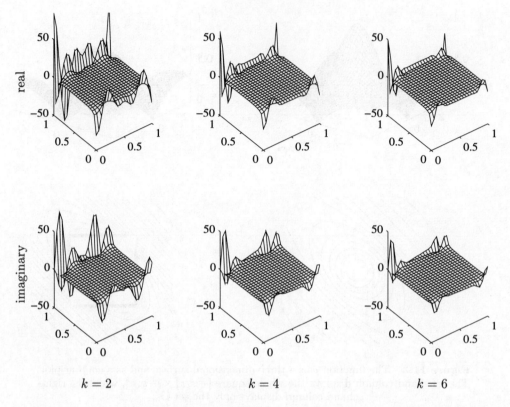

Figure 11.4   Real and imaginary components of $\varepsilon^{[k]}$ for $m = 20$ and $k = 2, 4, 6$.
Note the rapid elimination rate of the highly oscillatory components.

completely eliminated after only a few iterations. The non-oscillatory terms, however, are left and they account for the sedate and plodding rate of attenuation in the
asymptotic regime.

To rephrase this state of affairs, the Gauss–Seidel scheme is a *smoother:* its effect
after a few iterations is to filter out high frequencies from the 'signal'. This becomes
apparent in Figure 11.4, where the real and imaginary parts of $\varepsilon^{[k]}$ are displayed for
a 20×20 grid and $k = 2, 4, 6$. This is perhaps the place to emphasize that not all
iterative methods from Chapter 10 are smoothers; far from it. In Exercise 11.2, for
example, it is demonstrated that this attribute is absent from the Jacobi method.

Had this been all there were to it, the smoothing behaviour of Gauss–Seidel would
have been not much more than a mathematical curiosity. Suppose that we wished
to solve the five-point equations to a given tolerance $\delta > 0$. The fast attenuation
of the highly oscillatory components does not advance perceptibly the instant when
$\|\varepsilon^{[k]}\| < \delta$ (or, in a realistic computer program, $\|r^{[k]}\| < \delta$); it is the straggling non-
oscillatory wavenumbers that dictate the rate of convergence. Figure 10.5 does not
lie: Gauss–Seidel, in complete agreement with the theory of Chapter 10, will perform
just twice as well as Jacobi (which, according to Exercise 11.2, is not a smoother).
Fortunately, there is much more to the innocent phrase 'highly oscillatory components',

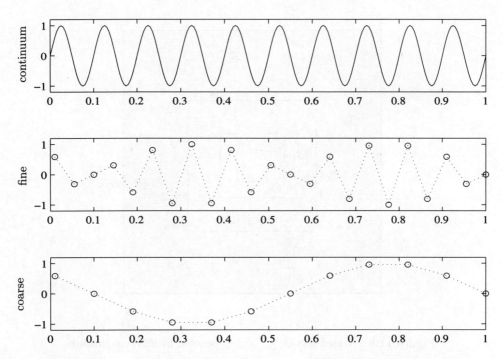

Figure 11.5   Now you see it, now you don't...: A highly oscillatory term and its restrictions to a fine and to a coarse grid.

and this forms our final clue about how to accelerate the Gauss–Seidel iteration.

Let us ponder for a moment the meaning of 'highly oscillatory terms'. A grid – any grid – is a window to the continuum, say $[0,1] \times [0,1]$. The continuum supports all possible frequencies and wavenumbers, but this is not the case with a grid. Suppose that the frequency is so high that a wave oscillates more than once between grid points – *this high oscillation will be invisible on the grid!* More precisely, observing the continuum through the narrow slits of the grid, we will, in all probability, register the wave as non-oscillatory. An example is presented in Figure 11.5 where, for simplicity, we have confined ourselves to a single dimension. The top graph displays the highly oscillatory wave $\sin 20\pi x$, $x \in [0,1]$. In the middle graph the signal has been sampled at 23 equidistant points, and this renders faithfully the oscillatory nature of the sinusoidal wave. However, in the bottom graph we have thrown away every second point. The new graph, with 12 points, completely misses the high frequencies!

The concept of a 'high oscillation' is, thus, a feature of a specific grid. This means that on grids of *different* spacing the Gauss–Seidel iteration attenuates rapidly *different* wavenumbers. Suppose that we coarsen a grid by taking out every second point, the outcome being a new square grid in $[0,1] \times [0,1]$ but with Δx replaced by $2\Delta x$. The range of former high frequencies $\mathbb{O}_0$ is no longer visible on the coarse grid. Instead, the new grid has its own range of high frequencies, on which Gauss–Seidel performs well – as far as the fine grid is concerned, these correspond to the wavenumbers

$$\mathbb{O}_1 := \left\{ (\theta, \psi) \; : \; \tfrac{1}{4}\pi \leq \max\{|\theta|, |\psi|\} \leq \tfrac{1}{2}\pi \right\}.$$

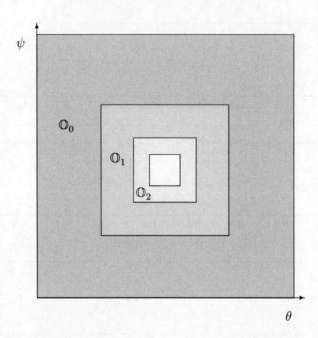

Figure 11.6 Nested sets $\mathbb{O}_s \subset [-\pi, \pi]$, denoted by different shading.

Needless to say, there is no need to stop with just a single coarsening. In general, we can cover the whole range of frequencies by a *hierarchy* of grids, embedded into each other, whose (grid-specific) high frequencies correspond, as far as the fine grid is concerned, to the sets

$$\mathbb{O}_s := \left\{ (\theta, \psi) \, : \, 2^{-s-1}\pi \leq \max\{|\theta|, |\psi|\} \leq 2^{-s}\pi \right\}, \qquad s = 1, 2, \ldots, \lfloor \log_2(m+1) \rfloor.$$

The sets $\mathbb{O}_s$ nest inside each other (see Figure 11.6) and their totality is the whole of $[-\pi, \pi] \times [-\pi, \pi]$. In the next section we describe a computational technique that sweeps across the sets $\mathbb{O}_s$, damping the highly oscillatory terms and using Gauss–Seidel in its 'fast' mode throughout the whole iteration.

11.2   The basic multigrid technique

Let us suppose for simplicity that $m = 2^s - 1$ and let us embed in our grid (to be thereby designated the *finest* grid) a hierarchy of successively coarser grids, as indicated in Figure 11.7.

The main idea behind the *multigrid* technique is to travel up and down the grid hierarchy, using Gauss–Seidel iterations to dampen the (locally) highly oscillating components of the error. *Coarsening* means that we are descending down the hierarchy to a coarser grid (in other words, getting rid of every other point), while *refinement* is the exact opposite, ascending from a coarser to a finer grid. Our goal is to solve tne five-point equations *on the finest grid* – the coarser grids are just a means to that end.

finest grid

coarsening

refinement

coarsest grid

Figure 11.7 Nested grids, from the finest to the coarsest.

In order to describe a multigrid algorithm we need to explain exactly how each coarsening or refinement step is performed, as well as to specify the exact strategy of how to start, when to coarsen, when to refine and when to terminate the whole procedure.

To describe refinement and coarsening it is enough to assume just two grids, one fine and one coarse. Suppose that we are solving the equation

$$A_f x_f = v_f \tag{11.4}$$

on the fine grid. Having performed a few Gauss–Seidel iterations, so as to smooth the high frequencies, we let $r_f := A_f x_f - v_f$ be the residual. This residual needs to be translated into the coarser grid. This is done by means of a *restriction matrix* R,

$$r_c = R r_f. \tag{11.5}$$

Remember the whole idea behind the multigrid technique: the vector r_f is constructed from low-frequency components (relative to the fine grid). Hence it makes sense to go on smoothing the coarsened residual r_c on the coarser grid.[3] To that end we set $v_c := -r_c$, and so solve

$$A_c x_c = -r_c. \tag{11.6}$$

The matrix A_c is, of course, the matrix of the original system (in our case, the matrix originating from the five-point scheme (7.16)) restricted to the coarser grid.

To move in the opposite direction, from coarse to fine, suppose that x_c is an approximate solution of (11.6), an outcome of Gauss–Seidel iterations on this and yet coarser grids. We translate x_c into the fine grid in terms of the *prolongation matrix* P, where

$$y_f = P x_c \tag{11.7}$$

and update the old value of x_f,

$$x_f^{\text{new}} = x_f^{\text{old}} + y_f. \tag{11.8}$$

Let us evaluate the residual r_f^{new} under the assumption that x_c is the exact solution of (11.6). Since

$$r_f^{\text{new}} = A_f x_f^{\text{new}} - v_f = A_f(x_f^{\text{old}} + y_f) - v_f,$$

(11.7) and (11.8) yield

$$r_f^{\text{new}} = r_f^{\text{old}} + A_f y_f = r_f^{\text{old}} + A_f P x_c.$$

Therefore, by (11.6),

$$r_f^{\text{new}} = r_f^{\text{old}} - A_f P A_c^{-1} r_c.$$

Finally, invoking (11.5), we deduce that

$$r_f^{\text{new}} = \left(I - A_f P A_c^{-1} R\right) r_f^{\text{old}}. \tag{11.9}$$

Thus, the sole contribution to the new residual comes from replacing a fine grid by a coarser one. Similar reasoning is valid even if x_c is an approximate solution of (11.6), provided that some bandwidths of wavenumbers have been eliminated in the course of the iteration. Moreover, suppose that (other) bandwidths of wavenumbers have been already filtered out of the residual r_f^{old}. Upon the update (11.8), the contribution of both bandwidths is restricted to the minor ill effects of the restriction and prolongation matrices.

Both the restriction and prolongation matrices are rectangular, but it is a very poor idea to execute them naively as matrix products. The proper procedure is to describe their effect on individual components of the grid, since this provides a convenient and cheap algorithm, as well as clarifying what are we trying to do in mathematical terms. Let $w_f = P w_c$, where $w_c = (w_{j,\ell}^c)_{j,\ell=1}^m$ (a subscript has just been promoted to a superscript, for notational convenience) and $w_f = (w_{j,\ell}^f)_{j,\ell=1}^{2m+1}$. The simplest course in restricting a grid is *injection*,

$$w_{j,\ell}^c = w_{2j,2\ell}^f, \qquad j, \ell = 1, 2, \ldots, m, \tag{11.10}$$

---

[3]To be exact, we have advanced an argument to justify this assertion for the error, rather than the residual. However, it is clear from Exercise 11.1 that the two assertions are equivalent. Of course, the residual, unlike the error, has an important virtue: we can calculate it without knowing the exact solution of the linear system....

but a popular alternative is *full weighting*

$$w^c_{j,\ell} = \tfrac{1}{4} w^f_{2j,2\ell} + \tfrac{1}{8}(w^f_{2j-1,2\ell} + w^f_{2j,2\ell-1} + w^f_{2j+1,2\ell} + w^f_{2j,2\ell+1}) + \tfrac{1}{16}(w^f_{2j-1,2\ell-1}$$
$$+ w^f_{2j+1,2\ell-1} + w^f_{2j-1,2\ell+1} + w^f_{2j+1,2\ell+1}), \qquad j,\ell = 1,2,\ldots,m. \qquad (11.11)$$

The latter can be rendered as a computational stencil (see Section 7.2) in the form

Why bother with (11.11), given the availability of the more natural injection (11.10)? The main reason is that in the latter case $R = \tfrac{1}{4} P^T$ for the prolongation P that we are just about to introduce in (11.12) (the factor of a quarter originates in the fourfold decrease in the number of grid points in coarsening), and this has important theoretical and practical advantages.

There is just one sensible way of prolonging a grid: *linear interpolation*. The exact equations are

$$w^f_{2j-1,2\ell-1} = w^c_{j,\ell}, \qquad j,\ell = 1,2,\ldots,m; \qquad (11.12)$$
$$w^f_{2j-1,2\ell} = \tfrac{1}{2}(w^c_{j,\ell} + w^c_{j,\ell+1}), \qquad j = 1,2,\ldots,m-1, \quad \ell = 1,2,\ldots,m;$$
$$w^f_{2j,2\ell-1} = \tfrac{1}{2}(w^c_{j,\ell} + w^c_{j+1,\ell}), \qquad j = 1,2,\ldots,m, \quad \ell = 1,2,\ldots,m-1;$$
$$w^f_{2j,2\ell} = \tfrac{1}{4}(w^c_{j,\ell} + w^c_{j,\ell+1} + w^c_{j+1,\ell} + w^c_{j+1,\ell+1}), \qquad j,\ell = 1,1,\ldots,m-1.$$

The values of $\boldsymbol{w}_f$ along the boundary are, of course, zero; recall that we are dealing with residuals!

Having learnt how to travel across the hierarchy of nested grids, we now need to specify an itinerary. There are many distinct multigrid strategies and here we mention just the simplest (and most popular), the *V-cycle*

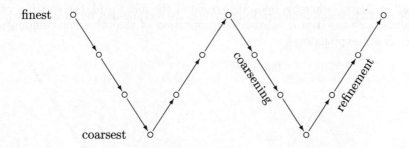

The whole procedure commences and ends at the finest grid. To start with, we stipulate an initial condition, we let $\boldsymbol{v}_f = \boldsymbol{b}$ (the original right-hand side of the linear system (11.1)), and we iterate a small number of times – n_r, say – with Gauss–Seidel.

Subsequently we evaluate the residual r_f, restrict it to the coarser grid, perform n_r further Gauss–Seidel iterations, evaluate the residual, restrict and so on, until we reach the coarsest grid, with just a single grid point, which we solve exactly. (In principle, it is possible to stop this procedure earlier, deciding that a 15×15 system, say, can be solved directly without further coarsening.) Having reached this stage, we have successively damped the influence of error components in the entire range of wavenumbers supported by the finest grid, except that a small amount of error might have been added by restriction.

Having reached the coarsest grid, we need to ascend all the way back to the finest. In each step we prolong, update the residual on the new grid and perform n_p Gauss–Seidel iterations to eliminate errors (corresponding to highly oscillatory wavenumbers on the grid in question) that might have been introduced by past prolongations.

Having returned to the finest grid, we have completed the V-cycle. It is now, and only now, that we check for convergence, by measuring the size of the residual vector. Provided that the error is below the required tolerance, the iteration is terminated; otherwise the V-cycle is repeated. This completes the description of the multigrid algorithm in its simplest manifestation.

11.3   The full multigrid technique

An obvious Achilles heel of all iterative methods is the choice of the starting vector $x^{[0]}$. Although the penalty for a wrong choice is not as drastic as in methods for nonlinear algebraic equations (see Chapter 6), it is nonetheless likely to increase the cost a great deal. By the same token, an astute choice of $x^{[0]}$ is bound to lead to considerable savings. So far, throughout Chapters 10 and 11, we have assumed that $x^{[0]} = 0$, a choice which is likely to be as good or as bad as many others.

The logic of multigrid – working in unison on a whole hierarchy of embedded grids – dictates a superior choice of starting value. Why not use an approximate solution from a coarser grid as the starting value on the finest? Of course, at the beginning of the iteration, exactly when the starting value is required, we have no solution available on the coarser grid, since the V-cycle iteration commences from the finest. The obvious remedy is to start from the coarsest grid and ascend by prolongation, performing n_p Gauss–Seidel iterations on each grid. Even better is the so-called *full multigrid*, whereby, upon its arrival at the finest grid, the starting value has been already cleansed of a substantial proportion of smooth error components. The pattern is self-explanatory:

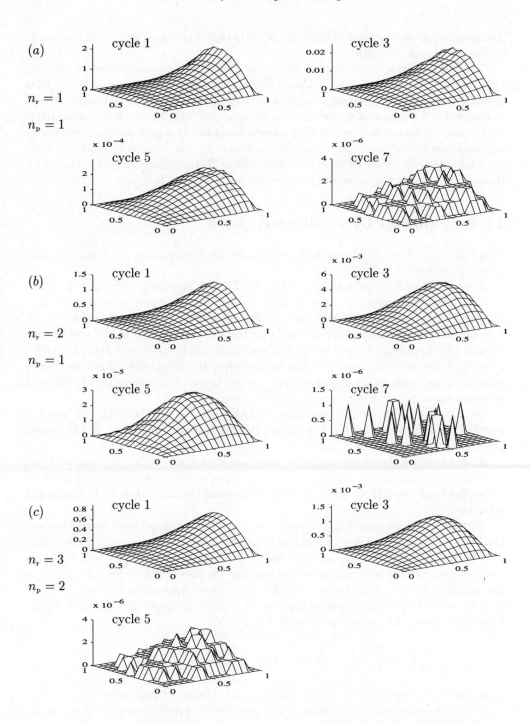

Figure 11.8   The V-cycle multigrid method for the Poisson equation (7.33).

The speed-up in convergence of the full multigrid technique, as evidenced in the results of Section 11.4, is spectacular.

The full multigrid combines two ideas: the first is the multigrid concept of using Gauss–Seidel, say, to smooth the highly oscillatory components by progressing from fine to coarse grids; the second is *nested iteration.* The latter uses estimates from a coarse grid as a starting value for an iteration on a fine grid. In principle, nested iteration can be used whenever an iterative scheme is applied in a grid, without necessarily any reference to multigrid. An example is provided by the solution of nonlinear algebraic equations by means of functional iteration or Newton–Raphson (cf. Chapter 6). However, it comes into its own in conjunction with multigrid.

11.4   Poisson by multigrid

This chapter is short on theory and, to remedy the situation, we have made it long on computational results.

Since Chapter 7 we have used a particular Poisson equation, the problem (7.33), as a yardstick to measure the behaviour of numerical methods, and we will continue this practice here. The finest grid used is always 63×63 (that is, with $\Delta x = \frac{1}{64}$. We measure the performance of the methods by the size of the error at the end of each V-cycle (disregarding, in the case of full multigrid, all but the 'complete V-cycles', from the finest to the coarsest grid and back again). It is likely that in practical error estimation residuals, rather than the error, are calculated. This might lead to different numbers but it will give the same qualitative picture.

We have tested three different choices of the pair (n_r, n_p) for both the 'regular' multigrid from Section 11.2 and the full multigrid technique, Section 11.3. The results are displayed in Figures 11.8 and 11.9.

Each figure displays three detailed iteration strategies: (a) $n_r = 1$, $n_p = 1$; (b) $n_r = 2$, $n_p = 1$; and (c) $n_r = 3$, $n_p = 2$. We have not printed the outcome of seven V-cycles (four in Figure 11.9), since the error is so small that it is likely to be a roundoff artefact.

To assess the cost of a single V-cycle, we disregard the expense of restriction and prolongation, counting just the number of smoothing (i.e., Gauss–Seidel) iterations. The latter are performed on grids of vastly different sizes, but this can be easily incorporated into our estimate by observing that the cost of Gauss–Seidel is linear in the number of grid points, hence a single coarsening decreases its operations' count by a factor of four. Let ϖ denote the cost of a single Gauss–Seidel iteration on the finest grid. Then the cost of one V-cycle is given by

$$\left(1 + \frac{1}{4} + \frac{1}{4^2} + \frac{1}{4^3} + \cdots\right)(n_r + n_p)\,\varpi \approx \tfrac{4}{3}(n_r + n_p)\,\varpi.$$

Remarkably, *the cost of a V-cycle is linear in the number of grid points on the finest grid!*[4] Incidentally, the initial phase of full multigrid is even cheaper: it 'costs' about $\frac{4}{9}(n_r + n_p)\,\varpi$.

---

[4]Our assumption is that all the calculations are performed in a serial, as distinct from a parallel, computer architecture. Otherwise the results are likely to be even more spectacular.

Figure 11.9 The full multigrid method for the Poisson equation (7.33).

There is no hiding the vastly superior performance of both the basic and the full versions of multigrid, in comparison with, say, the 'plain' Gauss–Seidel method. Compare Figure 10.5 with Figure 11.8, even though in the first we have $15^2 = 225$ equations, while the second comprises $63^2 = 3969$. To realize how much slower the 'plain' Gauss–Seidel method would have behaved with $m = 63$, let us compare the spectral radii of its iteration matrices (10.51): we find ≈ 0.96193976625564 for $m = 16$ and ≈ 0.99759236333610 for $m = 63$. Given that Gauss–Seidel spends almost all of its efforts in its asymptotic regime, this means that the number of iterations in Figure 10.5 needs to be multiplied by ≈ 16 to render comparison with Figures 11.8 and 11.9 more meaningful.

It is perhaps fairer to compare multigrid with SOR. The spectral radius of the latter's iteration matrix (for $m = 63$) is, *pace* (10.52), ≈ 0.90645470158276, and this is a great improvement upon Gauss–Seidel. Yet the residual after the eighth V-cycle of 'plain' multigrid (with $n_{\mathrm{r}} = n_{\mathrm{p}} = 1$) is $\approx 2.62 \times 10^{-5}$ and we need 243 SOR iterations to attain this value. (By comparison, Gauss–Seidel requires 6526 iterations to reduce the residual by a similar amount. Conjugate gradients, though, are marginally better than SOR, requiring just 179 iterations.)

Comparison of Figures 11.8 and 11.9 also confirms that, as expected, full multigrid further enhances the performance. The reason – and this should have been expected as well – is not a better rate of convergence but a superior starting value (on the finest grid): in case (a) both versions of multigrid attenuate the error by roughly a factor of ten per V-cycle. As a matter of interest, and in comparison with the previous paragraph, the residual of full multigrid (with $n_{\mathrm{r}} = n_{\mathrm{p}} = 1$) is $\approx 5.8510^{-6}$ after five V-cycles.

We conclude this 'iterative olympics' with a reminder that the errors in Figures 11.8 and 11.9 (and in Figures 10.5 and 10.7 also) display the departure of iterates from the solution of the five-point equations (7.16), not from the exact solution of the Poisson problem (7.33). Given that we are interested in solving the latter by means of the former, it makes little sense to iterate (with any method) beyond the theoretical accuracy of the five-point approximation. This is not as straightforward as it may seem, since (as we have already mentioned) practical convergence estimation employs residuals, rather than errors. Having said this, seeking a residual lower than 10^{-5} (for $m = 63$), is probably of no practical significance.

Comments and bibliography

The idea of using a hierarchy of grids has been around for a while, mainly in the context of nested iteration, but the first modern treatment of the multigrid technique was presented by Brandt (1977).

There exist a number of good introductory texts on multigrid techniques, e.g. Briggs (1987); Hackbusch (1985) and Wesseling (1992). Convergence and complexity (in the present context complexity means the estimation of computational cost) issues are addressed in the book of Bramble (1993) and in a survey by Yserentant (1993). It is important to emphasize that, although multigrid techniques can be introduced and explained in an elementary fashion, their convergence analysis is considerably challenging from a mathematical point of view.

Our treatment of multigrid has centred on just one version, the V-cycle, and we mention in passing that other strategies are perfectly viable and often preferable, e.g. the W-cycle:

finest

coarsest

The number of different strategies and implementations of multigrid is a source of major preoccupation to professionals, although it might be at times slightly baffling to other numerical analysts and to users of computational algorithms.

Gauss–Seidel is not the only smoother, although neither the Jacobi iteration nor SOR (with $\omega \neq 1$) possess this welcome property (see Exercise 11.2). An example of a smoother is a version of the incomplete LU factorization (not the Jacobi line relaxation from Section 10.1, though; see Exercise 11.3). Another example is *Jacobi over-relaxation* (JOR), an iterative scheme that is to Jacobi what SOR is to Gauss–Seidel, with a particular parameter value.

Multigrid methods would have been of little use had their applicability been restricted to the five-point equations in a square. Indeed, possibly the greatest virtue of multigrid is its versatility. Provided linear equations are specified in one grid and we can embed this into a hierarchy of nested grids of progressive coarseness, multigrid confers an advantage over a single-grid implementation of iterative methods. Indeed, we use the word 'grid' in a loose sense, since multigrid is, if at all, even more useful for finite elements than it is for finite differences!

Bramble, J.H. (1993), *Multigrid Methods,* Longman, Harlow, Essex.

Brandt, A. (1977), Multi-level adaptive solutions to boundary-value problems, *Mathematics of Computation* **31,** 333–390.

Briggs, W.L. (1987), *A Multigrid Tutorial,* SIAM, Philadelphia.

Hackbusch, W. (1985), *Multi-Grid Methods and Applications,* Springer-Verlag, Berlin.

Wesseling, P. (1992), *An Introduction to Multigrid Methods,* Wiley, Chichester.

Yserentant, H. (1993), Old and new convergence proofs for multigrid methods, *Acta Numerica* **2,** 285–326.

Exercises

11.1   Let $\hat{x}$ be the solution of (11.1), $\varepsilon^{[k]} := x^{[k]} - \hat{x}$ and $r^{[k]} := Ax^{[b]} - b$. Show that $r^{[k]} = A\varepsilon^{[k]}$. Further supposing that A is symmetric and that its eigenvalues reside in the interval $[\lambda_-, \lambda_+]$, prove that the inequality

$$\min\{|\lambda_-|, |\lambda_+|\}\|\varepsilon^{[k]}\| \leq \|r^{[k]}\| \leq \max\{|\lambda_-|, |\lambda_+|\}\|\varepsilon^{[k]}\|$$

holds in the Euclidean norm.

11.2   Apply the analysis of Section 11.1 to the Jacobi iteration (10.20) instead of the Gauss–Seidel iteration.

    a Finding an approximate recurrence relation for the Jacobi equivalent of the function $p^{[k]}(\theta, \psi)$, prove that the local attenuation of the wavenumber (θ, ψ) is approximately

$$\tilde{\rho}(\theta, \psi) = \tfrac{1}{2} |\cos\theta + \cos\psi| = \left| \cos\tfrac{1}{2}(\theta + \psi) \cos\tfrac{1}{2}(\theta - \psi) \right|.$$

    b Show that the best upper bound on $\tilde{\rho}$ in $\mathbb{O}$ is unity and hence that the Jacobi iteration does not smooth highly oscillatory terms.

11.3   Similarly to the last exercise, show that the line relaxation method from Section 10.4 is not a good smoother. You should prove that

$$\tilde{\rho}(\theta, \psi) = \frac{|\cos\psi|}{2 - \cos\theta}$$

and that it can attain unity in the set $\mathbb{O}$.

11.4   Assuming that $w_{j,\ell}^{\mathrm{f}} = g(j, \ell)$, where g is a linear function of both its arguments, and that $\boldsymbol{w}^{\mathrm{c}}$ has been obtained by the fully weighted restriction (11.11), prove that $w_{j,\ell}^{\mathrm{c}} = g(2j, 2\ell)$.

12

Fast Poisson solvers

12.1 TST matrices and the Hockney method

This chapter is concerned with yet another approach to the solution of linear equations that occur when the Poisson equation is discretized by finite differences. This approach is an alternative to the direct methods of Chapter 9 and to the iterative schemes of Chapters 10 and 11. We intend to present two techniques for the very fast approximation of $\nabla^2 u = f$, one in a rectangle and the other in a disc. These techniques share two features. Firstly, they originate in numerical solution of the Poisson equation – hence their sobriquet, *fast Poisson solvers*. Secondly, the secret of their efficacy rests in a clever use of the *fast Fourier transform* (FFT).

In the present section we assume that the Poisson equation (7.13) with Dirichlet boundary conditions (7.14) is solved in a rectangle with either the five-point formula (7.15) or the nine-point formula (7.28) (or, for that matter, the modified nine-point formula (7.32) – the matrix of the linear system does not depend on whether the nine-point method has been modified). In either case we assume that the linear equations have been assembled in *natural ordering*.

Suppose that the grid is $m_1 \times m_2$. The linear system $A\boldsymbol{x} = \boldsymbol{b}$ can be written in a block-TST form (recall from Section 10.2 that 'TST' stands for 'Toeplitz, symmetric, tridiagonal')

$$
\begin{bmatrix}
S & T & O & \cdots & O \\
T & S & T & \ddots & \vdots \\
O & \ddots & \ddots & \ddots & O \\
\vdots & \ddots & T & S & T \\
O & \cdots & O & T & S
\end{bmatrix}
\begin{bmatrix}
\boldsymbol{x}_1 \\
\boldsymbol{x}_2 \\
\vdots \\
\boldsymbol{x}_{m_2}
\end{bmatrix}
=
\begin{bmatrix}
\boldsymbol{b}_1 \\
\boldsymbol{b}_2 \\
\vdots \\
\boldsymbol{b}_{m_2}
\end{bmatrix},
\tag{12.1}
$$

where $\boldsymbol{x}_\ell$ and $\boldsymbol{b}_\ell$ correspond to the variables and to the portion of $\boldsymbol{b}$ along the ℓth column of the grid, respectively:

$$
\boldsymbol{x}_\ell =
\begin{bmatrix}
u_{1,\ell} \\
u_{2,\ell} \\
\vdots \\
u_{m_1,\ell}
\end{bmatrix},
\qquad
\boldsymbol{b}_\ell =
\begin{bmatrix}
b_{1,\ell} \\
b_{2,\ell} \\
\vdots \\
b_{m_1,\ell}
\end{bmatrix},
\qquad
\ell = 1, 2, \ldots, m_2.
$$

Both S and T are themselves $m_1 \times m_1$ TST matrices:

$$S = \begin{bmatrix} -4 & 1 & 0 & \cdots & 0 \\ 1 & -4 & 1 & \ddots & \vdots \\ 0 & \ddots & \ddots & \ddots & 0 \\ \vdots & \ddots & 1 & -4 & 1 \\ 0 & \cdots & 0 & 1 & -4 \end{bmatrix} \quad \text{and} \quad T = \begin{bmatrix} 1 & 0 & \cdots & 0 \\ 0 & 1 & \ddots & \vdots \\ \vdots & \ddots & \ddots & 0 \\ 0 & \cdots & 0 & 1 \end{bmatrix}$$

for the five-point formula and

$$S = \begin{bmatrix} -\frac{10}{3} & \frac{2}{3} & 0 & \cdots & 0 \\ \frac{2}{3} & -\frac{10}{3} & \frac{2}{3} & \ddots & \vdots \\ 0 & \ddots & \ddots & \ddots & 0 \\ \vdots & \ddots & \frac{2}{3} & -\frac{10}{3} & \frac{2}{3} \\ 0 & \cdots & 0 & \frac{2}{3} & -\frac{10}{3} \end{bmatrix} \quad \text{and} \quad T = \begin{bmatrix} \frac{2}{3} & \frac{1}{6} & 0 & \cdots & 0 \\ \frac{1}{6} & \frac{2}{3} & \frac{1}{6} & \ddots & \vdots \\ 0 & \ddots & \ddots & \ddots & 0 \\ \vdots & \ddots & \frac{1}{6} & \frac{2}{3} & \frac{1}{6} \\ 0 & \cdots & 0 & \frac{1}{6} & \frac{2}{3} \end{bmatrix}$$

in the case of the nine-point formula.

We rewrite (12.1) in the form

$$T\boldsymbol{x}_{\ell-1} + S\boldsymbol{x}_\ell + T\boldsymbol{x}_{\ell+1} = \boldsymbol{b}_\ell, \qquad \ell = 1, 2, \ldots, m_2, \tag{12.2}$$

where $\boldsymbol{x}_0, \boldsymbol{x}_{m_2+1} := \boldsymbol{0} \in \mathbb{R}^{m_1}$, and recall from Lemma 10.5 that the eigenvalues and eigenvectors of TST matrices are known and that all TST matrices of similar dimension share the same eigenvectors. In particular, according to (10.29) we have

$$S = QD_SQ, \qquad T = QD_TQ, \tag{12.3}$$

where

$$q_{j,\ell} = \sqrt{\frac{2}{m_1+1}} \sin\left(\frac{\pi j\ell}{m_1+1}\right), \qquad j, \ell = 1, 2, \ldots, m_1.$$

(Note that Q is both orthogonal and symmetric; thus, for example, $S = QD_SQ^{-1} = QD_SQ^T = QD_SQ$.) Both D_S and D_T are $m_1 \times m_1$ diagonal matrices whose diagonal components consist of the eigenvalues of S and T respectively, $\lambda_1^{(S)}, \lambda_2^{(S)}, \ldots, \lambda_{m_1}^{(S)}$ and $\lambda_1^{(T)}, \lambda_2^{(T)}, \ldots, \lambda_{m_1}^{(T)}$, say; cf. (10.28). We substitute (12.3) into (12.2) and multiply with $Q = Q^{-1}$ from the left. The outcome is

$$D_T\boldsymbol{y}_{\ell-1} + D_S\boldsymbol{y}_\ell + D_T\boldsymbol{y}_{\ell+1} = \boldsymbol{c}_\ell, \qquad \ell = 1, 2, \ldots, m_2, \tag{12.4}$$

where

$$\boldsymbol{y}_\ell := Q\boldsymbol{x}_\ell, \quad \boldsymbol{c}_\ell := Q\boldsymbol{b}_\ell, \qquad \ell = 1, 2, \ldots, m_2.$$

The crucial difference between (12.2) and (12.4) is that in the latter we have diagonal, rather than TST, matrices. To exploit this, we recall that $\boldsymbol{x}$ and $\boldsymbol{b}$ (and, indeed, the matrix A) have been obtained from a natural ordering of a rectangular grid by columns. Let us now reorder the $\boldsymbol{y}_\ell$ and the $\boldsymbol{c}_\ell$ *by rows*. Thus,

$$
\tilde{\boldsymbol{y}}_j := \begin{bmatrix} y_{j,1} \\ y_{j,2} \\ \vdots \\ y_{j,m_2} \end{bmatrix}, \quad \tilde{\boldsymbol{c}}_j := \begin{bmatrix} c_{j,1} \\ c_{j,2} \\ \vdots \\ c_{j,m_2} \end{bmatrix}, \quad j = 1, 2, \ldots, m_1.
$$

To derive linear equations that are satisfied by the $\tilde{\boldsymbol{y}}_j$, let us consider the first equation in each of the m_2 blocks in (12.4),

$$
\begin{aligned}
\lambda_1^{(S)} y_{1,1} + \lambda_1^{(T)} y_{1,2} &= c_{1,1}, \\
\lambda_1^{(T)} y_{1,\ell-1} + \lambda_1^{(S)} y_{1,\ell} + \lambda_1^{(T)} y_{1,\ell+1} &= c_{1,\ell}, \quad \ell = 2, 3, \ldots, m_2 - 1, \\
\lambda_1^{(T)} y_{1,m_2-1} + \lambda_1^{(S)} y_{1,m_2} &= c_{1,m_2},
\end{aligned}
$$

or, in a vector notation,

$$
\begin{bmatrix}
\lambda_1^{(S)} & \lambda_1^{(T)} & 0 & \cdots & & 0 \\
\lambda_1^{(T)} & \lambda_1^{(S)} & \lambda_1^{(T)} & \ddots & & \vdots \\
0 & \ddots & \ddots & \ddots & & 0 \\
\vdots & \ddots & \lambda_1^{(T)} & \lambda_1^{(S)} & \lambda_1^{(T)} \\
0 & \cdots & & 0 & \lambda_1^{(T)} & \lambda_1^{(S)}
\end{bmatrix} \tilde{\boldsymbol{y}}_1 = \tilde{\boldsymbol{c}}_1.
$$

Likewise, collecting together the jth equation from each block in (12.4) results in

$$
\Gamma_j \tilde{\boldsymbol{y}}_j = \tilde{\boldsymbol{c}}_j, \quad j = 1, 2, \ldots, m_1, \tag{12.5}
$$

where

$$
\Gamma_j := \begin{bmatrix}
\lambda_j^{(S)} & \lambda_j^{(T)} & 0 & \cdots & & 0 \\
\lambda_j^{(T)} & \lambda_j^{(S)} & \lambda_j^{(T)} & \ddots & & \vdots \\
0 & \ddots & \ddots & \ddots & & 0 \\
\vdots & \ddots & \lambda_j^{(T)} & \lambda_j^{(S)} & \lambda_j^{(T)} \\
0 & \cdots & & 0 & \lambda_j^{(T)} & \lambda_j^{(S)}
\end{bmatrix}, \quad j = 1, 2, \ldots, m_2.
$$

Hence, *switching from column-wise to row-wise ordering uncouples the* $(m_1 m_2) \times (m_1 m_2)$ *system* (12.4) *into* m_2 *systems, each of dimension* $m_1 \times m_1$.

The matrices Γ_j are also TST; hence their eigenvalues and eigenvectors are known, and this, in principle, can be used to solve (12.5). This, however, is a bad idea, since it is considerably easier to compute these linear systems by banded LU factorization (see Chapter 9). This costs just $\mathcal{O}(m_1)$ operations per system, altogether $\mathcal{O}(m_1 m_2)$ operations.

◇ **Counting operations**   Measuring the cost of numerical calculation is a
highly uncertain business. At the dawn of the computer era (not such a
long time ago!) it was usual to count multiplications, since they were significantly more expensive than additions or subtractions (divisions were even
more expensive, avoided if at all possible). As often in the computer business,
technological developments rendered this point of view obsolete. In modern
processors, operations like multiplication – and, for that matter, exponentia-
tion, square-rooting, the evaluation of logarithms and trigonometric functions
– are built into the hardware and can be performed exceedingly fast. Thus,
a more up-to-date measure of computational cost is a *flop,* an abbreviation
for *floating point operation.* A single flop is considered the equivalent of a
`FORTRAN` statement

        A(I) = B(I,J) * C(J) + D(J)

or alternative statements in `Pascal, C` or any other high-level computer language. Note that a flop involves a product, an addition and a number of
calculations of indices – none of these operations is free and the above statement, whose form is familiar even to the novice scientific programmer, combines them in a useful manner. The reader might have also encountered flops
as a yardstick of computer speed, in which case flops (kiloflops, megaflops,
gigaflops and perhaps, one day, teraflops) are measured per second.

Even flops, though, are of an increasingly uncertain provenance as measures
of computational cost, because modern computers – or, at least, the large
computers used for *macho* applications of scientific computing – typically involve parallel architectures of different configurations. This means that the
cost of computation varies between computers and no single number can provide a complete comparison. Matters are complicated further by the fact that
calculation is not the only time-consuming task of parallel, multi-processor
computers. The other is communication and message-passing among the different processors.

To make a long story short – and to avoid excessive departures from the
main theme of this book – we thereafter provide only the order of magnitude
(indicated by the $\mathcal{O}(\,\cdot\,)$ notation) of the cost of an algorithm. This can be
easily converted to a 'multiplication count' or to flops, but, for all intents and
purposes, the order of magnitude is illustrative enough. All our counts are in
serial architecture – parallel processing is likely to change everything! Having
said this, we hasten to add that, regardless of the underlying architecture,
most good methods remain good and most bad methods remain bad. It is
still better to use sparse LU factorization, say, for banded matrices, multigrid
still outperforms Gauss–Seidel and fast Poisson solvers are still... fast.   ◇

Having solved the tridiagonal systems, we end up with the vectors $\tilde{\boldsymbol{y}}_1, \tilde{\boldsymbol{y}}_2, \ldots, \tilde{\boldsymbol{y}}_{m_1}$,
which we 'translate' back to $\boldsymbol{y}_1, \boldsymbol{y}_2, \ldots, \boldsymbol{y}_{m_2}$ by rearranging rows to columns. Note
that this rearrangement is free of any computational cost since, in reality, we are (or, at
least, should) hold the information in the form of a $m_1 \times m_2$ array, corresponding to the
computational grid. Column-wise or row-wise natural ordering are purely notational
devices! Finally, we reverse the effects of the multiplication by Q and let $\boldsymbol{x}_\ell = Q\boldsymbol{y}_\ell$,

$\ell = 1, 2, \ldots, m_2$ (recall that $Q = Q^{-1}$).

Let us review briefly the stages in this, the *Hockney method*, estimating their computational cost.

(1) To start with, we have the vectors $\boldsymbol{b}_1, \boldsymbol{b}_2, \ldots, \boldsymbol{b}_{m_2} \in \mathbb{R}^{m_1}$ and we form the products $\boldsymbol{c}_\ell = Q\boldsymbol{b}_\ell$, $\ell = 1, 2, \ldots, m_2$. This costs $\mathcal{O}(m_1^2 m_2)$ operations.

(2) We rearrange columns into rows, i.e., $\boldsymbol{c}_\ell \in \mathbb{R}^{m_1}$, $\ell = 1, 2, \ldots, m_2$, into $\tilde{\boldsymbol{c}}_j \in \mathbb{R}^{m_2}$, $j = 1, 2, \ldots, m_1$. This is a purely a change of notation and is free of any computational cost.

(3) The tridiagonal systems $\Gamma_j \tilde{\boldsymbol{y}}_j = \tilde{\boldsymbol{c}}_j$, $j = 1, 2, \ldots, m_1$, are solved by banded LU factorization, and the cost of this procedure is $\mathcal{O}(m_1 m_2)$.

(4) We next rearrange rows into columns, i.e., $\tilde{\boldsymbol{y}}_j \in \mathbb{R}^{m_2}$, $j = 1, 2, \ldots, m_1$ into $\boldsymbol{y}_\ell \in \mathbb{R}^{m_1}$, $\ell = 1, 2, \ldots, m_2$. Again, this costs nothing.

(5) Finally, we find the solution of the discretized Poisson equation by forming the products $\boldsymbol{x}_\ell = Q\boldsymbol{y}_\ell$, $\ell = 1, 2, \ldots, m_2$, at the cost of $\mathcal{O}(m_1^2 m_2)$ operations.

Provided both m_1 and m_2 are large, matrix multiplications dominate the computational cost. Our first, trivial observation is that, the expense being $\mathcal{O}(m_1^2 m_2)$, it is a good policy to choose $m_1 \leq m_2$ (because of symmetry, we can always rotate the rectangle, in other words proceed from row to columns, and to rows again). However, a considerably more important observation is that the special form of the matrix Q can be exploited to make products of the form $\boldsymbol{s} = Q\boldsymbol{p}$, say, substantially cheaper than $\mathcal{O}(m_1^2)$. Because of (10.29), we have

$$s_\ell = c \sum_{j=1}^{m_1} p_j \sin\left(\frac{\pi j \ell}{m_1 + 1}\right) = c \operatorname{Im}\left[\sum_{j=0}^{m_1} p_j \exp\left(\frac{\pi \mathrm{i} j \ell}{m_1 + 1}\right)\right], \qquad \ell = 1, 2, \ldots, m_1,$$

(12.6)

where $c = [2/(m_1 + 1)]^{1/2}$ is a multiplicative constant. Such a product can be performed in a very efficient manner by using a fast Fourier transform, the theme of the next section.

An important remark is that in this section we have not used the fact that we are solving the Poisson equation! The crucial feature of the underlying linear system is that it is block-TST and each of the blocks is itself a TST matrix. There is nothing to prevent us from using the same approach for other equations that possess this structure, regardless of their origin. Moreover, it is an easy matter to extend this approach to, say, Poisson equations in three dimensions with Dirichlet conditions along the boundary of a parallelepiped: the matrix partitions into blocks of block-TST matrices and each such block-TST matrix is itself composed of TST matrices.

12.2 The fast Fourier transform

Let d be a positive integer and denote by Π_d the set of all complex sequences $\boldsymbol{x} = \{x_j\}_{j=-\infty}^{\infty}$ which are periodic with period d, i.e., which are such that $x_{j+d} = x_j$, $j \in \mathbb{Z}$.

It is an easy matter to demonstrate that Π_d is a linear space of dimension d over the complex numbers $\mathbb{C}$ (see Exercise 12.3).

Let

$$\omega_d := \exp\left(\frac{2\pi i}{d}\right)$$

denote the *primitive root of unity* of degree d. A *discrete Fourier transform (DFT)* is a linear mapping $\mathcal{F}_d$ defined for every $\boldsymbol{x} \in \Pi_d$ by

$$\boldsymbol{y} = \mathcal{F}_d \boldsymbol{x} \qquad \text{where} \qquad y_j = \frac{1}{d} \sum_{\ell=0}^{d-1} \omega_d^{-j\ell} x_\ell, \quad j \in \mathbb{Z}. \tag{12.7}$$

Lemma 12.1 *The DFT $\mathcal{F}_d$, as defined in (12.7), maps Π_d into itself. The mapping is invertible and*

$$\boldsymbol{x} = \mathcal{F}_d^{-1} \boldsymbol{y} \qquad \text{where} \qquad x_\ell = \sum_{j=0}^{d-1} \omega_d^{j\ell} y_j, \quad \ell \in \mathbb{Z}. \tag{12.8}$$

Moreover, $\mathcal{F}_d$ is an isomorphism of Π_d onto itself (A.1.4.2 and A.2.1.9).

Proof Since ω_d is a root of unity, it is true that $\omega_d^d = \omega_d^{-d} = 1$. Therefore it follows from (12.7) that

$$y_{j+d} = \frac{1}{d} \sum_{\ell=0}^{d-1} \omega_d^{-(j+d)\ell} x_\ell = \frac{1}{d} \sum_{\ell=0}^{d-1} \omega_d^{-j\ell} x_\ell = y_j, \qquad j \in \mathbb{Z},$$

and we deduce that $\boldsymbol{y} \in \Pi_d$. Therefore $\mathcal{F}_d$ indeed maps elements of Π_d into elements of Π_d.

To prove the stipulated form of the inverse, we denote $\boldsymbol{w} := \mathcal{F}_d \boldsymbol{x}$, where $\boldsymbol{x}$ has been defined in (12.8). Our first observation is that, provided $\boldsymbol{y} \in \Pi_d$, it is also true that $\boldsymbol{x} \in \Pi_d$ (just swap minus to plus in the above proof). Moreover, changing the order of summation,

$$w_m = \frac{1}{d} \sum_{\ell=0}^{d-1} \omega_d^{-m\ell} x_\ell = \frac{1}{d} \sum_{\ell=0}^{d-1} \omega_d^{-m\ell} \left(\sum_{j=0}^{d-1} \omega_d^{j\ell} y_j \right)$$

$$= \frac{1}{d} \sum_{j=0}^{d-1} \left(\sum_{\ell=0}^{d-1} \omega_d^{(j-m)\ell} \right) y_j, \qquad m \in \mathbb{Z}.$$

Within the parentheses is a geometric series that can be summed explicitly. If $j \neq m$, we have

$$\sum_{\ell=0}^{d-1} \omega_d^{(j-m)\ell} = \frac{1 - \omega_d^{(j-m)d}}{1 - \omega_d^{j-m}} = 0,$$

because $\omega_d^{sd} = 1$ for every $s \in \mathbb{Z} \setminus \{0\}$; whereas in the case $j = m$ we obtain

$$\sum_{\ell=0}^{d-1} \omega_d^{(j-m)d} = \sum_{\ell=0}^{d-1} 1 = d.$$

We thus conclude that $w_m = y_m$, $m \in \mathbb{Z}$, hence that $\boldsymbol{w} = \boldsymbol{y}$. Therefore, (12.8) indeed describes the inverse of $\mathcal{F}_d$.

The existence of an inverse for every $\boldsymbol{x} \in \Pi_d$ shows that $\mathcal{F}_d$ is an isomorphism. To conclude the proof and demonstrate that this DFT maps Π_d onto itself, we suppose that there exists a $\tilde{\boldsymbol{y}} \in \Pi_d$ that cannot be the destination of $\mathcal{F}_d\boldsymbol{x}$ for any $\boldsymbol{x} \in \Pi_d$. Let us define $\tilde{\boldsymbol{x}}$ in terms of (12.8). Then, as we have already observed, $\tilde{\boldsymbol{x}} \in \Pi_d$ and it follows from our proof that $\tilde{\boldsymbol{y}} = \mathcal{F}_d\tilde{\boldsymbol{x}}$. Therefore $\tilde{\boldsymbol{y}}$ is in the range of $\mathcal{F}_d$, in contradiction to our assumption, and we conclude that the DFT $\mathcal{F}_d$ is, indeed, an isomorphism of Π_d *onto* itself. ∎

It is of interest to mention an alternative proof that $\mathcal{F}_d$ is onto. According to a classical theorem from linear algebra, a linear mapping T from a finite-dimensional linear space V to itself is onto if and only if its kernel $\ker T$ consists just of the zero element of the space ($\ker T$ is the set of all $\boldsymbol{w} \in V$ such that $T\boldsymbol{w} = \boldsymbol{0}$; see A.1.4.2). Letting $\boldsymbol{x} \in \ker \mathcal{F}_d$, (12.7) yields

$$\sum_{\ell=0}^{d-1} \omega_d^{-j\ell} x_\ell = 0, \qquad j = 0, 1, \ldots, d-1. \tag{12.9}$$

This is a homogeneous linear system of d equations in the d unknowns $x_0, x_1, \ldots, x_{d-1}$. Its matrix is a *Vandermonde* matrix (A.1.2.5) and it is easy to prove that its determinant satisfies

$$\det \begin{bmatrix} 1 & \omega_d^{-1} & \cdots & \omega_d^{-(d-1)} \\ 1 & \omega_d^{-2} & \cdots & \omega_d^{-2(d-1)} \\ \vdots & \vdots & & \vdots \\ 1 & \omega_d^{-d} & \cdots & \omega_d^{-d(d-1)} \end{bmatrix} = \prod_{\ell=1}^{d} \prod_{j=0}^{\ell-1} (\omega_d^{-\ell} - \omega_d^{-j}) \neq 0.$$

Therefore the only possible solution of (12.9) is $x_0 = x_1 = \cdots = x_{d-1} = 0$ and we deduce that $\ker \mathcal{F}_d = \boldsymbol{0}$ and the mapping is onto Π_d.

◇ **Applications of the DFT**   It is difficult to overstate the importance of the discrete Fourier transform to a multitude of applications, ranging from numerical analysis to control theory, and from computer science to coding theory to time series analysis.... In the sequel we employ it to provide a fast solution to discretized differential equations but here we give another example that shows the power of the DFT: approximation of the *Fourier coefficients*, also known as *Fourier harmonics*, of a periodic function. This theme finds its application in the next section, as well as in numerous other places across applied mathematics, science and engineering.

Let g be an integrable complex-valued function, defined for all $x \in \mathbb{R}$, and periodic with period 2π (i.e., $g(x + 2\pi) = g(x)$ for all $x \in \mathbb{R}$). The *Fourier transform* of g is the sequence $\{\hat{g}_m\}_{m=-\infty}^{\infty}$, where

$$\hat{g}_m = \frac{1}{2\pi} \int_0^{2\pi} g(\tau) e^{-im\tau} \, d\tau, \qquad m \in \mathbb{Z}. \tag{12.10}$$

Fourier transforms feature in numerous branches of mathematical analysis
and its application: the library shelf labelled 'harmonic analysis' is, to a very
large extent, devoted to Fourier transforms and their ramifications. More to
the point, as far as the subject matter of this book is concerned, they are
crucial in the stability analysis of numerical methods for PDEs of evolution
(see Chapters 13 and 14).

The ubiquity of Fourier transforms in so many applications means that the
task of approximating (12.10) numerically is among the most important problems in scientific computing. The obvious approach is to pursue the logic of
Section 3.1 and replace integration by finite summation, i.e., by quadrature.
It is not difficult to prove that, provided $g(0) = g(2\pi)$, the best quadrature
nodes – our equivalent of the Gaussian quadrature nodes of Section 3.1 –
are equidistant along the interval $[0, 2\pi]$, and furthermore the corresponding
quadrature weights are all equal. In other words, given an integer $\nu \geq 1$, we
approximate

$$\hat{g}_m \approx \hat{h}_m := \frac{1}{\nu} \sum_{j=0}^{\nu-1} g\left(\frac{2\pi j}{\nu}\right) \exp\left(\frac{-2\pi \mathrm{i} mj}{\nu}\right) = \frac{1}{\nu} \sum_{j=0}^{\nu-1} \omega_\nu^{-mj} g_j, \qquad (12.11)$$

where $g_j = g(2\pi j/\nu)$, $j = 0, 1, \ldots, \nu - 1$. In a rare display of good mathematical fortune, virtue – Gaussian nodes – combines with ease of application:
equidistant nodes and equal weights.

Of course, (12.11) makes sense only for a finite set of indices m and we are
safe only if we restrict this to just ν values. Suppose for simplicity that ν
is even and let m vary in $\{-\nu/2 + 1, -\nu/2 + 2, \ldots, \nu/2\}$. Replacing m by
$\ell - \nu/2 + 1$, the identity $\omega_\nu^{\nu/2} = -1$ gives

$$\hat{h}_m = \frac{1}{\nu} \sum_{j=0}^{\nu-1} \omega_\nu^{-(\ell-\nu/2+1)j} g_j = \left(\frac{\omega_\nu^{\nu/2-1}}{\nu}\right) \sum_{j=0}^{\nu-1} \omega_\nu^{-\ell j} g_j$$

$$= -\omega_\nu^{-1} (\mathcal{F}_\nu \boldsymbol{g})_\ell, \qquad \ell = 0, 1, \ldots, \nu - 1,$$

where $\boldsymbol{g} = \{g_j\}_{j=-\infty}^\infty \in \Pi_\nu$ is obtained by letting $g_{j+s\nu} = g_j$ for every $j =
0, 1, \ldots, \nu - 1$ and integer s. In other words, (12.11) is nothing other but a
DFT!

Suppose that the *Fourier series* corresponding to g is absolutely convergent,
hence

$$g(x) = \sum_{k=-\infty}^{\infty} \hat{g}_k \mathrm{e}^{\mathrm{i} mx}, \qquad 0 \leq x \leq 2\pi. \qquad (12.12)$$

(This, incidentally, is the formula that reverses the action of the Fourier transform. Its representation of g as a linear combination of harmonics is perhaps
the most important reason for the popularity of Fourier series – and Fourier
transforms – in applications.) Letting $x = 2\pi j/\nu$, we obtain

$$g_j = \sum_{k=-\infty}^{\infty} \hat{g}_k \exp\left(\frac{2\pi jk}{\nu}\right), \qquad j = 0, 1, \ldots, \nu - 1,$$

and substitution into (12.11) yields

$$\hat{h}_m = \frac{1}{\nu} \sum_{j=0}^{\nu-1} \omega_\nu^{-mj} \left(\sum_{k=-\infty}^{\infty} \hat{g}_k \omega_\nu^{jk} \right) = \frac{1}{\nu} \sum_{k=-\infty}^{\infty} \hat{g}_k \sum_{j=0}^{\nu-1} \omega_\nu^{(k-m)j}.$$

We are justified in replacing the order of summation since we have assumed that the Fourier series converges absolutely. However, as we have already seen,

$$\sum_{j=0}^{\nu-1} \omega_\nu^{jr} = \begin{cases} 0, & r \bmod \nu \neq 0, \\ \nu, & r \bmod \nu = 0 \end{cases}$$

and we deduce that the error in our approximation of the Fourier coefficients $\hat{g}_m$ is

$$\hat{h}_m - \hat{g}_m = \sum_{\substack{j=-\infty \\ j \neq 0}}^{\infty} \hat{g}_{m+j\nu}. \tag{12.13}$$

Therefore, we can expect the error to decrease rapidly as a function of ν if the Fourier coefficients of g decay fast for large $|k|$.

The rate of decay of Fourier coefficients is known for a large variety of functions. In particular, suppose that (12.12) converges for all $z \in \mathbb{C}$ such that $0 \leq \operatorname{Re} z \leq 2\pi$, $|\operatorname{Im} z| \leq \alpha$ for some $\alpha > 0$. Then it is possible to show that

$$|\hat{g}_k| \leq c\, e^{-|k|\alpha}, \qquad k \in \mathbb{Z},$$

where $c > 0$ is a constant. Therefore, by dominating (12.13) with a geometric series we can deduce that

$$|\hat{h}_m - \hat{g}_m| \leq 2c \left(\frac{e^{-\nu\alpha}}{1 - e^{-\nu\alpha}} \right) \cosh m\alpha, \qquad |m| \leq \nu - 1.$$

Therefore, the error in replacing the Fourier integral by a DFT decays *exponentially* with ν. $\diamond$

On the face of it, the evaluation of the DFT (12.7) (or of its inverse) requires $\mathcal{O}(d^2)$ operations, since, by periodicity, it is obtained by multiplying a vector in $\mathbb{C}^d$ by a $d \times d$ complex matrix. It is one of the great wonders of computational mathematics, however, that this operation count can be reduced a very great deal. Let us assume for simplicity that $d = 2^n$, where n is a nonnegative integer. It is convenient to replace $\mathcal{F}_d$ by the mapping $\mathcal{F}_d^* := d\mathcal{F}_d$; clearly, if we can compute $\mathcal{F}_d^* \boldsymbol{x}$ cheaply, just $\mathcal{O}(d)$ operations will convert the result to $\mathcal{F}_d \boldsymbol{x}$.

Let us define, for every $\boldsymbol{x} \in \Pi_d$,

$$\boldsymbol{x}^{[e]} := \{x_{2j}\}_{j=-\infty}^{\infty} \qquad \text{and} \qquad \boldsymbol{x}^{[o]} := \{x_{2j+1}\}_{j=-\infty}^{\infty}.$$

Since $\boldsymbol{x}^{[e]}, \boldsymbol{x}^{[o]} \in \Pi_{d/2}$, we can make the mapppings

$$\boldsymbol{y}^{[e]} = \mathcal{F}_{d/2}^* \boldsymbol{x}^{[e]} \qquad \text{and} \qquad \boldsymbol{y}^{[o]} = \mathcal{F}_{d/2}^* \boldsymbol{x}^{[o]}.$$

Let $\boldsymbol{y} = \mathcal{F}_d \boldsymbol{x}$. Then, by (12.7),

$$
\begin{aligned}
y_j &= \sum_{\ell=0}^{d-1} \omega_d^{-j\ell} x_\ell = \sum_{\ell=0}^{2^n-1} \omega_{2^n}^{-j\ell} x_\ell \\
&= \sum_{\ell=0}^{2^{n-1}-1} \omega_{2^n}^{-2j\ell} x_{2\ell} + \sum_{\ell=0}^{2^{n-1}-1} \omega_{2^n}^{-j(2\ell+1)} x_{2\ell+1}, \qquad j = 0,1,\ldots,2^n-1.
\end{aligned}
$$

However,

$$
\omega_{2^n}^{2s} = \omega_{2^{n-1}}^{s}, \qquad s \in \mathbb{Z},
$$

therefore

$$
\begin{aligned}
y_j &= \sum_{j=0}^{2^{n-1}-1} \omega_{2^{n-1}}^{-j\ell} x_{2\ell} + \omega_{2^n}^{-j} \sum_{j=0}^{2^{n-1}-1} \omega_{2^{n-1}}^{-j\ell} x_{2\ell+1} \\
&= y_j^{[e]} + \omega_{2^n}^{-j} y_j^{[o]}, \qquad j = 0,1,\ldots,2^n-1. \tag{12.14}
\end{aligned}
$$

In other words, provided that $\boldsymbol{y}^{[e]}$ and $\boldsymbol{y}^{[o]}$ are already known, we can synthesize them into $\boldsymbol{y}$ in $\mathcal{O}(d)$ operations.

Incidentally – and before proceeding any further – we observe that the number of operations can be significantly reduced by exploiting the identity $\omega_{2s}^{-s} = -1$, $s \geq 1$. Hence, (12.14) yields

$$
\begin{aligned}
y_j &= y_j^{[e]} + \omega_{2^n}^{-j} y_j^{[o]}, \\
y_{j+2^{n-1}} &= y_{j+2^{n-1}}^{[e]} + \omega_{2^n}^{-j-2^{n-1}} y_{j+2^{n-1}}^{[o]} = y_j^{[e]} - \omega_{2^n}^{-j} y_j^{[o]},
\end{aligned}
\qquad j = 0,1,\ldots,2^{n-1}-1
$$

(recall that $\boldsymbol{y}^{[e]}, \boldsymbol{y}^{[o]} \in \Pi_{2^{n-1}}$, therefore $y_{j+2^{n-1}}^{[e]} = y_j^{[e]}$ etc.). In other words, to combine $\boldsymbol{y}^{[e]}$ and $\boldsymbol{y}^{[o]}$, we need to form just 2^{n-1} products $\omega_{2^{n-1}}^{-j} y_j^{[o]}$, subsequently adding or subtracting them, as required, from $y_j^{[e]}$ for $j = 0,1,\ldots,2^{n-1}$.

All this, needless to say, is based on the premise that $\boldsymbol{y}^{[e]}$ and $\boldsymbol{y}^{[o]}$ are known, which, as things stand, is false. Having said this, we can form, in a similar fashion to that above, $\boldsymbol{y}^{[e]}$ from $\mathcal{F}_{d/4}^* \boldsymbol{x}^{[ee]}, \mathcal{F}_{d/4}^* \boldsymbol{x}^{[eo]} \in \Pi_{d/4}$, where

$$
\boldsymbol{x}^{[ee]} = \{x_{4j}\}_{j=-\infty}^{\infty}, \qquad \boldsymbol{x}^{[eo]} = \{x_{4j+2}\}_{j=-\infty}^{\infty}.
$$

Likewise, $\boldsymbol{y}^{[o]}$ can be also obtained from two transforms of length $d/4$. This procedure can be iterated until we reach transforms of unit length, which, of course, are the variables themselves.

Practical implementation of this procedure, the famous *fast Fourier transform* (FFT), proceeds from the other end: we commence from 2^n transforms of length 1 and synthesize them to 2^{n-1} transforms of length 2. These are, in turn, combined into 2^{n-2} transforms of length 2^2, next into 2^{n-3} transforms of length 2^3 and so on, until we reach a single transform of length 2^n, the object of this whole exercise.

Assembling 2^{n-s+1} transforms of length 2^{s-1} into 2^{n-s} transforms of double the length costs $\mathcal{O}(2^{n-s} \times 2^s) = \mathcal{O}(d)$ operations. Since there are n such 'layers' the total

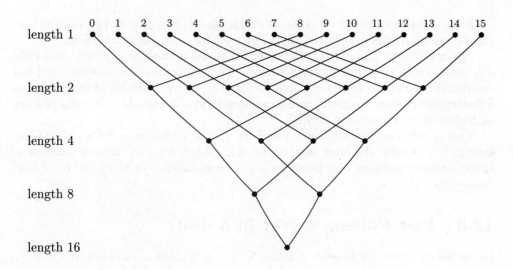

Figure 12.1   The assembly pattern in FFT for $d = 16 = 2^4$.

expense of the FFT is a multiple of $2^n n = d \log_2 d$ operations. For large values of d this results in a very significant saving, compared with naive matrix multiplication.

The order of assembly of one set of transforms into new transforms of twice the length is important – we do not just combine any two arbitrary 'strands'! The correct arrangement is displayed in Figure 12.1 in the case $n = 4$ and the general rule is obvious. It can be best understood by expressing the index ℓ in a binary representation, but we choose not to dwell further on that issue.

Having introduced FFTs, we use them to decrease the computational cost of the Hockney method from Section 12.1. Recall that the problematic part of the calculation consists of $2m_2$ products of the form (12.6). We recognize readily each such product as an inverse DFT in disguise,

$$s_\ell = c \operatorname{Im} \left(\sum_{j=0}^{m_1} p_j \omega_{2m_1+2}^{j\ell} \right), \qquad \ell = 1, 2, \ldots, m_1.$$

To bring this into a proper DFT form, we set $p_j = 0$, $j = m_1 + 1, m_1 + 2, \ldots, 2m_1 + 1$ and extend the range of ℓ to $\{0, 1, \ldots, 2m_1 + 1\}$. This yields

$$s_\ell = c \operatorname{Im} \left(\sum_{j=0}^{2m_1+1} p_j \omega_{2m_1+2}^{j\ell} \right), \qquad \ell = 0, 1, \ldots, 2m_1 + 1,$$

that is

$$\tilde{s} = c \operatorname{Im} \mathcal{F}_{2m_1+1}^{-1} \tilde{p},$$

where $\tilde{p}, \tilde{s} \in \Pi_{2m_1+2}$. Finally, we discard s_0 and $s_{m_1+1}, s_{m_1+2}, \ldots, s_{2m_1+1}$ and retain the remaining values as the result of our calculation.

Even when we calculate a system twice the size and employ complex arithmetic, this procedure results in substantial savings even for moderate values of m_1.[1]

The fast Fourier transform is instrumental to the performance of most fast Poisson solvers, and we return to this theme in the next section, in the context of fast solution of the Poisson equation in a disc. There are exceptions and, in the comments following this chapter, we briefly mention the method of *cyclic odd–even reduction and factorization,* which requires no FFT.

This is not the only application of FFT to the numerical solution of PDEs – indeed, it is hardly the most important. This distinction, in fairness, belongs to spectral methods, whose very power rests on the availability of FFTs and other 'fast transforms'.

12.3   Fast Poisson solver in a disc

Let us suppose that the Poisson equation $\nabla^2 u = g_0$ is given in the open unit disc

$$\mathbb{D} = \{(x, y) \in \mathbb{R}^2 : x^2 + y^2 < 1\},$$

together with Dirichlet boundary conditions along the unit circle,

$$u(\cos\theta, \sin\theta) = \phi(\theta), \qquad 0 \leq \theta \leq 2\pi, \tag{12.15}$$

where $\phi(0) = \phi(2\pi)$. It is convenient to translate the equation from Cartesian to polar coordinates. Thus, we let

$$\begin{aligned} v(r, \theta) &= u(r\cos\theta, r\sin\theta), \\ g(r, \theta) &= g_0(r\cos\theta, r\sin\theta), \end{aligned} \qquad 0 < r < 1, \quad 0 \leq \theta \leq 2\pi.$$

The form of ∇^2 in polar coordinates readily gives us

$$\frac{\partial^2 v}{\partial r^2} + \frac{1}{r}\frac{\partial v}{\partial r} + \frac{1}{r^2}\frac{\partial^2 v}{\partial \theta^2} = g, \qquad 0 < r < 1, \quad 0 \leq \theta \leq 2\pi. \tag{12.16}$$

The boundary conditions, however, are more delicate. Switching from Cartesian to polar means, in essence, that the disc $\mathbb{D}$ is replaced by the square

$$\widetilde{\mathbb{D}} = \{(r, \theta) : 0 < r < 1, \, 0 \leq \theta \leq 2\pi\}.$$

Unlike $\mathbb{D}$, which has just one boundary 'segment' – its circumference – the set $\widetilde{\mathbb{D}}$ boasts four portions of boundary and we need to allocate appropriate conditions at all of them.

The segment $r = 1$, $0 \leq \theta \leq 2\pi$, is the easiest, being the destination of the original boundary condition (12.15). Hence, we set

$$v(1, \theta) = \phi(\theta), \qquad 0 \leq \theta \leq 2\pi. \tag{12.17}$$

Next in order of difficulty are the line segments $0 < r < 1$, $\theta = 0$, and $0 < r < 1$, $\theta = 2\pi$. They both correspond to the same segment, namely $0 < x < 1$, $y = 0$, in the

---

[1] As a matter of fact, the calculation can be performed in real arithmetic, using the so-called *fast sine transform.*

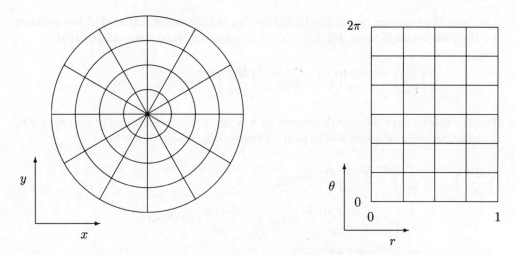

Figure 12.2   Computational grids in the disc $\mathbb{D}$ and the square $\widetilde{\mathbb{D}}$, associated by
the Cartesian-to-polar transformation.

original disc $\mathbb{D}$. The value of u on this segment is, of course, unknown, but, at the
very least, we know that it is the same whether we assign it to $\theta = 0$ or $\theta = 2\pi$. In
other words, we have the *periodic* boundary condition

$$v(r,0) = v(r,2\pi), \qquad 0 < r < 1. \tag{12.18}$$

Finally, we pay attention to the remaining portion of $\partial\widetilde{\mathbb{D}}$, namely $r = 0$, $0 \le \theta \le 2\pi$.
This whole line corresponds to just a single point in $\mathbb{D}$, namely the origin $x = y = 0$
(see Figure 12.2). Therefore, v is constant along that line or, to express it in a more
manageable form, we obtain the *Neumann* boundary condition

$$\frac{\partial}{\partial\theta}v(0,\theta) = 0, \qquad 0 \le \theta \le 2\pi. \tag{12.19}$$

We can approximate the solution of (12.16) with boundary conditions (12.17)–
(12.19) by inscribing a square grid into $\widetilde{\mathbb{D}}$, approximating $\partial v/\partial r$, $\partial^2 v/\partial r^2$ and $\partial^2 v/\partial\theta^2$
by central differences, say, at the grid points and taking adequate care of the boundary
conditions. The outcome is certainly preferable by far to imposing a square grid on
the original disc $\mathbb{D}$, a procedure that leads to excessively unpleasant equations at
near-boundary grid points. Having solved the Poisson equation in $\widetilde{\mathbb{D}}$, we can map the
outcome to the disc, i.e., to the concentric grid of Figure 12.2. This, however, is not
the fast solver that is the goal of this section. To calculate (12.16) considerably faster,
we again resort to FFTs.

We have already defined the *Fourier transform* (12.10) of an arbitrary complex-
valued, integrable function g in $\mathbb{R}$ that is periodic with period 2π. Because of the
periodic boundary condition (12.18), we can Fourier-transform $v(r, \cdot)$, and this results
in the sequence

$$\hat{v}_m(r) = \frac{1}{2\pi} \int_0^{2\pi} v(r,\theta)e^{-im\theta}\,d\theta, \qquad m \in \mathbb{Z}.$$

Our goal is to convert the PDE (12.16) into an infinite set of ODEs that are satisfied by the *Fourier coefficients* $\{v_m\}_{m=-\infty}^{\infty}$. It is easy to deduce from (12.10) that

$$\left(\widehat{\frac{\partial v}{\partial r}}\right)_m = v_m'(r) \quad \text{and} \quad \left(\widehat{\frac{\partial^2 v}{\partial r^2}}\right)_m = v_m''(r), \qquad m \in \mathbb{Z},$$

but the second derivative with respect to θ is less trivial – fortunately not by much! Integrating twice by parts and exploiting periodicity readily leads to

$$\begin{aligned}
\left(\widehat{\frac{\partial^2 v}{\partial \theta^2}}\right)_m &= \frac{1}{2\pi} \int_0^{2\pi} \frac{\partial^2 v(r, \theta)}{\partial \theta^2} e^{-im\theta} \, d\theta \\
&= \frac{1}{2\pi} \left[e^{-im\theta} \frac{\partial v(r, \theta)}{\partial \theta} \Big|_0^{2\pi} + im \int_0^{2\pi} \frac{\partial v(r, \theta)}{\partial \theta} e^{-im\theta} \, d\theta \right] \\
&= \frac{im}{2\pi} \int_0^{2\pi} \frac{\partial v(r, \theta)}{\partial \theta} e^{-im\theta} \, d\theta \\
&= \frac{im}{2\pi} \left[e^{-im\theta} v(r, \theta) \Big|_0^{2\pi} + im \int_0^{2\pi} v(r, \theta) e^{-im\theta} \, d\theta \right] = -m^2 \hat{v}_m(r), \qquad m \in \mathbb{Z}.
\end{aligned}$$

We now multiply (12.16) by $e^{-im\theta}$, integrate from 0 to 2π and divide by 2π. The outcome is the ODE

$$\hat{v}_m'' + \frac{1}{r} \hat{v}_m' - \frac{m^2}{r^2} \hat{v}_m = \hat{g}_m, \tag{12.20}$$

which is obeyed by each Fourier coefficient $\hat{v}_m$, $m \in \mathbb{Z}$, for $0 < r < 1$. The right-hand side is the mth Fourier coefficient of the inhomogeneous term g.

What are the boundary conditions associated with the differential equation (12.20)? Firstly, the periodic conditions (12.18), having played their role in validating the use of the Fourier transform, disappear in tandem with the variable θ. Secondly, the right-hand condition (12.17) translates at once into

$$\hat{v}_m(1) = \hat{\phi}_m \qquad m \in \mathbb{Z}. \tag{12.21}$$

Finally, we must render the natural condition (12.19) in the language of Fourier coefficients. This presents more of a challenge.

We commence by using the inverse Fourier transform formula (12.12), which allows us to synthesize v from its coefficients:

$$v(r, \theta) = \sum_{m=-\infty}^{\infty} \hat{v}_m(r) e^{im\theta}, \qquad 0 \le r \le 1, \qquad 0 \le \theta \le 2\pi.$$

Next, we differentiate with respect to θ and set $r = 0$. Comparison with (12.19) yields

$$i \sum_{m=-\infty}^{\infty} m \hat{v}_m(0) e^{im\theta} \equiv 0, \qquad 0 \le \theta \le 2\pi.$$

The trigonometric functions $e^{im\theta}$, $m \in \mathbb{Z}$, are linearly independent. Therefore, whenever their linear combination vanishes identically, all its components must be zero.[2] The outcome is the boundary condition

$$\hat{v}_m(0) = 0, \qquad m \in \mathbb{Z} \setminus \{0\}. \tag{12.22}$$

This leaves us with just a single missing item of information, namely the boundary value for the zeroth harmonic. The Cartesian-to-polar transformation,

$$v(r, \theta) = u(r \cos \theta, r \sin \theta)$$

implies

$$\frac{\partial v(0, \theta)}{\partial r} = \cos \theta \, \frac{\partial u(0,0)}{\partial x} + \sin \theta \, \frac{\partial u(0,0)}{\partial y}.$$

Therefore

$$\hat{v}'(0) = \frac{1}{2\pi} \int_0^{2\pi} \frac{\partial v(0, \theta)}{\partial r} \, \mathrm{d}\theta = \frac{1}{2\pi} \int_0^{2\pi} \cos \theta \, \mathrm{d}\theta \frac{\partial u(0,0)}{\partial x} + \frac{1}{2\pi} \int_0^{2\pi} \sin \theta \, \mathrm{d}\theta \frac{\partial u(0,0)}{\partial y} = 0,$$

since

$$\int_0^{2\pi} \cos \theta \, \mathrm{d}\theta, \int_0^{2\pi} \sin \theta \, \mathrm{d}\theta = 0.$$

We thus obtain the missing boundary condition,

$$\hat{v}'(0) = 0. \tag{12.23}$$

The crucial fact about the ODEs (12.20) (with the initial conditions (12.21)–(12.23)) is that the Fourier transform *uncouples* the harmonics – the equation for each $\hat{v}_m$ is a two-point boundary-value problem, whose solution is independent of all other Fourier coeffcients!

The two-point boundary problem (12.20) can be solved easily by finite differences (see Exercise 12.4 for an alternative, using the finite element method). Thus, we choose a positive integer d and cover the interval $[0, 1]$ with d sub-intervals of length $\Delta r = 1/d$. Adopting the usual notation from Chapters 1–6, $\hat{v}_{m,k}$ denotes an approximation to $\hat{v}_m(k\Delta r)$. Employing the simplest central difference approximation,

$$\hat{v}'_m(k\Delta r) \approx \frac{1}{2\Delta r} \left(\hat{v}_{m,k+1} - \hat{v}_{m,k-1} \right),$$

$$\hat{v}''_m(k\Delta r) \approx \frac{1}{(\Delta r)^2} \left(\hat{v}_{m,k+1} - 2\hat{v}_{m,k} + \hat{v}_{m,k-1} \right),$$

the ODE (12.20) leads to the difference equation

$$\frac{1}{(\Delta r)^2} \left(\hat{v}_{m,k+1} - 2\hat{v}_{m,k} + \hat{v}_{m,k-1} \right) + \frac{2}{2k(\Delta r)^2} \left(\hat{v}_{m,k+1} - \hat{v}_{m,k-1} \right) - \frac{m^2}{k^2(\Delta r)^2} \hat{v}_{m,k} = \hat{g}_{m,k},$$

---

[2]The argument is *slightly* more complicated, since the standard theorem about linear independence refers to a finite number of components. For reasons of brevity, we take for granted the generalization to linear combinations having an infinite number of components.

where $\hat{g}_{m,k} = \hat{g}_m(k\Delta r)$ and k ranges in $\{1, 2, \ldots, d-1\}$. Rearranging terms, we arrive at

$$\left(1 - \frac{1}{2k}\right)\hat{v}_{m,k-1} - \left(2 + \frac{m^2}{k^2}\right)\hat{v}_{m,k} + \left(1 + \frac{1}{2k}\right)\hat{v}_{m,k+1} = (\Delta r)^2\hat{g}_{m,k}, \quad k = 1, \ldots, d-1.$$

$$(12.24)$$

We complement (12.24) with boundary values. Thus, (12.21) yields

$$\hat{v}_{m,d} = \hat{\phi}_m, \qquad m \in \mathbb{Z}$$

and (12.22) results in

$$\hat{v}_{m,0} = 0, \qquad m \in \mathbb{Z} \setminus \{0\}.$$

Finally, we use forward differences to approximate (12.23): for example,

$$-\tfrac{5}{4}\hat{v}_{0,0} + \tfrac{3}{2}\hat{v}_{0,1} - \tfrac{1}{4}\hat{v}_{0,2} = 0.$$

The outcome is a tridiagonal linear system for every $m \neq 0$ and an almost-tridiagonal system for $m = 0$, with just one 'rogue' element outside a three-diagonal band.[3] Such systems can be solved with minimal effort by sparse LU factorization, see Chapter 9.

The practical computation of (12.24) should be confined, needless to say, to a finite subset of $m \in \mathbb{Z}$, for example $-m^* + 1 \leq m \leq m^*$. Any discussion of the side-effects of such truncation is way beyond the scope of this volume; we are concerned mainly with the speed of approximation of periodic functions by truncated Fourier series. We just point out that, provided both g and ϕ are sufficiently smooth, good accuracy is attainable with relatively modest values of m^*.

Let us turn our attention to the nuts and bolts of a fast Poisson solver in the disc $\mathbb{D}$. We commence by choosing a positive integer n and let $m^* = 2^{n-1}$. Next we approximate $\{\hat{g}_m\}_{m=-m^*+1}^{m^*}$ and $\{\hat{\phi}_m\}_{m=-m^*+1}^{m^*}$ with FFTs, a task that carries a computational price tag of $\mathcal{O}(m^* \log_2 m^*)$ operations. As we have commented in Section 12.2, the error in such a procedure is very small and, provided that g and ϕ are sufficiently well behaved, it decays at an exponential speed as a function of m^*.

Having calculated the two transforms and chosen a positive integer d, we proceed to solve the linear systems (12.24) for the relevant range of m by employing sparse LU factorization. The total cost of this procedure is $\mathcal{O}(dm^*)$ operations.

Finally, having evaluated $\hat{v}_{m,k}$ for $-m^* + 1 \leq m \leq m^*$ and $k = 1, \ldots, d-1$, we employ $d - 1$ inverse FFTs to produce values of v on a $d \times (2m^*)$ square grid, or, alternatively, on a concentric grid in $\mathbb{D}$ (see Figure 12.2). Specifically, we use Fourier coefficients to reconstruct the function, in line with the formula (12.12),

$$u\left(k\Delta r \cos(\pi\ell/m^*), k\Delta r \sin(\pi\ell/m^*)\right) = v(k\Delta x, \omega_{2m^*}^\ell) = \sum_{m=-m^*+1}^{m^*} \hat{v}_{m,k}\omega_{2m^*}^{\ell m}$$

for $k = 1, 2, \ldots, d-1$ and $m = 0, 1, \ldots, 2m^* - 1$.[4] This is the most computationally intense part of the algorithm and the cost totals $\mathcal{O}(dm^* \log_2 m^*)$. It is comparable,

---

[3] Of course, we have $d + 1$ unknowns and a matching number of equations when $m = 0$.

[4] We do not need to perform an inverse FFT for $d = 0$ since, obviously, $u(0,0) = v(0,\theta) = \hat{v}_{0,0}$.

though, with the expense of the Hockney method from Section 12.1 (which, of course, acts in a different geometry) and very modest indeed in comparison with other computational alternatives.

Why is the Poisson problem in a disc so important as to deserve a section all its own? According to the *conformal mapping theorem*, given any simply connected open set $\mathbb{B} \subset \mathbb{C}$ with a sufficiently smooth boundary, there exists an analytic, univalent (that is, one-to-one) function χ that maps $\mathbb{B}$ onto the complex unit disc and $\partial\mathbb{B}$ onto the unit circle. (There are many such functions but to attain uniqueness it is enough, for example, to require that an arbitrary $z_0 \in \mathbb{B}$ is mapped into the origin with a positive derivative.) Such a function χ is called a *conformal mapping* of $\mathbb{B}$ on the complex unit disc.

Suppose that the Poisson equation

$$\nabla^2 w = f \tag{12.25}$$

is given for all $(x, y) \in \mathbb{B}^\star$, the natural projection of $\mathbb{B}$ on $\mathbb{R}^2$, where

$$(x, y) \in \mathbb{B}^\star \qquad \text{if and only if} \qquad x + \mathrm{i}y \in \mathbb{B}.$$

We accompany (12.25) with Dirichlet boundary conditions $w = \psi$ across $\partial\mathbb{B}^\star$.

Provided that a conformal mapping χ from $\mathbb{B}$ onto the complex unit disc is known, it is possible to translate the problem of numerically solving (12.25) into the unit disc. Being one-to-one, χ possesses an inverse $\eta = \chi^{-1}$.

We let

$$u(x, y) = w\left(\mathrm{Re}\,\eta(x + \mathrm{i}y), \mathrm{Im}\,\eta(x + \mathrm{i}y)\right), \qquad (x, y) \in \mathrm{cl}\,\mathbb{D}.$$

Therefore

$$\frac{\partial^2 u}{\partial x^2} = \left(\frac{\partial^2 \mathrm{Re}\,\eta}{\partial x^2}\right)\frac{\partial w}{\partial x} + \left(\frac{\partial^2 \mathrm{Im}\,\eta}{\partial x^2}\right)\frac{\partial w}{\partial y} + \left(\frac{\partial \mathrm{Re}\,\eta}{\partial x}\right)^2\frac{\partial^2 w}{\partial x^2} + \left(\frac{\partial \mathrm{Im}\,\eta}{\partial x}\right)^2\frac{\partial^2 w}{\partial y^2}$$

$$\frac{\partial^2 u}{\partial y^2} = \left(\frac{\partial^2 \mathrm{Re}\,\eta}{\partial y^2}\right)\frac{\partial w}{\partial x} + \left(\frac{\partial^2 \mathrm{Im}\,\eta}{\partial y^2}\right)\frac{\partial w}{\partial y} + \left(\frac{\partial \mathrm{Re}\,\eta}{\partial y}\right)^2\frac{\partial^2 w}{\partial x^2} + \left(\frac{\partial \mathrm{Im}\,\eta}{\partial y}\right)^2\frac{\partial^2 w}{\partial y^2}$$

and we deduce that

$$\nabla^2 u = \left[\left(\frac{\partial \mathrm{Re}\,\eta}{\partial x}\right)^2 + \left(\frac{\partial \mathrm{Re}\,\eta}{\partial y}\right)^2\right]\frac{\partial^2 w}{\partial x^2} + \left[\left(\frac{\partial \mathrm{Im}\,\eta}{\partial x}\right)^2 + \left(\frac{\partial \mathrm{Im}\,\eta}{\partial y}\right)^2\right]\frac{\partial^2 w}{\partial y^2}$$

$$+ (\nabla^2 \mathrm{Re}\,\eta)\frac{\partial w}{\partial x} + (\nabla^2 \mathrm{Im}\,\eta)\frac{\partial w}{\partial y}. \tag{12.26}$$

The function η is the inverse of a univalent analytic function, hence it is itself analytic and obeys the *Cauchy–Riemann* equations

$$\frac{\partial \mathrm{Re}\,\eta}{\partial x} = \frac{\partial \mathrm{Im}\,\eta}{\partial y}, \qquad \frac{\partial \mathrm{Re}\,\eta}{\partial y} = -\frac{\partial \mathrm{Im}\,\eta}{\partial x}, \qquad (x, y) \in \mathrm{cl}\,\mathbb{D}.$$

We conclude that

$$\frac{\partial^2 \mathrm{Re}\,\eta}{\partial y^2} = \frac{\partial}{\partial y}\left(-\frac{\partial \mathrm{Im}\,\eta}{\partial x}\right) = -\frac{\partial}{\partial x}\left(\frac{\partial \mathrm{Im}\,\eta}{\partial y}\right) = -\frac{\partial^2 \mathrm{Re}\,\eta}{\partial x^2},$$

consequently $\nabla^2 \operatorname{Re} \eta = 0$. Likewise $\nabla^2 \operatorname{Im} \eta = 0$ and we deduce the familiar theorem that the real and imaginary parts of an analytic function are *harmonic* (i.e., they obey the Laplace equation).

Another outcome of the Cauchy–Riemann equations is that

$$\left(\frac{\partial \operatorname{Re} \eta}{\partial x} \right)^2 + \left(\frac{\partial \operatorname{Re} \eta}{\partial y} \right)^2 = \left(\frac{\partial \operatorname{Im} \eta}{\partial x} \right)^2 + \left(\frac{\partial \operatorname{Im} \eta}{\partial y} \right)^2 := \kappa(x, y),$$

say. Therefore, substitution in (12.26), in tandem with the Poisson equation (12.25), yields

$$
\begin{aligned}
\nabla^2 u &= \kappa(x, y)\, \nabla^2 w\, (\operatorname{Re} \eta(x + iy), \operatorname{Im} \eta(x + iy)) \\
&= \kappa(x, y) f\left(\operatorname{Re} \eta(x + iy), \operatorname{Im} \eta(x + iy) \right) := g_0(x, y), \qquad (x, y) \in \mathbb{D}
\end{aligned}
$$

and we are back to the Poisson equation in a unit disc! The boundary condition is $u(x, y) = \psi(\operatorname{Re} \eta(x + iy), \operatorname{Im} \eta(x + iy))$, $(x, y) \in \partial \mathbb{D}$.

Even if χ is unknown, all is not lost, since there are very effective numerical methods for its approximation. Their efficacy is based – again – on a clever use of the FFT technique. Moreover, approximate maps χ and η are typically expressible as DFTs, and this means that, having solved the equation in a unit disc, we return to $\mathbb{B}^\star$ with FFT.... The outcome is a numerical solution on a curved grid, the image of the concentric grid under the function η (see Figure 12.3).

Comments and bibliography

The secret of the fast solvers from Sections 12.1 and 12.3 is the fast Fourier transform (FFT). As we have already remarked in Section 12.2, the fast Fourier transform is an almost miraculous computational device, ubiquitous in a very wide range of applications. Arguably, no other computational algorithm has ever changed the practice of science and engineering as much as this rather simple trick – already implicit in the writings of Gauss, discovered by Lanczos (and forgotten) and, at a more opportune moment, rediscovered in Cooley and Tuckey in 1965.

To present a small and roundabout item in evidence, the author has run a computerized search on the bibliographic data base at California Institute of Technology. This came up with 1272 papers that have been published in the scientific literature in the last four years and that contain 'FFT' in their title or abstract. This is, of course, a gross underestimate, since the FFT is used in so many subjects as a matter of course that it is no longer mentioned in the abstract. In contrast, just 32 papers include the phrase 'Fourier transform' without the adjective 'fast'....

Henrici's review (1979) of the FFT and its mathematical applications is a must for every open-minded applied mathematician. This survey is comprehensive, readable and inspiring – if you plan to read just one mathematical paper this year, Henrici's review will be a very rewarding choice!

As far as fast Poisson solvers are concerned, Golub's survey (1971) and Pickering's monograph (1986) are probably the most comprehensive, although some methods appear also in Henrici's survey and elsewhere.

The name 'Poisson solver' is frequently a misnomer. While the rationale behind Section 12.3 is intimately linked with the Laplace operator, this is not the case with the Hockney

Figure 12.3 Conformal mappings and induced grids for three subsets of $\mathbb{R}^2$. The corresponding mapppings are $z/(\frac{3}{2} - z^2)$, $z(4 + z^2)^{1/2}$ and $(2 + z^3)^{1/2}/(\frac{7}{4} + z)^{3/2}$.

method of Section 12.1 and, indeed, with many other 'fast Poisson solvers'. A more appropriate name, in line with the comments that conclude Section 12.1, would be 'fast block-TST solvers'. For example, see Exercise 12.2 for an application of a fast block-TST solver to the Helmholtz equation.

We conclude these remarks with a brief survey of a fast Poisson (or, again, a fast block-TST) solver that does not employ the FFT: *cyclic odd–even reduction and factorization.*

The starting point for our discussion is the block-TST equations (12.2), where both S and T are $m_1 \times m_1$ matrices. Neither S nor T need be TST, but we assume that they commute (an assumption which, of course, is certainly true when they are TST). We assume that $m_2 = 2^n$ for some $n \geq 1$. For every $\ell = 1, 2, \ldots, 2^{n-1}$ we multiply the $(2\ell - 1)$th equation by T, the (2ℓ)th equation by S and the $(2\ell + 1)$th equation by T. Therefore

$$
\begin{aligned}
T^2 \boldsymbol{x}_{2\ell-2} + TS\boldsymbol{x}_{2\ell-1} + T^2 \boldsymbol{x}_{2\ell} &= T\boldsymbol{b}_{2\ell-1}, \\
- ST\boldsymbol{x}_{2\ell-1} - S^2 \boldsymbol{x}_{2\ell} - ST\boldsymbol{x}_{2\ell+1} &= -S\boldsymbol{b}_{2\ell}, \\
T^2 \boldsymbol{x}_{2\ell} + TS\boldsymbol{x}_{2\ell+1} + T^2 \boldsymbol{x}_{2\ell+2} &= T\boldsymbol{b}_{2\ell+1}
\end{aligned}
$$

and summation, in tandem with $ST = TS$, results in

$$
T^2 \boldsymbol{x}_{2(\ell-1)} + (2T^2 - S^2)\boldsymbol{x}_{2\ell} + T^2 \boldsymbol{x}_{2(\ell+1)} = T(\boldsymbol{b}_{2\ell-1} + \boldsymbol{b}_{2\ell+1}) - S\boldsymbol{b}_{2\ell}, \qquad \ell = 1, 2, \ldots, 2^{n-1}.
$$
$$(12.27)$$

The linear system (12.27) is also block-TST and it possesses exactly half the number of blocks of (12.2). Provided that its solution is known, we can easily recover the missing components by solving the $m_1 \times m_1$ linear systems

$$S \boldsymbol{x}_{2\ell-1} = \boldsymbol{b}_{2\ell-1} - T(\boldsymbol{x}_{2\ell-2} + \boldsymbol{x}_{2\ell}), \qquad \ell = 1, 2, \ldots, 2^{n-1}.$$

Our choice of $m_2 = 2^n$ already gives the game away – we continue by repeating this procedure again and again, each time reducing the size of the system. Thus, we let

$$S^{[0]} := S, \qquad S^{[r+1]} := 2(T^{[r]})^2 - (S^{[r]})^2,$$
$$T^{[0]} := T, \qquad T^{[r+1]} := (T^{[r]})^2,$$
$$\boldsymbol{b}_\ell^{[0]} := \boldsymbol{b}, \qquad \boldsymbol{b}_\ell^{[r+1]} = T^{[r]}(\boldsymbol{b}_{\ell-2^r}^{[r]} + \boldsymbol{b}_{\ell+2^r}^{[r]}) - S^{[r]} \boldsymbol{b}_\ell^{[r]}$$

and recover the missing components by iterating backwards:

$$S^{[r-1]} \boldsymbol{x}_{j2^r - 2^{r-1}} = \boldsymbol{b}_{j2^r - 2^{r-1}}^{[r-1]} - T^{[r-1]}(\boldsymbol{x}_{j2^r} + \boldsymbol{x}_{(j-1)2^r}), \qquad j = 1, 2, \ldots, 2^{n-r-1}.$$

We hasten to warn that, as presented, the method is ill conditioned, since each 'iteration' $S^{[r]} \to S^{[r+1]}$, $T^{[r]} \to T^{[r+1]}$ not only destroys sparsity but also produces matrices that are progressively less amenable for numerical manipulation. It is possible to stabilize this algorithm, but this is outside the scope of this brief survey.

There exists an intriguing common thread in multigrid methods, the FFT technique and the cyclic odd–even reduction and factorization method. All these factorize their 'medium' – whether a grid, a sequence or a system of equations – into a hierarchy of nested subsystems. This procedure is increasingly popular in modern computational mathematics and other examples are provided by wavelets and by the method of hierarchical bases in the finite element method.

Golub, G.H. (1971), Direct methods for solving elliptic difference equations, in *Symposium on the Theory of Numerical Analysis* (J.L. Morris, editor), Lecture Notes in Mathematics 193, Springer-Verlag, Berlin.

Henrici, P. (1979), Fast Fourier methods in computational complex analysis, *SIAM Review* **21**, 481–527.

Pickering, M. (1986), *An Introduction to Fast Fourier Transform Methods for Partial Differential Equations, with Applications*, Research Studies Press, Herts.

Exercises

12.1   Show how to modify the Hockney method to evaluate numerically the solution of the Poisson equation in the three-dimensional cube

$$\{(x_1, x_2, x_3) : 0 \le x_1, x_2, x_3 \le 1\}$$

with Dirichlet boundary conditions.

12.2   The Helmholtz equation

$$\nabla^2 u + \lambda u = g$$

is given in a rectangle in $\mathbb{R}^2$, accompanied by Dirichlet boundary conditions. Amend the Hockney method so that it can provide fast numerical solution of this equation.

12.3 Prove that Π_d satisfies all the axioms of a linear space from Appendix section A.2.1.1. Find a basis of Π_d, thereby demonstrating that $\dim \Pi_d = d$.

12.4* An alternative to solving (12.20) by the finite difference equations (12.24) is the finite element method.

   a Find a variational problem whose Euler–Lagrange equation is (12.20).

   b Formulate explicitly the Ritz equations.

   c Discuss the choice of finite element functions that are likely to produce an accuracy similar to the finite difference method (12.24).

12.5 The Poisson integral formula

$$v(r,\theta) = \frac{1}{\pi} \int_0^{2\pi} \left[\frac{1 - r^2}{1 - 2r\cos(\theta - \tau) + r^2} \right] g(\tau) \, d\tau$$

confers an alternative to the natural boundary condition (12.23).

   a Find explicitly the value of $v(0, \theta)$.

   b Deduce the value of $\hat{v}_0(0)$. [*Hint: Express $v(0, \theta)$ as a linear combination of Fourier coefficients, in line with* (12.12).]

12.6 Amend the fast Poisson solver from Section 12.3 to approximate the solution of $\nabla^2 u = g_0$ in the unit disc, but with (12.15) replaced by the Neumann boundary condition

$$\frac{\partial u(\cos\theta, \sin\theta)}{\partial x} \cos\theta + \frac{\partial u(\cos\theta, \sin\theta)}{\partial y} \sin\theta = \phi(\theta), \qquad 0 \le \theta \le 2\pi,$$

where $\phi(0) = \phi(2\pi)$ and

$$\int_0^{2\pi} \phi(\theta) \, d\theta = 0.$$

12.7 Describe a fast Poisson solver for the Poisson equation $\nabla^2 u = g_0$, given with Dirichlet boundary conditions in the annulus

$$\{(x, y) : \rho < x^2 + y^2 < 1\},$$

where $\rho \in (0, 1)$ is given.

12.8* Generalize the fast Poisson solver from Section 12.3 from the unit disc to the three-dimensional cylinder

$$\{(x_1, x_2, x_3) : x_1^2 + x_2^2 < 1, \ 0 < x_3 < 1\}.$$

You may assume Dirichlet boundary conditions.

PART III

Partial differential equations of evolution

13

The diffusion equation

13.1 A simple numerical method

It is often useful to classify partial differential equations into two kinds: *steady-state* equations, where all the variables are spatial, and *evolutionary* equations, which combine differentiation with respect to space and to time. We have already seen some examples of steady-state equations, namely the Poisson equation and the biharmonic equation. Typically, equations of this type describe physical phenomena whose behaviour depends on the minimization of some quantity, e.g. potential energy, and they are ubiquitous in mechanics and elasticity theory.[1] Evolutionary equations, however, model systems that undergo change as a function of time and they are important *inter alia* in the description of wave phenomena, thermodynamics, diffusive processes and population dynamics.

It is usual in the theory of PDEs to distinguish between *elliptic, parabolic* and *hyperbolic* equations. We do not wish to pursue here this formalism – or even provide the requisite definitions – except for remarking that elliptic equations are of the steady-state type whilst both parabolic and hyperbolic PDEs are evolutionary. A brief explanation of this distinction rests in the different kind of characteristics admitted by the three types of equations.

Evolutionary differential equations are, in a sense, reminiscent of ODEs. Indeed, one can view ODEs as evolutionary equations without space variables. We will see in what follows that there are many similarities between the numerical treatment of ODEs and of evolutionary PDEs and that, in fact, one of the most effective means to compute the latter is by approximate conversion to an ODE system. However, this similarity is deceptive. The numerical solution of evolutionary PDEs requires us to discretize both in time and in space and, in a successful algorithm, these two procedures are not independent. The concepts underlying the numerical analysis of PDEs of evolution might sound familiar but they are often surprisingly more intricate and subtle than the comparable concepts from Chapters 1–3.

Our first example of an evolutionary equation is the *diffusion equation,*

$$\frac{\partial u}{\partial t} = \frac{\partial^2 u}{\partial x^2}, \qquad 0 \leq x \leq 1, \quad t \geq 0, \tag{13.1}$$

also known as the *heat conduction equation.* The function $u = u(x, t)$ is accompanied

---

[1] This minimization procedure can be often rendered in the language of the theory of variations, and this provides a bridge to the material of Chapter 8.

by two kinds of 'side conditions', namely an *initial condition*

$$u(x, 0) = g(x), \qquad 0 \le x \le 1 \tag{13.2}$$

and the *boundary conditions*

$$u(0, t) = \varphi_0(t), \quad u(1, t) = \varphi_1(t), \qquad t \ge 0 \tag{13.3}$$

(of course, $g(0) = \varphi_0(0)$ and $g(1) = \varphi_1(0)$). As its name implies, (13.1) models diffusive and thermodynamic phenomena.

The equation (13.1) is the simplest form of a diffusion equation and it can be generalized in several ways:

- by allowing more *spatial variables,* giving

$$\frac{\partial u}{\partial t} = \nabla^2 u, \tag{13.4}$$

  where $u = u(x, y, t)$, say;

- by adding to (13.1) a *forcing term* f, giving

$$\frac{\partial u}{\partial t} = \frac{\partial^2 u}{\partial x^2} + f, \tag{13.5}$$

  where $f = f(x, t)$;

- by allowing a variable *diffusion coefficient* a, giving

$$\frac{\partial u}{\partial t} = \frac{\partial}{\partial x} \left[a(x) \frac{\partial u}{\partial x} \right], \tag{13.6}$$

  where $a = a(x)$ is a differentiable function such that $0 < a(x) < \infty$ for all $x \in [0, 1]$;

- by letting x range in an arbitrary interval of $\mathbb{R}$. The most important special case is the *Cauchy problem,* where $-\infty < x < \infty$ and the boundary conditions (13.3) are replaced by the requirement that the solution $u(\,\cdot\,, t)$ is square-integrable for all t, i.e.,

$$\int_{-\infty}^{\infty} [u(x, t)]^2 \, dx < \infty, \qquad t \ge 0.$$

Needless to say, we can combine several such generalizations.

We commence from the most elementary framework but will address ourselves hereafter to various generalizations. Our intention being to approximate (13.1) by finite differences, we choose a positive integer d and inscribe into the strip

$$\{(x, t) : x \in [0, 1], \ t \ge 0\}$$

a rectangular grid

$$\{(\ell \Delta x, n \Delta t), \quad \ell = 0, 1, \dots, d + 1, \ n \ge 0\},$$

where $\Delta x = 1/(d+1)$. The approximant of $u(\ell\Delta x, n\Delta t)$ is denoted by u_ℓ^n. Observe that in the latter, n is a superscript, not a power – we employ this notation to establish a firm and clear distinction between space and time, a central *leitmotif* in the numerical analysis of evolutionary equations.

Let us replace the second spatial derivative and the first temporal derivative respectively by the central difference

$$\frac{\partial^2 u(x,t)}{\partial x^2} \approx \frac{1}{(\Delta x)^2}[u(x-\Delta x, t) - 2u(x,t) + u(x+\Delta x, t)] + \mathcal{O}((\Delta x)^2), \qquad \Delta x \to 0,$$

and the forward difference

$$\frac{\partial u(x,t)}{\partial t} = \frac{1}{\Delta t}[u(x, t+\Delta t) - u(x,t)] + \mathcal{O}((\Delta t)), \qquad \Delta t \to 0.$$

Substitution into (13.1) and multiplication by Δt results in the *Euler method*

$$u_\ell^{n+1} = u_\ell^n + \mu(u_{\ell-1}^n - 2u_\ell^n + u_{\ell+1}^n), \qquad \ell = 1, 2, \ldots, d, \; n = 0, 1, \ldots, \qquad (13.7)$$

where the ratio

$$\mu = \frac{\Delta t}{(\Delta x)^2}$$

is important enough to be given a name all of its own, the *Courant number*.

To launch the recursive procedure (13.7) we use the initial condition (13.2), setting

$$u_\ell^0 = g(\ell\Delta x), \qquad \ell = 1, 2, \ldots, d.$$

Note that the calculation of (13.7) for $\ell = 1$ and $\ell = d$ requires us to invoke the boundary values from (13.3), namely $u_0^n = \varphi_0(n\Delta t)$ and $u_{d+1}^n = \varphi_1(n\Delta t)$ respectively.

How accurate is the method (13.7)? In line with our definiton of the order of a numerical scheme for ODEs, we observe that

$$\frac{u(x, t+\Delta t) - u(x,t)}{\Delta t} - \frac{u(x-\Delta x, t) - 2u(x,t) + u(x+\Delta x, t)}{(\Delta x)^2} = \mathcal{O}((\Delta x)^2, \Delta t)$$

$$(13.8)$$

for $\Delta x, \Delta t \to 0$. Let us assume that Δx and Δt approach zero in such a manner that μ stays constant – it will be seen later that this assumption makes perfect sense! Therefore $\Delta t = \mu(\Delta x)^2$ and (13.8) becomes

$$\frac{u(x, t+\Delta t) - u(x,t)}{\Delta t} - \frac{u(x-\Delta x, t) - 2u(x,t) + u(x+\Delta x, t)}{(\Delta x)^2} = \mathcal{O}((\Delta x)^2)$$

for $\Delta x \to 0$. We say that the Euler method (13.7) is of *order* two.[2]

The concept of order is important in studying how well a finite difference scheme models a continuous differential equation but – as was the case with ODEs in Chapter 2

---

[2]There is some room for confusion here, since for ODE methods an error of $\mathcal{O}(h^{p+1})$ meant order p. The reason for the present definition of order will be clear in the course of the proof of Theorem 13.1.

– our main concern is *convergence*, not order. We say that (13.7) (or, for that matter, any other finite difference method) is convergent if, given any $t^* > 0$, it is true that

$$\lim_{\Delta x \to 0} \left[\lim_{\ell \to x/\Delta x} \left(\lim_{n \to t/\Delta t} u_\ell^n \right) \right] = u(x, t) \qquad \text{for all} \qquad x \in [0, 1], \quad t \in [0, t^*].$$

As before, $\mu = \Delta t / (\Delta x)^2$ is kept constant.

Theorem 13.1 *If $\mu \leq \frac{1}{2}$ then the method (13.7) is convergent.*

Proof Let $t^* > 0$ be an arbitrary constant and define

$$e_\ell^n := u_\ell^n - u(\ell \Delta x, n \Delta t), \qquad \ell = 0, 1, \ldots, d+1, \quad n = 0, 1, \ldots, n_{\Delta t},$$

where $n_{\Delta t} = \lfloor t^*/\Delta t \rfloor = \lfloor t^*/(\mu(\Delta x)^2) \rfloor$ is the rightmost endpoint of the range of n. The definition of convergence can be expressed in the terminology of the variables e_ℓ^n as

$$\lim_{\Delta x \to 0} \left[\max_{\ell = 0, 1, \ldots, d+1} \left(\max_{n = 0, 1, \ldots, n_{\Delta t}} |e_\ell^n| \right) \right] = 0.$$

Letting

$$\eta_n := \max_{\ell = 0, 1, \ldots, d+1} |e_\ell^n|, \qquad n = 0, 1, \ldots, n_{\Delta t},$$

we rewrite this as

$$\lim_{\Delta x \to 0} \left(\max_{n = 0, 1, \ldots, n_{\Delta t}} \eta_n \right) = 0. \tag{13.9}$$

Since

$$u_\ell^{n+1} = u_\ell^n + \mu(u_{\ell-1}^n - 2u_\ell^n + u_{\ell-1}^n),$$
$$\tilde{u}_\ell^{n+1} = \tilde{u}_\ell^n + \mu[\tilde{u}_{\ell-1}^n - 2\tilde{u}_\ell^n + \tilde{u}_{\ell+1}^n] + \mathcal{O}((\Delta x)^4),$$
$$\ell = 0, 1, \ldots, d+1, \quad n = 0, 1, \ldots, n_{\Delta t} - 1,$$

where $\tilde{u}_\ell^n = u(\ell \Delta x, n \Delta t)$, subtraction results in

$$e_\ell^{n+1} = e_\ell^n + \mu(e_{\ell-1}^n - 2e_\ell^n + u_{\ell+1}^n) + \mathcal{O}((\Delta x)^4),$$
$$\ell = 0, 1, \ldots, d+1, \quad n = 0, 1, \ldots, n_{\Delta t} - 1.$$

Similarly to the proof of Theorem 1.1, we may now deduce that, provided u is sufficiently smooth (as it will be, provided that the initial and boundary conditions are so, but we choose not to elaborate this point further), there exists a constant $c > 0$, independent of Δx, such that, for every $\ell = 0, 1, \ldots, d+1$,

$$|e_\ell^{n+1} - e_\ell^n - \mu(e_{\ell-1}^n - 2e_\ell^n + e_{\ell+1}^n)| \leq c(\Delta x)^4,$$
$$\ell = 0, 1, \ldots, d+1, \quad n = 0, 1, \ldots, n_{\Delta t} - 1.$$

Therefore, by the triangle inequality and the definition of η^n,

$$\begin{aligned}
|e_\ell^{n+1}| &\leq \left| e_\ell^n + \mu(e_{\ell-1}^n - 2e_\ell^n + e_{\ell+1}^n) \right| + c(\Delta x)^4 \\
&\leq \mu|e_{\ell-1}^n| + |1 - 2\mu| \, |e_\ell^n| + \mu|e_{\ell+1}^n| + c(\Delta x)^4 \\
&\leq (2\mu + |1 - 2\mu|)\eta^n + c(\Delta x)^4, \qquad n = 0, 1, \ldots, n_{\Delta t} - 1.
\end{aligned}$$

Because $\mu \leq \frac{1}{2}$, we thus deduce that

$$\eta^{n+1} = \max_{\ell=0,1,\ldots,d+1} |e_\ell^{n+1}| \leq \eta^n + c(\Delta x)^4, \quad n = 0, 1, \ldots, n_{\Delta t} - 1.$$

Thus, by induction

$$\eta^{n+1} \leq \eta^n + c(\Delta x)^4 \leq \eta^{n-1} + 2c(\Delta x)^4 \leq \eta^{n-2} + 3c(\Delta x)^4 \leq \cdots$$

and we conclude that

$$\eta^n \leq \eta^0 + nc(\Delta x)^4, \qquad n = 0, 1, \ldots, n_{\Delta t}.$$

Since $\eta^0 = 0$ (because initial conditions at the grid points match for the exact and the discretized equation) and $n(\Delta x)^2 = n\Delta t/\mu \leq t^*/\mu$, we deduce that

$$\eta^n \leq \frac{ct^*}{\mu}(\Delta x)^2, \qquad n = 0, 1, \ldots, n_{\Delta t}.$$

Therefore $\lim_{\Delta x \to 0} \eta^n = 0$ for all n, and comparison with (13.9) completes the proof of convergence. ∎

◇ **A numerical example**   Let us consider the diffusion equation (13.1) with the initial and boundary conditions

$$g(x) = \sin \tfrac{1}{2}\pi x + \tfrac{1}{2}\sin 2\pi x, \qquad 0 \leq x \leq 1, \tag{13.10}$$

$$\varphi_0(t) \equiv 0, \qquad \varphi_1(t) = e^{-\pi^2 t/4}, \qquad t \geq 0,$$

respectively. Its exact solution is, incidentally,

$$u(x,t) = e^{-\pi^2 t/4} \sin \tfrac{1}{2}\pi x + \tfrac{1}{2}e^{-4\pi^2 t} \sin 2\pi x, \qquad 0 \leq x \leq 1, \quad t \geq 0.$$

Figure 13.1 displays the error in the solution of this equation by the Euler method (13.7) with two choices of μ, one just within the interval $(0, \frac{1}{2}]$ and one outside. It is evident that, while for $\mu = \frac{1}{2}$ the solution looks right, the second choice of the Courant number soon deteriorates into complete nonsense.

A different aspect of the solution is highlighted in Figure 13.2, where μ is kept constant (and within a 'safe' range), while the size of the spatial grid is doubled. The error can be observed to be roughly divided by four with each doubling of d, a behaviour entirely consistent with our statement that (13.7) is a second-order method. ◇

Two important remarks are in order. Firstly, unless a method converges, it should not be used – the situation is similar to the numerical analysis of ODEs, with one important exception, as follows. An ODE method is either convergent or not, whereas a method

Figure 13.1   The error in the solution of the diffusion equation (13.10) by the
Euler method (13.7) with $d = 20$ and two choices of μ, 0.5 and 0.509.

for evolutionary PDEs (e.g. for the diffusion equation (13.1)) possesses a parameter
μ and it is entirely possible that it converges only for some values of μ. (We will see
later examples of methods that converge for all $\mu > 0$ and, in Exercise 13.13, a method
which diverges for all $\mu > 0$.)

Secondly, 'keeping μ constant' means in practice that, each time we refine Δx, we
need to amend Δt so that the quotient $\mu = \Delta t/(\Delta x)^2$ remains constant. This implies
that Δt is likely to be considerably smaller than Δx; for example, $d = 20$ and $\mu = \frac{1}{2}$
yields $\Delta x = \frac{1}{20}$ and $\Delta t = \frac{1}{800}$, leading to very large computational cost.[3] For example,
the lower right-hand surface in Figure 13.2 was produced with

$$\Delta x = \frac{1}{160} \quad \text{and} \quad \Delta t = \frac{1}{64000}.$$

Much of the effort associated with designing and analysing numerical methods for
the diffusion equation is invested in circumventing such restrictions and attaining
convergence in regimes of Δx and Δt that are of a more comparable size.

---

[3]To be fair, we have not yet proved that $\mu > \frac{1}{2}$ is bound to lead to a loss of convergence, although
Figure 13.1 certainly seems to indicate that this is likely. The proof of the necessity of $\mu \leq \frac{1}{2}$ is
deferred to Section 13.5.

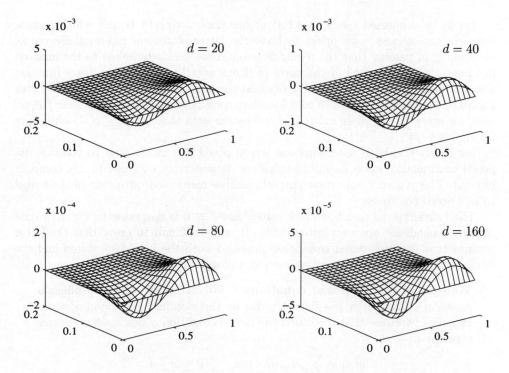

Figure 13.2   The numerical error in the solution of the diffusion equation (13.10) by the Euler method (13.7) with $d = 20, 40, 80, 160$ and $\mu = \frac{2}{5}$.

13.2   Order, stability and convergence

In the present section we wish to discuss the numerical solution by finite differences of a general linear PDE of evolution,

$$\frac{\partial u}{\partial t} = \mathcal{L}u + f, \qquad \boldsymbol{x} \in \mathcal{U}, \quad t \geq 0, \tag{13.11}$$

where $\mathcal{U} \subset \mathbb{R}^s$, $u = u(\boldsymbol{x}, t)$, $f = f(t, \boldsymbol{x})$ and $\mathcal{L}$ is a *linear differential operator*,

$$\mathcal{L} = \sum_{i_1 + i_2 + \cdots + i_s \leq r} a_{i_1, i_2, \ldots, i_s} \frac{\partial^{i_1 + i_2 + \cdots + i_s}}{\partial x_1^{i_1} \partial x_2^{i_2} \cdots \partial x_s^{i_s}}.$$

The equation (13.11) is, as usual, provided with an initial value $u(\boldsymbol{x}, 0) = g(\boldsymbol{x})$, $\boldsymbol{x} \in \mathcal{U}$, as well as appropriate boundary conditions.

We express the solution of (13.11) in the form $u(\boldsymbol{x}, t) = \mathcal{E}(t)g(\boldsymbol{x})$, where $\mathcal{E}$ is the *evolution operator* of $\mathcal{L}$. In other words, $\mathcal{E}(t)$ takes an initial value and maps it to the solution at time t. As an aside, note that $\mathcal{E}(0) = \mathcal{I}$, the identity operator, and that $\mathcal{E}(t_1 + t_2) = \mathcal{E}(t_1)\mathcal{E}(t_2) = \mathcal{E}(t_2)\mathcal{E}(t_1)$ for all $t_1, t_2 \geq 0$. An operator with the latter two properties is called a *semigroup*.

Let $\mathbb{H}$ be a normed space (A.2.1.4) of functions acting in $\mathcal{U}$ and which possess sufficient smoothness – we prefer to leave the latter statement intentionally vague, mentioning in passing that the requisite smoothness is closely linked to the analysis in Chapter 8. Denote by $|\cdot|$ the norm of $\mathbb{H}$ and recall (A.2.1.8) that every function norm induces an operator norm. We say that the equation (13.11) is *well posed* (with regard to the space $\mathbb{H}$) if, *letting both boundary conditions and the forcing term f equal zero,* for every $t^* > 0$ there exists $0 < c(t^*) < \infty$ such that $|\mathcal{E}(t)| \leq c(t^*)$ uniformly for all $0 \leq t \leq t^*$.

Intuitively speaking, if an equation is well posed this means that its solution depends continuously upon its initial value and is uniformly bounded in any compact interval. This is a very important property and we restrict our attention in the sequel to well-posed equations.

The restriction to zero boundary values and $f \equiv 0$ is not essential for such continuous dependence upon an initial value. It is not difficult to prove that the latter remains true (for well-posed equations) provided both the boundary values and the forcing term are themselves continuous and uniformly bounded.

$\diamond$ **Well-posed and ill-posed equations**   We commence with the diffusion equation (13.1). It is possible to prove by the standard technique of separation of variables that, provided the initial condition g possesses a Fourier expansion,

$$g(x) = \sum_{m=1}^{\infty} \alpha_m \sin \pi m x, \qquad 0 \leq x \leq 1,$$

the solution of (13.1) can be written explicitly in the form

$$u(x,t) = \sum_{m=1}^{\infty} \alpha_m e^{-\pi^2 m^2 t} \sin \pi m x, \qquad 0 \leq x \leq 1, \quad t \geq 0. \qquad (13.12)$$

Note that u does indeed obey zero boundary conditions.

Suppose that $|\cdot|$ is the familiar Euclidean norm,

$$|f| = \left\{ \int_0^1 [f(x)]^2 \, dx \right\}^{1/2}.$$

Then, according to (13.12) (and allowing ourselves to exchange the order of summation and integration)

$$|\mathcal{E}(t)g|^2 = \int_0^1 [u(x,t)]^2 \, dx = \int_0^1 \left[\sum_{m=1}^{\infty} \alpha_m e^{-\pi^2 m^2 t} \sin \pi m x \right]^2 dx$$

$$= \sum_{m=1}^{\infty} \sum_{j=1}^{\infty} \alpha_m \alpha_j e^{-\pi^2 (m^2 + j^2) t} \int_0^1 \sin \pi m x \sin \pi j x \, dx.$$

But

$$\int_0^1 \sin \pi m x \sin \pi j x \, dx = \begin{cases} \frac{1}{2}, & m = j, \\ 0, & \text{otherwise;} \end{cases}$$

consequently

$$|\mathcal{E}(t)g|^2 = \frac{1}{2}\sum_{m=1}^{\infty} \alpha_m^2 e^{-2\pi^2 m^2 t} \leq \frac{1}{2}\sum_{m=1}^{\infty} \alpha_m^2 = |g|^2.$$

Therefore $|\mathcal{E}(t)| \leq 1$ for every $t \geq 0$ and we deduce that (13.1) is well posed. Another example of a well posed equation is provided by the *advection equation*

$$\frac{\partial u}{\partial t} = \frac{\partial u}{\partial x},$$

which, for simplicity, we define for all $x \in \mathbb{R}$; therefore, there is no need to specify boundary conditions although, of course, we still need to define u at $t = 0$. We will encounter this equation time and again in Chapter 14.

The exact solution of the advection equation is a unilateral shift, $u(x,t) = g(x+t)$ (verify this!). Therefore, employing again the Euclidean norm,

$$|\mathcal{E}g|^2 = \int_{-\infty}^{\infty} [g(x+t)]^2 \, \mathrm{d}x = \int_{-\infty}^{\infty} [g(x)]^2 \, \mathrm{d}x = |g|^2$$

and the equation is well posed.

For an example of an ill-posed problem we resort to the 'reversed-time' diffusion equation

$$\frac{\partial u}{\partial t} = -\frac{\partial^2 u}{\partial x^2}.$$

Its solution, obtained by separation of variables, is almost identical to (13.12), except that we need to replace a decaying exponential by an increasing one. Therefore

$$|\mathcal{E}(t)\sin \pi m x| = e^{\pi^2 m^2 t} |\sin \pi m x|, \qquad m = 1, 2, \ldots,$$

and it is easy to ascertain that $|\mathcal{E}|$ is unbounded.

That the 'reversed-time' diffusion equation is ill posed is intimately linked to one of the main principles of thermodynamics, namely that it is impossible to tell the thermal history of an object from its present temperature distribution. $\diamond$

There are, basically, two avenues toward the design of finite difference schemes for the PDE (13.11). Firstly, we can replace the derivatives with respect to each of the variables $t, x_1, x_2, \ldots, x_s$, by finite differences. The outcome is a linear recurrence relation that allows us to advance from $t = n\Delta t$ to $t = (n+1)\Delta t$; the method (13.7) is a case in point. Arranging all the components at the time level $n\Delta t$ in a vector $\boldsymbol{u}_{\Delta x}^n$, we can write a general *full discretization* (FD) of (13.11) in the form

$$\boldsymbol{u}_{\Delta x}^{n+1} = A_{\Delta x}\boldsymbol{u}_{\Delta x}^n + \boldsymbol{k}_{\Delta x}^n, \qquad n = 0, 1, \ldots, \tag{13.13}$$

where the vector $\boldsymbol{k}_{\Delta x}^n$ contains the contributions of the forcing term f and the influence of the boundary values. The elements of the matrix $A_{\Delta x}$ and of the vector $\boldsymbol{k}_{\Delta x}^n$ may depend upon Δx and the *Courant number* $\mu = \Delta t/(\Delta x)^r$ (recall that r is the largest order of differentiation in $\mathcal{L}$).

◇ **The Euler method as FD**   The Euler method (13.7) can be written in the form (13.13) with

$$
A_{\Delta x} =
\begin{bmatrix}
1-2\mu & \mu & 0 & \cdots & & 0 \\
\mu & 1-2\mu & \mu & & \ddots & \vdots \\
0 & \ddots & \ddots & \ddots & & 0 \\
\vdots & & \ddots & \mu & 1-2\mu & \mu \\
0 & & \cdots & 0 & \mu & 1-2\mu
\end{bmatrix},
\qquad (13.14)
$$

and $k_{\Delta x}^n \equiv 0$. It might appear that neither $A_{\Delta x}$ nor $k_{\Delta x}^n$ depend upon Δx and that we have followed the bad counsel of excessive mathematical nitpicking. Not so! Δx expresses itself via the dimension of the space, since $(d+1)\Delta x = 1$. ◇

Let us denote the exact solution of (13.11) at the time level $n\Delta t$, arranged into a similar vector, by $\tilde{u}_{\Delta x}^n$. We say that the FD method (13.13) is of *order p* if, for every initial condition,

$$
\tilde{u}_{\Delta x}^{n+1} - A_{\Delta x}\tilde{u}_{\Delta x}^n - k_{\Delta x}^n = \mathcal{O}\!\left((\Delta x)^{p+r}\right), \qquad \Delta x \to 0, \qquad (13.15)
$$

for all $n \geq 0$ and if there exists at least one initial condition g for which the $\mathcal{O}((\Delta x)^{p+r})$ term on the right-hand side does not vanish. The reason why the exponent $p+r$, rather than p, features in the definition, is nontrivial and it will be justified in Lemma 13.2.

We equip the underlying linear space with the Euclidean norm

$$
\|g_{\Delta x}\|_{\Delta x} = \left(\Delta x \sum |g_j|^2\right)^{1/2},
$$

where the sum ranges over all the grid points.

◇ **Why the factor Δx?**   Before we progress further, this is the place to comment on the presence of the mysterious factor $(\Delta x)^{1/2}$ in our definition of the vector norm. Recall that in the present context vectors approximate functions and suppose that $g_{\Delta x,\ell} = g(\ell\Delta x)$, $\ell = 1, 2, \ldots, d$, where $(d+1)\Delta x = 1$. Provided that g is square-integrable and letting Δx tend to zero in the Riemann sum, it follows from elementary calculus that

$$
\lim_{\Delta x \to 0} \|g_{\Delta x}\|_{\Delta x} = \left\{\int_0^1 [g(x)]^2\,\mathrm{d}x\right\}^{1/2} = |g|.
$$

Thus, scaling by $(\Delta x)^{1/2}$ provides for continuous passage from a vector to a function norm.   ◇

A method is *convergent* if for every initial condition and all $t^* > 0$ it is true that

$$
\lim_{\Delta x \to 0}\left(\max_{n=0,1,\ldots,\lfloor t^*/\Delta t\rfloor} \|u_{\Delta x}^n - \tilde{u}_{\Delta x}^n\|_{\Delta x}\right) = 0.
$$

We stipulate that the Courant number is constant as $\Delta x, \Delta t \to 0$, hence $\Delta t = \mu(\Delta x)^r$.

Lemma 13.2 *Let $\|A_{\Delta x}\|_{\Delta x} \leq 1$ and suppose that the order condition (13.15) holds. Then, for every $t^* > 0$, there exists $c = c(t^*) > 0$ such that*

$$\|\boldsymbol{u}^n - \tilde{\boldsymbol{u}}^n\|_{\Delta x} \leq c(\Delta x)^p, \qquad n = 0, 1, \ldots, \lfloor t^*/\Delta t \rfloor,$$

for all sufficiently small $\Delta x > 0$.

Proof We subtract (13.15) from (13.13), therefore

$$e_{\Delta x}^{n+1} = A_{\Delta x} e_{\Delta x}^n + \mathcal{O}\left((\Delta x)^{p+r}\right), \qquad \Delta x \to 0,$$

where $e_{\Delta x}^n := \boldsymbol{u}_{\Delta x}^n - \tilde{\boldsymbol{u}}_{\Delta x}^n$, $n \geq 0$. The residuals $e_{\Delta x}^n$ obey zero initial conditions as well as zero boundary conditions. Hence, and provided that $\Delta x > 0$ is small and the solution of the differential equation is sufficiently smooth (the latter condition depends solely on the requisite smoothness of the initial condition), there exists $c = c(t^*)$ such that

$$\|e_{\Delta x}^{n+1} - A_{\Delta x} e_{\Delta x}^n\|_{\Delta x} \leq c(\Delta x)^{p+r}, \qquad n = 0, 1, \ldots, n_{\Delta t} - 1,$$

where, as in the proof of Theorem 13.1, $n_{\Delta t} := \lfloor t^*/\Delta t \rfloor$. We deduce that

$$\|e_{\Delta x}^{n+1}\|_{\Delta x} \leq \|A_{\Delta x}\|_{\Delta x} \times \|e_{\Delta x}^n\|_{\Delta x} + c(\Delta x)^{p+r}, \qquad n = 0, 1, \ldots, n_{\Delta t} - 1;$$

induction, in tandem with $e_{\Delta x}^0 = \boldsymbol{0}_{\Delta x}$, then readily yields

$$\|e_{\Delta x}^n\|_{\Delta x} \leq c\left(1 + \|A_{\Delta x}\|_{\Delta x} + \cdots + \|A_{\Delta x}\|_{\Delta x}^{n-1}\right)(\Delta x)^{p+r}, \qquad n = 0, 1, \ldots, n_{\Delta t}.$$

Since $\|A_{\Delta x}\|_{\Delta x} \leq 1$, this leads to

$$\|e_{\Delta x}^n\|_{\Delta x} \leq cn(\Delta x)^{p+r} \leq c\frac{t^*}{\Delta t}(\Delta x)^{p+r} = \frac{ct^*}{\mu}(\Delta x)^p, \qquad n = 0, 1, \ldots, n_{\Delta t}$$

and the proof of the lemma is complete. ∎

We mention in passing that, with little extra effort, the condition $\|A_{\Delta x}\|_{\Delta x} \leq 1$ in the above lemma can be replaced by the weaker condition that the underlying method is stable – although, of course, we have not yet said what is meant by stability! This is a good moment to introduce this important concept.

Let us suppose that $f \equiv 0$ and that the boundary conditions are continuous and uniformly bounded. Reiterating that $\Delta t = \mu(\Delta x)^r$, we say that (13.13) is *stable* if for every t^* there exists a constant $c(t^*) > 0$ such that

$$\|\boldsymbol{u}_{\Delta x}^n\|_{\Delta x} \leq c(t^*), \qquad n = 0, 1, \ldots, \lfloor t^*/\Delta t \rfloor, \quad \Delta x \to 0. \tag{13.16}$$

Suppose further that the boundary values vanish. In that case $\boldsymbol{k}_{\Delta x}^n \equiv \boldsymbol{0}_{\Delta x}$, the solution of (13.13) is $\boldsymbol{u}_{\Delta x}^n = A_{\Delta x}^n \boldsymbol{u}_{\Delta x}^0$, $n = 0, 1, \ldots$ and (13.16) becomes equivalent to

$$\lim_{\Delta x \to 0} \left(\max_{n=0,1,\ldots,\lfloor t^*/\Delta t \rfloor} \|A_{\Delta x}^n\|_{\Delta x} \right) \leq c(t^*). \tag{13.17}$$

Needless to say, (13.16) and (13.17) each require that the Courant number be kept constant.

An important feature of both order and stability is that they are not attributes of any single numerical scheme (13.13) but of the totality of such schemes as $\Delta x \to 0$. This distinction, to which we will return in Chapter 14, is crucial to the understanding of stability.

Before we make use of this concept, it is only fair to warn the reader that *there is no connection between the present concept of stability and the notion of A-stability from Chapter 4*. Mathematics is replete with diverse concepts bearing the identical sobriquet 'stability' and a careful mathematician should always verify whether a casual reference to 'stability' has to do with stable ultrafilters in logic, with stable fluid flow or, perhaps, with (13.17).

The purpose of our definition of stability is the following theorem. Without much exaggeration, it can be singled out as the lynchpin of the modern numerical theory of evolutionary PDEs.

Theorem 13.3 (The Lax equivalence theorem)   *Provided that the linear evolutionary PDE* (13.11) *is well posed, the fully discretized numerical method* (13.13) *is convergent if and only if it is stable and of order* $p \geq 1$.   ∎

The last theorem plays a similar role to the Dahlquist equivalence theorem (Theorem 2.2) in the theory of multistep numerical methods for ODEs. Thus, on the one hand the concept of convergence might be the central goal of our analysis, but it is, in general, difficult to verify from basic definitions – Theorem 13.1 is almost in the nature of the exception that proves the rule! On the other hand, it is easy to derive the order and, as we will see in Sections 13.4 and 13.5 and Chapter 14, a number of powerful techniques are available to determine whether a given method (13.13) is stable. Exactly as in Theorem 2.2, we thus replace an awkward, analytic requirement by more manageable algebraic conditions.

Discretizing all the derivatives in line with the recursion (13.13) is not the only possible – or, indeed, useful – approach to the numerical solution of evolutionary PDEs. An alternative technique follows by subjecting only spatial derivatives to finite difference discretization. This procedure, which we term *semi-discretization* (SD), converts a PDE into a system of coupled ODEs. Using a similar notation to that for FD schemes, in particular 'stretching' grid points into a long vector, we write an archetypical SD method in the form

$$\boldsymbol{v}'_{\Delta x} = P_{\Delta x} \boldsymbol{v}_{\Delta x} + \boldsymbol{h}_{\Delta x}(t), \qquad t \geq 0, \tag{13.18}$$

where $\boldsymbol{h}_{\Delta x}$ consists of the contribution of the forcing term and of the boundary values. Note that the use of a prime to denote a derivative is unambiguous: since we have replaced all spatial derivatives by differences, only the temporal derivative is left.

Needless to say, having derived (13.18) we next solve the ODEs, putting into use the theory of Chapters 1–6. Although the outcome is a full discretization – at the end of the day, both spatial and temporal variables are discretized – it is, in general, simpler to derive an SD scheme first and then apply to it the considerable apparatus of numerical ODE methods. Moreover, instead of using finite differences to discretize in space, there is nothing to prevent us from employing finite elements (via the Galerkin approach) or other means, e.g. spectral methods. Only limitations of space (no pun intended) prevent us from debating these issues further.

The method (13.18) is occasionally termed, mainly in the more traditional numerical literature, as *the method of lines,* to reflect the fact that each component of $\boldsymbol{v}_{\Delta x}$ describes a variable along the line $t \geq 0$. To prevent confusion and for assorted æsthetic reasons we will not use this name.

The concepts of order, convergence and stability can be easily generalized to the SD framework. Denoting by $\tilde{\boldsymbol{v}}_{\Delta x}(t)$ the vector of exact solutions of (13.11) at the (spatial) grid points, we say that the method (13.18) is of *order p* if for all initial conditions it is true that

$$\tilde{\boldsymbol{v}}'_{\Delta x}(t) - P_{\Delta x}\tilde{\boldsymbol{v}}_{\Delta x}(t) - \boldsymbol{h}_{\Delta x}(t) = \mathcal{O}((\Delta x)^p), \qquad \Delta x \to 0, \quad t \geq 0, \qquad (13.19)$$

and if the error is precisely $\mathcal{O}((\Delta x)^p)$ for some initial condition. It is *convergent* if for every initial condition and all $t^* > 0$ it is true that

$$\lim_{\Delta x \to 0} \left(\max_{t \in [0, t^*]} \|\boldsymbol{v}_{\Delta x}(t) - \tilde{\boldsymbol{v}}_{\Delta x}(t)\|_{\Delta x} \right) = 0.$$

The semi-discretized method is *stable* if, whenever $f \equiv 0$ and the boundary values are uniformly bounded, for every $t^* > 0$ there exists a constant $c(t^*) > 0$ such that

$$\|\boldsymbol{v}_{\Delta x}(t)\|_{\Delta x} \leq c(t^*), \qquad t \in [0, t^*]. \qquad (13.20)$$

Suppose that the boundary values vanish, in which case $\boldsymbol{h}_{\Delta x} \equiv \boldsymbol{0}_{\Delta x}$ and the solution of (13.18) is

$$\boldsymbol{v}_{\Delta x}(t) = \mathrm{e}^{tP_{\Delta x}}\boldsymbol{v}_{\Delta x}(0), \qquad t \geq 0.$$

We note that the exponential of an arbitrary square matrix B is defined by means of the Taylor series

$$\mathrm{e}^B = \sum_{k=0}^{\infty} \frac{1}{k!} B^k,$$

which always converges (see Exercise 13.4). Therefore, (13.20) becomes equivalent to

$$\lim_{\Delta x \to 0} \left(\max_{t \in [0, t^*]} \|\mathrm{e}^{tP_{\Delta x}}\|_{\Delta x} \right) \leq c(t^*). \qquad (13.21)$$

Theorem 13.4 (The Lax equivalence theorem for SD schemes) *Provided that the linear evolutionary PDE (13.11) is well posed, the semi-discretized numerical method (13.18) is convergent if and only if it is stable and of order $p \geq 1$.* ∎

Approximating a PDE by an ODE is, needless to say, only half of the computational job and the effect of the best semi-discretization can be undone by an inappropriate choice of an ODE solver for the equations (13.18). We will return to this issue later.

The two equivalence theorems establish a firm bedrock and a starting point for a proper theory of discretized PDEs of evolution. It is easy to discretize a PDE and to produce numbers, but only methods that adhere to the conditions of these theorems confer the necessary confidence for us to regard such numbers with a modicum of trust.

13.3   Numerical schemes for the diffusion equation

We have already seen one method for the equation (13.1), namely the Euler scheme
(13.7). In the present section we follow a more systematic route toward the design
of numerical methods – semi-discretized and fully discretized alike – for the diffusion
equation (13.1) and some of its generalizations.

Let

$$
v'_\ell = \frac{1}{(\Delta x)^2} \sum_{k=-\alpha}^{\beta} a_k v_{\ell+k}, \qquad \ell = 1, 2, \ldots, d, \tag{13.22}
$$

be a general SD scheme for (13.1). Note, incidentally, that, unless $\alpha, \beta \leq 1$, we
need somehow to provide additional information in order to implement (13.22). For
example, if $\alpha = 2$ we could require a value for v_{-1} (which is not provided by the
boundary conditions (13.3)). Alternatively, we need to replace the first equation
in (13.22) by a different scheme. This procedure is akin to boundary effects for the
Poisson equation (see Chapter 7) and, more remotely, to the requirement for additional
starting values to launch a multistep method (see Chapter 2).

Now set

$$
a(z) := \sum_{k=-\alpha}^{\beta} a_k z^k, \qquad z \in \mathbb{C}.
$$

Theorem 13.5   *The SD method (13.22) is of order p if and only if there exists a
constant $c \neq 0$ such that*

$$
a(z) = (\ln z)^2 + c(z-1)^{p+2} + \mathcal{O}\left(|z-1|^{p+3}\right), \qquad z \to 1. \tag{13.23}
$$

Proof   We employ the terminology of finite difference operators from Section 7.1,
except that we add to each operator a subscript that denotes the variable. For example,
$\mathcal{D}_t = \mathrm{d}/\mathrm{d}t$, whereas $\mathcal{E}_x$ stands for the shift operator along the x-axis. Letting

$$
\tilde{v}_\ell = u(\ell \Delta x, \cdot), \qquad \ell = 0, 1, \ldots, d+1,
$$

we can thus write the error in the form

$$
\tilde{e}_\ell := \tilde{v}'_\ell - \frac{1}{(\Delta x)^2} \sum_{k=-\alpha}^{\beta} a_k \tilde{v}_{\ell+k} = \left[\mathcal{D}_t - \frac{1}{(\Delta x)^2} a(\mathcal{E}_x) \right] \tilde{v}_\ell, \qquad \ell = 1, 2, \ldots, d.
$$

Recall that the function $\tilde{v}_\ell$ is the solution of the diffusion equation (13.1) at $x = \ell \Delta x$.
In other words,

$$
\mathcal{D}_t \tilde{v}_\ell = \frac{\partial u(\ell \Delta x, t)}{\partial t} = \frac{\partial^2 u(\ell \Delta x, t)}{\partial x^2} = \mathcal{D}_x^2 \tilde{v}_\ell, \qquad \ell = 1, 2, \ldots, d.
$$

Consequently,

$$
\tilde{e}_\ell = \left[\mathcal{D}_x^2 - \frac{1}{(\Delta x)^2} a(\mathcal{E}_x) \right] \tilde{v}_\ell, \qquad \ell = 1, 2, \ldots, d.
$$

According to (7.2), however, it is true that

$$\mathcal{D}_x = \frac{1}{\Delta x} \ln \mathcal{E}_x,$$

allowing us to deduce that

$$\tilde{e} = \frac{1}{(\Delta x)^2} \left[(\ln \mathcal{E}_x)^2 - a(\mathcal{E}_x) \right] \tilde{v}, \tag{13.24}$$

where $\tilde{e} = [\, \tilde{e}_1 \quad \tilde{e}_2 \quad \cdots \quad \tilde{e}_d \,]^T$.

Since, formally, $\mathcal{E}_x = \mathcal{I} + \mathcal{O}(\Delta x)$, it follows that (13.23) is equivalent to

$$[(\ln \mathcal{E}_x)^2 - a(\mathcal{E}_x)]\tilde{v} = c(\Delta x)^{p+2}\mathcal{D}_x^{p+2}\tilde{v} + \mathcal{O}\big((\Delta x)^{p+3}\big), \qquad \Delta x \to 0,$$

provided that the solution u of (13.1) is sufficiently smooth. In particular, substitution in (13.24) gives

$$\tilde{e} = c(\Delta x)^p \mathcal{D}_x^{p+2}\tilde{v} + \mathcal{O}\big((\Delta x)^{p+1}\big), \qquad \Delta x \to 0.$$

It now follows from (13.19) that the SD scheme (13.22) is indeed of order p. ∎

◇ **Examples of SD methods**   Our first example is

$$v_\ell' = \frac{1}{(\Delta x)^2}(v_{\ell-1} - 2v_\ell + v_{\ell+1}), \qquad \ell = 1, 2, \ldots, d. \tag{13.25}$$

In this case $a(z) = z^{-1} - 2 + z$ and, to derive the order, we let $z = \mathrm{e}^{\mathrm{i}\theta}$; hence

$$a(\mathrm{e}^{\mathrm{i}\theta}) = \mathrm{e}^{-\mathrm{i}\theta} - 2 + \mathrm{e}^{\mathrm{i}\theta} = -4 \sin^2 \tfrac{1}{2}\theta = -\theta^2 + \tfrac{1}{12}\theta^4 + \cdots, \qquad \theta \to 0,$$

while $(\ln \mathrm{e}^{\mathrm{i}\theta})^2 = (\mathrm{i}\theta)^2 = -\theta^2$. Therefore (13.25) is of order two.

Bearing in mind that $a(\mathcal{E}_x)$ is nothing other than a finite difference approximant of $(\Delta x \mathcal{D}_x)^2$, the form of (13.25) is not very surprising, once we write it in the language of finite difference operators:

$$v_\ell' = \frac{1}{(\Delta x)^2}\Delta_{0,x}^2 v_\ell, \qquad \ell = 1, 2, \ldots, d. \tag{13.26}$$

Likewise, we can use (7.8) as a starting point for the SD scheme

$$v_\ell' = \frac{1}{(\Delta x)^2}\left(\Delta_{0,x}^2 - \tfrac{1}{12}\Delta_{0,x}^4\right) v_\ell \tag{13.27}$$

$$= -\tfrac{1}{12}v_{\ell-2} + \tfrac{4}{3}v_{\ell-1} - \tfrac{5}{2}v_\ell + \tfrac{4}{3}v_{\ell+1} - \tfrac{1}{12}v_{\ell+2}, \qquad \ell = 1, 2, \ldots, d,$$

where, needless to say, at $\ell = 1$ and $\ell = d$ a special 'fix' might be required to cover for the missing values. It is left to the reader in Exercise 13.5 to verify that (13.27) is of order four.

Both (13.25) and (13.27) are constructed using central differences and their coefficients display an obvious spatial symmetry. We will see in Section 13.4 that this state of affairs confers an important advantage. Other things being equal, we prefer such schemes and this is the rule for equation (13.1). In Chapter 14, though, while investigating different equations we will encounter a situation where 'other things' are not equal. ◇

The method (13.22) can be easily amended to cater for (13.4), the diffusion equation in several space variables, and it can withstand the addition of a forcing term. Examples are the counterpart of (13.25) in a square,

$$v'_{k,\ell} = \frac{1}{(\Delta x)^2}(v_{k-1,\ell} + v_{k,\ell-1} + v_{k+1,\ell} + v_{k,\ell+1} - 4v_{k,\ell}), \qquad k,\ell = 1,2,\ldots,d, \quad (13.28)$$

(unsurprisingly, the term on the right is the five-point discretization of the Laplacian ∇^2), and an SD scheme for (13.5), the diffusion equation with a forcing term,

$$v'_\ell = \frac{1}{(\Delta x)^2}(v_{\ell-1} - 2v_\ell + v_{\ell+1}) + f_\ell(t), \qquad \ell = 1,2,\ldots,d. \quad (13.29)$$

Both (13.28) and (13.29) are second-order discretizations.

Extending (13.22) to the case of a variable diffusion coefficient, e.g. to equation (13.6), is equally easy if done correctly. We extend (13.25) by replacing (13.26) with

$$v'_\ell = \frac{1}{(\Delta x)^2}\Delta_{0,x}\left(a_\ell \Delta_{0,x} v_\ell\right), \qquad \ell = 1,2,\ldots,d,$$

where $a_\gamma = a(\kappa \Delta x)$, $\kappa \in [0, d+1]$. The outcome is

$$\begin{aligned}
v'_\ell &= \frac{1}{(\Delta x)^2}\Delta_{0,x}[a_\ell(v_{\ell+1/2} - v_{\ell-1/2})] \\
&= \frac{1}{(\Delta x)^2}[a_{\ell-1/2}v_{\ell-1} - (a_{\ell-1/2} + a_{\ell+1/2})v_\ell + a_{\ell+1/2}v_{\ell+1}], \qquad \ell = 1,2,\ldots,d
\end{aligned}$$
$$(13.30)$$

and it involves solely the values of v on the grid. Again, it is easy to prove that, subject to the requisite smoothness of a, the method is second order.

The derivation of FD schemes can proceed along two distinct avenues, which we explore in the case of the 'plain-vanilla' diffusion equation (13.1). Firstly, we may combine the SD scheme (13.22) with an ODE solver. Three ODE methods are of sufficient interest in this context to merit special mention.

- The Euler method (1.4), that is

$$\boldsymbol{y}_{n+1} = \boldsymbol{y}_n + \Delta t \boldsymbol{f}(n\Delta t, \boldsymbol{y}_n),$$

yields the similarly named *Euler* scheme

$$u_\ell^{n+1} = u_\ell^n + \mu \sum_{k=-\alpha}^{\beta} a_k u_{\ell+k}^n, \qquad \ell = 1,2,\ldots,d, \quad n \geq 0. \quad (13.31)$$

- An application of the trapezoidal rule (1.9),

$$\boldsymbol{y}_{n+1} = \boldsymbol{y}_n + \tfrac{1}{2}\Delta t[\boldsymbol{f}(n\Delta t, \boldsymbol{y}_n) + \boldsymbol{f}((n+1)\Delta t, \boldsymbol{y}_{n+1})]$$

results, after minor manipulation, in the *Crank–Nicolson* scheme

$$u_\ell^{n+1} - \tfrac{1}{2}\mu \sum_{k=-\alpha}^{\beta} a_k u_{\ell+k}^{n+1} = u_\ell^n + \tfrac{1}{2}\mu \sum_{k=-\alpha}^{\beta} a_k u_{\ell+k}^n, \qquad \ell = 1,2,\ldots,d, \quad n \geq 0.$$
$$(13.32)$$

Unlike (13.31), the Crank-Nicolson method is *implicit* – to advance the recursion by a single step, we need to solve a system of linear equations.

- The explicit midpoint rule

$$\boldsymbol{y}_{n+2} = \boldsymbol{y}_n + 2\Delta t \boldsymbol{f}((n+1)\Delta t, \boldsymbol{y}_{n+1})$$

(see Exercise 2.5), in tandem with (13.22), yields the *leapfrog* method

$$u_\ell^{n+2} = 2\mu \sum_{k=-\alpha}^{\beta} a_k u_{\ell+k}^{n+1} + u_\ell^n, \qquad \ell = 1, 2, \ldots, \quad n \geq 1. \tag{13.33}$$

The leapfrog scheme is *multistep* (specifically, two-step). This is not very surprising, given that the explicit midpoint method itself requires two steps.

Suppose that the SD scheme is of order p_1 and the ODE solver of order p_2. Hence, the contribution of semi-discretization to the error is $\Delta t\, \mathcal{O}((\Delta x)^{p_1}) = \mathcal{O}((\Delta x)^{p_1+2})$, while the ODE solver adds $\mathcal{O}((\Delta t)^{p_2+1}) = \mathcal{O}((\Delta x)^{2p_2+2})$. Altogether, according to (13.15), the order of the FD method is thus

$$p = \min\{p_1, 2p_2\} \tag{13.34}$$

(see also Exercise 13.6).

◇ **FD from SD**   Let us marry the SD scheme (13.25) with the ODE solvers (13.29)–(13.31). In the first instance this yields the Euler method (13.7). Since, *pace* Theorem 13.5, $p_1 = 2$ and, of course, $p_2 = 1$, we deduce from (13.34) that the order is two – a result that is implicit in the proof of Theorem 13.1.

Putting (13.25) into (13.32) yields the Crank–Nicolson scheme

$$-\tfrac{1}{2}\mu u_{\ell-1}^{n+1} + (1+\mu)u_\ell^{n+1} - \tfrac{1}{2}\mu u_{\ell+1}^{n+1} = \tfrac{1}{2}\mu u_{\ell-1}^n + (1-\mu)u_\ell^n + \tfrac{1}{2}\mu u_{\ell+1}. \tag{13.35}$$

Since $p_1 = 2$ (the trapezoidal rule is second order – see Section 1.3) and $p_2 = 2$, we have order two. The superior order of the trapezoidal rule has not helped in improving the order of Crank-Nicolson as compared with Euler's method (13.7). Bearing in mind that (13.35) is, as well as everything else, implicit, it is fair to query why should we bother with it in the first place. The one-word answer, which will be discussed at length in Sections 13.4 and 13.5, is 'stability'.

The explicit midpoint rule is also of order two, and so is the order of the leapfrog scheme

$$u_\ell^{n+2} = 2\mu(u_{\ell-1}^{n+1} - 2u_\ell^{n+1} + u_{\ell+1}^{n+1}) + u_{\ell-1}^n. \tag{13.36}$$

Similar reasoning can be applied to more general versions of the diffusion equation and to the SD schemes (13.26)–(13.28).   ◇

An alternative technique in designing FD schemes follows similar logic to Theorems 2.1 and 13.5, identifying the order of a method with the order of approximation to a certain function. In line with (13.22), we write a general FD scheme for the diffusion equation (13.1) in the form

$$\sum_{k=-\gamma}^{\delta} b_k(\mu)u_{\ell+k}^{n+1} = \sum_{k=-\alpha}^{\beta} c_k(\mu)u_{\ell+k}^n, \qquad \ell = 1, 2, \ldots, d, \quad n \geq 0, \qquad (13.37)$$

where, as before, a different type of discretization might be required near the boundary if $\max\{\alpha, \beta, \gamma, \delta\} \geq 2$. We assume that the identity

$$\sum_{k=-\gamma}^{\delta} b_k(\mu) \equiv 1 \qquad (13.38)$$

holds and that $b_{-\gamma}, b_\delta, c_{-\alpha}, c_\beta \not\equiv 0$. Otherwise the coefficients $b_k(\mu)$ and $c_k(\mu)$ are, for the time being, arbitrary. If $\gamma = \delta = 0$ then (13.37) is explicit, otherwise the method is implicit.

We set

$$\tilde{a}(z, \mu) := \frac{\sum_{k=-\alpha}^{\beta} c_k(\mu)z^k}{\sum_{k=-\gamma}^{\delta} b_k(\mu)z^k}, \qquad z \in \mathbb{C}, \quad \mu > 0.$$

Theorem 13.6   *The method (13.37) is of order p if and only if*

$$\tilde{a}(z, \mu) = e^{\mu(\ln z)^2} + c(\mu)(z-1)^{p+2} + \mathcal{O}\left(|z-1|^{p+3}\right), \qquad z \to 1, \qquad (13.39)$$

where $c \not\equiv 0$.

Proof   The argument is similar to the proof of Theorem 13.5, hence we just present its outline. Thus, applying (13.37) to the exact solution, we obtain

$$\tilde{e}_\ell^n = \sum_{k=-\gamma}^{\delta} b_k(\mu)\tilde{u}_{\ell+k}^{n+1} - \sum_{k=-\alpha}^{\beta} c_k(\mu)\tilde{u}_{\ell+k}^n$$

$$= \left[\mathcal{E}_t \sum_{k=-\gamma}^{\delta} b_k(\mu)\mathcal{E}_x^k - \sum_{k=-\alpha}^{\beta} c_k(\mu)\mathcal{E}_x^k \right] \tilde{u}_\ell^n.$$

We deduce from the differential equation (13.1) and the finite difference calculus in Section 7.1 that

$$\mathcal{E}_t = e^{(\Delta t)\mathcal{D}_t} = e^{\mu(\Delta x \mathcal{D}_x)^2} = e^{\mu(\ln \mathcal{E}_x)^2},$$

and this renders $\tilde{e}_\ell^n$ into the language of $\mathcal{E}_x$,

$$\tilde{e}_\ell^n = \left[e^{\mu(\ln \mathcal{E}_x)^2} \sum_{k=-\gamma}^{\delta} b_k(\mu)\mathcal{E}_x^k - \sum_{k=-\alpha}^{\beta} c_k(\mu)\mathcal{E}_x^k \right] \tilde{u}_\ell^n.$$

Next, we conclude from (13.39), $\mathcal{E}_x = \mathcal{I} + \mathcal{O}(\Delta x)$ and the normalization condition (13.38) that

$$e^{\mu(\ln \mathcal{E}_x)^2} \sum_{k=-\gamma}^{\delta} b_k(\mu)\mathcal{E}_x^k - \sum_{k=-\alpha}^{\beta} c_k(\mu)\mathcal{E}_x^k = \mathcal{O}((\Delta x)^{p+2})$$

and comparison with (13.15) completes the proof. ∎

◇ **FD from the function** $\tilde{a}$ We commence by revisiting methods that have already been presented in this chapter. As we have seen earlier in this section, it is a useful practice to let $z = e^{i\theta}$, so that $z \to 1$ is replaced by $\theta \to 0$.
In the case of the Euler method (13.7) we have

$$\tilde{a}(z, \mu) = 1 + \mu(z^{-1} - 2 + z);$$

therefore

$$\tilde{a}(e^{i\theta}) = 1 - 4\mu \sin^2 \tfrac{1}{2}\theta = 1 - \mu\theta^2 + \tfrac{1}{12}\mu\theta^4 + \mathcal{O}(\theta^6)$$
$$= e^{-\mu\theta^2} + \mathcal{O}(\theta^4), \qquad \theta \to 0,$$

and we deduce order two from (13.39).
For the Crank–Nicolson method (13.35) we have

$$\tilde{a}(z, \mu) = \frac{1 + \tfrac{1}{2}\mu(z^{-1} - 2 + z)}{1 - \tfrac{1}{2}\mu(z^{-1} - 2 + z)}$$

(note that (13.38) is satisfied), hence

$$\tilde{a}(e^{i\theta}) = \frac{1 - 2\mu \sin^2 \tfrac{1}{2}\theta}{1 + 2\mu \sin^2 \tfrac{1}{2}\theta} = 1 - \mu\theta^2 + \left(\tfrac{1}{3}\mu + \tfrac{1}{4}\mu^2\right)\theta^4 + \mathcal{O}(\theta^6)$$
$$= e^{-\mu\theta^2} + \mathcal{O}(\theta^4), \qquad \theta \to 0.$$

Again, we obtain order two.
The leapfrog method (13.36) does not fit into the framework of Theorem 13.6, but it is not difficult to derive order conditions along the lines of (13.39) for two step methods, a task left to the reader.   ◇

Using the approach of Theorems 13.5 and 13.6, it is possible to express order conditions as a problem in approximation in the two-dimensional case also. This is not so, however, for a variable diffusion coefficient; the quickest practical route toward FD schemes for (13.6) lies in combining the SD method (13.30) with, say, the trapezoidal rule.

13.4   Stability analysis I: Eigenvalue techniques

Throughout this section we will restrict our attention, mainly for the sake of simplicity and brevity, to the case of zero boundary conditions. Therefore, the relevant stability requirements are (13.17) and (13.21) for FD and SD schemes respectively.

A real, square matrix B is *normal* if $BB^T = B^T B$ (A.1.2.5). Important special cases are symmetric and skew-symmetric matrices.

Two properties of normal matrices are of interest to the material of this section. Firstly, every $d \times d$ normal matrix B possesses a complete set of unitary eigenvectors; in other words, the eigenvectors of B span a d-dimensional linear space and $\bar{w}_j^T w_\ell = 0$ for any two distinct eigenvectors $w_j, w_\ell \in \mathbb{C}^d$ (A.1.5.3). Secondly, all normal matrices B obey the identity $\|B\| = \rho(B)$, where $\| \cdot \|$ is the Euclidean norm and ρ is the spectral radius. The proof is easy and we leave it to the reader (see Exercise 13.10).

We denote the usual Euclidean inner product by $\langle \cdot, \cdot \rangle$, hence

$$\langle x, y \rangle = x^T y, \qquad x, y \in \mathbb{R}^d. \tag{13.40}$$

Theorem 13.7   *Let us suppose that the matrix $A_{\Delta x}$ is normal for every sufficiently small $\Delta x > 0$ and that there exists $\nu \geq 0$ such that*

$$\rho(A_{\Delta x}) \leq e^{\nu \Delta t}, \qquad \Delta x \to 0. \tag{13.41}$$

Then the FD method (13.13) is stable.[4]

Proof   We choose an arbitrary $t^* > 0$ and, as before, let $n_{\Delta t} := t^*/\Delta t$. Hence it is true for every vector $w_{\Delta x} \neq 0_{\Delta x}$ that

$$\begin{aligned}
\|A_{\Delta x}^n w_{\Delta x}\|_{\Delta x}^2 &= \langle A_{\Delta x}^n w_{\Delta x}, A_{\Delta x}^n w_{\Delta x} \rangle_{\Delta x} \\
&= \langle w_{\Delta x}, (A_{\Delta x}^n)^T A_{\Delta x}^n w_{\Delta x} \rangle_{\Delta x}, \qquad n = 0, 1, \ldots, n_{\Delta t}.
\end{aligned}$$

Note that we have used here the identity $\langle Bx, y \rangle = \langle x, B^T y \rangle$, which follows at once from (13.40).

It is trivial to verify by induction, using the normalcy of $A_{\Delta x}$, that

$$(A_{\Delta x}^n)^T A_{\Delta x}^n = (A_{\Delta x}^T A_{\Delta x})^n, \qquad n = 0, 1, \ldots, n_{\Delta t}.$$

Therefore, by the triangle inequality (A.1.3.3) and the definition of a matrix norm (A.1.3.4), we have

$$\begin{aligned}
\|A_{\Delta x}^n w_{\Delta x}\|_{\Delta x}^2 &= \langle w_{\Delta x}, (A_{\Delta x}^T A_{\Delta x})^n w_{\Delta x} \rangle_{\Delta x} \\
&\leq \|w_{\Delta x}\|_{\Delta x} \times \|(A_{\Delta x}^T A_{\Delta x})^n w_{\Delta x}\|_{\Delta x} \\
&\leq \|w_{\Delta x}\|_{\Delta x}^2 \times \|(A_{\Delta x}^T A_{\Delta x})^n\|_{\Delta x} \\
&\leq \|w_{\Delta x}\|_{\Delta x}^2 \times \|A_{\Delta x}\|_{\Delta x}^{2n}
\end{aligned}$$

for $n = 0, 1, \ldots, n_{\Delta t}$. Recalling that $A_{\Delta x}$ is normal, hence that its norm and spectral radius coincide, we deduce from (13.41) the inequality

$$\frac{\|A_{\Delta x}^n w_{\Delta x}\|_{\Delta x}}{\|w_{\Delta x}\|_{\Delta x}} \leq [\rho(A_{\Delta x})]^n \leq e^{\nu n \Delta t} \leq e^{\nu t^*}, \qquad n = 0, 1, \ldots, n_{\Delta t}. \tag{13.42}$$

---

[4]We recall that $\Delta t \to 0$ as $\Delta x \to 0$ and that the Courant number remains constant.

The crucial observation about (13.42) is that it holds *uniformly* for $\Delta x \to 0$. Since by the definition of a matrix norm

$$\|A_{\Delta x}^n\|_{\Delta x} = \max_{\boldsymbol{w}_{\Delta x} \neq \boldsymbol{0}_{\Delta x}} \frac{\|A_{\Delta x}^n \boldsymbol{w}_{\Delta x}\|_{\Delta x}}{\|\boldsymbol{w}_{\Delta x}\|_{\Delta x}},$$

it follows that (13.17) is satisfied by $c(t^*) = e^{\nu t^*}$ and the method (13.13) is stable. $\blacksquare$

It is of interest to consider an alternative proof of the theorem, since it highlights the role of normalcy and clarifies why, in its absence, the condition (13.41) may not be sufficient for stability. Suppose, thus, that $A_{\Delta x}$ has a complete set of eigenvectors but is not necessarily normal. We can factorize $A_{\Delta x}$ as $V_{\Delta x} D_{\Delta x} V_{\Delta x}^{-1}$, where $V_{\Delta x}$ is the matrix of the eigenvectors, while $D_{\Delta x}$ is a diagonal matrix of eigenvalues (A.1.5.4). It follows that, for every $n = 0, 1, \ldots, n_{\Delta t}$,

$$\begin{aligned}
\|A_{\Delta x}^n\| &= \|(V_{\Delta x} D_{\Delta x} V_{\Delta x}^{-1})^n\|_{\Delta x} = \|V_{\Delta x} D_{\Delta x}^n V_{\Delta x}^{-1}\|_{\Delta x} \\
&\leq \|V_{\Delta x}\|_{\Delta x} \times \|D_{\Delta x}^n\|_{\Delta x} \times \|V_{\Delta x}^{-1}\|_{\Delta x}.
\end{aligned}$$

The matrix $D_{\Delta x}$ is diagonal and its diagonal components, $d_{j,j}$, say, are the eigenvalues of $A_{\Delta x}$. Therefore

$$\|D_{\Delta x}^n\|_{\Delta x} = \max |d_{j,j}^n| = (\max |d_{j,j}|)^n = [\rho(A_{\Delta x})]^n$$

and we deduce that

$$\|A_{\Delta x}^n\| \leq \kappa_{\Delta x} [\rho(A_{\Delta x})]^n, \tag{13.43}$$

where

$$\kappa_{\Delta x} := \|V_{\Delta x}\|_{\Delta x} \times \|V_{\Delta x}^{-1}\|_{\Delta x}$$

is the *spectral condition number* of the matrix $V_{\Delta x}$.

On the face of it, we could have continued from (13.43) in a manner similar to the proof of Theorem 13.7, thereby proving the inequality

$$\|A_{\Delta x}^n\|_{\Delta x} \leq \kappa_{\Delta x} e^{\nu t^*}, \qquad n = 0, 1, \ldots, n_{\Delta t}.$$

This looks deceptively like a proof of stability without assuming normalcy in the process. The snag, of course, is in the number $\kappa_{\Delta x}$: as Δx tends to zero, it is entirely possible that $\kappa_{\Delta x}$ becomes infinite! However, provided $A_{\Delta x}$ is normal, its eigenvectors are orthogonal, therefore $\|V_{\Delta x}\|_{\Delta x}$, $\|V_{\Delta x}^{-1}\|_{\Delta x} \equiv 1$ for all Δx (A.1.3.4) and we can indeed use (13.43) to construct an alternative proof of the theorem.

Using the same approach as in Theorem 13.7, we can prove a stability condition for SD schemes with normal matrices.

Theorem 13.8   *Let the matrix $P_{\Delta x}$ be normal for every sufficiently small $\Delta x > 0$. If there exists $\eta \in \mathbb{R}$ such that*

$$\text{Re}\, \lambda \leq \eta \qquad \text{for every} \qquad \lambda \in \sigma(P_{\Delta x}) \qquad \text{and} \qquad \Delta x \to 0 \tag{13.44}$$

then the SD method (13.18) is stable.

Proof   Let $t^* > 0$ be given. Because of the normalcy of $P_{\Delta x}$, it follows along similar lines to the proof of Theorem 13.7 that

$$
\begin{aligned}
\| e^{tP_{\Delta x}} \boldsymbol{w}_{\Delta x} \|_{\Delta x}^2 &= \langle e^{tP_{\Delta x}} \boldsymbol{w}_{\Delta x}, \, e^{tP_{\Delta x}} \boldsymbol{w}_{\Delta x} \rangle_{\Delta x} = \langle \boldsymbol{w}_{\Delta x}, \, (e^{tP_{\Delta x}})^T e^{tP_{\Delta x}} \boldsymbol{w}_{\Delta x} \rangle_{\Delta x} \\
&= \langle \boldsymbol{w}_{\Delta x}, \, e^{tP_{\Delta x}^T} e^{tP_{\Delta x}} \boldsymbol{w}_{\Delta x} \rangle_{\Delta x} = \langle \boldsymbol{w}_{\Delta x}, \, e^{t(P_{\Delta x}^T + P_{\Delta x})} \boldsymbol{w}_{\Delta x} \rangle_{\Delta x} \\
&\leq \| \boldsymbol{w}_{\Delta x} \|_{\Delta x}^2 \times \| e^{t(P_{\Delta x}^T + P_{\Delta x})} \|_{\Delta x} = \| \boldsymbol{w}_{\Delta x} \|_{\Delta x}^2 \, \rho(e^{t(P_{\Delta x}^T + P_{\Delta x})}) \\
&= \| \boldsymbol{w}_{\Delta x} \|_{\Delta x}^2 \max \left\{ e^{2t\mathrm{Re}\,\lambda} : \lambda \in \sigma(P_{\Delta x}) \right\} \leq \| \boldsymbol{w}_{\Delta x} \|_{\Delta x}^2 e^{2\eta t^*}, \quad t \in [0, t^*].
\end{aligned}
$$

We leave it to the reader to verify that for all normal matrices B and for $t \geq 0$ it is true that

$$
(e^{tB})^T = e^{tB^T} \qquad e^{tB^T} e^{tB} = e^{t(B^T + B)}
$$

(note that the second identity might fail unless B is normal!) and that

$$
\sigma(e^{t(B^T + B)}) = \{ e^{2t\mathrm{Re}\,\lambda}, \ \lambda \in \sigma(P_{\Delta x}) \}.
$$

We deduce stability from the definition (13.21).   ■

The *spectral abscissa* of a square matrix B is the real number

$$
\tilde{\alpha}(B) := \max \left\{ \mathrm{Re}\,\lambda \, : \, \lambda \in \sigma(B) \right\}.
$$

We can rephrase Theorem 13.8 by requiring $\tilde{\alpha}(P_{\Delta x}) \leq \eta$ for all $\Delta x \to 0$.

The great virtue of Theorems 13.7 and 13.8 is that they reduce the task of determining stability to that of locating the eigenvalues of a normal matrix. Even better, to establish stability it is often sufficient to bound the spectral radius or the spectral abscissa. According to a broad principle that we have mentioned in Chapter 2, we replace the analytic – and difficult – stability conditions (13.17) and (13.21) by algebraic requirements.

◇ **Eigenvalues and stability of methods for the diffusion equation**   The matrix associated with the SD method (13.25) is (in the natural ordering of grid points, from left to right) TST and, according to Lemma 10.5, its eigenvalues are $-4\sin^2[\pi\ell/(d+1)]$, $\ell = 1, 2, \ldots, d$. Therefore

$$
\tilde{\alpha}(P_{\Delta x}) = -4\sin^2(\pi\Delta x) \leq 0, \qquad \Delta x > 0
$$

(recall that $(d+1)\Delta x = 1$) and the method is stable.

Next we consider Euler's FD scheme (13.7). The matrix $A_{\Delta x}$ is again TST and its eigenvalues are $1 - 4\mu\sin^2[\pi\ell/(d+1)]$, $\ell = 1, 2, \ldots, d$. Therefore

$$
\rho(A_{\Delta x}) \equiv |1 - 4\mu|, \qquad \Delta x > 0.
$$

Consequently, (13.41) is satisfied by $\nu = 0$ for $\mu \leq \frac{1}{2}$, whereas no ν will do for $\mu > \frac{1}{2}$. This, in tandem with the Lax equivalence theorem (Theorem 13.3) and our observation from Section 13.3 that (13.7) is a second-order method, provides a brief alternative proof of Theorem 13.1.

The next candidate for our attention is the Crank–Nicolson scheme (13.35), which we also render in a vector form. Disregarding for a moment our assumption that the forcing term and boundary contributions vanish, we have

$$A_{\Delta x}^{[+]} u_{\Delta x}^{n+1} = A_{\Delta x}^{[-]} u_{\Delta x}^n + \tilde{k}_{\Delta x}^n, \qquad n \geq 0,$$

where the matrices $A_{\Delta x}^{[\pm]}$ are TST, while the vector $\tilde{k}_{\Delta x}^n$ contains the contribution of both forcing and boundary terms. Therefore $A_{\Delta x} = A_{\Delta x}^{[+]^{-1}} A_{\Delta x}^{[-]}$ and $k_{\Delta x}^n = A_{\Delta x}^{[+]^{-1}} \tilde{k}_{\Delta x}^n$. (We insist on the presence of forcing terms $\tilde{k}_{\Delta x}^n$, before eliminating them for the sake of stability analysis, since this procedure illustrates how to construct the form (13.13) for general implicit FD schemes.) According to Lemma 10.5, TST matrices of the same dimension share the same set of eigenvectors. Moreover, these eigenvactors span the whole space, consequently $A_{\Delta x}^{\pm} = V_{\Delta x} D_{\Delta x}^{[\pm]} V_{\Delta x}^{-1}$, where $V_{\Delta x}$ is the matrix of the eigenvectors and $D_{\Delta x}^{[\pm]}$ are diagonal. Therefore

$$A_{\Delta x} = V_{\Delta x} D_{\Delta x}^{[+]^{-1}} D_{\Delta x}^{[-]} V_{\Delta x}^{-1}$$

and the eigenvalues of the quotient matrix of two TST matrices – itself, in general, not TST – are the quotient of the eigenvalues of $A_{\Delta x}^{[\pm]}$. Employing again Lemma 10.5, we write down explicitly the eigenvalues of the latter,

$$\sigma(A_{\Delta x}^{[\pm]}) = \left\{ 1 \pm 2\mu \sin^2\left(\frac{\pi \ell}{d+1}\right) : \ell = 1, 2, \ldots, d \right\},$$

hence

$$\sigma(A_{\Delta x}) = \left\{ \frac{1 - 2\mu \sin^2[\pi \ell/(d+1)]}{1 + 2\mu \sin^2[\pi \ell/(d+1)]} : \ell = 1, 2, \ldots, d \right\}$$

and we deduce that

$$\rho(A_{\Delta x}) = \frac{|1 - 2\mu \sin^2(\pi \Delta x)|}{1 + 2\mu \sin^2(\pi \Delta x)} \leq 1.$$

Therefore the Crank–Nicolson scheme is stable *for all* $\mu > 0$.

All three aforementioned examples make use of TST matrices, but this technique is, unfortunately, of limited scope. Consider, for example, the SD scheme (13.30) for the variable diffusion coefficient PDE (13.6). Writing this in the form (13.18), we obtain for $(\Delta x)^2 P_{\Delta x}$ the following matrix:

$$
\begin{bmatrix}
-a_{-1/2} - a_{1/2} & a_{1/2} & 0 & \cdots & & 0 \\
a_{1/2} & -a_{1/2} - a_{3/2} & a_{3/2} & & \ddots & \vdots \\
0 & \ddots & \ddots & \ddots & & 0 \\
\vdots & \ddots & & a_{d-3/2} & -a_{d-3/2} - a_{d-1/2} & a_{d-1/2} \\
0 & \cdots & 0 & & a_{d-1/2} & -a_{d-1/2} - a_{d+1/2}
\end{bmatrix}.
$$

Clearly, in general $P_{\Delta x}$ is not a Toeplitz matrix and we are not allowed to use Lemma 10.5. However, it is symmetric, hence normal, and we are within the conditions of Theorem 13.8.

Although we cannot find the eigenvalues of $P_{\Delta x}$, we exploit the Gerschgorin criterion (Lemma 7.3) to derive enough information about their location to prove stability. Since $a(x) > 0$, $x \in [0,1]$, it follows at once that all the Gerschgorin discs $\mathbb{S}_i$, $i = 1, 2, \ldots, d$, lie in the closed complex left half-plane. Therefore $\tilde{\alpha}(P_{\Delta x}) \leq 0$, hence stability. $\diamond$

13.5   Stability analysis II: Fourier techniques

We commence this section by assuming that we are solving an evolutionary PDE given (in a single spatial dimension) for all $x \in \mathbb{R}$ and that in place of boundary conditions, say, (13.3) we impose the requirement that the function $u(\,\cdot\,, t)$ is square-integrable in $\mathbb{R}$ for all $t \geq 0$. As we have already mentioned in Section 13.1, this is known as the *Cauchy* problem.

The technique of this section is valid whenever a Cauchy problem for an arbitrary linear PDE of evolution with constant coefficients is solved by a method – either SD or FD – that employs an identical formula at each grid point. For simplicity, however, we restrict ourselves here to the diffusion equation and to the general SD and FD schemes (13.22) and (13.37) respectively (except that the range of ℓ now extends all across $\mathbb{Z}$). The reader should bear in mind, however, that special properties of the diffusion equation – except in the narrowest technical sense, e.g. with regard to the power of Δx in (13.22) and (13.37) – are never used in our exposition. This makes for entirely straightforward generalization.

The definition of stability depends on the underlying norm and throughout this section we consider exclusively the Euclidean norm over bi-infinite sequences. The set $\ell[\mathbb{Z}]$ is the linear space of all complex sequences, indexed over the integers, that are bounded in the Euclidean vector norm. In other words,

$$\boldsymbol{w} = \{w_m\}_{m=-\infty}^{\infty} \in \ell_2[\mathbb{Z}] \qquad \text{if and only if} \qquad \|\boldsymbol{w}\| := \left(\sum_{m=-\infty}^{\infty} |w_m|^2 \right)^{1/2} < \infty.$$

Note that throughout the present section we omit the factor $(\Delta x)^{1/2}$ in the definition of the Euclidean norm, mainly to unclutter the notation and to bring it into line with the standard terminology of Fourier analysis. We also allow ourselves the liberating convention of dropping the subscripts $_{\Delta x}$ in our formulae, the reason being that Δx no longer expresses the reciprocal of the number of grid points – which is infinite for all Δx. The only influence of Δx on the underlying equations is expressed in the multiplier $(\Delta x)^{-2}$ for SD equations and – most importantly – in the spacing of the grid along which we are sampling the initial condition g.

We let $L[0, 2\pi]$ denote the set of all complex, square-integrable functions in $[0, 2\pi]$, equipped with the Euclidean function norm:

$$w \in L[0, 2\pi] \qquad \text{if and only if} \qquad \|w\| = \left[\frac{1}{2\pi} \int_0^{2\pi} |w(\theta)| \, \mathrm{d}\theta \right]^{1/2} < \infty.$$

Solutions of either (13.22) or (13.37) 'live' in $\ell[\mathbb{Z}]$ (remember that the index ranges across all integers!), consequently we phrase their stability in terms of the norm in that space. As it turns out, however, it is considerably more convenient to investigate stability in $L[0, 2\pi]$. The opportunity to abandon $\ell[\mathbb{Z}]$ in favour of $L[0, 2\pi]$ is conferred by the *Fourier transform*. We have already encountered a similar concept in Chapters 11 and 12 in a different context. For our present purpose, we choose a different definition to that in (12.9) and (12.10), letting

$$\hat{w}(\theta) = \sum_{m=-\infty}^{\infty} w_m \mathrm{e}^{-\mathrm{i}m\theta}, \qquad \boldsymbol{w} = (w_m)_{m=-\infty}^{\infty} \in \ell[\mathbb{Z}]. \tag{13.45}$$

Lemma 13.9   *The mapping* (13.45) *takes* $\ell[\mathbb{Z}]$ *onto* $L[0, 2\pi]$. *It is an* isomorphism *(i.e., a one-to-one mapping) and its inverse is given by*

$$w_m = \frac{1}{2\pi} \int_0^{2\pi} \hat{w}(\theta) \mathrm{e}^{\mathrm{i}m\theta} \, \mathrm{d}\theta, \qquad m \in \mathbb{Z}, \quad \hat{w} \in L[0, 2\pi]. \tag{13.46}$$

Moreover, (13.45) *is an* isometry:

$$\|\hat{w}\| = \|\boldsymbol{w}\|, \qquad \boldsymbol{w} \in \ell[\mathbb{Z}]. \tag{13.47}$$

Proof   We combine the proof of $\hat{w} \in L[0, 2\pi]$ (hence, that (13.45) indeed takes $\ell[\mathbb{Z}]$ to $L[0, 2\pi]$) with the proof of (13.47), by evaluating the norm of $\hat{w}$:

$$
\begin{aligned}
\|\hat{w}\|^2 &= \frac{1}{2\pi} \int_0^{2\pi} \sum_{m=-\infty}^{\infty} \sum_{j=-\infty}^{\infty} w_m \bar{w}_j \mathrm{e}^{\mathrm{i}(j-m)\theta} \, \mathrm{d}\theta \\
&= \frac{1}{2\pi} \sum_{m=-\infty}^{\infty} \sum_{j=-\infty}^{\infty} w_m \bar{w}_j \int_0^{2\pi} \mathrm{e}^{\mathrm{i}(j-m)\theta} \, \mathrm{d}\theta = \sum_{m=-\infty}^{\infty} |w_m|^2 = \|\boldsymbol{w}\|^2.
\end{aligned}
$$

Note our use of the identity

$$\frac{1}{2\pi} \int_0^{2\pi} \mathrm{e}^{\mathrm{i}k\theta} \, \mathrm{d}\theta = \begin{cases} 1, & k = 0, \\ 0, & \text{otherwise,} \end{cases} \qquad k \in \mathbb{Z}.$$

The argument required to prove that the mapping $\boldsymbol{w} \mapsto \hat{w}$ is an isomorphism onto $L[0, 2\pi]$ and that its inverse is given by (13.46) is an almost exact replica of the proof of Lemma 12.1.   ∎

We will call $\hat{w}$ the *Fourier transform* of $\boldsymbol{w}$. This is at variance with the terminology of Chapter 12 – by rights, we should call $\hat{w}$ the *inverse* Fourier transform of $\boldsymbol{w}$. The present usage, however, has the advantage of brevity.

The isomorphic isometry of the Fourier transform is perhaps the great secret of its importance in a wide range of applications. A Euclidean norm typically measures the energy of physical systems and a major consequence of (13.47) is that, while travelling back and forth between $\ell[\mathbb{Z}]$ and $L[0, 2\pi]$ by means of the Fourier transform and its inverse, the energy stays intact.

We commence our analysis with the SD scheme (13.22), recalling that the index ranges across all $\ell \in \mathbb{Z}$. We multiply the equation by $e^{-i\ell\theta}$ and sum over ℓ; the outcome is

$$\frac{\partial \hat{v}(\theta, t)}{\partial t} = \sum_{\ell=-\infty}^{\infty} v'_m e^{-i\ell\theta} = \frac{1}{(\Delta x)^2} \sum_{\ell=-\infty}^{\infty} \sum_{k=-\alpha}^{\beta} a_k v_{\ell+k} e^{-i\ell\theta}$$

$$= \frac{1}{(\Delta x)^2} \sum_{k=-\alpha}^{\beta} a_k \sum_{\ell=-\infty}^{\infty} v_{\ell+k} e^{-i\ell\theta} = \frac{1}{(\Delta x)^2} \sum_{k=-\alpha}^{\beta} a_k \sum_{\ell=-\infty}^{\infty} v_\ell e^{-i(\ell-k)\theta}$$

$$= \frac{1}{(\Delta x)^2} \sum_{k=-\alpha}^{\beta} a_k e^{ik\theta} \sum_{\ell=-\infty}^{\infty} v_\ell e^{-i\ell\theta} = \frac{a(e^{i\theta})}{(\Delta x)^2} \hat{v}(\theta, t),$$

where the function $a(\,\cdot\,)$ has been defined in Section 13.3. The crucial step in the above argument is the shift of the index from ℓ to $\ell - k$ without changing the endpoints of the summation, a trick that explains why we require that ℓ should range across all the integers.

We have just proved that the Fourier transform $\hat{v} = \hat{v}(\theta, t)$ obeys, as a function of t, the linear ODE

$$\frac{\partial \hat{v}}{\partial t} = \frac{a(e^{i\theta})}{(\Delta x)^2} \hat{v}, \qquad t \geq 0, \quad \theta \in [0, 2\pi].$$

with the initial condition $\hat{v}(\theta, 0) = \hat{g}$, where $g_m = u(m\Delta x, 0)$, $m \in \mathbb{Z}$, is the projection on the grid of the initial condition of the PDE. The solution of the ODE can be written down explicitly:

$$\hat{v}(\theta, t) = \hat{g}(\theta) \exp\left[\frac{a(e^{i\theta})t}{(\Delta x)^2}\right], \qquad t \geq 0, \quad \theta \in [0, 2\pi]. \tag{13.48}$$

Suppose that

$$\operatorname{Re} a(e^{i\theta}) \leq 0, \qquad \theta \in [0, 2\pi]. \tag{13.49}$$

In that case it follows from (13.48) that

$$\|\hat{v}\|^2 = \frac{1}{2\pi} \int_0^{2\pi} |\hat{g}(\theta)|^2 \exp\left[\frac{2\operatorname{Re} a(e^{i\theta})}{(\Delta x)^2}\right] d\theta \leq \frac{1}{2\pi} \int_0^{2\pi} |\hat{g}(\theta)|^2 \, d\theta = \|\hat{g}\|^2,$$

Therefore, according to (13.47),

$$\|v(t)\| \leq \|v(0)\|$$

for all possible initial conditions $v \in \ell[\mathbb{Z}]$. We thus conclude that, according to (13.20), the method is stable.

Next, we consider the case when the condition (13.49) is violated, in other words, when there exists $\theta_0 \in [0, 2\pi]$ such that $\operatorname{Re} a(e^{i\theta_0}) > 0$. Since $a(e^{i\theta})$ is continuous in θ, there exist $\varepsilon > 0$ and $0 \leq \theta_- < \theta_+ < 2\pi$ such that

$$\operatorname{Re} a(e^{i\theta}) > \varepsilon, \qquad \theta \in [\theta_-, \theta_+].$$

We choose an initial condition g such that $\hat{g}$ is a characteristic function of the interval $[\theta_-, \theta_+]$,

$$\hat{g}(\theta) = \begin{cases} 1, & \theta \in [\theta_-, \theta_+], \\ 0, & \text{otherwise} \end{cases}$$

(it is possible to choose g to be square-integrable – why?). It follows from (13.48) that

$$\|\hat{v}\|^2 = \frac{1}{2\pi} \int_{\theta_-}^{\theta_+} \exp\left[\frac{2\operatorname{Re} a(e^{i\theta})t}{(\Delta x)^2}\right] d\theta \geq \frac{1}{2\pi} \int_{\theta_-}^{\theta_+} \exp\left[\frac{2\varepsilon t}{(\Delta x)^2}\right] d\theta$$

$$= \frac{\theta_+ - \theta_-}{2\pi} \exp\left[\frac{2\varepsilon t}{(\Delta x)^2}\right].$$

Therefore $\|\hat{v}\|$ cannot be uniformly bounded for $t \in [0, t^*]$ (regardless of the size of $t^* > 0$) as $\Delta x \to 0$. We will again exploit isometry to argue that (13.22) is unstable.

Theorem 13.10   *The SD method* (13.22), *when applied to a Cauchy problem, is stable if and only if the inequality* (13.49) *is obeyed.* ∎

Fourier analysis can be applied with similarly telling effect to the FD scheme (13.37) – again, with $\ell \in \mathbb{Z}$. The argument is almost identical, hence we present it with greater brevity.

Theorem 13.11   *The FD method* (13.37), *when applied to a Cauchy problem, is stable for a specific value of the Courant number* μ *if and only if*

$$|\tilde{a}(e^{i\theta}, \mu)| \leq 1, \qquad \theta \in [0, 2\pi], \tag{13.50}$$

where

$$\tilde{a}(z, \mu) = \frac{\sum_{k=-\alpha}^{\beta} c_k(\mu) z^k}{\sum_{k=-\gamma}^{\delta} b_k(\mu) z^k}, \qquad z \in \mathbb{C}.$$

Proof   We multiply both sides of (13.37) by $e^{-i\ell\theta}$ and sum over $\ell \in \mathbb{Z}$. The outcome is the recursive relationship

$$\hat{u}^{n+1} = \tilde{a}(e^{i\theta}, \mu)\hat{u}^n, \qquad n \geq 0,$$

between the Fourier transforms in adjacent time levels. Iterating this recurrence results in the explicit formula

$$\hat{u}^n = [\tilde{a}(e^{i\theta}, \mu)]^n \hat{u}^0, \qquad n \geq 0,$$

where, of course, $\hat{u}^0 = \hat{g}$. Therefore

$$\|\boldsymbol{u}^n\|^2 = \|\hat{u}^n\|^2 = \frac{1}{2\pi} \int_0^{2\pi} [\tilde{a}(e^{i\theta})]^n \hat{u}^0(\theta) \, d\theta, \qquad n \geq 0. \tag{13.51}$$

If (13.50) is satisfied we deduce from (13.51) that

$$\|\boldsymbol{u}^n\| \leq \|\boldsymbol{u}^0\|, \qquad n \geq 0.$$

Stability follows from (13.16) by virtue of isometry.

The course of action when (13.50) fails is identical to our analysis of SD methods. We have $\varepsilon > 0$ such that $|\tilde{a}(e^{i\theta}, \mu)| \geq 1 + \varepsilon$ for all $\theta \in [\theta_-, \theta_+]$. Picking $\hat{u}^0 = \hat{g}$ as the characteristic function of the interval $[\theta_-, \theta_+]$, we exploit (13.51) to argue that

$$\|\boldsymbol{u}^n\|^2 = \|\hat{u}^n\|^2 = \frac{1}{2\pi} \int_0^{2\pi} |\tilde{a}(e^{i\theta}, \mu)|^{2n} g(\theta)\, d\theta \geq \frac{\theta_+ - \theta_-}{2\pi}(1 + \varepsilon)^n, \qquad n \geq 0.$$

This concludes the proof of instability.   ∎

◇ **The Fourier technique in practice**   It is trivial to use Theorem 13.10 to prove that the SD method (13.25) is stable, since $a(e^{i\theta}) = -4\sin^2 \frac{1}{2}\theta$. Let us attempt a more ambitious goal, the fourth-order SD scheme (13.27). We have

$$
\begin{aligned}
a(e^{i\theta}) &= -\tfrac{1}{12}e^{-2i\theta} + \tfrac{4}{3}e^{-i\theta} - \tfrac{5}{2} + \tfrac{4}{3}e^{i\theta} - \tfrac{1}{12}e^{2i\theta} \\
&= -\tfrac{7}{3} + \tfrac{8}{3}\cos\theta - \tfrac{1}{3}\cos^2\theta = -\tfrac{1}{3}(1 - \cos\theta)(7 - \cos\theta) \leq 0
\end{aligned}
$$

for all $\theta \in [0, 2\pi]$, hence stability.

Whenever the Fourier technique can be put to work, the results are easy to come by, and this is also true with regard to FD schemes. The Euler method (13.7) yields

$$\tilde{a}(e^{i\theta}, \mu) = 1 - 4\mu\sin^2 \tfrac{1}{2}\theta, \qquad \theta \in [0, 2\pi],$$

and it is trivial to deduce that (13.50) implies stability if and only if $\mu \leq \frac{1}{2}$. Likewise, for the Crank–Nicolson method we have

$$\tilde{a}(e^{i\theta}, \mu) = \frac{1 - 2\mu\sin^2 \frac{1}{2}\theta}{1 + 2\mu\sin^2 \frac{1}{2}\theta} \in [-1, 1], \qquad \theta \in [0, 2\pi],$$

hence stability for all $\mu > 0$.

Let us set ourselves a fairer challenge. Solving the SD scheme (13.25) with the Adams–Bashforth method (2.6) results in the second-order, two-step scheme

$$u_\ell^{n+2} = u_\ell^{n+1} + \tfrac{3}{2}\mu(u_{\ell-1}^{n+1} - 2u_\ell^{n+1} + u_{\ell+1}^{n+1}) - \tfrac{1}{2}(u_{\ell-1}^n - 2u_\ell^n + u_{\ell+1}^n). \quad (13.52)$$

It is not difficult to extend the Fourier technique to multistep methods. We multiply (13.52) by $e^{-i\ell\theta}$ and sum for all $\ell \in \mathbb{Z}$; the outcome is the three-term recurrence relation

$$\hat{u}^{n+2} - \left(1 - 6\mu\sin^2 \tfrac{1}{2}\theta\right)\hat{u}^{n+1} - 2\mu\left(\sin^2 \tfrac{1}{2}\theta\right)\hat{u}^n = 0, \qquad n \geq 0. \quad (13.53)$$

The general solution of the difference equation (13.53) is

$$\hat{u}^n = q_-(\theta)[\omega_-(\theta)]^n + q_+(\theta)[\omega_+(\theta)]^n, \qquad n = 0, 1, \ldots,$$

where $\omega_\pm$ are zeros of the characteristic equation

$$\omega^2 - \left(1 - 6\mu\sin^2 \tfrac{1}{2}\theta\right)\omega - 2\mu\sin^2 \tfrac{1}{2}\theta = 0$$

and $q_\pm$ depend on the starting values (see Section 4.4 for solution of comparable difference equations). As before, the condition for stability is uniform boundedness of the set $\{\|\hat{u}^n\|\}_{n=0,1\ldots}$, since this implies that the vectors

$\{\|\boldsymbol{u}^n\|\}_{n=0,1,\ldots}$ are uniformly bounded. Evidently, the Fourier transforms are uniformly bounded for all $q_\pm$ if and only if $|\omega_\pm(\theta)| \leq 1$ for all $\theta \in [0, 2\pi]$ and, whenever $|\omega_\pm(\theta)| = 1$ for some θ, the two zeros are distinct – in other words, the root condition all over again!

We use Lemma 4.9 to verify the root condition and this, after some trivial, yet tedious algebra, results in the stability condition $\mu \leq \frac{2}{5}$.

Another example of a two-step method, the leapfrog scheme (13.36), features in Exercise 13.13. $\diamond$

The scope of the Fourier technique can be generalized in several directions. The easiest is from one to several spatial dimensions and requires a multivariate counterpart of the Fourier transform (13.45).

More interesting is the relaxation of the ban on boundary conditions in finite time – after all, most physical objects subjected to mathematical modelling possess finite size! In Chapter 14 we will mention briefly periodic boundary conditions, which lend themselves to the same treatment as the Cauchy problem. Here, though, we address ourselves to the Dirichlet boundary conditions (13.3), which are more characteristic of parabolic equations. Without going into any proofs, we simply state that, provided that an SD or an FD method uses just one point from the right and one from the left (in other words, $\max\{\alpha, \beta, \gamma, \delta\} = 1$), the scope of the Fourier technique extends to finite intervals. Thus, the outcome of the Fourier analysis for the SD (13.25), the Euler method (13.7), the Crank–Nicolson scheme (13.35) and, indeed, the Adams–Bashforth two-step FD (13.52) extends *in toto* to Dirichlet boundary conditions, but this is not the case with the fourth-order SD scheme (13.27). There, everything depends on our treatment of the 'missing' values near the boundary. This is an important subject – admittedly more important in the context of hyperbolic differential equations, the theme of Chapter 14 – which requires a great deal of advanced mathematics and is well outside the scope of this book.

This section will not be complete without mentioning a remarkable connection, which might have already caught the eye of a vigilant reader. Let us consider a simple example, the 'basic' SD (13.25). The Fourier condition for stability is

$$\operatorname{Re} a(e^{i\theta}) = -4 \sin^2 \tfrac{1}{2}\theta \leq 0, \qquad \theta \in [0, 2\pi],$$

while the eigenvalue condition is nothing other than

$$\operatorname{Re} a(\omega_{d+1}^\ell) = -4\sin^2\left(\frac{\pi\ell}{d+1}\right) \leq 0, \qquad \ell = 1, 2, \ldots, d,$$

where $\omega_d = \exp[2i\pi/(d+1)]$ is the dth root of unity. A similar connection exists for Euler's method and Crank–Nicolson. Before we get carried away, it is perhaps in order to clarify that this coincidence is restricted, at least in the context of the Cauchy problem, to methods that are constructed from TST matrices.

13.6 Splitting

Even the stablest and the most heavily analysed method must be, ultimately, run on a computer. This can be even more expensive for PDEs of evolution than for the Poisson equation; in a sense, using an implicit method for (13.1) in two spatial dimensions, say, and with a forcing term is equivalent to solving a Poisson equation in every time step. Needless to say, by this stage we know full well that effective solution of the diffusion equation calls for implicit schemes; otherwise, we would need to advance with such a miniscule step Δt as to render the whole procedure unrealistically expensive.

The emphasis on two (or more) space dimensions is important, since in one dimension the algebraic equations originating in the Crank–Nicolson scheme, say, are tridiagonal (cf. (13.32)) and can be easily solved with banded LU factorization from Chapter 9.

We restrict our analysis to the diffusion equation

$$\frac{\partial u}{\partial t} = \nabla(a\nabla u), \qquad 0 \le x, y \le 1, \tag{13.54}$$

where the diffusion coefficient $a = a(x, y)$ is bounded and positive in $[0, 1] \times [0, 1]$.

The starting point of our discussion is an extension of the SD equations (13.29) to two dimensions,

$$
\begin{aligned}
v'_{k,\ell} = \frac{1}{(\Delta x)^2} &\Big[a_{k-1/2,\ell} v_{k-1,\ell} + a_{k,\ell-1/2} v_{k,\ell-1} + a_{k+1/2,\ell} v_{k+1,\ell} + a_{k,\ell+1/2} v_{k,\ell+1} \\
&\quad - (a_{k-1/2,\ell} + a_{k,\ell-1/2} + a_{k+1/2,\ell} + a_{k,\ell+1/2}) v_{k,\ell} \Big] \\
&+ h_{k,\ell}, \qquad k, \ell = 1, \dots, d,
\end{aligned}
$$

where $h_{k,\ell}$ includes the contribution of the boundary values. (We could have also added a forcing term without changing the equation materially.) We commence by assuming that $h_{k,\ell} = 0$ for all $k, \ell = 1, 2, \dots, d$ and, employing natural ordering, write the method in a vector form,

$$v' = \frac{1}{(\Delta x)}(B_x + B_y)v, \qquad t \ge 0, \qquad v(0) \text{ given.} \tag{13.55}$$

Here B_x and B_y are $d^2 \times d^2$ matrices that contain the contribution of the differentiation in the x- and y- variables, respectively. In other words, B_y is a block-diagonal matrix and its diagonal is constructed from the tridiagonal $d \times d$ matrices

$$
\begin{bmatrix}
-(b_{1/2} + b_{3/2}) & b_{3/2} & 0 & \cdots & & 0 \\
b_{3/2} & -(b_{3/2} + b_{5/2}) & b_{5/2} & \ddots & & \vdots \\
0 & \ddots & \ddots & \ddots & & 0 \\
\vdots & \ddots & & b_{d-3/2} & -(b_{d-3/2} + b_{d-1/2}) & b_{d-1/2} \\
0 & \cdots & & 0 & b_{d-1/2} & -(b_{d-1/2} + b_{d+1/2})
\end{bmatrix},
$$

where $b_\ell = a_{k,\ell}$ and $k = 1, 2, \dots, d$. The matrix B_x contains all the remaining terms. A crucial observation is that its sparsity pattern is also block-diagonal, with tridiagonal blocks, *provided that the grid is ordered by rows, rather than by columns.*

Letting $v^n := v(n\Delta t)$, $n \ge 0$, the solution of (13.55) can be written explicitly as

$$v^{n+1} = e^{\mu(B_x + B_y)} v^n, \qquad n \ge 0. \tag{13.56}$$

It might be remembered that an exponential of a matrix has been already defined in Section 13.2 (see also Exercise 13.4). To solve (13.56) we may discretize the exponential by means of the *Padé approximant* $\hat{r}_{1/1}$ (Theorem 4.5). The outcome,

$$\boldsymbol{u}^{n+1} = \hat{r}_{1/1}(\mu(B_x+B_y))\boldsymbol{u}^n = \left[I - \tfrac{1}{2}\mu(B_x + B_y)\right]^{-1}\left[I + \tfrac{1}{2}\mu(B_x + B_y)\right]\boldsymbol{u}^n, \qquad n \geq 0,$$

is nothing other than the Crank–Nicolson method (in two dimensions) in disguise. Advancing the solution by a single time step is tantamount to solving a linear algebraic system by use of the matrix $I - \tfrac{1}{2}\mu(B_x + B_y)$, a task which can be quite expensive, even with the fast methods of Chapter 11, when repeated for a large number of steps.

An exponential, however, is a very special function. In particular, we are all aware of the identity $\mathrm{e}^{z_1+z_2} = \mathrm{e}^{z_1}\mathrm{e}^{z_2}$, where $z_1, z_2 \in \mathbb{C}$. Were this identity true for matrices, so that

$$\mathrm{e}^{t(Q+S)} = \mathrm{e}^{tQ}\mathrm{e}^{tS}, \qquad t \geq 0, \tag{13.57}$$

for all square matrices Q and S of equal dimension, we could replace Crank–Nicolson by

$$\begin{aligned}
\boldsymbol{u}^{n+1} &= \hat{r}_{1/1}(\mu B_x)\hat{r}_{1/1}(\mu B_y)\boldsymbol{u}^n \tag{13.58}\\
&= \left(I - \tfrac{1}{2}\mu B_x\right)^{-1}\left(I + \tfrac{1}{2}\mu B_x\right)\left(I - \tfrac{1}{2}\mu B_y\right)^{-1}\left(I + \tfrac{1}{2}\mu B_y\right)\boldsymbol{u}^n, \qquad n \geq 0.
\end{aligned}$$

The implementation of (13.58) has a great advantage over the unadulterated form of Crank–Nicolson. We need to solve two linear systems to advance one step, but the second matrix, $I - \tfrac{1}{2}\mu B_y$, is tridiagonal, whilst the first, $I - \tfrac{1}{2}\mu B_x$, can be converted into a tridiagonal form by reordering the grid by rows. Hence, (13.58) can be solved by sparse LU factorization in $\mathcal{O}(d^2)$ operations!

Unfortunately, in general the identity (13.57) is false. Thus, let $[Q, P] := QS - SQ$ be the *commutator* of Q and S. Since

$$\begin{aligned}
\mathrm{e}^{tQ}\mathrm{e}^{tS} - \mathrm{e}^{t(Q+S)} &= \left(I + tQ + \tfrac{1}{2}t^2Q^2 + \cdots\right)\left(I + tS + \tfrac{1}{2}t^2S^2 + \cdots\right) \tag{13.59}\\
&\quad - \left\{I + t(Q + S) + \tfrac{1}{2}t^2(Q + S)^2 + \cdots\right\} = \tfrac{1}{2}t^2[S, Q] + \mathcal{O}(t^3),
\end{aligned}$$

we deduce that (13.57) cannot be true unless Q and S commute. If $a \equiv 1$ and (13.54) reduces to (13.4) then $[B_x, B_y] = O$ (see Exercise 13.14) and we are fully justified in using (13.58), but this is not the case when the diffusion coefficient is allowed to vary.

As with every good policy, the rule that mathematical injunctions must always be followed has its exceptions. For instance, were we to disregard for a moment the breakdown in commutativity and use (13.58) with a variable diffusion coefficient a, the difference would hardly register. The reason is explained in Figure 13.3, where we plot (in the top graph) the error $\|\mathrm{e}^{\mu B_x}\mathrm{e}^{\mu B_y} - \mathrm{e}^{\mu(B_x+B_y)}\|$ for a specific variable diffusion coefficient. Evidently, loss of commutativity does not necessarily cause a damaging loss of accuracy. Part of the reason is that, according to (13.59), $\mathrm{e}^{\mu B_x}\mathrm{e}^{\mu B_y} - \mathrm{e}^{\mu(B_x+B_y)} = \mathcal{O}(\mu^2)$; but this hardly explains the phenomenon, since we are interested in large values of μ. Another clue is that, provided B_x, B_y and B_x+B_y have their eigenvalues in the left half-plane, all exponents vanish for $\mu \to \infty$ (cf. Section 4.1). More justification is provided in Exercise 13.16. In any case, the error in the *splitting*

$$\mathrm{e}^{\mu B_x}\mathrm{e}^{\mu B_y} \approx \mathrm{e}^{\mu(B_x+B_y)}$$

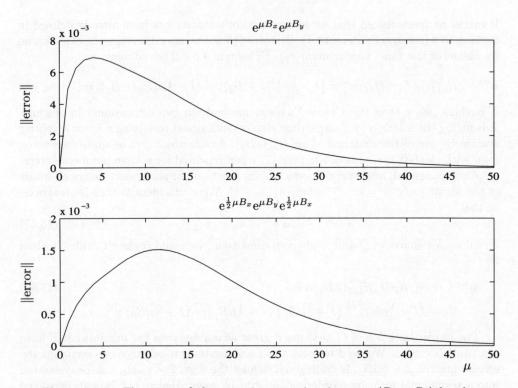

Figure 13.3   The norm of the error in approximating $\exp \mu(B_x + B_y)$ by the 'naive' splitting $\mathrm{e}^{\mu B_x}\mathrm{e}^{\mu B_y}$ and by the Strang splitting $\mathrm{e}^{\frac{1}{2}\mu B_x}\mathrm{e}^{\mu B_y}\mathrm{e}^{\frac{1}{2}\mu B_x}$ for $a(x,y) = 1 + \frac{1}{4}(x - y)$ and $d = 10$.

is sufficiently small to justify the use of (13.58) even in the absence of commutativity.

Even better is the *Strang splitting*

$$\mathrm{e}^{\frac{1}{2}\mu B_x}\mathrm{e}^{\mu B_y}\mathrm{e}^{\frac{1}{2}\mu B_x} \approx \mathrm{e}^{\mu(B_x+B_y)}.$$

It can be observed from Figure 13.3 that it produces a smaller error – Exercise 13.15 is devoted to proving that this is $\mathcal{O}(\mu^3)$.

In general, splitting presents an affordable alternative to the 'full' Crank–Nicolson and the technique can be generalized to other PDEs of evolution and diverse computational schemes.

We have assumed in our exposition that boundary conditions vanish, but this is not strictly necessary. Let us add to (13.54) nonzero boundary conditions and, possibly, a forcing term. Thus, in place of (13.55) we now have

$$\boldsymbol{v}' = \frac{1}{(\Delta x)^2}(B_x + B_y)\boldsymbol{v} + \boldsymbol{h}(t), \quad t \geq 0, \quad \boldsymbol{v}(0) \text{ given},$$

an equation whose explicit solution is

$$\boldsymbol{v}^{n+1} = \mathrm{e}^{\mu(B_x+B_y)}\boldsymbol{v}^n + \Delta t \int_0^1 \mathrm{e}^{(1-\tau)\mu(B_x+B_y)}\boldsymbol{h}((n+\tau)\Delta t)\,\mathrm{d}\tau, \quad n \geq 0.$$

We replace the integral using the trapezoidal rule – a procedure whose error is within the same order of magnitude as that of the original SD scheme. The outcome is

$$\tilde{v}^{n+1} = e^{\mu(B_x+B_y)}\tilde{v}^n + \tfrac{1}{2}\Delta t[e^{\mu(B_x+B_y)}h(n\Delta t) + h((n+1)\Delta t)]$$
$$= e^{\mu(B_x+B_y)}\left[\tilde{v}^n + \tfrac{1}{2}(\Delta t)h(n\Delta t)\right] + \tfrac{1}{2}\Delta t h((n+1)\Delta), \qquad n \geq 0,$$

and we form an FD scheme by splitting the exponential and approximating it with the $\hat{r}_{1/1}$ Padé approximant.

Comments and bibliography

Numerical theory for PDEs of evolution is sometimes presented in a deceptively simple way. On the face of it, nothing could be more straightforward: discretize all spatial derivatives by finite differences and apply a reputable ODE solver, without paying heed to that fact that, actually, one is attempting to solve a PDE. This nonsense has, unfortunately, taken root in many textbooks and lecture courses, which, not to mince words, propagate shoddy mathematics and poor numerical practice. Reputable literature is surprisingly scarce, considering the importance and the depth of the subject. The main source and a good reference to much of the advanced theory is the monograph of Richtmyer & Morton (1967). Other surveys of finite differences that get stability and convergence right are Gekeler (1984), Hirsch (1988) and Mitchell & Griffiths (1980), while the slim volume of Gottlieb & Orszag (1977), whose main theme is entirely different, contains a great deal of useful material applicable to the stability of finite difference schemes for evolutionary PDEs.

Both the eigenvalue approach and the Fourier technique are often – and confusingly – termed in the literature 'the von Neumann method'. While paying due homage to John von Neumann, who originated both techniques, we prefer a more descriptive and less ambiguous terminology.

Modern stability theory ranges far and wide beyond the exposition of this chapter. A useful technique, the *energy method,* will be introduced in Chapter 14. Perhaps the most significant effort in generalizing the framework of stability theory has been devoted to the matter of boundary conditions in the Fourier technique. This is perhaps more significant in the context of hyperbolic equations, the theme of Chapter 14; our only remark here is that these are *very* deep mathematical waters. The original reference, not for the faint-hearted, is Gustaffson *et al.* (1972).

Another interesting elaboration on the theme of stability is connected to the *Kreiss matrix theorem,* its generalizations and applications (van Dorsselaer *et al.,* 1993).

The finite difference technique can be extended (and transformed out of recognition) by the following approach. Suppose that we semi-discretize the equation (13.1), say, at the ℓth grid point, where $\ell \in \{1, 2, \ldots, d\}$. The main idea is to approximate the spatial derivatives with a formula that uses ℓ points to the left and $d+1-\ell$ points to the right. We then obtain

$$v'_\ell = \frac{1}{(\Delta x)^2} \sum_{k=0}^{d+1} a_{\ell,k} v_k, \qquad \ell = 1, 2, \ldots, d \tag{13.60}$$

and each equation couples all grid points. Needless to say, the coefficients in the linear combination are different for each ℓ, but they can be evaluated quite easily and cheaply. The semi-discretization method (13.60) is a special case of a *pseudospectral* method (Fornberg & Sloan, 1994) – more precisely, it yields a formula identical to a method that can be derived by a pseudospectral approach.

The good news is that the SD formula (13.60) can be made accurate to order d (for d even). The bad news is that we need to solve in each step an ODE system with a dense matrix – but, given the high order of the method, this is not much of a problem, since we can typically choose d smaller by an order of magnitude than the number of grid points of a standard SD approximation. Worse news than this is that, alas, (13.60) is unstable. Not all is lost, however! This pseudospectral method can be rendered stable by choosing a non-equidistant grid, denser near the endpoints than at the centre of the interval (Fornberg & Sloan, 1994).

The splitting algorithms of Section 13.6 are a popular means to solve multivariate PDEs of evolution. An alternative, first pioneered by electrical engineers and subsequently adopted and enhanced by numerical analysts, is *waveform relaxation*. Like splitting, it is concerned with effective solution of the ODEs that occur in the course of semi-discretization. Let us suppose that an SD method can be written in the form

$$v' = \frac{1}{(\Delta x)^2} Pv + h(t), \quad t \geq 0, \qquad v(0) = v_0,$$

and that we can express P as a sum of two matrices, $P = Q + S$, say, so that it is easy to solve linear ODE systems with the matrix Q; for example, Q might be diagonal (*Jacobi* waveform relaxation) or lower triangular (*Gauss–Seidel* waveform relaxation). We replace the ODE system by the recursion

$$\left(v^{[k+1]}\right)' = \frac{1}{(\Delta x)^2} Qv^{[k+1]} + \frac{1}{(\Delta x)^2} Sv^{[k]} + h(t), \quad t \geq 0, \qquad v^{[k+1]} = v_0, \qquad k = 0, 1, \ldots,$$

(13.61)

where $v^{[k+1]}(0) \equiv v_0$. In each kth iteration we apply a standard ODE solver, e.g. a multistep method, to (13.61), until the procedure converges to our satisfaction.[5] This idea might appear to be very strange indeed – to replace a single ODE by an infinite (in principle) system of such equations. However, solving the original, unamended ODE by conversion into an algebraic system and employing an iterative procedure from Chapters 10 and 11 also replaces a single equation by an infinite recursion....

There exists a respectable theory that predicts the rate of convergence of (13.61) with k, similar in spirit to the convergence theory from Sections 10.2 and 10.3. An important advantage of waveform relaxation is that it can easily be programmed in a way that takes full advantage of parallel computer architectures, and it can also be combined with multigrid techniques (Vandewalle, 1993).

van Dorsselaer, J.L.M., Kraaijevanger, J.F.B.M. and Spijker, M.N. (1993), Linear stability analysis in the numerical solution of initial value problems, *Acta Numerica* **2**, 199–237.

Fornberg, B. and Sloan, D.M. (1994), A review of pseudospectral methods for solving partial differential equations, *Acta Numerica* **3**, 203–268.

Gekeler, E. (1984), *Discretization Methods for Stable Initial Value Problems*, Springer-Verlag, Berlin.

Gottlieb, D. and Orszag, S.A. (1977), *Numerical Analysis of Spectral Methods: Theory and Applications*, SIAM, Philadelphia,

---

[5]This brief description does little justice to a complicated procedure. For example, for 'intermediate' values of k there is no need to solve the implicit equations with high precision, and this leads to substantial savings (Vandewalle, 1993).

Gustafsson, B., Kreiss, H.-O. and Sundström, A. (1972), Stability theory of difference approximations for mixed initial boundary value problems, *Mathematics of Computation* **26**, 649–686.

Hirsch, C. (1988), *Numerical Computation of Internal and External Flows, Vol. I: Fundamentals of Numerical Discretization*, Wiley, Chichester.

Mitchell, A.R. and Griffiths, D.F. (1980), *The Finite Difference Method in Partial Differential Equations*, Wiley, London.

Richtmyer, R.D. and Morton, K.W. (1967), *Difference Methods for Initial-Value Problems*, Interscience, New York.

Vandewalle, S. (1993), *Parallel Multigrid Waveform Relaxation for Parabolic Problems*, B.G. Teubner, Stuttgart.

Exercises

13.1 Extend the method of proof of Theorem 13.1 to furnish a direct proof that the Crank–Nicolson method (13.32) converges.

13.2 Let

$$
u_\ell^{n+1} = u_m^n + \mu(u_{\ell-1}^n - 2u_\ell^n + u_{\ell+1}^n) - b\frac{\mu}{2}\Delta x\{u_{\ell+1}^n - u_{\ell-1}^n\}
$$

be an FD scheme for the *convection–diffusion* equation

$$
\frac{\partial u}{\partial t} = \frac{\partial^2 u}{\partial x^2} - b\frac{\partial u}{\partial x}, \qquad 0 \le x \le 1, \quad t \ge 0,
$$

where $b > 0$ is given. Prove from first principles that the method converges. [*You can take for granted that the convection–diffusion equation is well posed.*]

13.3 Let

$$
c(t) := \|u(\,\cdot\,,t)\| = \left\{ \int_0^1 [u(x,t)]^2 \, dx \right\}^{1/2}, \qquad t \ge 0,
$$

be the Euclidean norm of the exact solution of the diffusion equation (13.1) with zero boundary conditions.

a Prove that $c'(t) \le 0$, $t \ge 0$, hence $c(t) \le c(0)$, $t \ge 0$, thereby deducing an alternative proof that (13.1) is well posed.

b Let $\boldsymbol{u}^n = (u_\ell^n)_{\ell=0}^{d+1}$ be the Crank–Nicolson solution (13.32) and define

$$
c^n := \|\boldsymbol{u}^n\|_{\Delta x} = \left(\Delta x \sum_{\ell=0}^{d+1} |u_\ell^n|^2 \right)^{1/2}, \qquad n \ge 0,
$$

as the discretized counterpart of the function c. Demonstrate that

$$
(c^{n+1})^2 = (c^n)^2 - \frac{\Delta t}{2\Delta x} \sum_{\ell=1}^{d} \left(u_\ell^{n+1} + u_\ell^n - u_{\ell-1}^{n+1} - u_{\ell-1}^n \right)^2.
$$

Consequently $c_n \le c_0$, $n \ge 0$ and this furnishes yet another proof that the Crank–Nicolson method is stable. [*This is an example of the energy method, which we will encounter again in Chapter 14.*]

13.4   The exponential of a $d \times d$ matrix B is defined by the Taylor series

$$e^B = \sum_{k=0}^{\infty} \frac{1}{k!} B^k.$$

a Prove that the series converges and that

$$\|e^B\| \le e^{\|B\|}.$$

[*This particular result does not depend on the choice of a norm and you should be able to prove it directly from the definition of a norm in A.1.3.3.*]

b Suppose that $B = VDV^{-1}$, where V is nonsingular. Prove that

$$e^{tB} = Ve^{tD}V^{-1}, \qquad t \ge 0.$$

Deduce that, provided B has distinct eigenvalues $\lambda_1, \lambda_2, \ldots, \lambda_d$, there exist $d \times d$ matrices $E_1, E_2, \ldots, E_d$ such that

$$e^{tB} = \sum_{m=1}^{d} e^{t\lambda_m} E_m, \qquad t \ge 0.$$

c Prove that the solution of the linear ODE system

$$\boldsymbol{y}' = B\boldsymbol{y}, \quad t \ge t_0, \qquad \boldsymbol{y}(t_0) = \boldsymbol{y}_0,$$

is

$$\boldsymbol{y}(t) = e^{(t-t_0)B}\boldsymbol{y}_0, \qquad t \ge 0.$$

d Generalize the result from **c**, proving that the explicit solution of

$$\boldsymbol{y}' = B\boldsymbol{y} + \boldsymbol{p}(t), \quad t \ge t_0, \qquad \boldsymbol{y}(t_0) = \boldsymbol{y}_0,$$

is

$$\boldsymbol{y}(t) = e^{(t-t_0)B}\boldsymbol{y}_0 + \int_{t_0}^{t} e^{(t-\tau)B}\boldsymbol{p}(\tau)\,\mathrm{d}\tau, \qquad t \ge t_0.$$

e Let $\|\cdot\|$ be the Euclidean norm and let B be a normal matrix. Prove that $\|e^{tB}\| \le e^{t\tilde{\alpha}(B)}$, $t \ge 0$, where $\tilde{\alpha}(\cdot)$, the *spectral abscissa*, was defined in Section 13.4.

13.5   Prove that the SD method (13.27) is of order four.

13.6   Suppose that an SD scheme of order p_1 for the PDE (13.11) is computed with an ODE solver of order p_2, and that this results in an FD method (possibly multistep). Show that this method is of order $\min\{p_1, rp_2\}$.

13.7 The diffusion equation (13.1) is solved by the fully discretized scheme

$$u_\ell^{n+1} - \tfrac{1}{2}(\mu - \zeta)\left(u_{\ell-1}^{n+1} - 2u_\ell^{n+1} + u_{\ell+1}^{n+1}\right) = u_\ell^n + \tfrac{1}{2}(\mu + \zeta)\left(u_{\ell-1}^n - 2u_\ell^n + u_{\ell+1}^n\right),$$

$$(13.62)$$

where ζ is a given constant. Prove that (13.62) is a second-order method for all $\zeta \neq \tfrac{1}{6}$, while for the choice $\zeta = \tfrac{1}{6}$ (the *Crandall* method) it is of order four.

13.8 Determine the order of the SD method

$$v_\ell' = \frac{1}{(\Delta x)^2}\left(\tfrac{11}{12}v_{\ell-1} - \tfrac{5}{3}v_\ell + \tfrac{1}{2}v_{\ell+1} + \tfrac{1}{3}v_{\ell+2} - \tfrac{1}{12}v_{\ell+3}\right)$$

for the diffusion equation (13.1). Is it stable? [*Hint: Express the function* $\operatorname{Re} a(\mathrm{e}^{\mathrm{i}\theta})$ *as a cubic polynomial in* $\cos\theta$.]

13.9 The SD scheme (13.30) for the diffusion equation with a variable coefficient (13.6) is solved by means of the Euler method.

a Write down the fully discretized equations.

b Prove that the FD method is stable, provided that $\mu \leq 1/(2a_{\min})$, where $a_{\min} = \min\{a(x) : 0 \leq x \leq 1\} > 0$.

13.10 Let B be a $d \times d$ *normal* matrix and let $\boldsymbol{y} \in \mathbb{C}^d$ be an arbitrary vector such that $\|\boldsymbol{y}\| = 1$ (in the Euclidean norm).

a Prove that there exist numbers $\alpha_1, \alpha_2, \ldots, \alpha_d$ such that $\boldsymbol{y} = \sum_{k=1}^d \alpha_k \boldsymbol{w}_k$, where $\boldsymbol{w}_1, \boldsymbol{w}_2, \ldots, \boldsymbol{w}_d$ are the eigenvectors of B. Express $\|\boldsymbol{y}\|^2$ explicitly in terms of α_k, $k = 1, 2, \ldots, d$.

b Let $\lambda_1, \lambda_2, \ldots, \lambda_d$ be the eigenvalues of B, $B\boldsymbol{w}_k = \lambda_k \boldsymbol{w}_k$, $k = 1, 2, \ldots, d$. Prove that

$$\|B\boldsymbol{y}\|^2 = \sum_{k=1}^d |\alpha_k \lambda_k|^2.$$

c Deduce that $\|B\| = \rho(B)$.

13.11 Apply the Fourier stability technique to the FD scheme

$$
\begin{aligned}
u_\ell^{n+1} &= \tfrac{1}{2}(2 - 5\mu + 6\mu^2)u_\ell^n + \tfrac{2}{3}\mu(2 - 3\mu)(u_{\ell-1}^n + u_{\ell+1}^n) \\
&\quad - \tfrac{1}{12}\mu(1 - 6\mu)(u_{\ell-2}^n + u_{\ell+2}^n), \qquad \ell \in \mathbb{Z}.
\end{aligned}
$$

You should find that stability occurs if and only if $0 \leq \mu \leq \tfrac{2}{3}$. [*We have not specified which equation – if any – the scheme is supposed to solve, but this, of course, has no bearing on the question of stability.*]

13.12 Investigate the stability of the FD scheme (13.62) (see Exercise 13.7) for different values of ζ by both the eigenvalue and the Fourier technique.

13.13* Prove that the leapfrog scheme (13.33) for the diffusion equation is unstable for every choice of $\mu > 0$. [*An experienced student of mathematics will not be surprised to hear that this was the first-ever discretization method for the diffusion equation to be published in the scientific literature. Sadly, it is still occasionally used by the unwary – those who forget the history of mathematics are condemned to repeat its mistakes. . . .*]

13.14* Prove that the matrices B_x and B_y from Section 13.6 commute when $a \equiv$ constant. [*Hint: Employ the techniques from Section 12.1 to factorize these matrices and demonstrate that they share the same eigenvalues.*]

13.15 Prove that
$$e^{\frac{1}{2}tQ}e^{tS}e^{\frac{1}{2}tQ} = e^{t(Q+S)} + \mathcal{O}(t^3), \qquad t \to 0,$$
for any $d \times d$ matrices Q and S, thereby establishing the order of the *Strang splitting*.

13.16* Let $E(t) := e^{tQ}e^{tS}$, $t \geq 0$, where Q and S are $d \times d$ matrices.

    a Prove that
$$E' = (Q + S)E + [e^{tQ}, S]e^{tS}, \qquad t \geq 0.$$

    b Using the explicit formula from Exercise 13.4**d** – or otherwise – show that
$$E(t) = e^{t(Q+S)} + \int_0^t e^{(t-\tau)(Q+S)}[e^{\tau Q}, S]e^{\tau S}\,d\tau, \qquad t \geq 0.$$

    c Let P, Q and $P + Q$ be symmetric, negative-definite matrices. Prove that
$$\|e^{tQ}e^{tS} - e^{t(Q+S)}\| \leq 2\|S\| \int_0^t \exp\{(t - \tau)\tilde{\alpha}(Q + S) + \tau[\tilde{\alpha}(Q) + \tilde{\alpha}(S)]\}\,d\tau$$
for $t \geq 0$, where $\tilde{\alpha}(\,\cdot\,)$, the spectral abscissa, has been defined in Section 13.4. [*Hint: Use the estimate from Exercise 13.4***e***.*]

14

Hyperbolic equations

14.1 Why the advection equation?

Much of the discussion in this section is centred upon the *advection equation*

$$\frac{\partial u}{\partial t} + \frac{\partial u}{\partial x} = 0, \qquad 0 \le x \le 1, \quad t \ge 0, \tag{14.1}$$

which is specified in tandem with the initial value

$$u(x,0) = g(x), \qquad 0 \le x \le 1, \tag{14.2}$$

as well as the boundary condition

$$u(0,t) = \varphi_0(t), \qquad t \ge 0, \tag{14.3}$$

where $g(0) = \varphi_0(0)$.

The first and most natural question pertaining to any mathematical construct should not be 'how?' (the knee-jerk reaction of many a trained mathematical mind) but 'why?'. This is a particularly poignant remark with regard to an equation whose exact solution is both well-known and trivial,

$$u(x,t) = \begin{cases} g(x-t), & t \le x, \\ \varphi_0(t-x), & x \le t. \end{cases} \tag{14.4}$$

Note that (14.4) can be verified at once by direct differentiation and that it makes clear why the single boundary condition (14.3) is sufficient.

There are three reasons why numerical study of the advection equation is of interest. Firstly, by its very simplicity, (14.1) affords an insight into a multitude of computational phenomena that are specific to *hyperbolic PDEs*. It is a fitting counterpart of the linear ODE $y' = \lambda y$ that was so fruitful in Chapter 4 in elucidating the behaviour of ODE solvers for stiff equations. Secondly, various generalizations of (14.1) lead to PDEs that are crucial in many applications for which we require in practice numerical solutions: for example the advection equation with a variable coefficient,

$$\frac{\partial u}{\partial t} + \tau(x)\frac{\partial u}{\partial x} = 0, \qquad 0 \le x \le 1, \quad t \ge 0, \tag{14.5}$$

the advection equation in two dimensions,

$$\frac{\partial u}{\partial t} + \frac{\partial u}{\partial x} + \frac{\partial u}{\partial y} = 0, \qquad 0 \le x, y \le 1, \quad t \ge 0, \tag{14.6}$$

and the *wave equation*

$$\frac{\partial^2 u}{\partial t^2} = \frac{\partial^2 u}{\partial x^2}, \qquad -1 \le x \le 1, \quad t \ge 0. \tag{14.7}$$

Equations (14.5)–(14.7) need to be equipped with appropriate initial and boundary conditions and we will address this problem later. Here we just mention the connection that takes us from (14.1) to (14.6). Consider two coupled advection equations, specifically

$$\begin{aligned} \frac{\partial u}{\partial t} + \frac{\partial v}{\partial x} &= 0, \\ \frac{\partial v}{\partial t} + \frac{\partial u}{\partial x} &= 0, \end{aligned} \qquad 0 \le x \le 1, \quad t \ge 0. \tag{14.8}$$

It follows that

$$\frac{\partial^2 u}{\partial t^2} = \frac{\partial}{\partial t}\left(\frac{\partial u}{\partial t}\right) = \frac{\partial}{\partial t}\left(-\frac{\partial v}{\partial x}\right) = -\frac{\partial}{\partial x}\left(\frac{\partial v}{\partial t}\right) = -\frac{\partial}{\partial x}\left(-\frac{\partial u}{\partial x}\right) = \frac{\partial^2 u}{\partial x^2}$$

and so u obeys the wave equation. The system (14.8) is a special case of the vector advection equation

$$\frac{\partial \boldsymbol{u}}{\partial t} + A\frac{\partial \boldsymbol{u}}{\partial x} = \boldsymbol{0}, \qquad 0 \le x \le 1, \quad t \ge 0, \tag{14.9}$$

where the matrix A is diagonalizable and has only real eigenvalues.

The third, and perhaps the most interesting, reason why the humble advection equation is so interesting leads us into the realm of nonlinear hyperbolic equations, ubiquitous in wave theory and in quantum mechanics, e.g. the *Burgers equation*

$$\frac{\partial u}{\partial t} + \frac{1}{2}\frac{\partial u^2}{\partial x^2} = 0, \qquad -\infty < x < \infty, \quad t \ge 0 \tag{14.10}$$

and the *Korteweg–de-Vries* equation

$$\frac{\partial u}{\partial t} + \kappa\frac{\partial u}{\partial x} + \frac{3\kappa}{4\eta}\frac{\partial u^2}{\partial x} + \frac{\kappa\eta^2}{6}\frac{\partial^3 u}{\partial x^3} = 0, \qquad -\infty < x < \infty, \quad t \ge 0, \tag{14.11}$$

whose name is usually abbreviated to KdV. Both display a wealth of nonlinear phenomena of a kind that we have not encountered elsewhere in this volume.

Figure 14.1 displays the evolution of the solution of the Burgers equation (14.10) in the interval $0 \le x \le 2\pi$ with the periodic boundary condition $u(2\pi, t) = u(0, t)$, $t \ge 0$. The initial condition is $g(x) = \frac{5}{2} + \sin x$, $0 \le x \le 2\pi$ and, as t increases from the origin, $g(x)$ is transported with unit speed to the right – as we can expect from the original advection equation – and simultaneously evolves into a function with an increasingly sharper profile which, after a while, looks (and is!) discontinuous. The same picture emerges even more vividly from Figure 14.2, where six 'snapshots' of the

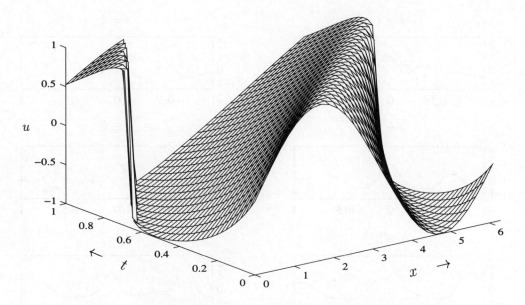

Figure 14.1 The solution of the Burgers equation (14.10) with periodic boundary conditions and $u(x,0) = \frac{5}{2} + \sin x$, $x \in [0, 2\pi)$.

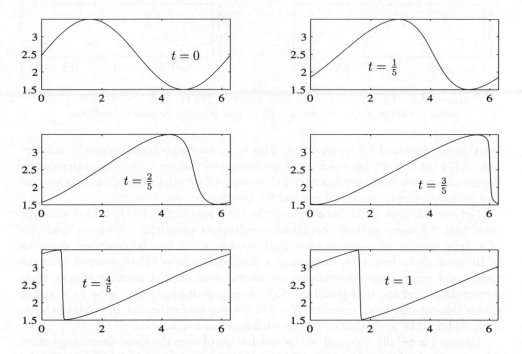

Figure 14.2 The solution of the Burgers equation from Figure 14.1 at times $t = i/5$, $i = 1, 2, \ldots, 5$.

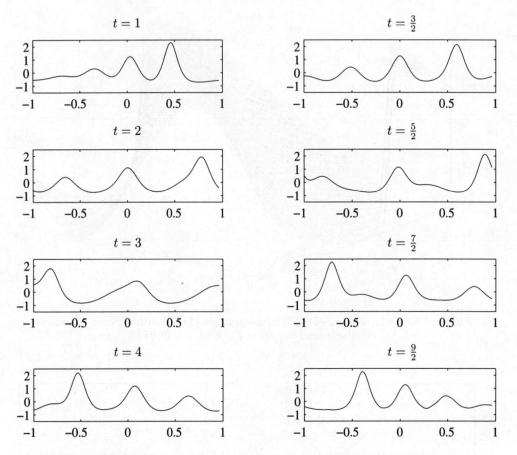

Figure 14.3 The solution of the KdV equation (14.11) with $\kappa = \frac{1}{10}$, $\eta = \frac{1}{5}$, an initial condition $g(x) = \cos \pi x$, $x \in [0, 1)$, and periodic boundary conditions.

solution are displayed for increasing t. This is a remarkable phenomenon, characteristic of (14.10) and similar *nonlinear hyperbolic conservation laws:* a smooth solution degenerates into a discontinuous one. In Section 14.5 we briefly explain this behaviour and present a simple numerical method for the Burgers equation.

Not less intricate is the behaviour of the KdV equation (14.11). It is possible to show that, for every periodic boundary condition, a nontrivial solution is made up of a finite number of *active modes.* Such modes, which can be described in terms of Riemann theta functions, interact in a nonlinear fashion. They move at different speeds and, upon colliding, coalesce yet emerge after a brief delay to resume their former shape and speed of travel. A 'KdV movie' is displayed in Figure 14.3, and it makes this concept more concrete. We will not pursue further the interesting theme of modelling KdV and other equations with such *soliton* solutions.

Having – hopefully – argued to the satisfaction of even the most discerning reader why numerical schemes for the humble advection equation might be of interest, the task of motivating the present chapter is not yet complete. For, have we not just spent a whole chapter deliberating in some detail how to discretize evolutionary PDEs by

finite differences and discussing issues of stability and implementation? According
to this comforting point of view, we just need to employ finite difference operators
to construct a numerical method, evaluate an eigenvalue or two to prove stability...
and the task of computing the solution of (14.1) will be complete. Nothing could be
further from the truth!

To convince a sceptical reader (and all good readers ought to be sceptical!) that
hyperbolic equations require a subtly different approach, we prove a theorem. Its
statement might sound at first quite incredible – as, of course, it is. Having carefully
studied Chapter 13, the reader should be adequately equipped to verify – or reject –
the veracity of the following statement.

'Theorem' $1 = 2$.

Proof We construct the simplest possible genuine finite difference method for
(14.1) by replacing the time derivative by forward differences and the space derivative
by backward differences. The outcome is the *Euler scheme*

$$u_\ell^{n+1} = u_\ell^n - \mu(u_\ell^n - u_{\ell-1}^n) = \mu u_{\ell-1}^n + (1-\mu)u_\ell^n, \quad \ell = 1, 2, \ldots, d, \quad n \geq 0, \quad (14.12)$$

where

$$\mu = \frac{\Delta t}{\Delta x}$$

is the Courant number in this case. Assuming for even greater simplicity that the
boundary value φ_0 is identically zero, we pose the question, what is the set of all
numbers μ that bring about stability?

We address ourselves to this problem by two different techniques, based on eigenvalue analysis (Section 13.4) and on Fourier transforms (Section 13.5) respectively.
Firstly, we write (14.12) in a vector form,

$$\boldsymbol{u}^{n+1} = A\boldsymbol{u}^n, \qquad n \geq 0, \qquad \text{where} \qquad A = \begin{bmatrix} 1-\mu & 0 & \cdots & \cdots & 0 \\ \mu & 1-\mu & \ddots & & \vdots \\ 0 & \ddots & \ddots & \ddots & \vdots \\ \vdots & \ddots & \mu & 1-\mu & 0 \\ 0 & \cdots & 0 & \mu & 1-\mu \end{bmatrix}.$$

Since A is lower triangular, its eigenvalues all equal $1 - \mu$. Requiring $|1 - \mu| \leq 1$ for
stability, we thus deduce that

$$\text{stability} \quad \Longleftrightarrow \quad \mu \in (0, 2]. \tag{14.13}$$

Next we turn our attention to the Fourier approach. Multiplying (14.12) by $\mathrm{e}^{-i\ell\theta}$
and summing over ℓ we easily deduce that the Fourier transform obeys the recurrence

$$\hat{u}^{n+1} = [\mu \mathrm{e}^{-i\theta} + (1-\mu)]\hat{u}^n, \qquad \theta \in [0, 2\pi], \quad n \geq 0.$$

Therefore, the Fourier stability condition is

$$|1 - \mu(1 - \mathrm{e}^{-i\theta})| \leq 1, \qquad \theta \in [0, 2\pi].$$

Straightforward algebra renders this in the form

$$1 - |1 - \mu(1 - e^{-i\theta})|^2 = 4\mu(1 - \mu)\sin^2 \tfrac{1}{2}\theta \ge 0, \qquad \theta \in [0, 2\pi],$$

hence $\mu(1 - \mu) \ge 0$ and we conclude that

$$\text{stability} \quad \Longleftrightarrow \quad \mu \in (0, 1]. \tag{14.14}$$

Comparison of (14.13) with (14.14) proves the assertion of the theorem. ∎

Before we get carried away by the last theorem, it is fair to give the game away and confess that it is, after all, just a prank. It is a prank with a point, though; more precisely, three points. Firstly, its 'proof' is entirely consistent with several books of numerical analysis. Secondly, it is the author's experience that a fair share of professional numerical analysts fail to spot exactly what is wrong. Thirdly, although a rebuttal of the proof should be apparent upon careful reading of Chapter 13, it affords us an opportunity to emphasize the very different ground rules that apply for hyperbolic PDEs and their discretizations.

Which part of the proof is wrong? In principle, both, except that the second part can be mended with relative ease, while the first is based on a blunder, pure and simple.

The Fourier stability technique from Section 13.5 is based on the assumption that we are analysing a Cauchy problem: the range of x is the whole real axis and there are no boundary conditions except for the requirement that the solution is square-integrable. In the proof of the theorem, however, we have stipulated that (14.1) holds in $[0, 1]$ with zero boundary conditions at $x = 0$. This can be easily amended and we can convert this equation into a Cauchy problem without changing the nonzero portion of the solution of (14.12). To that end let us define

$$u_\ell^n = 0, \qquad \ell \in \mathbb{Z} \setminus \{1, 2, \dots, d\}, \tag{14.15}$$

and let the index ℓ in (14.12) range in $\mathbb{Z}$, rather than $\{1, 2, \dots, d\}$. We denote the new solution sequence by $\breve{u}^n$ and claim that $\|\breve{u}^n\| \ge \|u^n\|$, $n \ge 0$. This is obvious from the following diagram, describing the flow of information in the scheme (14.12). This diagram also clarifies why only a left-hand side boundary condition is required to implement this method. We denote by '$\bullet$' a point that belongs to the original grid and let '$\circ$' stand for any value that we have set to zero, whether as a consequence of letting $\varphi_0 = 0$ or of (14.15) or of the FD scheme. Finally, let '$\circledast$' be any value of u_ℓ^n for $\ell \ge d + 1$ that is rendered nonzero by (14.12).

The arrows denote the flow of information – from each u_ℓ^n to u_ℓ^{n+1} and to $u_{\ell+1}^n$ – and we can see at once that padding u^0 with zeros does not introduce any changes in the solution for $\ell = 1, 2, \ldots, d$. Therefore, $\|\check{\boldsymbol{u}}^n\| \geq \|\boldsymbol{u}^n\|$ for all $n \geq 0$ and we have thereby deduced the stability of the original problem by Fourier analysis.

Before advancing further, we should perhaps use this opportunity to comment that often it is more natural to solve (14.1) in $x \in [0, \infty)$ (or, more specifically, in $x \in [0, 1+t)$, $t \geq 0$), since typically it is of interest to follow the wave-like phenomena modelled by hyperbolic PDEs throughout their evolution and at all their destinations. Thus, it is $\check{\boldsymbol{u}}^n$, rather than $\boldsymbol{u}^n$, that should be measured for the purposes of stability analysis, in which case (14.14) is valid (as an 'if and only if' statement) by virtue of Theorem 13.11.

Unlike (14.14), the stability condition (14.13) is false. Recall from Section 13.4 that using eigenvalues and spectral radii to prove stability is justified only if the underlying matrix A is *normal,* which is not so in the present case. It might seem that this clear injunction should be enough to deter anybody from 'proving' stability by eigenvalues. Unfortunately, most students of numerical analysis are weaned on the diffusion equation (where all reasonable finite difference schemes are symmetric) and then given a brief treatment of the wave equation (where, as we will see in Section 14.4, all reasonable finite difference schemes are skew-symmetric). Sooner or later the limited scope of eigenvalue analysis is likely to be forgotten. . ..

It is easy to convince ourselves that the present matrix A is not normal by verifying that $AA^T \neq AA^T$. It is of interest, however, to go back to the theme of Section 13.4 and see what exactly it is that goes wrong with this matrix. The purpose of stability analysis is to deduce uniform bounds on *norms,* while eigenvalue analysis delivers *spectral radii.* Had A been normal, it would have been true that $\rho(A) = \|A\|$ and, in greater generality, $\rho(A^n) = \|A^n\|$. Let us estimate the norm of $A = A_{\Delta x}$ for (14.12), demonstrating that it is consistent with the Fourier estimate rather than the eigenvalue estimate.

In greater generality, we let

$$
S_d = \begin{bmatrix} s & 0 & \cdots & & \cdots & 0 \\ q & s & \ddots & & & \vdots \\ 0 & \ddots & \ddots & \ddots & & \vdots \\ \vdots & & \ddots & q & s & 0 \\ 0 & \cdots & & 0 & q & s \end{bmatrix}
$$

be a *bidiagonal* $d \times d$ matrix and assume that $s, q \neq 0$. (Letting $s = 1 - \mu$ and $q = \mu$ recovers the matrix A.) To evaluate $\|S_d\|$ (in the usual Euclidean norm) we recall (A.1.5.2) that $\|B\| = [\rho(B^T B)]^{1/2}$ for any real square matrix B. Let us thus form the product

$$
S_d^T S_d = \begin{bmatrix} s^2 & sq & 0 & \cdots & & 0 \\ sq & s^2 + q^2 & sq & & \ddots & \vdots \\ 0 & \ddots & \ddots & \ddots & & 0 \\ \vdots & & \ddots & sq & s^2 + q^2 & sq \\ 0 & & \cdots & 0 & sq & s^2 + q^2 \end{bmatrix}.
$$

This is almost a TST matrix – just a single rogue element prevents us from applying Lemma 10.5 to determine its norm! Instead, we take a more roundabout approach. Firstly, it readily follows from the Gerschgorin criterion (Lemma 7.3) that

$$\|S_d\|^2 = \rho(S_d^T S_d) \leq \max\{s^2 + |sq|, \ s^2 + q^2 + 2|sq|\} = (|s| + |q|)^2. \tag{14.16}$$

Secondly, set $w_{d,\ell} := (\operatorname{sgn} s/q)^{\ell-1}$, $\ell = 1, 2, \ldots, d$ and $\boldsymbol{w}_d = (w_{d,\ell})_{\ell=1}^d$. Since

$$S_d^T S_d \boldsymbol{w}_d = \begin{bmatrix} s^2 + |sq| \\ (|s| + |q|)^2 \\ \vdots \\ (|s| + |q|)^2 \\ s^2 + |sq| + q^2 \end{bmatrix} \boldsymbol{w}_d = (|s| + |q|)^2 \boldsymbol{w}_d - \begin{bmatrix} |sq| + q^2 \\ 0 \\ \vdots \\ 0 \\ w_{d,d}|sq| \end{bmatrix},$$

it follows from the definition of a matrix norm (A.1.3.4) that

$$\|S_d\|^2 = \|S_d^T S_d\| = \max_{\boldsymbol{y} \neq \boldsymbol{0}} \frac{\|S_d^T S_d \boldsymbol{y}\|}{\|\boldsymbol{y}\|} \geq \frac{\|S_d^T S_d \boldsymbol{w}_d\|}{\|\boldsymbol{w}_d\|} = (|s| + |q|)^2 + \mathcal{O}\left(d^{-1/2}\right), \quad d \to \infty.$$

Comparison with (14.16) demonstrates that

$$\|S_d\| = |s| + |q| + \mathcal{O}\left(d^{-1/2}\right), \qquad d \to \infty$$

which, returning to the matrix A, becomes

$$\|A\| = |1 - \mu| + \mu + \mathcal{O}\left((\Delta x)^{1/2}\right), \qquad \Delta x \to 0.$$

Hence $|1 - \mu| + \mu \leq 1$ is sufficient for stability, as we have already deduced by Fourier analysis.

Hopefully, we have made the case that numerical solution of hyperbolic equations deserves further elaboration and effort.

14.2   Finite differences for the advection equation

We are concerned with semi-discretizations of the form

$$v'_\ell + \frac{1}{\Delta x} \sum_{k=-\alpha}^{\beta} a_k v_{k+\ell} = 0 \tag{14.17}$$

and with fully discretized schemes

$$\sum_{k=-\gamma}^{\delta} b_k(\mu) u_{\ell+k}^{n+1} = \sum_{k=-\alpha}^{\beta} c_k(\mu) u_{\ell+k}^n, \qquad n \geq 0, \tag{14.18}$$

where

$$\sum_{k=-\gamma}^{\delta} b_k(\mu) \equiv 1$$

(compare with (13.22) and (13.37) respectively), when applied to the advection equation (14.1).

As soon as we contemplate stability, (14.1) will be augmented by boundary conditions; we plan to devote most of our attention to the Cauchy problem and to *periodic* boundary conditions. For the time being we focus on the *order* of (14.17) and that of (14.18), a task for which it is not yet necessary to specify the exact range of ℓ.

Theorem 14.1 *The SD method* (14.17) *is of order p if and only if*

$$a(z) := \sum_{k=-\alpha}^{\beta} a_k z^k = \ln z + c(z-1)^{p+1} + \mathcal{O}\big(|z-1|^{p+2}\big), \qquad z \to 1, \qquad (14.19)$$

where $c \neq 0$, while the FD scheme (14.18) *is of order p for Courant number $\mu = \Delta t/\Delta x$ if and only if there exists $c(\mu) \neq 0$ such that*

$$\tilde{a}(z,\mu) := \frac{\sum_{k=-\alpha}^{\beta} c_k(\mu) z^k}{\sum_{k=-\gamma}^{\delta} b_k(\mu) z^k} = z^{-\mu} + c(\mu)(z-1)^{p+1} + \mathcal{O}\big(|z-1|^{p+2}\big), \qquad z \to 1.$$

$$(14.20)$$

Proof   Our analysis is similar to the material of Section 13.3 and, if anything, easier. Letting $\tilde{v}_\ell(t) := u(\ell\Delta x, t)$ and $\tilde{u}_\ell^n := u(\ell\Delta x, n\Delta t)$ stand for the exact solution at the grid points, we have

$$\tilde{v}_\ell' + \frac{1}{\Delta x} \sum_{k=-\alpha}^{\beta} a_k \tilde{v}_{k+\ell} = \left(\mathcal{D}_t + \frac{1}{\Delta x} \sum_{k=-\alpha}^{\beta} a_k \mathcal{E}_x^k \right) \tilde{v}_\ell.$$

As far as the exact solution of the advection equation is concerned, we have $\mathcal{D}_t = -\mathcal{D}_x$ and, by (7.1), $\mathcal{D}_x = (\Delta x)^{-1} \ln \mathcal{E}_x$. Therefore

$$\tilde{v}_\ell' + \frac{1}{\Delta x} \sum_{k=-\alpha}^{\beta} a_k \tilde{v}_{k+\ell} = -\frac{1}{\Delta x} [\ln \mathcal{E}_x - a(\mathcal{E}_x)] v_\ell$$

and we deduce that (14.19) is necessary and sufficient for order p similarly to the proof of Theorem 13.5.

The order condition for FD schemes is based on the same argument and its derivation proceeds along the lines of Theorem 13.6. Thus, briefly,

$$\sum_{k=-\gamma}^{\delta} b_k(\mu) \tilde{u}_{\ell+k}^{n+1} - \sum_{k=-\alpha}^{\beta} c_k(\mu) \tilde{u}_{\ell+k}^{n} = \left[\mathcal{E}_t \sum_{k=-\gamma}^{\delta} b_k(\mu) \mathcal{E}_x^k - \sum_{k=-\alpha}^{\beta} c_k(\mu) \mathcal{E}_x^k \right] \tilde{u}_\ell$$

while, by (14.1) and Section 7.1,

$$\mathcal{E}_t = e^{(\Delta t)\mathcal{D}_t} = e^{-(\Delta t)\mathcal{D}_x} = e^{-\mu(\Delta x)\mathcal{D}_x} = e^{-\mu\mathcal{E}_x} = \mathcal{E}_x^{-\mu}.$$

Hence

$$\sum_{k=-\gamma}^{\delta} b_k(\mu) \tilde{u}_{\ell+k}^{n+1} - \sum_{k=-\alpha}^{\beta} c_k(\mu) \tilde{u}_{\ell+k}^{n} = \left[\mathcal{E}_x^{-\mu} \sum_{k=-\gamma}^{\delta} b_k(\mu) \mathcal{E}_x^k - \sum_{k=-\alpha}^{\beta} c_k(\mu) \mathcal{E}_x^k \right] \tilde{u}_\ell.$$

This and the normalization of the denominator of $\tilde{a}$ are now used to complete the proof that the pth-order condition is indeed (14.20).   ∎

◇ **Examples of methods and their order**   It is possible to show that, given any $\alpha, \beta \geq 0$, $\alpha + \beta \geq 1$, there exists for the advection equation a unique SD method of order $\alpha + \beta$ and that no other method may attain this bound. The coefficients of such a method are not difficult to derive explicitly, a task that is deferred to Exercise 14.2. Here we present four such schemes for future consideration. In each case we specify the function a.

$$\alpha = 1, \quad \beta = 0: \qquad a(z) = -z^{-1} + 1; \tag{14.21}$$

$$\alpha = 0, \quad \beta = 1: \qquad a(z) = z - 1; \tag{14.22}$$

$$\alpha = 1, \quad \beta = 1: \qquad a(z) = -\tfrac{1}{2}z^{-1} + \tfrac{1}{2}z; \tag{14.23}$$

$$\alpha = 3, \quad \beta = 1: \qquad a(z) = -\tfrac{1}{12}z^{-3} + \tfrac{1}{2}z^{-2} - \tfrac{3}{2}z^{-1} + \tfrac{5}{6} + \tfrac{1}{4}z. \tag{14.24}$$

To demonstrate the power of Theorem 14.1 we address ourselves to the most complicated method above, verifying that the order of (14.24) is indeed four. Letting $z = \mathrm{e}^{\mathrm{i}\theta}$, (14.19) becomes equivalent to

$$a(\mathrm{e}^{\mathrm{i}\theta}) = \mathrm{i}\theta + \mathcal{O}\left(\theta^{p+1}\right), \qquad \theta \to 0.$$

In our case (14.24)

$$
\begin{aligned}
a(\mathrm{e}^{\mathrm{i}\theta}) &= -\tfrac{1}{12}\mathrm{e}^{-3\mathrm{i}\theta} + \tfrac{1}{2}\mathrm{e}^{-2\mathrm{i}\theta} - \tfrac{3}{2}\mathrm{e}^{-\mathrm{i}\theta} + \tfrac{5}{6} + \tfrac{1}{4}\mathrm{e}^{\mathrm{i}\theta} \\
&= -\tfrac{1}{12}\left(1 - 3\mathrm{i}\theta - \tfrac{9}{2}\theta^2 + \tfrac{9}{2}\mathrm{i}\theta^3 + \tfrac{27}{8}\theta^4\right) + \tfrac{1}{2}\left(1 - 2\mathrm{i}\theta - 2\theta^2 + \tfrac{4}{3}\mathrm{i}\theta^3 + \tfrac{2}{3}\theta^4\right) \\
&\quad - \tfrac{3}{2}\left(1 - \mathrm{i}\theta - \tfrac{1}{2}\theta^2 + \tfrac{1}{6}\mathrm{i}\theta^3 + \tfrac{1}{24}\theta^4\right) + \tfrac{5}{6} \\
&\quad + \tfrac{1}{4}\left(1 + \mathrm{i}\theta - \tfrac{1}{2}\theta^2 - \tfrac{1}{6}\theta^3 + \tfrac{1}{24}\theta^4\right) + \mathcal{O}\left(\theta^5\right) \\
&= \mathrm{i}\theta + \mathcal{O}\left(\theta^5\right), \qquad \theta \to 0,
\end{aligned}
$$

hence the method is of order four. It is substantially easier to check that the orders of both (14.21) and (14.22) are unity and that (14.23) is a second-order scheme.

As was the case with the diffusion equation in Chapter 13, the easiest technique in the design of FD schemes is the combination of an SD method with an ODE solver (typically, of at least the same order, cf. Exercise 13.6). Thus, pairing (14.21) with the Euler method (1.4) results in the FD scheme (14.12), which we have already encountered in Section 14.1 in somewhat strange circumstances. The marriage of (1.4) and (14.22) yields

$$u_\ell^{n+1} = u_\ell^n - \mu(u_{\ell+1}^n - u_\ell^n) = (1+\mu)u_\ell^n - \mu u_{\ell+1}^n, \qquad n \geq 0, \tag{14.25}$$

a method that looks very similar to (14.12) – but, as we will see later, is quite different.

The SD scheme (14.23) is of order two and we consider two popular schemes that are obtained when it is combined with second-order ODE schemes. Our first example is the *Crank–Nicolson method*

$$-\tfrac{1}{4}\mu u_{\ell-1}^{n+1} + u_\ell^{n+1} + \tfrac{1}{4}\mu u_{\ell+1}^{n+1} = \tfrac{1}{4}\mu u_{\ell-1}^n + u_\ell^n - \tfrac{1}{4}\mu u_{\ell+1}^n, \qquad n \geq 0, \tag{14.26}$$

which originates in the trapezoidal rule. Although we can deduce directly from Exercise 13.6 that it is of order two, we can also prove it by using (14.20). To that end, we again exploit the substitution $z = e^{i\theta}$. Since

$$\tilde{a}(e^{i\theta}, \mu) = \frac{\frac{1}{4}\mu e^{-i\theta} + 1 - \frac{1}{4}\mu e^{i\theta}}{-\frac{1}{4}\mu e^{-i\theta} + 1 + \frac{1}{4}\mu e^{i\theta}} = \frac{1 - \frac{1}{2}i\mu \sin\theta}{1 + \frac{1}{2}i\mu \sin\theta}$$

$$= \left(1 - \tfrac{1}{2}i\mu \sin\theta\right)\left(1 - \tfrac{1}{2}i\mu \sin\theta - \tfrac{1}{4}\mu^2 \sin^2\theta + \tfrac{1}{8}i\mu^3 \sin^3\theta + \cdots\right)$$

$$= 1 - i\mu\theta - \tfrac{1}{2}\mu^2\theta^2 + \tfrac{1}{12}i\mu(2 + 3\mu^2)\theta^3 + \mathcal{O}(\theta^4), \qquad \theta \to 0,$$

and

$$e^{-i\mu\theta} = 1 - i\mu\theta - \tfrac{1}{2}\mu^2\theta^2 + \tfrac{1}{6}i\mu^3\theta^3 + \mathcal{O}(\theta^4), \qquad \theta \to 0,$$

we deduce that

$$\tilde{a}(e^{i\theta}, \mu) = e^{i\mu\theta} + \tfrac{1}{12}i\mu(2 + \mu^2)\theta^3 + \mathcal{O}(\theta^4), \qquad \theta \to 0.$$

Thus, Crank–Nicolson is a second-order scheme.

Another ubiquitous scheme originates when (14.23) is combined with the explicit midpoint rule from Exercise 2.5. An outcome, the *leapfrog method*, uses two steps,

$$u_\ell^{n+2} = \mu(u_{\ell-1}^{n+1} - u_{\ell+1}^{n+1}) + u_\ell^n, \qquad n \geq 0 \tag{14.27}$$

(cf. (13.33)). Although we have addressed ourselves in Theorem 14.1 to one-step FD methods, a generalization to two steps is easy: since

$$e^{-2i\mu\theta} - \mu(e^{-i\theta} - e^{i\theta})e^{-i\mu\theta} - 1 = -\tfrac{1}{3}i\mu(1 - \mu^2)\theta^3 + \mathcal{O}(\theta^4), \qquad \theta \to 0,$$

this method is also of order two. Note that the error constant $c(\mu) = -\frac{1}{3}i\mu(1 - \mu^2)$ vanishes at $\mu = 1$ and so the method is of superior order for this value of the Courant number μ. An explanation of this phenomenon is the theme of Exercise 14.3.

Not all interesting FD methods can be derived easily from semi-discretized schemes; an example is the *angled derivative method*

$$u_\ell^{n+2} = (1 - 2\mu)(u_\ell^{n+1} - u_{\ell-1}^{n+1}) + u_{\ell-1}^n, \qquad n \geq 0. \tag{14.28}$$

It can be proved that the method is of order two, a task that we relegate to Exercise 14.4. Exercise 14.4 also includes the order analysis of the *Lax–Wendroff scheme*

$$u_\ell^{n+1} = \tfrac{1}{2}\mu(1 + \mu)u_{\ell-1}^n + (1 - \mu^2)u_\ell^n - \tfrac{1}{2}\mu(1 - \mu)u_{\ell+1}^n, \qquad n \geq 0. \tag{14.29}$$

$\diamond$

Proceeding next to the stability analysis of the advection equation (14.1) and paying heed to the lesson of the 'theorem' from Section 14.1, we choose not to use eigenvalue techniques.[1] Our standard tool in the reminder of this section is Fourier analysis.

---

[1]Eigenvalues retain a marginal role, since some methods yield normal matrices; cf. Exercise 14.5.

It is of little surprise, thus, that we commence by considering the Cauchy problem, where the initial condition is given on the whole real line. Not much change is needed in the theory of Section 13.5 – as far as stability analysis is concerned, the exact identity of the PDE is irrelevant! The one obvious exception is that, since the space derivative in (14.1) is on the left-hand side, the inequality in the stability condition for SD schemes needs to be reversed. Without further ado, we thus formulate an equivalent of Theorems 13.10 and 13.11 appropriate to the current discussion.

Theorem 14.2   *The SD method* (14.17) *is stable (for a Cauchy problem) if and only if*

$$\operatorname{Re} a(e^{i\theta}) \geq 0, \qquad \theta \in [0, 2\pi], \tag{14.30}$$

where the function a is defined in (14.19). *Likewise, the FD method* (14.18) *is stable (for a Cauchy problem) for a given Courant number* $\mu \in \mathbb{R}$ *if*

$$|\tilde{a}(e^{i\theta}, \mu)| \leq 1, \qquad \theta \in [0, 2\pi], \tag{14.31}$$

the function $\tilde{a}$ *is defined in* (14.20).   ∎

It is easy to observe that (14.21) and (14.23) are stable, while (14.22) is not. This is an important point, intimately connected to the lack of symmetry in the advection equation. Since the exact solution is a unilateral shift, each value is transported to the right at a constant speed. Hence, numerical methods have a privileged direction and it is popular to choose schemes – whether SD or FD – that employ more points to the left than to the right of the current point, a practice known under the name of *upwinding*. Both (14.21) and (14.24) are upwind schemes, (14.23) is symmetric and (14.22), a downwind scheme, seeks information at a wrong venue.

Being upwind is not a guarantee of stability, but it certainly helps. Thus, (14.24) is stable, since

$$\begin{aligned}
\operatorname{Re} a(e^{i\theta}) &= \operatorname{Re}\left(-\tfrac{1}{12}e^{-3i\theta} + \tfrac{1}{2}e^{-2i\theta} - \tfrac{3}{2}e^{-i\theta} + \tfrac{5}{6} + \tfrac{1}{4}e^{i\theta}\right) \\
&= -\tfrac{1}{12}\cos 3\theta + \tfrac{1}{2}\cos 2\theta - \tfrac{5}{4}\cos\theta + \tfrac{5}{6} \\
&= -\tfrac{1}{12}(4\cos^3\theta - 3\cos\theta) + \tfrac{1}{2}(2\cos^2\theta - 1) - \tfrac{5}{4}\cos\theta + \tfrac{5}{6} \\
&= -\tfrac{1}{3}\cos^3\theta + \cos^2\theta - \cos\theta + \tfrac{1}{3} \\
&= \tfrac{1}{3}(1 - \cos\theta)^2 \geq 0, \qquad \theta \in [0, 2\pi].
\end{aligned}$$

Fully discretized schemes lend themselves to Fourier analysis just as easily. We have already seen that the Euler method (14.12) is stable for $\mu \in (0, 1]$. The outlook is more grim with regard to the downwind scheme (14.25) and, indeed,

$$|\tilde{a}(e^{i\theta}, \mu)|^2 = |1 + \mu - \mu e^{i\theta}|^2 = 1 + 4\mu(1 + \mu)\sin^2 \tfrac{1}{2}\theta, \qquad \theta \in [0, 2\pi],$$

exceeds unity for every $\theta \neq 0, 2\pi$ and $\mu > 0$. Before we discard this method, however, let us pause for a while and recall the vector equation (14.9). Suppose that $A = VDV^{-1}$, where D is diagonal. The elements along the diagonal of D are real since, as we have already mentioned, $\sigma(A) \subset \mathbb{R}$, but they might be negative or positive (in particular, in the important case of the wave equation (14.8), one is positive and the other negative). Letting $\boldsymbol{w}(x,t) := V^{-1}\boldsymbol{u}(x,t)$, the equation (14.9) factorizes into

$$\frac{\partial \boldsymbol{w}}{\partial t} + D\frac{\partial \boldsymbol{w}}{\partial x} = \boldsymbol{0}, \qquad t \geq 0,$$

hence into

$$\frac{\partial w_k}{\partial t} + \lambda_k \frac{\partial w_k}{\partial x} = 0, \qquad t \geq 0, \qquad k = 1,2,\ldots,m, \tag{14.32}$$

where m is the dimension of $\boldsymbol{u}$ and $\lambda_1, \lambda_2, \ldots, \lambda_m$ are the eigenvalues of A (and form the diagonal of D). A similar transformation can be applied to a numerical method, replacing $\boldsymbol{u}^n$ by, say, $\boldsymbol{w}^n$, and it is obvious that the two solution sequences are uniformly bounded (or otherwise) in norm for the same values of μ. Let us suppose that $\mathcal{M} \subseteq \mathbb{R}$ is the set of all numbers (positive, negative or zero) such that $\mu \in \mathcal{M}$ implies that an FD method is stable for the equation (14.1). If we wish to apply this FD scheme to (14.9), it follows from (14.32) that we require

$$\lambda_k \mu \in \mathcal{M}, \qquad k = 1, 2, \ldots, m. \tag{14.33}$$

We recognize a situation familiar from Section 4.2, the interval $\mathcal{M}$ playing a role similar to the linear stability domain of an ODE solver. In most cases of interest, $\mathcal{M}$ is a closed interval, which we denote by $[\mu_-, \mu_+]$.

Provided all eigenvalues are positive, (14.33) merely rescales μ by $\rho(A)$. If they are all negative, the method (14.25) becomes stable (for appropriate values of μ), while (14.12) loses its stability. More interesting, though, is the situation, as in (14.8), when some eigenvalues are positive and others negative, since then, unless $\mu_- < 0$ and $0 < \mu_+$, no value of μ may coexist with stability. Both (14.12) and (14.21) fail in this situation, but this is not the case with Crank–Nicolson, since

$$|\tilde{a}(\mathrm{e}^{\mathrm{i}\theta}, \mu)|^2 = \left|\frac{1 - \frac{1}{2}\mathrm{i}\sin\theta}{1 + \frac{1}{2}\mathrm{i}\sin\theta}\right| \equiv 1, \qquad \theta \in [0, 2\pi],$$

hence we have stability for all $\mu \in (-\infty, \infty)$! Another example is the Lax–Wendroff scheme, whose explicit form confers important advantages in comparison with Crank–Nicolson. Since

$$\begin{aligned}
|\tilde{a}(\mathrm{e}^{\mathrm{i}\theta}, \mu)|^2 &= \left|\tfrac{1}{2}\mu(1+\mu)\mathrm{e}^{-\mathrm{i}\theta} + (1-\mu^2) - \tfrac{1}{2}\mu(1-\mu)\mathrm{e}^{\mathrm{i}\theta}\right|^2 \\
&= |1 - \mu(1-\cos\theta) - \mathrm{i}\mu\sin\theta|^2 \\
&= 1 - 4\mu^2(1-\mu^2)\sin^4\tfrac{1}{2}\theta, \qquad \theta \in [0, 2\pi],
\end{aligned}$$

we obtain $\mu_- = -1$, $\mu_+ = 1$.

Periodic boundary conditions, our next theme, are important in the context of the wave-like phenomena that are typically described by hyperbolic PDEs. Thus, let us complement the advection equation (14.1) with, say, the boundary condition

$$u(0,t) = u(1,t), \qquad t \geq 0. \tag{14.34}$$

The exact solution is no longer (14.4) but is instead periodic in t: the initial condition is transported to the right with unit speed but, as soon as it disappears through $x = 1$, it reappears from the other end, hence $u(x,1) = g(x)$, $0 \leq x \leq 1$.

To emphasize the difference between Dirichlet and periodic boundary conditions we write the Lax–Wendroff scheme (14.29) in a matrix form, $\boldsymbol{u}^{n+1} = A\boldsymbol{u}^n$, say. Assuming (zero) Dirichlet conditions, we have

$$
A = A^{[\mathrm{D}]} := \begin{bmatrix}
1-\mu^2 & \frac{1}{2}\mu(\mu-1) & 0 & \cdots & & 0 \\
\frac{1}{2}\mu(1+\mu) & 1-\mu^2 & \frac{1}{2}\mu(\mu-1) & \ddots & & \vdots \\
0 & \ddots & \ddots & \ddots & & 0 \\
\vdots & & \ddots & \frac{1}{2}\mu(1+\mu) & 1-\mu^2 & \frac{1}{2}\mu(\mu-1) \\
0 & & \cdots & 0 & \frac{1}{2}\mu(1+\mu) & 1-\mu^2
\end{bmatrix},
$$

while periodic boundary conditions yield

$$
A = A^{[\mathrm{p}]} := \begin{bmatrix}
1-\mu^2 & \frac{1}{2}\mu(\mu-1) & 0 & \cdots & & 0 & \frac{1}{2}\mu(1+\mu) \\
\frac{1}{2}\mu(1+\mu) & 1-\mu^2 & \frac{1}{2}\mu(\mu-1) & \ddots & & \vdots & 0 \\
0 & \frac{1}{2}\mu(1+\mu) & \ddots & \ddots & & 0 & \vdots \\
\vdots & & 0 & \ddots & & \frac{1}{2}\mu(\mu-1) & 0 \\
0 & & \vdots & \ddots & \frac{1}{2}\mu(1+\mu) & 1-\mu^2 & \frac{1}{2}\mu(\mu-1) \\
\frac{1}{2}\mu(\mu-1) & 0 & \cdots & & 0 & \frac{1}{2}\mu(1+\mu) & 1-\mu^2
\end{bmatrix}.
$$

The reason for the discrepancies in the top right-hand and lower left-hand corners is that, in the presence of periodic boundary conditions, each time we need a value from outside the set $\{0, 1, \ldots, d-1\}$ at one end, we borrow it from the other end.[2]

The difference between $A^{[\mathrm{D}]}$ and $A^{[\mathrm{p}]}$ does seem minor – just two entries in what are likely to be very large matrices. However, as we will see soon, these two matrices could hardly be more dissimilar in their properties. In particular, while stability analysis with Dirichlet boundary conditions, a subject to which we will have returned briefly by the end of this section, is very intricate, periodic boundary conditions surrender their secrets much more easily. In fact, we have the unexpected comfort of both eigenvalue and Fourier analysis being absolutely straightforward in the periodic case!

The matrix $A^{[\mathrm{p}]}$ is a special case of a *circulant* – the latter being a $d \times d$ matrix C whose jth row, $j = 2, 3, \ldots, d$, is a 'right-rotated' $(j-1)$th row,

$$
C = C(\boldsymbol{\kappa}) = \begin{bmatrix}
\kappa_0 & \kappa_1 & \kappa_2 & \cdots & \kappa_{d-1} \\
\kappa_{d-1} & \kappa_0 & \kappa_1 & \cdots & \kappa_{d-2} \\
\kappa_{d-2} & \kappa_{d-1} & \kappa_0 & \cdots & \kappa_{d-3} \\
\vdots & & & & \vdots \\
\kappa_1 & \kappa_2 & \kappa_3 & \cdots & \kappa_0
\end{bmatrix}; \tag{14.35}
$$

specifically, $\kappa_0 = 1-\mu^2$, $\kappa_1 = \frac{1}{2}\mu(\mu-1)$, $\kappa_2 = \cdots = \kappa_{d-2} = 0$ and $\kappa_{d-1} = \frac{1}{2}\mu(1+\mu)$.

---

[2]We have just tacitly adopted the convention that the unknowns in a periodic problem are the points with spatial coordinates $\ell\Delta x$, $\ell = 0, 1, \ldots, d-1$, where $\Delta x = 1/d$. This makes for a somewhat less unwieldy notation.

Lemma 14.3   *The eigenvalues of $C(\boldsymbol{\kappa})$ are $\kappa(\omega_d^j)$, $j = 0, 1, \ldots, d-1$, where*

$$\kappa(z) := \sum_{\ell=0}^{d-1} \kappa_\ell z^\ell, \qquad z \in \mathbb{C}$$

and $\omega_d = \exp(2\pi \mathrm{i}/d)$ is the dth primitive root of unity. To each $\lambda_j = \kappa(\omega_d^j)$ there corresponds the eigenvector

$$\boldsymbol{w}_j = \begin{bmatrix} 1 \\ \omega_d^j \\ \omega_d^{2j} \\ \vdots \\ \omega_d^{(d-1)j} \end{bmatrix}, \qquad j = 0, 1, \ldots, d-1.$$

Proof   We show directly that $C(\boldsymbol{\kappa})\boldsymbol{w}_j = \lambda_j \boldsymbol{w}_j$ for all $j = 0, 1, \ldots, d-1$. To that end we observe in (14.35) that the mth row of $C(\boldsymbol{\kappa})$ is

$$\begin{bmatrix} \kappa_{d-m} & \kappa_{d-m+1} & \cdots & \kappa_{d-1} & \kappa_0 & \kappa_1 & \cdots & \kappa_{d-m-1} \end{bmatrix},$$

hence the mth component of $C(\boldsymbol{\kappa})\boldsymbol{w}_j$ is

$$\sum_{\ell=0}^{d-1} c_{m,\ell} w_{j,\ell} = \sum_{\ell=0}^{m-1} \kappa_{d-m+\ell} \omega_d^{j\ell} + \sum_{\ell=m}^{d-1} \kappa_{\ell-m} \omega_d^{j\ell}.$$

Let us replace the summation indices on the right by $\ell' = d - m + \ell$, $\ell = 1, 2, \ldots, m-1$ and $\ell' = \ell - m$, $\ell = m, m+1, \ldots, d-1$ respectively. Since $\omega_d^d = 1$, the outcome (dropping the prime from the index ℓ') is

$$\sum_{\ell=0}^{d-1} c_{m,\ell} w_{j,\ell} = \sum_{\ell=d-m}^{d-1} \kappa_\ell \omega_d^{j(\ell-d+m)} + \sum_{\ell=0}^{d-1-m} \kappa_\ell \omega_d^{j(\ell+m)}$$

$$= \left(\sum_{\ell=0}^{d-1} \kappa_\ell \omega_d^{j\ell} \right) \omega_d^{jm} = \lambda_j w_{j,m}, \qquad m = 0, 1, \ldots, d-1.$$

We conclude that the $\boldsymbol{w}_j$ are indeed eigenvectors corresponding to the eigenvalues $\kappa(\omega_d^j)$, $j = 0, 1, \ldots, d-1$, respectively. ∎

The lemma has several interesting consequences. For example, since the matrix of the eigenvectors is *exactly* the inverse discrete Fourier transform (12.8), the theory of Section 12.2 demonstrates that multiplying an arbitrary $d \times d$ circulant by a vector can be executed very fast by two FFTs. More interestingly from our point of view, the eigenvectors of $C(\boldsymbol{\kappa})$ do not depend at all on $\boldsymbol{\kappa}$: all $d \times d$ circulants share the same eigenvectors, hence *all such matrices commute*.

The matrix of eigenvectors,

$$\begin{bmatrix} \boldsymbol{w}_0 & \boldsymbol{w}_1 & \cdots & \boldsymbol{w}_{d-1} \end{bmatrix},$$

is *unitary* since it is trivial to prove that

$$\langle \boldsymbol{w}_j, \boldsymbol{w}_\ell \rangle = \bar{\boldsymbol{w}}_j^T \boldsymbol{w}_\ell = 0, \qquad j, \ell = 0, 1, \ldots, d-1, \quad j \neq \ell.$$

Therefore each and every circulant is *normal* (A.1.2.5). An alternative proof is left to Exercise 14.9. As we already know from Theorems 13.7 and 13.8, the stability of finite difference schemes with normal matrices can be completely specified in terms of their eigenvalues. Since these eigenvalues have been fully described in Lemma 14.3, stability analysis becomes almost as easy as painting by numbers. Thus, for Lax–Wendroff,

$$\kappa_0 = 1 - \mu^2, \qquad \kappa_1 = \tfrac{1}{2}\mu(\mu-1), \qquad \kappa_{d-1} = \tfrac{1}{2}\mu(\mu+1)$$

$$\Rightarrow \quad \lambda_j = (1 - \mu^2) + \tfrac{1}{2}\mu(\mu-1)\omega_d^j + \tfrac{1}{2}\mu(\mu+1)\omega_d^{(d-1)j}$$

$$= (1 - \mu^2) + \tfrac{1}{2}\mu(\mu-1)\exp(2\pi\mathrm{i}j/d) + \tfrac{1}{2}\mu(\mu+1)\exp(-2\pi\mathrm{i}j/d)$$

$$= \tilde{a}(\exp(2\pi\mathrm{i}j/d), \mu), \qquad j = 0, \ldots, d-1,$$

where $\tilde{a}$ has been already encountered in the context of both order and Fourier stability analysis.

There is nothing special about the Lax–Wendroff scheme – it is the presence of periodic boundary conditions that makes the whole difference. The identity

$$\lambda_j = \tilde{a}(\exp(2\pi\mathrm{i}j/d), \mu), \qquad j = 0, 1, \ldots, d-1$$

is valid for all FD methods (14.18). Letting $d \to \infty$, we can now use Theorem 13.7 (or similar analysis, in tandem with Theorem 13.8, in the case of SD schemes) to extend the scope of Theorem 14.2 to the realm of periodic boundary conditions.

Theorem 14.4 *Let us assume the periodic boundary conditions* (14.34). *The SD method* (14.17) *is stable subject to the inequality* (14.30), *and the FD scheme* (14.18) *is stable subject to the inequality* (14.31). ∎

An alternative route to Theorem 14.4 proceeds via Fourier analysis with a *discrete Fourier transform*. It is identical in both content and consequences; as far as circulants are concerned, the main difference between Fourier and eigenvalue analysis is just a matter of terminology.

The stability analysis for a Dirichlet boundary problem is considerably more complicated and the conditions of Theorem 14.2, say, are necessary but often far from sufficient. The Euler method (14.12) is a double exception. Firstly, the Fourier conditions are both necessary and sufficient for stability. Secondly, the statement in the previous sentence can be proved by elementary means (cf. Section 14.1). In general, even if (14.30) or (14.31) are sufficient to attain stability, the proof is far from elementary.

Let us consider the solution of (14.1) with the initial condition $g(x) = \sin 8\pi x$, $x \in [0, 1]$, and the Dirichlet boundary condition $\varphi_0(t) = -\sin 8\pi t$, $t \geq 0$ – its exact solution, acording to (14.4), is $u(x, t) = \sin 8\pi(x - t)$, $x \in [0, 1]$, $t \geq 0$.

To illustrate the difficulty of stability analysis in the presence of the Dirichlet boundary conditions, we solve this equation with the leapfrog method (14.27), the first step having been evaluated with the Lax–Wendroff scheme (14.29). However,

Figure 14.4 A leapfrog solution of (14.1) with Dirichlet boundary conditions.

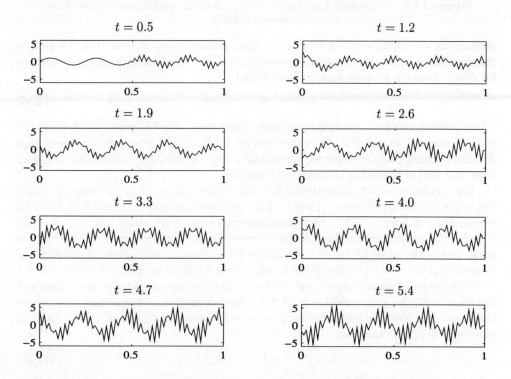

Figure 14.5 Evolution of u^n for the leapfrog method with $d = 80$.

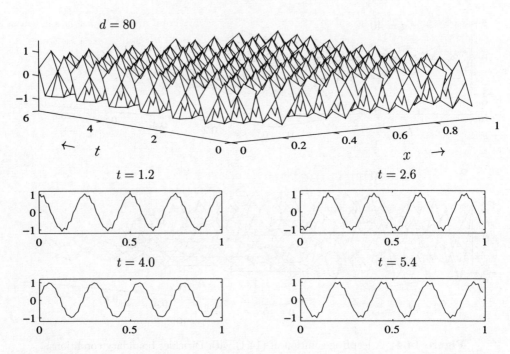

Figure 14.6 A leapfrog solution of (14.1) with the stable *artificial boundary condition* (14.37).

attempting to implement leapfrog, it is soon realized that a vital item of data is missing: since there is no boundary condition at $x = 1$, (14.27) cannot be executed for $\ell = d$. Instead, we have simply substituted

$$u_d^{n+1} = 0, \qquad n \geq 0, \tag{14.36}$$

which seems a safe bet – what can be more stable than zero?! Figure 14.4 displays the solution for $d = 40$ and $d = 80$ in the interval $t \in [0, 6]$ and it is quite apparent that it looks nothing like the expected sinusoidal curve. Worse, the solution deteriorates when the grid is refined, a hallmark of instability.

The mechanism that causes instability and deterioration of the solution is indeed the rogue 'boundary scheme' (14.36). This perhaps becomes more evident upon an examination of Figure 14.5, which displays snapshots of $\boldsymbol{u}^n$ along intervals of equal length. The oscillatory overlay on the (correct) sinusoidal curve at $t = 0.5$ gives the game away: the substitution (14.36) allows for increasing oscillations that enter the interval $[0, 1]$ at $x = 1$ and travel leftwards, *in the wrong direction*.

This amazing sensitivity to the choice of just one point (whose value does not influence at all the *exact* solution in $[0, 1)$) is further emphasized in Figure 14.6, where the very same leapfrog has been used to solve an identical equation, except that, in place of (14.36), we have used

$$u_d^{n+1} = u_{d-1}^n \qquad n \geq 0. \tag{14.37}$$

Like Wordsworth's 'Daffodils', the sinusoidal curves of Figure 14.6 are 'stretch'd in a never-ending line', perfectly aligned and stable. This is already apparent from a cursory inspection of the solution, while the four 'snapshots' are fully consistent with numerical error of $\mathcal{O}((\Delta x)^{-2})$, which we can expect from a second-order convergent scheme.

The general rules governing stability in the presence of boundaries are far too complicated for an introductory text and they require sophisticated mathematical machinery. Our simple example demonstrates that adding boundaries is a genuine issue, not simply a matter of mathematical nitpicking, and that a wrong choice of a 'boundary fix' might well corrupt a stable scheme.

14.3   The energy method

Both the eigenvalue technique and Fourier analysis are, as we have amply demonstrated, of limited scope. Sooner or later – sooner if we set our mind on solving nonlinear PDEs – we are bound to come across a numerical scheme that defies both. The one means left is the recourse of the desperate, the *energy method*. The truth of the matter is that, far from being a coherent technique, the energy method is essentially a brute force approach toward proving the stability conditions (13.16) or (13.20) by direct manipulation of the underlying scheme.

We demonstrate the energy method by a single example, namely numerical solution of the variable-coefficient advection equation (14.5) with the zero boundary condition $\varphi_0 \equiv 0$ by the SD scheme

$$v'_\ell = \frac{\tau_\ell}{2\Delta x}(v^n_{\ell-1} - v^n_{\ell+1}), \qquad \ell = 1, 2, \ldots, d, \quad t \geq 0, \tag{14.38}$$

where $\tau_\ell := \tau(\ell\Delta x)$, $\ell = 1, 2, \ldots, d$. Being a generalization of (14.23), this method is of order two, but our current goal is to investigate its stability.

Fourier analysis is out of the question. The whole point about this technique is that it requires exactly the same difference scheme at every ℓ and a variable function τ_ℓ renders this impossible. It takes more effort to demonstrate that the eigenvalue technique is not up to the task either. The matrix of the SD system (14.38) is

$$P_{\Delta x} = \frac{1}{2\Delta x} \begin{bmatrix} 0 & -\tau_1 & 0 & \cdots & 0 \\ \tau_2 & 0 & -\tau_2 & \ddots & \vdots \\ 0 & \ddots & \ddots & \ddots & 0 \\ \vdots & \ddots & \tau_{d-1} & 0 & -\tau_{d-1} \\ 0 & \cdots & 0 & \tau_d & 0 \end{bmatrix},$$

where $(d+1)\Delta x = 1$. It is an easy, yet tedious, task to prove that, subject to τ being twice differentiable, the matrix $P_{\Delta x}$ cannot be normal for $\Delta x \to 0$ unless τ is a constant. The proof is devoid of intrinsic interest and has no connection with our discussion; hence let us, without further ado, take this result for granted. This means that we cannot use eigenvalues to deduce stability.

Let us assume that the function τ obeys the *Lipschitz condition*

$$|\tau(x) - \tau(y)| \leq \lambda|x - y|, \qquad x, y \in [0, 1], \tag{14.39}$$

for some constant $\lambda \geq 0$. Recall from Chapter 13 that we are measuring the magnitude of $v_{\Delta x}$ in the Euclidean norm

$$\|w_{\Delta x}\|_{\Delta x} = \left(\Delta x \sum_{\ell=1}^{d} w_\ell^2 \right)^{1/2}.$$

Differentiating $\|v_{\Delta x}\|_{\Delta x}^2$ yields

$$\frac{\mathrm{d}}{\mathrm{d}t}\|v_{\Delta x}\|_{\Delta x}^2 = \Delta x \frac{\mathrm{d}}{\mathrm{d}t}\sum_{\ell=1}^{d} v_\ell^2 = 2\Delta x \sum_{\ell=1}^{d} v_\ell v_\ell'.$$

Substituting v_ℓ by the SD equations (14.38) and changing the order of summation, we obtain

$$\frac{\mathrm{d}}{\mathrm{d}t}\|v_{\Delta x}\|_{\Delta x}^2 = \sum_{\ell=1}^{d} \tau_\ell v_\ell (v_{\ell-1} - v_{\ell+1}) = \sum_{\ell=0}^{d-1} \tau_{\ell+1} v_\ell v_{\ell+1} - \sum_{\ell=1}^{d} \tau_\ell v_\ell v_{\ell+1}$$

$$= \sum_{\ell=1}^{d} (\tau_{\ell+1} - \tau_\ell) v_\ell v_{\ell+1}.$$

Note that we have used the zero boundary condition $v_0 = 0$. Observe next that the Lipschitz condition (14.39) implies

$$|\tau_{\ell+1} - \tau_\ell| = |\tau((\ell+1)\Delta x) - \tau(\ell \Delta x)| \leq \lambda \Delta x,$$

therefore

$$\frac{\mathrm{d}}{\mathrm{d}t}\|v_{\Delta x}\|_{\Delta x}^2 \leq \left| \sum_{\ell=1}^{d} (\tau_{\ell+1} - \tau_\ell) v_\ell v_{\ell+1} \right| \leq \sum_{\ell=1}^{d} |\tau_{\ell+1} - \tau_\ell| \times |v_\ell v_{\ell+1}| \leq \lambda \Delta x \sum_{\ell=1}^{d} |v_\ell v_{\ell+1}|.$$

Finally, we resort to the *Cauchy–Schwarz inequality* (A.1.3.1) to deduce that

$$\frac{\mathrm{d}}{\mathrm{d}t}\|v_{\Delta x}\|_{\Delta x}^2 \leq \lambda \Delta x \left(\sum_{\ell=1}^{d} v_\ell^2 \right)^{1/2} \times \left(\sum_{\ell=1}^{d} v_{\ell+1}^2 \right)^{1/2} \leq \lambda \|v_{\Delta x}\|_{\Delta x}^2. \qquad (14.40)$$

It follows at once from (14.40) that

$$\|v_{\Delta x}(t)\|_{\Delta x}^2 \leq \mathrm{e}^{\lambda t} \|v_{\Delta x}(0)\|_{\Delta x}^2, \qquad t \in [0, t^*].$$

Since

$$\lim_{\Delta x \to 0} \|v_{\Delta x}(0)\|_{\Delta x} = \|g\| < \infty,$$

where g is the initial condition (recall our remark on Riemann sums in Section 13.2!), it is possible to bound $\|v_{\Delta x}(0)\|_{\Delta x} \leq c$, say, uniformly for sufficiently small Δx, thereby deducing the inequality

$$\|v_{\Delta x}(t)\|_{\Delta x}^2 \leq c^2 \mathrm{e}^{\lambda t^*}, \qquad t \in [0, t^*].$$

This is precisely what is required for (13.20), the definition of stability for SD schemes, and we thus deduce that the scheme (14.38) is stable.

14.4 The wave equation

As we have already noted in Section 14.1, the wave equation can be expressed as a system of advection equations (14.8). At least in principle, this enables us to exploit the theory of Sections 14.2 and 14.3 to produce finite difference schemes for the wave equation. Unfortunately, we soon encounter two practical snags. Firstly, the wave equation is equipped with two initial conditions, namely

$$u(x,0) = g_0(x), \qquad \frac{\partial u(x,0)}{\partial t} = g_1(x), \qquad 0 \leq x \leq 1, \tag{14.41}$$

and, typically, two Dirichlet boundary conditions,

$$u(0,t) = \varphi_0(t), \quad u(1,t) = \varphi_1(t), \qquad t \geq 0 \tag{14.42}$$

(we will return later to the issue of boundary conditions). Rendering (14.41) and (14.42) in the terminology of a vector advection equation (14.9) makes for strange-looking conditions that needlessly complicate the exposition. The second difficulty comes to light as soon as we attempt to generalize the SD method (14.23), say, to cater for the system (14.9). On the face of it, nothing could be easier: just replace $(\Delta x)^{-1}$ by $(\Delta x)^{-1}A$, thereby obtaining

$$\boldsymbol{v}'_\ell + \frac{1}{2\Delta x}A(\boldsymbol{v}_{\ell+1} - \boldsymbol{v}_{\ell-1}) = \mathbf{0}. \tag{14.43}$$

This is entirely reasonable so far as a general matrix A is concerned. However, choosing

$$A = \begin{bmatrix} 0 & 1 \\ 1 & 0 \end{bmatrix},$$

converts (14.43) into

$$v'_{1,\ell} = -\frac{1}{2\Delta x}(v_{2,\ell+1} - v_{2,\ell-1}),$$

$$v'_{2,\ell} = -\frac{1}{2\Delta x}(v_{1,\ell+1} - v_{1,\ell-1}).$$

According to Section 14.1, $v_\ell := v_{1,\ell}$ approximates the solution of the wave equation. Further differentiation helps us to eliminate the second coordinate,

$$v''_\ell = -\frac{1}{2\Delta x}(v'_{2,\ell+1} - v'_{2,\ell-1}) = -\frac{1}{2\Delta x}\left[-\frac{1}{2\Delta x}(v_{\ell+2} - v_\ell) + \frac{1}{2\Delta x}(v_\ell - v_{\ell-2})\right],$$

and results in the SD scheme

$$v''_\ell = \frac{1}{4(\Delta x)^2}(v_{\ell-2} - 2v_\ell + v_{\ell+2}). \tag{14.44}$$

Although (14.44) is a second-order scheme, it makes very little sense. There is absolutely no good reason to make v''_ℓ depend on $v_{\ell\pm2}$ rather than on $v_{\ell\pm1}$ and we have at least one powerful incentive for the latter course – it is likely to make the numerical error significantly smaller. A simple trick can sort this out: replace (14.43) by the formal scheme

$$v'_\ell + \frac{1}{\Delta x} A(v_{\ell+1/2} - v_{\ell-1/2}).$$

In general this is nonsensical, but for the present matrix A the outcome is the SD scheme

$$v''_\ell = \frac{1}{(\Delta x)^2}(v_{\ell-1} - 2v_\ell + v_{\ell+1}), \tag{14.45}$$

which is of exactly the right form. An alternative route leading to (14.45) is to discretize the second spatial derivative using central differences, exactly as we have done in Chapters 7 and 12.

Choosing to follow the path of analytic expansion, along the lines of Sections 13.3 and 14.2, we consider the general SD method

$$v''_\ell = \frac{1}{(\Delta x)^2} \sum_{k=-\alpha}^{\beta} a_k v_{\ell+k}. \tag{14.46}$$

The only difference from (13.22) and (14.7) is that this scheme possesses a second time derivative, hence being of the right form to satisfy both of the initial conditions (14.41).

Letting $\tilde{v}_\ell(t) := u(\ell\Delta x, t)$, $t \geq 0$, and engaging without any further ado in the already-familiar calculus of finite difference operators, we deduce from (14.8) that

$$\tilde{v}''_\ell - \frac{1}{(\Delta x)^2} \sum_{k-\alpha}^{\beta} a_k \tilde{v}_{\ell+k} = \frac{1}{(\Delta x)^2} \left[(\ln \mathcal{E}_x)^2 - \sum_{k=-\alpha}^{\beta} a_k \mathcal{E}_x^k \right] \tilde{v}_\ell.$$

Therefore (14.46) is of order p if and only if

$$a(z) := \sum_{k=-\alpha}^{\beta} a_k z^k = (\ln z)^2 + c(z-1)^{p+2} + \mathcal{O}\big(|z-1|^{p+3}\big), \qquad z \to 1, \tag{14.47}$$

for some $c \neq 0$.

It is an easy matter to verify that both (14.44) and (14.45) are of order two. A little more effort is required to demonstrate that the SD scheme

$$v''_\ell = \frac{1}{(\Delta x)^2} \left(-\tfrac{1}{12}v_{\ell-2} + \tfrac{4}{3}v_{\ell-1} - \tfrac{5}{2}v_\ell + \tfrac{4}{3}v_{\ell+1} - \tfrac{1}{12}v_{\ell+2} \right)$$

is fourth-order.

Our next step consists of discretizing the ODE system (14.46) and it affords us an opportunity to discuss the numerical solution of second-order ODEs with greater generality. Consider thus the equations

$$z'' = f(t, z), \quad t \geq t_0, \qquad z(t_0) = z_0, \quad z'(t_0) = z'_0, \tag{14.48}$$

where f is a given function. Note that we can easily cast the semi-discretized scheme (14.46) in this form.

The easiest means to solve (14.48) numerically is conversion into a first-order system of twice the number of variables. It can be verified at once that, subject to the substitution $\boldsymbol{y}_1(t) := \boldsymbol{z}(t)$, $\boldsymbol{y}_2(t) := \boldsymbol{z}'(t)$, $t \geq t_0$, (14.48) is equivalent to the ODE system

$$\begin{aligned} \boldsymbol{y}_1' &= \boldsymbol{y}_2, \\ \boldsymbol{y}_2' &= \boldsymbol{f}(t, \boldsymbol{y}_1), \end{aligned} \qquad t \geq t_0, \qquad (14.49)$$

with the initial condition

$$\boldsymbol{y}_1(t_0) = \boldsymbol{z}_0, \qquad \boldsymbol{y}_2(t_0) = \boldsymbol{z}_0'.$$

On the face of it, we may choose any ODE scheme from Chapters 1–3 and apply it to (14.49). The outcome can be surprising....

Suppose, thus, that (14.49) is solved with Euler's method (1.4), hence, in the notation of Chapters 1–6,

$$\begin{aligned} \boldsymbol{y}_{1,n+1} &= \boldsymbol{y}_{1,n} + h\boldsymbol{y}_{2,n}, \\ \boldsymbol{y}_{2,n+1} &= \boldsymbol{y}_{2,n} + h\boldsymbol{f}(t_n, \boldsymbol{y}_{1,n}), \end{aligned} \qquad n \geq 0.$$

According to the first equation,

$$\boldsymbol{y}_{2,n} = \frac{1}{h}(\boldsymbol{y}_{1,n+1} - \boldsymbol{y}_{1,n}).$$

Substituting this twice (once with n and once with $n+1$) into the second equation allows us to eliminate $\boldsymbol{y}_{2,n}$ altogether:

$$\frac{1}{h}(\boldsymbol{y}_{1,n+2} - \boldsymbol{y}_{1,n+1}) = \frac{1}{h}(\boldsymbol{y}_{1,n+1} - \boldsymbol{y}_{1,n}) + h\boldsymbol{f}(t_n, \boldsymbol{y}_{1,n}).$$

The outcome is the two-step explicit method

$$\boldsymbol{z}_{n+2} - 2\boldsymbol{z}_{n+1} + \boldsymbol{z}_n = h^2\boldsymbol{f}(t_n, \boldsymbol{z}_n), \qquad n \geq 0. \qquad (14.50)$$

A considerably more clever approach is to solve the first set of equations in (14.49) with the backward Euler method (1.15), while retaining the usual Euler method for the second set. This yields

$$\begin{aligned} \boldsymbol{y}_{1,n+1} &= \boldsymbol{y}_{1,n} + h\boldsymbol{y}_{2,n+1}, \\ \boldsymbol{y}_{2,n+1} &= \boldsymbol{y}_{2,n} + h\boldsymbol{f}(t_n, \boldsymbol{y}_{1,n}), \end{aligned} \qquad n \geq 0.$$

Substitution of

$$\boldsymbol{y}_{2,n+1} = \frac{1}{h}(\boldsymbol{y}_{1,n+1} - \boldsymbol{y}_{1,n})$$

in the second equation and a shift in the index results in the *Störmer* method

$$\boldsymbol{z}_{n+2} - 2\boldsymbol{z}_{n+1} + \boldsymbol{z}_n = h^2\boldsymbol{f}(t_{n+1}, \boldsymbol{z}_{n+1}), \qquad n \geq 0. \qquad (14.51)$$

Although we have used the backward Euler method in its construction, the Störmer method is explicit.

A numerical method for the ODE (14.49) is of *order p* if substitution of the exact solution results in a perturbation of $\mathcal{O}(h^{p+2})$. As can be expected, the method (14.50) is of order one – after all, it is nothing other than the Euler scheme. However, as far as (14.51) is concerned, symmetry and a subtle interplay between the forward and backward Euler methods mean that its order is increased and its performance improved significantly compared with what we might have expected from naive premises. Let $\tilde{z}_n = z(t_n)$, $n \geq 0$, be the exact solution of (14.48). Substitution of this in (14.51), expansion about the point t_{n+1} and the differential equation (14.48) yield

$$
\begin{aligned}
&\tilde{z}_{n+2} - 2\tilde{z}_{n+1} + \tilde{z}_n - h^2 \boldsymbol{f}(t_{n+1}, \tilde{z}_{n+1}) \\
&= \left[\tilde{z}_{n+1} + h\tilde{z}'_{n+1} + \tfrac{1}{2}h^2 \tilde{z}''_{n+1} + \tfrac{1}{6}h^3 \tilde{z}'''_{n+1} + \mathcal{O}(h^4) \right] - 2\tilde{z}_{n+1} \\
&\quad + \left[\tilde{z}_{n+1} - h\tilde{z}'_{n+1} + \tfrac{1}{2}h^2 \tilde{z}''_{n+1} - \tfrac{1}{6}h^3 \tilde{z}'''_{n+1} + \mathcal{O}(h^4) \right] - h^2 \tilde{z}''_{n+1} \\
&= \mathcal{O}(h^4)
\end{aligned}
$$

and thus we see that the Störmer method is of order two.

Both the methods (14.50) and (14.51) are two-step and explicit. Their implementation is likely to entail a very similar expense. Yet, (14.51) is of order two, while (14.50) is just first-order – yet another example of a free lunch in numerical mathematics.[3]

Applying Störmer's method (14.51) to the semi-discretized scheme (14.45) results in the following two-step FD recursion, the *leapfrog* method,

$$
u_\ell^{n+2} - 2u_\ell^{n+1} + u_\ell^n = \mu^2 (u_{\ell-1}^{n+1} - 2u_\ell^{n+1} + u_{\ell+1}^{n+1}), \qquad n \geq 0, \tag{14.52}
$$

where $\mu = \Delta t / \Delta x$.

Being composed of a second-order space discretization and a second-order approximation in time, (14.52) is itself a second-order method. To analyse its stability we commence by assuming a Cauchy problem and proceeding as in our investigation of the Adams–Bashforth-like method (13.52) in Section 13.5. Relocating to Fourier space, straightforward manipulation results in

$$
\hat{u}^{n+2} - 2(1 - 2\mu^2 \sin^2 \tfrac{1}{2}\theta)\hat{u}^{n+1} + \hat{u}^n = 0, \qquad \theta \in [0, 2\pi].
$$

All solutions of this three-term recurrence are uniformly bounded (and the underlying leapfrog method is stable) if and only if the zeros of the quadratic

$$
\omega^2 - 2(1 - 2\mu^2 \sin^2 \tfrac{1}{2}\theta)\omega + 1 = 0
$$

both reside in the closed unit disc for all $\theta \in [0, 2\pi]$. Although we could now use Lemma 7.3, it is perhaps easier to write the zeros explicitly,

$$
\omega_\pm = 1 - 2\mu \sin^2 \tfrac{1}{2}\theta \pm 2i\mu \sin \tfrac{1}{2}\theta \left(1 - \mu^2 \sin^2 \tfrac{1}{2}\theta \right)^{1/2}.
$$

Provided $0 < \mu \leq 1$, both ω_+ and ω_- are of unit modulus, while if μ exceeds unity then so does the magnitude of one of the zeros. We deduce that the leapfrog method is stable for all $0 < \mu \leq 1$ insofar as the Cauchy problem is concerned. However, similarly to the Euler method for the advection equation in Section 14.1, we are

---

[3]To add insult to injury, (14.50) leads to an *unstable* FD scheme for all $\mu > 0$ (cf. Exercise 14.12).

allowed to infer from Cauchy to Dirichlet. For, suppose that the wave equation (14.7) is given for $0 \leq x \leq 1$ with zero boundary conditions $\varphi_0, \varphi_1 \equiv 0$. Since the leapfrog scheme (14.52) is explicit and each u_ℓ^{n+2} is coupled to just the nearest neighbours on the spatial grid, it follows that, as long as $u_0^n, u_{d+1}^n \equiv 0$, $n \geq 0$, we can assign arbitrary values to u_ℓ^n for $\ell \leq -1$ and $\ell \geq d+1$ without any influence upon $u_1^n, u_2^n, \ldots, u_d^n$. In particular, we can embed a Dirichlet problem into a Cauchy one by padding with zeros, without amending the Euclidean norm of $\boldsymbol{u}^n$. Thus we have stability in a Dirichlet setting for $0 < \mu \leq 1$.

Launching the first iteration of the leapfrog method we first need to derive the vector $\boldsymbol{u}^1$ by other means. Recall that both u and $\partial u/\partial t$ are specified along $x = 0$, and this can be exploited in the derivation of a second-order approximant at $t = \Delta t$. Let $\breve{u}_\ell^0 = g_1(\ell\Delta x)$, $\ell = 1, 2, \ldots, d$. Expanding about $(\ell\Delta x, 0)$ in a Taylor series, we have

$$u(\ell\Delta x, \Delta t) = u(\ell\Delta x, 0) + \Delta t \frac{\partial u(\ell\Delta x, 0)}{\partial t} + \tfrac{1}{2}(\Delta t)^2 \frac{\partial^2 u(\ell\Delta x, 0)}{\partial t^2} + \mathcal{O}\big((\Delta t)^3\big).$$

Substituting intial values and the second derivative from the SD scheme (14.45) results in

$$u_\ell^1 = u_\ell^0 + (\Delta t)\breve{u}_\ell^0 + \tfrac{1}{2}\mu^2(u_{\ell-1}^0 - 2u_\ell^0 + u_{\ell+1}^0), \qquad \ell = 1, 2, \ldots, d, \qquad (14.53)$$

a scheme whose order is consistent with the leapfrog method (14.52).

The Dirichlet boundary conditions (14.42) are not the only interesting means for determining the solution of the wave equation. It is a well-known peculiarity of hyperbolic differential equations that, for every $t_0 \geq 0$, an initial condition along an interval $[x_-, x_+]$, say, determines uniquely the solution for all (x, t) in a set $\mathbb{D}_{x_-, x_+}^{t_0} \subset \mathbb{R} \times [t_0, \infty)$, the *domain of dependence*. For example, it is easy to deduce from (14.4) that the domain of dependence of the advection equation is the parallelogram

$$\mathbb{D}_{x_-, x_+}^{t_0} = \{(x, t) : t \geq t_0,\ x_- + t - t_0 \leq x \leq x_+ + t - t_0\}.$$

The domain of dependence of the wave equation, also known as the *Monge cone*, is the triangle

$$\mathbb{D}_{x_-, x_+}^{t_0} = \big\{(x, t) : t_0 \leq t \leq t_0 + \tfrac{1}{2}(x_+ - x_-),\ x_- + t - t_0 \leq x \leq x_+ - t + t_0\big\}.$$

In other words, provided that we specify $u(x, t_0)$ and $\partial u(x, t_0)/\partial t$ in $[x_-, x_+]$, the solution of (14.6) can be uniquely determined in the Monge cone without any need for boundary conditions.

This dependence of hyperbolic PDEs on local data has two important implications in their numerical analysis. Firstly, consider an explicit FD scheme of the form

$$u_\ell^{n+1} = \sum_{k=-\alpha}^{\beta} c_k(\mu)u_{\ell+k}^n, \qquad n \geq 0, \qquad (14.54)$$

where $c_{-\alpha}(\mu), c_\beta(\mu) \neq 0$. Remember that each u_j^m approximates the solution at $(j\Delta x, m\mu\Delta x)$ and suppose that for sufficiently many ℓ and n in the region of interest it is true that

$$\big(\ell\Delta x, (n+1)\mu\Delta x\big) \notin \mathbb{D}_{(\ell-\alpha)\Delta x, (\ell+\beta)\Delta x}^{n\mu\Delta x}.$$

As $\Delta x \to 0$, points that we wish to determine persistently stay outside their domain of dependence, hence the numerical solution cannot converge there. In other words, a necessary condition for the stability of explicit FD schemes (and, for that matter, SD schemes) for hyperbolics is that μ should be small enough, that the new point 'almost always' fits into the domain of dependence of the 'footprint'.[4] This is the celebrated *Courant–Friedrichs–Lewy* condition, usually known under its acronym, the CFL condition.

As an example, let us consider again the method (14.12) for the advection equation. Comparing the flow of information (see page 312) with the shape of the domain of dependence, we obtain

where the domain of dependence is enclosed between the parallel dotted lines. We deduce at once that stability requires $\mu \leq 1$, which, of course, we already know. The CFL condition, however, is more powerful than that! Thus, suppose that in an explicit method for the advection equation u_ℓ^{n+s} depends on u_ℓ^{n+i} and $u_{\ell+1}^{n+i}$ for $i = 0, 1, \ldots, s - 1$. No matter how we choose the coefficients, the above geometrical argument demonstrates at once that stability is inconsistent with $\mu > 1$.

In the case of the wave equation and the leapfrog method (14.52), the diagram is

and, again, we need $\mu \leq 1$, otherwise the method overruns the Monge cone.

Suppose that the wave equation is specified with the initial conditions (14.41) but without boundary conditions. Since

$$\mathbb{D}_{0,1}^0 = \left\{ (x,t) : 0 \leq t \leq \tfrac{1}{2}, \ t \leq x \leq 1 - t \right\},$$

it makes sense to use the leapfrog method (14.52), in tandem with the starting method (14.53), to derive the solution there. Each new point depends on its immediate neighbours at the previous time level, and this, together with the value of $\mu \in (0, 1]$, restricts the portion of $\mathbb{D}_{0,1}^0$ that can be reached with the leapfrog method. This is illustrated in the following diagram, for $\mu = \tfrac{3}{5}$.

---

[4]We do not propose to elaborate on the meaning of 'almost always' here. It is enough to remark that for both the advection equation and the wave equation the condition is either obeyed for all ℓ and n or violated for all of them – a clear enough distinction.

Here, the reachable points are shown as solid and we can observe that they leave a portion of the Monge cone out of reach of the numerical method.

14.5 The Burgers equation

The Burgers equation (14.10) is the simplest *nonlinear hyperbolic conservation law* and its generalization gives us the *Euler equations* of inviscid, compressible fluid dynamics. We have already seen in Figures 14.1 and 14.2 that its solution displays strange behaviour and might generate discontinuities from an arbitrarily smooth initial condition. Before we take up the challenge of numerically solving it, let us first devote some attention to the analytic properties of the solution.

◇ **Why analysis?**   The assembling of analytic information before even considering computation is a hallmark of good numerical analysis, while the ugly instinct of 'discretizing everything in sight and throwing it on the nearest computer' is the worst kind of advice. Partial differential equations are complicated constructs and, as soon as nonlinearities are allowed, they may exhibit a great variety of difficult phenomena. Before we even pose the question how well is a numerical algorithm performing, we need to formulate more precisely what 'well' means!

In reality, it is a two-way traffic between analysis and computation since, when it comes to truly complicated equations, the best way for a pure mathematician to guess what should be proved is by using numerical experimentation. Recent advances in the understanding of nonlinear behaviour in PDEs have been not the work of narrow specialists in hermetically sealed intellectual compartments but the outcome of collaboration between mathematical analysts, computational experts, applied mathematicians and even experimenters in their laboratories. There is little doubt that future advances will increasingly depend on a greater permeability of discipline boundaries.   ◇

Throughout this section we assume a Cauchy problem,

$$u(x,0) = g(x), \quad -\infty < x < \infty, \quad \text{where} \quad \int_{-\infty}^{\infty} [g(x)]^2 \, \mathrm{d}x < \infty$$

and g is differentiable, since this allows us to disregard boundaries and simplify the notation. As a matter of fact, introducing Dirichlet boundary conditions leaves most of our conclusions unchanged.

Figure 14.7 Characteristics for $g(x) = e^{-5x^2}$ and shock formation.

Let us choose $(x, t) \in \mathbb{R} \times [0, \infty)$ and consider the algebraic equation

$$x = \xi + g(\xi)t. \tag{14.55}$$

Supposing that a unique solution $\xi = \xi(x, t)$ exists for all (x, t) in a suitable subset of $\mathbb{R} \times [0, \infty)$, we set there

$$w(x, t) := g(\xi(x, t)).$$

Therefore

$$\begin{aligned}
\frac{\partial w}{\partial x} &= g'(\xi)\frac{\partial \xi}{\partial x}, \\
\frac{\partial w}{\partial t} &= g'(\xi)\frac{\partial \xi}{\partial t}
\end{aligned} \tag{14.56}$$

(it can be easily proved that, provided the solution of (14.55) is unique, the function ξ is differentiable with respect to x and t). The partial derivatives of ξ can be readily obtained by differentiating (14.55):

$$1 = \frac{\partial \xi}{\partial x} + g'(\xi)t\frac{\partial \xi}{\partial x} \quad \Rightarrow \quad \frac{\partial \xi}{\partial x} = \frac{1}{1 + g'(\xi)t},$$

$$0 = \frac{\partial \xi}{\partial t} + g'(\xi)t\frac{\partial \xi}{\partial t} + g(\xi) \quad \Rightarrow \quad \frac{\partial \xi}{\partial t} = -\frac{g(\xi)}{1 + g'(\xi)t}.$$

Substituting in (14.56) gives

$$\frac{\partial w}{\partial t} + \frac{1}{2}\frac{\partial w^2}{\partial x} = \left[\frac{\partial \xi}{\partial t} + g(\xi)\frac{\partial \xi}{\partial x}\right]g'(\xi) = \left[-\frac{g(\xi)}{1 + g'(\xi)t} + \frac{g(\xi)}{1 + g'(\xi)t}\right]g'(\xi) = 0,$$

thus proving that the function w obeys the Burgers equation.

Since the solution of (14.55) for $t = 0$ is $\xi = x$, it follows that $w(x, 0) = g(x)$. In other words, the function w obeys both the correct equation and the required initial conditions, hence within its domain of definition it coincides with u. (We have just used – without a proof – the uniqueness of the solution of (14.10). In fact, and as we are about to see, the solution need not be unique for general (x, t), but it is so within the domain of definition of w.)

characteristics

Figure 14.8 Characteristics for a step function and formation of a rarefaction region.

Let us reconsider our conclusion that $u(x,t) = g(\xi(x,t))$ in a new light. We choose an arbitrary $\xi \in \mathbb{R}$. It follows from (14.55) that for every $0 \leq t < t_\xi$, say, it is true that $u(\xi + g(\xi)t, t) = g(\xi)$. In other words, at least for a short while, the solution of the Burgers equation is constant along a straight line.

We have already seen an example of similar behaviour in the case of the advection equation, since, according to (14.4), its solution is constant along straight lines of slope $+1$. The crucial difference is that for the Burgers equation the slopes of the straight lines depend on an initial condition g and, in general, vary with ξ. This immediately creates a problem: what if the straight lines collide? At such a point of collision, which might occur for an arbitrarily small $t \geq 0$, we cannot assign a value to the solution – it has a discontinuity there. Figure 14.7 displays the straight lines (under their proper name, *characteristics*) for the function $g(x) = \mathrm{e}^{-5x^2}$. The formation of a discontinuity is apparent and it is easy to ascertain (cf. Exercise 14.16) that it starts to develop from the very beginning at the point $x = 0$.

Another illustration of how discontinuities develop in the solution of the Burgers equation is presented in Figures 14.1 and 14.2.

A discontinuity that originates in a collision of characteristics is called a *shock* and its position is determined by the requirement that characteristics must always flow *into* a shock and never emerge from it. Representing the position of a shock at time t by $\eta(t)$, say, it is not difficult to derive from elementary geometric considerations the *Rankine–Hugoniot* condition

$$\eta'(t) = \tfrac{1}{2}(u_\mathrm{L} + u_\mathrm{R}), \tag{14.57}$$

where u_L and u_R are the values 'carried' by the characteristics to the left and the right of the shock respectively.

No sooner have we explained the mechanism of shock formation than another problem comes to light. If discontinuities are allowed, it is possible for characteristics to depart from each other, leaving a void which is reached by none of them. Such a situation is displayed in Figure 14.8, where the initial condition is already a discontinuous step function.

A domain that is left alone by the characteristics is called a *rarefaction region*. (The terminology of 'shocks' and 'rarefactions' originates in the shock-tube problem of gas dynamics.) For valid physical reasons, it is important to fill such a rarefaction region by imposing the *entropy condition*

$$\frac{1}{2}\frac{\partial u^2}{\partial t} + \frac{1}{3}\frac{\partial u^3}{\partial x} \leq 0. \tag{14.58}$$

The origin of (14.58) is the Burgers equation with *artificial viscosity*,

$$\frac{\partial u}{\partial t} + \frac{1}{2}\frac{\partial u^2}{\partial x} = \nu\frac{\partial^2 u}{\partial x^2},$$

where $\nu > 0$ is small. The addition of the parabolic term $\nu\partial^2 u/\partial x^2$ causes dissipation and the solution is smooth, with neither shocks nor rarefaction regions. Letting $\nu \to 0$, it is possible to derive the inequality (14.58). More importantly, it is possible to prove that, subject to the Rankine–Hugoniot condition (14.57) and the entropy condition (14.58), the Burgers equation possesses a unique solution.

There are many numerical schemes to solve the Burgers equation but we restrict our exposition to perhaps the simplest algorithm that takes on board the special structure of (14.10) – the *Godunov* method.

The main idea behind this approach is to approximate locally the solution by a piecewise-constant function. Since, as we will see soon, the Godunov method amends the step size, we can no longer assume that Δt is constant. Instead, we denote by Δt_n the step that takes us from the nth to the $(n+1)$th time level, $n \geq 0$, and let $u_\ell^n \approx u(\ell\Delta x, t_n)$, where $t_0 = 0$ and $t_{n+1} = t_n + \Delta t_n$, $n \geq 0$.

We let

$$u_\ell^0 = \frac{1}{\Delta x}\int_{(\ell-1/2)\Delta x}^{(\ell+1/2)\Delta x} g(x)\,\mathrm{d}x \tag{14.59}$$

for all ℓ-values of interest. Supposing that the u_ℓ^n are known, we construct a piecewise-constant function $w^{[n]}(\cdot, t_n)$ by letting it equal u_ℓ^n in each interval $\mathbb{I}_\ell := (x_{\ell-1/2}, x_{\ell+1/2}]$ and evaluate the exact solution of this so-called *Riemann problem* ahead of $t = t_n$. The idea is to let each interval $\mathbb{I}_\ell$ 'propagate' in the direction determined by its characteristics.

Let us choose a point (x, t), $t \geq t_n$. There are three possibilities.

(1)   There exists a unique ℓ such that the point is reached by a characteristic from $\mathbb{I}_\ell$. Since characteristics propagate constant values, the solution of the Riemann problem at this point is u_ℓ^n.

(2)   There exists a unique ℓ such that the point is reached by characteristics from the intervals $\mathbb{I}_\ell$ and $\mathbb{I}_{\ell+1}$. In this case, as the two intervals 'propagate' in time, they are separated by a shock. It is trivial to verify from (14.57) that the shock advances along a straight line that commences at $(\ell + \frac{1}{2})\Delta x$ and whose slope is the average of the slopes in $\mathbb{I}_\ell$ and $\mathbb{I}_{\ell+1}$ – in other words, it is $\frac{1}{2}(u_\ell^n + u_{\ell+1}^n)$. Let us denote this line by ρ_ℓ.[5] The value at (x, t) is u_ℓ^n if $x < \rho_\ell(t)$ and $u_{\ell+1}^n$ if $x > \rho_\ell(t)$. (We disregard the case when $x = \rho_\ell(t)$ and the point resides on the shock, since it makes absolutely no difference to the algorithm.)

---

[5]Not every ρ_ℓ is a shock, but this makes no difference to the method.

the segments ρ_ℓ

Figure 14.9   A piecewise-constant approximation, showing the line segments ρ_ℓ (solid) and the first vertical line to collide with ρ_ℓ (dotted).

(3)   Characteristics from more than two intervals reach the point (x, t). In this case we cannot assign a value to the point.

Simple geometrical considerations demonstrate that case 3, which we must avoid, occurs (for some x) for $t > \check{t}$, where $\check{t} > t_n$ is the lowest solution of the equation $\rho_\ell(t) = \rho_{\ell+1}(t)$ for some ℓ. This becomes obvious upon an examination of Figure 14.9. Let us consider the vertical lines rising from the points $(\ell + \frac{1}{2})\Delta x$. Unless the original solution is identically zero, sooner or later one such line is bound to hit one of the segments ρ_j. We let $\check{t}$ be the time at which the first such encounter takes place, choose $t_{n+1} \in (t_n, \check{t}]$ and set $\Delta t_n = t_{n+1} - t_n$. Since $t_{n+1} \in (t_n, \check{t}]$ (cf. Figure 14.9), cases (1) and (2) can be used to construct a unique solution $w^{[n]}(x, t)$ for all $t_n \le t \le t_{n+1}$. We choose the u_ℓ^{n+1} as averages of $w^{[n]}(\,\cdot\,, t_{n+1})$ along the intervals $\mathbb{I}_\ell$,

$$u_\ell^{n+1} = \frac{1}{\Delta x} \int_{(\ell-1/2)\Delta x}^{(\ell+1/2)\Delta x} w^{[n]}(x, t_{n+1})\, \mathrm{d}x. \tag{14.60}$$

Our description of the Godunov method is complete, except for an important remark: the integral in (14.60) can be calculated with great ease. Disregarding shocks, the function $w^{[n]}$ obeys the Burgers equation for $t \in [t_n, t_{n+1}]$. Therefore, integrating in t,

$$\frac{\partial w^{[n]}}{\partial t} + \frac{1}{2}\frac{\partial [w^{[n]}]^2}{\partial x} = 0 \quad \Rightarrow \quad w^{[n]}(x, t_{n+1}) = w^{[n]}(x, t_n) - \frac{1}{2}\int_{t_n}^{t_{n+1}} \frac{\partial [w^{[n]}(x, t)]^2}{\partial x}\, \mathrm{d}t.$$

Substitution in (14.60) results in

$$u_\ell^{n+1} = \frac{1}{\Delta x} \int_{(\ell-1/2)\Delta x}^{(\ell+1/2)\Delta x} \left\{ w^{[n]}(x, t_n) - \frac{1}{2}\int_{t_n}^{t_{n+1}} \frac{\partial [w^{[n]}(x, t)]^2}{\partial x}\, \mathrm{d}t \right\} \mathrm{d}x.$$

Since the u_ℓ^n have been obtained by an averaging procedure as given in (14.60) (this is the whole purpose of (14.59)), we have, after an interchange of the order of integration,

$$
u_\ell^{n+1} = u_\ell^n - \frac{1}{2\Delta x} \int_{t_n}^{t_{n+1}} \int_{(\ell-1/2)\Delta x}^{(\ell+1/2)\Delta x} \frac{\partial [w^{[n]}(x,t)]^2}{\partial x} \, dx \, dt
$$

$$
= u_\ell^n - \frac{1}{2\Delta x} \int_{t_n}^{t_{n+1}} \left\{ \left[w^{[n]} \left((\ell + \tfrac{1}{2})\Delta x, t \right) \right]^2 - \left[w^{[n]} \left((\ell - \tfrac{1}{2})\Delta x, t \right) \right]^2 \right\} \, dt.
$$

Let us now recall our definition of t_{n+1}. No vertical line segments $((\ell + \tfrac{1}{2})\Delta x, t)$, $t \in [t_n, t_{n+1}]$, may cross the discontinuities ρ_j, therefore the value of $w^{[n]}$ across each such segment is constant – equalling either u_ℓ^n or $u_{\ell+1}^n$ (depending on the slope of ρ_ℓ: if it points rightwards it is u_ℓ^n, otherwise $u_{\ell+1}^n$). Let us denote this value by $\chi_{\ell+1/2}$; then

$$
u_\ell^{n+1} = u_\ell^n - \tfrac{1}{2}\mu_n(\chi_{\ell+1/2}^2 - \chi_{\ell-1/2}^2), \tag{14.61}
$$

where

$$
\mu_n := \frac{\Delta t_n}{\Delta x}.
$$

The Godunov method is a first-order approximant to the solution of the Burgers equation, since the only error that we have incurred comes from replacing the values along each step by piecewise-constant approximants. It satisfies the Rankine–Hugoniot condition and it is possible to prove that it is stable. However, more work, outside the scope of this exposition, is required to ensure that the entropy condition is obeyed as well.

It is possible to generalize the Godunov method to more complicated nonlinear hyperbolic conservation laws

$$
\frac{\partial u}{\partial t} + \frac{\partial f(u)}{\partial x} = 0,
$$

where f is a general differentiable function, as well as to systems of such equations. In each case we obtain a recursion of the form (14.61), except that the definition of the *flux* $\chi_{\ell+1/2}$ needs to be amended and is slightly more complicated. In the special case $f(u) = u$ we are back to the advection equation and the Godunov method (14.61) becomes the familiar scheme (14.12) with $0 < \mu \leq 1$.

Comments and bibliography

Fluid and gas dynamics, relativity theory, quantum mechanics, aerodynamics – this is just a partial list of subjects that need hyperbolic PDEs to describe their mathematical foundations. Such equations – the Euler equations of inviscid, compressible flow, the Schrödinger equation of quantum wave mechanics, Einstein's equations of general relativity etc. – are nonlinear and generally multivariate and multidimensional, and their numerical solution presents a formidable challenge. This perhaps explains the major effort that has gone into the computation of hyperbolic PDEs in the last few decades. A bibliographical journey through the hyperbolic landscape might commence with texts on their theory, mainly in a nonlinear setting – thus Drazin & Johnson (1988) on solitons, Lax (1973) and LeVeque (1992) on conservation laws and Whitham (1974) for a general treatment of wave theory. The next

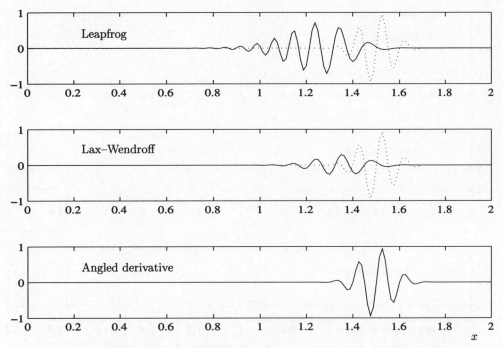

Figure 14.10 Numerical wave propagation by three FD schemes. The dotted line presents the position of the exact solution at time $t = 1.5$. We have used $\Delta x = \frac{1}{80}$ and $\mu = \frac{1}{2}$.

destination might be the classic volume of Richtmyer and Morton (1967), still the best all-round volume on the foundations of the numerical treatment of evolutionary PDEs, followed by more specialized sources: LeVeque (1992) and Hirsch (1988) on numerical conservation laws. Finally, there is an abundance of texts on themes that bear some relevance to the subject matter: Gustaffson *et al.*, (1972) on the influence of boundary conditions on stability; Trefethen (1992) on an alternative treatment, by means of pseuodospectra, of the issue of numerical stability in the absence of normalcy; Iserles & Nørsett (1991) on how to derive optimal schemes for the advection equation, a task that bears a striking similarity to some of the themes from Chapter 4; and Davis (1979) on circulant matrices.

As soon as we concern ourselves with computational wave mechanics (which, in a way, is exactly what the numerical solution of hyperbolic PDEs is all about), there are additional considerations besides order and stability. In Figure 14.10 we display a numerical solution of the advection equation (14.1) with the initial condition

$$g(x) = \mathrm{e}^{-100(x-1/2)^2} \sin 20\pi x, \qquad -\infty \leq x \leq \infty. \qquad (14.62)$$

The function g is a *wave packet* – a highly oscillatory wave, modulated by a sharply-decaying exponential so that, for all intents and purposes, it vanishes outside a small support. The exact solution of (14.1) and (14.62) at time t is the function g, unilaterally translated right-wards by a distance t. In Figure 14.10 we can observe what happens to the wave packet under the influence of discretization by three *stable* FD schemes. Firstly, the leapfrog scheme evidently moves the wave packet at the wrong speed, distorting it in the process. The energy of the packet – that is, the Euclidean norm of the solution – stays constant, as it does in the exact leapfrog is a *conservative* method. However, the energy is transported at the

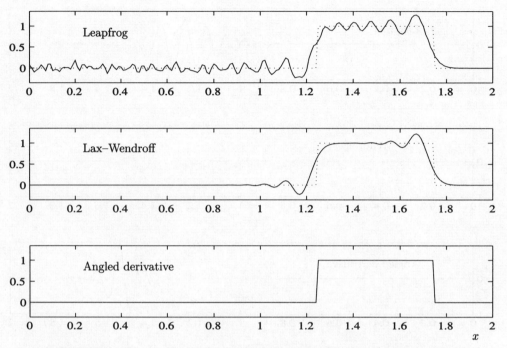

Figure 14.11 Numerical shock propagation by three FD schemes. The dotted line presents the position of the exact solution at time $t = 1.5$. We have used $\Delta x = \frac{1}{100}$ and $\mu = \frac{1}{2}$.

solution: wrong speed. This wrong speed of propagation depends on the wavenumber: the higher the oscillation (in comparison with Δx – recall from Chapter 11 that frequencies larger than $\pi/\Delta x$ are 'invisible' on a grid scale), the more false the reading and, for sufficiently high frequencies, a wave can be transported in the wrong direction altogether. Of course, we can always decrease Δx so as to render the frequency of any particular wave small on the grid scale, although this, obviously, increases the cost of computation. Unfortunately, if the initial condition is discontinuous then its Fourier transform (i.e., its decomposition as a linear combination of periodic 'waves') contains all frequencies that are 'visible' in a grid and this cannot be changed by decreasing Δx; see Figure 14.11.

The behaviour of the Lax–Wendroff method as shown in Figure 14.10 poses another difficulty. Not only does the wave packet lag somewhat; the main problem is that it has almost disappeared! Its energy has decreased by about a factor of three and this is hardly acceptable. Lax–Wendroff is *dissipative*, rather than conservative. The dissipation is governed by the size of Δx and it disappears as $\Delta x \to 0$, yet it might be highly problematic in some applications.

Unlike either leapfrog or Lax–Wendroff, the angled derivative method (14.28) displays the correct qualitative behaviour – no dissipation, no dispersion and, at least up to plotting accuracy, high frequencies are transported at the right speed (Trefethen, 1982).

A similar picture emerges from Figure 14.11, where we have displayed the evolution of the piecewise-constant function

$$g(x) = \begin{cases} 1, & \frac{1}{4} \leq x < \frac{3}{4}, \\ 0, & \text{otherwise.} \end{cases}$$

Leapfrog emerges the worst, both degrading the shock front and transporting some waves too slowly or, even worse, in the wrong direction. Lax–Wendroff is somewhat better: although it also smoothes the sharp shock front, the 'missing mass' is simply dissipated, rather than reappearing in the wrong place. But, unarguably, the angled derivative method emerges again the clear winner. (Lest there is an impression that the latter method is *always* bound to transport wavenumbers correctly, we hasten to note that no method is perfect – but some methods are better than others!)

By this stage, we should be well aware why the correct propagation of shock fronts is so important. Figure 14.11 reaffirms a principle that underlay much of the discussion of hyperbolic PDEs: methods should follow characteristics.

In the particular context of conservation laws, 'following characteristics' means upwinding. This is easy for the advection equation but becomes a more formidable task when characteristics change direction, e.g. for the Burgers equation (14.10). Seen in this light, the Godunov method from Section 14.5 is all about the local determination of the upwind direction. Another popular choice of an upwinding technique is the use of *Engquist–Osher switches*

$$f_-(y) := [\min\{y, 0\}]^2, \quad f_+(y) := [\max\{y, 0\}]^2, \quad y \in \mathbb{R},$$

to form the SD scheme

$$u'_\ell + \frac{1}{\Delta x}\left[\Delta_+ f_-(u_\ell) + \Delta_- f_+(u_\ell)\right] = 0.$$

If $u_{\ell-1}, u_\ell$ and $u_{\ell+1}$ are all positive and the characteristics propagate rightwards, we have $\Delta_+ f_-(u_\ell) = 0$ and $\Delta_- f_+(u_\ell) = [u_\ell]^2 - [u_{\ell-1}]^2$, while if all three values are negative then $\Delta_+ f_-(u_\ell) = [u_{\ell+1}]^2 - [u_\ell]^2$ and $\Delta_- f_+(u_\ell) = 0$. Again, the scheme determines an upwind direction on a local basis.

Numerical methods for nonlinear conservation laws are among the great success stories of modern numerical analysis. They come in many shapes and sizes – finite differences, finite elements, particle methods, spectral methods, ..., but all good schemes have in common elements of upwinding and attention to shock propagation.

Davis, P.J. (1979), *Circulant Matrices*, Wiley, New York.

Drazin, P.G. and Johnson, R.S. (1988), *Solitons: An Introduction*, Cambridge University Press, Cambridge.

Gustafsson, B., Kreiss, H.-O. and Sundström, A. (1972), 'Stability theory of difference approximations for mixed initial boundary value problems', *Mathematics of Computation* **26**, 649–686.

Hirsch, C. (1988), *Numerical Computation of Internal and External Flows, Vol. I: Fundamentals of Numerical Discretization*, Wiley, Chichester.

Iserles, A. and Nørsett, S.P. (1991), *Order Stars*, Chapman & Hall, London.

Lax, P.D. (1973), *Hyperbolic Systems of Conservation Laws and the Mathematical Theory of Shock Waves*, SIAM, Philadelphia.

LeVeque, R.J. (1992), *Numerical Methods for Conservation Laws*, Birkhäuser, Basel.

Richtmyer, R.D. and Morton, K.W. (1967), *Difference Methods for Initial-Value Problems*, Interscience, New York.

Trefethen, L.N. (1982), 'Group velocity in finite difference schemes', *SIAM Review* **24**, 113–136.

Trefethen, L.N. (1992), 'Pseudospectra of matrices', in *Numerical Analysis 1991* (D.F. Griffiths & G.A. Watson, editors), Longman, London, 234–266.

Whitham, G. (1974), *Linear and Nonlinear Waves,* Wiley, New York.

Exercises

14.1   In Section 14.1 we have proved that the Fourier stability analysis of the Euler method (14.12) for the advection equation can be translated from the real axis (i.e. where it is a Cauchy problem) to the interval $[0,1]$ and a zero boundary condition (14.3). Can we use the same technique of proof to analyse the stability of the Euler method (13.7) for the diffusion equation (13.1) with zero Dirichlet boundary conditions (13.3)?

14.2*   Let α and β be nonnegative integers, $\alpha + \beta \geq 1$, and set

$$p_{\alpha,\beta}(z) = \sum_{\substack{k=-\alpha \\ k \neq 0}}^{\beta} \breve{a}_k z^k + q_{\alpha,\beta}, \qquad z \in \mathbb{C},$$

where

$$\breve{a}_k = \frac{(-1)^{k-1}}{k} \frac{\alpha!\beta!}{(\alpha+k)!(\beta-k)!}, \qquad k = -\alpha, -\alpha+1, \ldots, \beta, \quad k \neq 0,$$

and the constant $q_{\alpha,\beta}$ is such that $p_{\alpha,\beta}(1) = 0$.

a Evaluate $p'_{\alpha,\beta}$, proving that

$$p'_{\alpha,\beta}(z) = \frac{1}{z} + \mathcal{O}\big(|z-1|^{\alpha+\beta}\big), \qquad z \to 1.$$

b Integrate the last expression, thereby demonstrating that

$$p_{\alpha,\beta}(z) = \ln z + \mathcal{O}\big(|z-1|^{\alpha+\beta+1}\big), \qquad z \to 1.$$

c Determine the order of the SD scheme

$$u'_\ell + \frac{1}{\Delta x} \left(\sum_{k=-\alpha}^{-1} \breve{a}_k u_{\ell+k} + q_{\alpha,\beta} u_\ell + \sum_{k=1}^{\beta} \breve{a}_k u_{\ell+k} \right) = 0$$

for the advection equation (14.1).

14.3   Show that the leapfrog method (14.27) recovers the exact solution of the advection equation when the Courant number μ equals unity. [*It should be noted that this is of little or no relevance to the solution of the system* (14.9) *or nonlinear equations.*]

14.4 Analyse the order of the following FD methods for the advection equation.

   a The angled derivative scheme (14.28).

   b The Lax–Wendroff scheme (14.29).

   c The *Lax–Friedrichs* scheme

$$u_\ell^{n+1} = \tfrac{1}{2}(1-\mu)u_{\ell-1}^n + \tfrac{1}{2}(1+\mu)u_{\ell+1}^n, \qquad n \geq 0.$$

14.5 Carefully justifying your arguments, use the eigenvalue technique from Section 13.4 to prove that the SD scheme (14.23) is stable.

14.6 Find a third-order upwind SD method (14.17) with $\alpha = 3$, $\beta = 0$ and prove that it is unstable for the Cauchy problem.

14.7 Determine the range of Courant numbers μ for which

   a the leapfrog scheme (14.27); and

   b the angled derivative scheme (14.28)

   are stable for the Cauchy problem. Use the method of proof from Section 14.1 to show that the result for the angled derivative scheme can be generalized from the Cauchy problem to a zero Dirichlet boundary condition.

14.8 Find the order of the *box* method

$$(1-\mu)u_{\ell-1}^{n+1} + (1+\mu)u_\ell^{n+1} = (1+\mu)u_{\ell-1}^n + (1-\mu)u_\ell^n, \qquad n \geq 0.$$

   Determine the range of Courant numbers μ for which the method is stable.

14.9 Show that the transpose of a circulant matrix is itself a circulant and use this to prove that every circulant matrix is normal.

14.10 Find the order of the method

$$u_{j,\ell}^{n+1} = \tfrac{1}{2}\mu(\mu-1)u_{j+1,\ell+1}^n + (1-\mu^2)u_{j,\ell}^n + \tfrac{1}{2}\mu(\mu+1)u_{j-1,\ell-1}^n, \qquad n \geq 0,$$

   for the bivariate advection equation (14.6). Here $\Delta x = \Delta y$, $\Delta t = \mu \Delta x$ and $u_{j,\ell}^n$ approximates $u(j\Delta x, \ell\Delta x, n\Delta t)$.

14.11 Determine the order of the *Numerov* method

$$z_{n+2} - 2z_{n+1} + z_n = \tfrac{1}{12}h^2 \left[\boldsymbol{f}(t_{n+2}, \boldsymbol{z}_{n+2}) \right.$$
$$\left. + 10\boldsymbol{f}(t_{n+1}, \boldsymbol{z}_{n+1}) + \boldsymbol{f}(t_n, \boldsymbol{z}_n) \right], \qquad n \geq 0,$$

   for the second-order ODE system (14.48).

14.12 Application of the method (14.50) to the second-order ODE system (14.45) results in a two-step FD scheme for the wave equation. Prove that this method is unstable for all $\mu > 0$ by two alternative techniques:

a by Fourier analysis; and

b directly from the CFL condition.

14.13   Determine the order of the scheme

$$u_\ell^{n+2} - u_\ell^{n+1} + u_\ell^n = \tfrac{1}{12}\mu^2(-u_{\ell-2}^{n+1} + 16u_{\ell-1}^{n+1} - 30u_\ell^{n+1} + 16u_{\ell+1}^{n+1} - u_{\ell+2}^{n+1}),$$

$$n \geq 0,$$

for the solution of the wave equation (14.8) and find the range of Courant numbers μ for which the method is stable for the Cauchy problem.

14.14   Let $\tilde{u}_\ell^n = u(\ell\Delta x, n\Delta x)$, $\ell = 1, 2, \ldots, d+1$, $n \geq 0$, where $(d+1)\Delta x = 1$ and u is the solution of the wave equation (14.8) with the initial conditions (14.41) and the boundary conditions (14.42).

a Prove the identity

$$\tilde{u}_\ell^{n+2} - \tilde{u}_\ell^n = \tilde{u}_{\ell+1}^{n+1} - \tilde{u}_{\ell-1}^{n+1}, \qquad \ell = 1, 2, \ldots, d, \quad n \geq 0.$$

[*Hint: You might use without proof the d'Alembert solution of the wave equation*

$$u(x,t) = g(x - t) + h(x + t)$$

for all $0 \leq x \leq 1$ and $t \geq 0$.]

b Suppose that the wave equation is solved using the FD method (14.52) with Courant number $\mu = 1$ and denote $e_\ell^n := u_\ell^n - \tilde{u}_\ell^n$, $\ell = 0, 1, \ldots, d+1$, $n \geq 0$. Prove by induction that

$$e_\ell^n = \sum_{j=0}^{n-1} (-1)^j e_{\ell+n-2j-1}^1, \qquad \ell = 1, 2, \ldots, d, \quad n \geq 1,$$

where we let $e_\ell^1 = 0$ for $\ell \notin \{1, 2, \ldots, d\}$.

c The leapfrog method cannot be used to obtain u_ℓ^1 and, instead, we use the scheme (14.53). Assuming that it is known that $|e_\ell^1| \leq \varepsilon$, $\ell = 1, 2, \ldots, d$, prove that

$$|e_\ell^n| \leq \min\{n, \lfloor \tfrac{1}{2}(d+1)\rfloor\}\varepsilon, \qquad n \geq 0. \tag{14.63}$$

[*Naive considerations provide a geometrically increasing upper bound on the error, but (14.63) demonstrates that it is much too large and that the increase in the error is at most linear.*]

14.15   Prove that it is impossible for any explicit FD method (14.54) for the advection equation to be convergent for $\mu > \alpha$.

14.16   Let g be a continuous function and suppose that $x_0 \in \mathbb{R}$ is its strict (local) maximum. The Burgers equation (14.10) is solved with the initial condition $u(x,0) = g(x)$, $x \in \mathbb{R}$. Prove that a shock propagates from the point x_0.

14.17 Show that the entropy condition is satisfied by a solution of the Burgers equation as an equality in any portion of $\mathbb{R} \times [0, \infty)$ that is neither a shock nor a rarefaction region. [*Hint: Multiply the equation by u.*]

14.18* Amend the Godunov method from Section 14.5 to solve the advection equation (14.1) rather than the Burgers equation. Prove that the result is the Euler scheme (14.12) with μ automatically restricted to the stable range $(0, 1]$.

Bluffer's guide to useful mathematics

This is not a review of undergraduate mathematics *or a distillation of the wisdom of many lecture courses into a few pages. Certainly, nobody should use it to understand new material. Mathematics is not learnt from crib-sheets and brief compendia but by careful study of definitions, theorems and – most importantly, perhaps – proofs, by elucidating the intuition behind ideas and grasping the interconnectedness between what might seem disparate concepts at first glance. There are no shortcuts and no cherry-tasting knowledge capsules to help you along your path. . . .*

A conscious attempt has been made throughout the volume not to take for granted any knowledge that an advanced mathematics undergraduate is unlikely to possess. If we need it, we explain it. However, every book has to start from somewhere.

Unless you have basic knowledge of the first two years of university or college mathematics, this appendix will not help you and, indeed, this is the wrong book for you. However, it is not unusual for students to attend a lecture course, study material, absorb it, pass an exam with flying colours – and yet, a year or two later, a concept is perhaps not entirely forgotten but resides so deep in the recesses of memory that it cannot be used here and now. In these circumstances a virtuous reader consults another textbook or perhaps her lecture notes. A less virtuous reader usually means to do so – not just yet – but in the meantime plunges ahead with a decreased level of comprehension. This appendix has been written in recognition of the poverty and scarcity of virtue.

While trying to read a mathematical textbook, nothing can be worse than gradually losing the thread, progressively understanding less and less. This can happen either because the reader fails to understand the actual material – and the fault may well rest with the author – or when she encounters unfamiliar mathematical constructs.

If in this volume you occasionally come across a mathematical concept and, for the life of you, simply cannot recall exactly what it means (or perhaps are not sure of the finer details of its definition), do glance in this appendix – you might find it here! However, if these glances become a habit, rather than an exception, perhaps you had better use a proper textbook!

There are two sections to this appendix, one on linear algebra and the second on analysis. Neither is complete – they both endeavour to answer possible queries arising from this book, rather than providing a potted summary of a subject.

There is nothing on basic calculus. Unless you are familiar with calculus then, I am afraid, you are trying to dance the samba before you can walk.

A.1   Linear algebra

A.1.1   Vector spaces

A.1.1.1   A *vector space* or a *linear space* $\mathbb{V}$ over the field $\mathbb{F}$ (which in our case will be either $\mathbb{R}$, the reals, or $\mathbb{C}$, the complex numbers) is a set of elements closed with respect to addition, i.e.,

$$x_1, x_2 \in \mathbb{V} \qquad \text{implies} \qquad x_1 + x_2 \in \mathbb{V},$$

and multiplication by a *scalar*, i.e.,

$$\alpha \in \mathbb{F},\ x \in \mathbb{V} \qquad \text{implies} \qquad \alpha x \in \mathbb{V}.$$

Addition obeys the axioms of an abelian group: $x_1 + x_2 = x_2 + x_1$ for all $x_1, x_2 \in \mathbb{V}$ (commutativity); $(x_1 + x_2) + x_3 = x_1 + (x_2 + x_3)$, $x_1, x_2, x_3 \in \mathbb{V}$ (associativity); there exists an element $0 \in \mathbb{V}$ (the zero element) such that $x + 0 = x$, $x \in \mathbb{V}$; and for every $x_1 \in \mathbb{V}$ there exists an element $x_2 \in \mathbb{V}$ (the inverse) such that $x_1 + x_2 = 0$ (we write $x_2 = -x_1$). Multiplication by a scalar is also commutative: $\alpha(\beta x) = (\alpha\beta)x$ for all $\alpha, \beta \in \mathbb{F}$, $x \in \mathbb{V}$, and multiplication by the unit element of $\mathbb{F}$ leaves $x \in \mathbb{V}$ intact, $1x = x$. Moreover, addition and multiplication by a scalar are linked by the distributive laws $\alpha(x_1 + x_2) = \alpha x_1 + \alpha x_2$ and $(\alpha + \beta)x_1 = \alpha x_1 + \beta x_1$, $\alpha, \beta \in \mathbb{F}$, $x_1, x_2 \in \mathbb{V}$.

The elements of $\mathbb{V}$ are sometimes called *vectors*.

If $\mathbb{V}_1 \subseteq \mathbb{V}_2$, where both $\mathbb{V}_1$ and $\mathbb{V}_2$ are vector spaces, we say that $\mathbb{V}_1$ is a *subspace* of $\mathbb{V}_2$.

A.1.1.2   The vectors $x_1, x_2, \ldots, x_m \in \mathbb{V}$ are *linearly independent* if

$$\exists\, \alpha_1, \alpha_2, \ldots, \alpha_m \in \mathbb{F} \quad \text{such that} \quad \sum_{\ell=1}^{m} \alpha_\ell x_\ell = 0 \quad \Longrightarrow \quad \alpha_1, \alpha_2, \ldots, \alpha_m = 0.$$

A vector space $\mathbb{V}$ is of *dimension* $\dim \mathbb{V} = d$ if there exist d linearly independent elements $y_1, y_2, \ldots, y_d \in \mathbb{V}$ such that for every $x \in \mathbb{V}$ we can find scalars $\beta_1, \beta_2, \ldots, \beta_d \in \mathbb{F}$ such that

$$x = \sum_{\ell=1}^{d} \beta_\ell y_\ell.$$

The set $\{y_1, y_2, \ldots, y_d\} \subset \mathbb{V}$ is then said to be a *basis* of $\mathbb{V}$. In other words, all the elements of $\mathbb{V}$ can be expressed by forming linear combinations of its basis elements.

A.1.1.3   The vector space $\mathbb{R}^d$ (over $\mathbb{F} = \mathbb{R}$, the reals) consists of all real d-tuples

$$x = \begin{bmatrix} x_1 \\ x_2 \\ \vdots \\ x_d \end{bmatrix},$$

with addition and multiplication by a scalar defined by

$$
\begin{bmatrix} x_1 \\ x_2 \\ \vdots \\ x_d \end{bmatrix} + \begin{bmatrix} y_1 \\ y_2 \\ \vdots \\ y_d \end{bmatrix} = \begin{bmatrix} x_1 + y_1 \\ x_2 + y_2 \\ \vdots \\ x_d + y_d \end{bmatrix} \quad \text{and} \quad \alpha \begin{bmatrix} x_1 \\ x_2 \\ \vdots \\ x_d \end{bmatrix} = \begin{bmatrix} \alpha x_1 \\ \alpha x_2 \\ \vdots \\ \alpha x_d \end{bmatrix}
$$

respectively. It is of dimension d with a *canonical* basis

$$
\boldsymbol{e}_1 = \begin{bmatrix} 1 \\ 0 \\ 0 \\ \vdots \\ 0 \end{bmatrix}, \quad \boldsymbol{e}_2 = \begin{bmatrix} 0 \\ 1 \\ 0 \\ \vdots \\ 0 \end{bmatrix}, \quad \ldots, \quad \boldsymbol{e}_d = \begin{bmatrix} 0 \\ 0 \\ \vdots \\ 0 \\ 1 \end{bmatrix}
$$

of *unit vectors*.

Likewise, letting $\mathbb{F} = \mathbb{C}$ we obtain the vector space $\mathbb{C}^d$ of all complex d-tuples, with similarly defined operations of addition and multiplication by a scalar. It is again of dimension d and possesses an identical basis $\{\boldsymbol{e}_1, \boldsymbol{e}_2, \ldots, \boldsymbol{e}_d\}$ of unit vectors.

In what follows, unless explicitly stated, we restrict our review to $\mathbb{R}^d$. Transplantation to $\mathbb{C}^d$ is straightforward.

A.1.2 Matrices

A.1.2.1 A *matrix* A is an $d \times n$ array of real numbers,

$$
A = \begin{bmatrix} a_{1,1} & a_{1,2} & \cdots & a_{1,n} \\ a_{2,1} & a_{2,2} & \cdots & a_{2,n} \\ \vdots & \vdots & & \vdots \\ a_{d,1} & a_{d,2} & \cdots & a_{d,n} \end{bmatrix}.
$$

It is said to have d *rows* and n *columns*. The addition and multiplication by a scalar of two $d \times n$ matrices are defined by

$$
\begin{bmatrix} a_{1,1} & a_{1,2} & \cdots & a_{1,n} \\ a_{2,1} & a_{2,2} & \cdots & a_{2,n} \\ \vdots & \vdots & & \vdots \\ a_{d,1} & a_{d,2} & \cdots & a_{d,n} \end{bmatrix} + \begin{bmatrix} b_{1,1} & b_{1,2} & \cdots & b_{1,n} \\ b_{2,1} & b_{2,2} & \cdots & b_{2,n} \\ \vdots & \vdots & & \vdots \\ b_{d,1} & b_{d,2} & \cdots & b_{d,n} \end{bmatrix} = \begin{bmatrix} a_{1,1} + b_{1,1} & a_{1,2} + b_{1,2} & \cdots & a_{1,n} + b_{1,n} \\ a_{2,1} + b_{2,1} & a_{2,2} + b_{2,2} & \cdots & a_{2,n} + b_{2,n} \\ \vdots & \vdots & & \vdots \\ a_{d,1} + b_{d,1} & a_{d,2} + b_{d,2} & \cdots & a_{d,n} + b_{d,n} \end{bmatrix}
$$

and

$$
\alpha \begin{bmatrix} a_{1,1} & a_{1,2} & \cdots & a_{1,n} \\ a_{2,1} & a_{2,2} & \cdots & a_{2,n} \\ \vdots & \vdots & & \vdots \\ a_{d,1} & a_{d,2} & \cdots & a_{d,n} \end{bmatrix} = \begin{bmatrix} \alpha a_{1,1} & \alpha a_{1,2} & \cdots & \alpha a_{1,n} \\ \alpha a_{2,1} & \alpha a_{2,2} & \cdots & \alpha a_{2,n} \\ \vdots & \vdots & & \vdots \\ \alpha a_{d,1} & \alpha a_{d,2} & \cdots & \alpha a_{d,n} \end{bmatrix}.
$$

Given an $m \times d$ matrix A and an $d \times n$ matrix B, the product $C = AB$ is the $m \times n$ matrix

$$
C = \begin{bmatrix}
c_{1,1} & c_{1,2} & \cdots & c_{1,n} \\
c_{2,1} & c_{2,2} & \cdots & c_{2,n} \\
\vdots & \vdots & & \vdots \\
c_{m,1} & c_{m,2} & \cdots & c_{m,n}
\end{bmatrix},
$$

where

$$
c_{i,j} = \sum_{\ell=1}^{d} a_{i,\ell} b_{\ell,j}, \qquad i = 1, 2, \ldots, m, \quad j = 1, 2, \ldots, n.
$$

Any $x \in \mathbb{R}^d$ is itself an $d \times 1$ matrix. Hence, the *matrix–vector product* $y = Ax$, where A is $m \times d$, is an element of $\mathbb{R}^m$ such that

$$
y_i = \sum_{\ell=1}^{d} a_{i,\ell} x_\ell, \qquad i = 1, 2, \ldots, m.
$$

In other words, any $m \times d$ matrix A is a *linear transformation* that maps $\mathbb{R}^d$ to $\mathbb{R}^m$.

A.1.2.2 The dth *identity matrix* is the $d \times d$ matrix

$$
I = \begin{bmatrix}
1 & 0 & \cdots & 0 \\
0 & 1 & \ddots & \vdots \\
\vdots & \ddots & \ddots & 0 \\
0 & \cdots & 0 & 1
\end{bmatrix}.
$$

It is true that $IA = A$ and $BI = B$ for any $d \times n$ matrix A and $m \times d$ matrix B respectively.

A.1.2.3 A matrix is *square* if it is $d \times d$ for some $d \geq 1$. The *determinant* of a square matrix A can be defined by induction. If $d = 1$, so that A is simply a real number, $\det A = A$. Otherwise

$$
\det A = \sum_{j=1}^{d} (-1)^{d+j} a_{d,j} \det A_j,
$$

where $A_1, A_2, \ldots, A_d$ are $(d-1) \times (d-1)$ matrices given by

$$
A_j = \begin{bmatrix}
a_{1,1} & \cdots & a_{1,j-1} & a_{1,j+1} & \cdots & a_{1,d} \\
a_{2,1} & \cdots & a_{2,j-1} & a_{2,j+1} & \cdots & a_{2,d} \\
\vdots & & \vdots & \vdots & & \vdots \\
a_{d-1,1} & \cdots & a_{d-1,j-1} & a_{d-1,j+1} & \cdots & a_{d-1,d}
\end{bmatrix}, \qquad j = 1, 2, \ldots, d.
$$

An alternative means of defining $\det A$ is as

$$
\det A = \sum_{\sigma} (-1)^{|\sigma|} \prod_{j=1}^{d} a_{j,\sigma(j)},
$$

where the summation is carried across all $d!$ *permutations* σ of the numbers $1, 2, \ldots, d$. The *parity* $|\sigma|$ of the permutation σ is the minimal number of two-term exchanges that are needed to convert it to the unit permutation $i = (1, 2, \ldots, d)$.

Provided that $\det A \neq 0$, the $d \times d$ matrix A possesses a unique *inverse* A^{-1}; this is a $d \times d$ matrix such that $A^{-1}A = AA^{-1} = I$. An explicit definition of A^{-1} is

$$A^{-1} = \frac{\text{adj } A}{\det A},$$

where the (i, j)th component $b_{i,j}$ of the $d \times d$ *adjunct matrix* adj A is

$$b_{i,j} = (-1)^{i+j} \det \begin{bmatrix} a_{1,1} & \cdots & a_{1,j-1} & a_{1,j+1} & \cdots & a_{1,d} \\ \vdots & & \vdots & \vdots & & \vdots \\ a_{i-1,1} & \cdots & a_{i-1,j-1} & a_{i-1,j+1} & \cdots & a_{i-1,d} \\ a_{i+1,1} & \cdots & a_{i+1,j-1} & a_{i+1,j+1} & \cdots & a_{i+1,d} \\ \vdots & & \vdots & \vdots & & \vdots \\ a_{d,1} & \cdots & a_{d,j-1} & a_{d,j+1} & \cdots & a_{d,d} \end{bmatrix}, \quad i, j = 1, 2, \ldots, d.$$

A matrix A such that $\det A \neq 0$ is *nonsingular;* otherwise it is *singular*.

A.1.2.4   The *transpose* A^T of a $d \times n$ matrix A is an $n \times d$ matrix such that

$$A^T = \begin{bmatrix} a_{1,1} & a_{2,1} & \cdots & a_{n,1} \\ a_{1,2} & a_{2,2} & \cdots & a_{n,2} \\ \vdots & \vdots & & \vdots \\ a_{1,d} & a_{2,d} & \cdots & a_{n,d} \end{bmatrix}.$$

A.1.2.5   A square $d \times d$ matrix A is

- *diagonal* if $a_{j,\ell} = 0$ for every $j \neq \ell$, $j, \ell = 1, 2, \ldots, d$.

- *symmetric* if $A^T = A$;

- *Hermitian* or *self-adjoint* if A is complex and $\bar{A}^T = A$;

- *skew-symmetric* (or *anti-symmetric*) if $A^T = -A$;

- *skew-Hermitian* if A is complex and $\bar{A}^T = -A$;

- *orthogonal* if $A^T A = I$, the identity matrix, which is equivalent to

$$\sum_{\ell=1}^{d} a_{\ell,i} a_{\ell,j} = \begin{cases} 1, & i = j, \\ 0, & i \neq j, \end{cases} \quad i, j = 1, 2, \ldots, d.$$

Note that in this case $A^{-1} = A^T$ and that A^T is also orthogonal;

- *unitary* if A is complex and $\bar{A}^T A = I$;

- a *permutation matrix* if all its elements are either 0 or 1 and there is exactly one 1 in each row and column. The matrix–vector product $A\boldsymbol{x}$ permutes the elements of $\boldsymbol{x} \in \mathbb{R}^d$, while $A^T\boldsymbol{y}$ causes an inverse permutation – therefore $A^T = A^{-1}$ and A is orthogonal;

- *tridiagonal* if $a_{i,j} = 0$ for every $|i - j| \geq 2$, in other words

$$A = \begin{bmatrix} a_{1,1} & a_{1,2} & 0 & \cdots & & 0 \\ a_{2,1} & a_{2,2} & a_{2,3} & \ddots & & \vdots \\ 0 & \ddots & \ddots & \ddots & & 0 \\ \vdots & \ddots & & a_{d-1,d-2} & a_{d-1,d-1} & a_{d-1,d} \\ 0 & \cdots & & 0 & a_{d,d-1} & a_{d,d} \end{bmatrix};$$

- a *Vandermonde matrix* if

$$A = \begin{bmatrix} 1 & \xi_1 & \xi_1^2 & \cdots & \xi_1^{d-1} \\ 1 & \xi_2 & \xi_2^2 & \cdots & \xi_2^{d-1} \\ \vdots & \vdots & \vdots & & \vdots \\ 1 & \xi_d & \xi_d^2 & \cdots & \xi_d^{d-1} \end{bmatrix}$$

for some $\xi_1, \xi_2, \ldots, \xi_d \in \mathbb{C}$. It is true in this case that

$$\det A = \prod_{i=2}^{d} \prod_{j=1}^{i-1} (\xi_i - \xi_j),$$

hence a Vandermonde matrix is nonsingular if and only if $\xi_1, \xi_2, \ldots, \xi_d$ are distinct numbers;

- *normal* if $A^T A = A A^T$. Symmetric, skew-symmetric and orthogonal matrices are all normal.

A.1.3   Inner products and norms

A.1.3.1   An *inner product*, also known as a *scalar product*, is a function $\langle \cdot, \cdot \rangle :$ $\mathbb{R}^d \times \mathbb{R}^d \to \mathbb{R}$ with the following properties.

(1)   *Nonnegativity:* $\langle \boldsymbol{x}, \boldsymbol{x} \rangle \geq 0$ for every $\boldsymbol{x} \in \mathbb{R}^d$ and $\langle \boldsymbol{x}, \boldsymbol{x} \rangle = 0$ if and only if $\boldsymbol{x} = \boldsymbol{0}$, the *zero vector*.

(2)   *Linearity:* $\langle \alpha \boldsymbol{x} + \beta \boldsymbol{y}, \boldsymbol{z} \rangle = \alpha \langle \boldsymbol{x}, \boldsymbol{z} \rangle + \beta \langle \boldsymbol{y}, \boldsymbol{z} \rangle$ for all $\alpha, \beta \in \mathbb{R}$ and $\boldsymbol{x}, \boldsymbol{y}, \boldsymbol{z} \in \mathbb{R}^d$.

(3)   *Symmetry:* $\langle \boldsymbol{x}, \boldsymbol{y} \rangle = \langle \boldsymbol{y}, \boldsymbol{x} \rangle$ for all $\boldsymbol{x}, \boldsymbol{y} \in \mathbb{R}^d$.

(4)   *The Cauchy–Schwarz inequality:* For every $\boldsymbol{x}, \boldsymbol{y} \in \mathbb{R}^d$ it is true that

$$|\langle \boldsymbol{x}, \boldsymbol{y} \rangle| \leq [\langle \boldsymbol{x}, \boldsymbol{x} \rangle]^{1/2} [\langle \boldsymbol{y}, \boldsymbol{y} \rangle]^{1/2}.$$

A particular example is the *Euclidean* (or ℓ_2) inner product

$$\langle x, y \rangle = x^T y, \qquad x, y \in \mathbb{R}^d.$$

In the case of the complex-valued vector space $\mathbb{C}^d$, the symmetry axiom of the inner product needs to be replaced by $\langle x, y \rangle = \overline{\langle y, x \rangle}$, $x, y \in \mathbb{C}^d$, where the bar denotes conjugation, while the complex Euclidean inner product is

$$\langle x, y \rangle = \bar{x}^T y, \qquad x, y \in \mathbb{C}^d.$$

A.1.3.2   Two vectors $x, y \in \mathbb{R}^d$ such that $\langle x, y \rangle = 0$ are said to be *orthogonal*. A basis $\{y_1, y_2, \ldots, y_d\}$ constructed from orthogonal vectors is called an *orthogonal basis*, or, if $\langle y_\ell, y_\ell \rangle = 1$, $\ell = 1, 2, \ldots, d$, an *orthonormal basis*. The canonical basis $\{e_1, e_2, \ldots, e_d\}$ is orthonormal with respect to the Euclidean norm.

A.1.3.3   A vector *norm* is a function $\| \cdot \| : \mathbb{R}^d \to \mathbb{R}$ that obeys the following axioms.

(1)   *Nonnegativity:* $\|x\| \geq 0$ for every $x \in \mathbb{R}^d$ and $\|x\| = 0$ if and only if $x = \mathbf{0}$.

(2)   *Rescaling:* $\|\alpha x\| = |\alpha| \|x\|$ for every $\alpha \in \mathbb{R}$ and $x \in \mathbb{R}^d$.

(3)   *The triangle inequality:* $\|x + y\| \leq \|x\| + \|y\|$ for every $x, y \in \mathbb{R}^d$.

Any inner product $\langle \cdot, \cdot \rangle$ induces a norm $\|x\| = [\langle x, x \rangle]^{1/2}$, $x \in \mathbb{R}^d$. In particular, the Euclidean inner product induces the ℓ_2 norm, also known as the *Euclidean*, the *least squares* norm or *energy* norm, $\|x\| = (x^T x)^{1/2}$, $x \in \mathbb{R}^d$ (or $\|x\| = (\bar{x}^T x)^{1/2}$, $x \in \mathbb{C}^d$).

Not every vector norm is induced by an inner product. Well-known and useful examples are the ℓ_1 norm (also known as the *Manhattan* norm)

$$\|x\| = \sum_{\ell=1}^d |x_\ell|, \qquad x \in \mathbb{R}^d,$$

and the ℓ_∞ norm (also known as the *Chebyshev* norm, the *uniform* norm, the *max* norm or the *sup* norm),

$$\|x\| = \max_{\ell=1,2,\ldots,d} |x_\ell|, \qquad x \in \mathbb{R}^d.$$

A.1.3.4   Every vector norm $\| \cdot \|$ acting on $\mathbb{R}^d$ can be extended to a norm on $d \times d$ matrices, the *induced matrix norm*, by letting

$$\|A\| = \max_{\substack{x \in \mathbb{R}^d \\ x \neq 0}} \frac{\|Ax\|}{\|x\|} = \max_{\substack{x \in \mathbb{R}^d \\ \|x\|=1}} \|Ax\|.$$

It is always true that

$$\|Ax\| \leq \|A\| \times \|x\|, \qquad x \in \mathbb{R}^d$$

and
$$\|AB\| \leq \|A\| \times \|B\|,$$
where both A and B are $d \times d$ matrices.

The Euclidean norm of an orthogonal matrix always equals unity.

A.1.3.5   A $d \times d$ symmetric matrix A is said to be *positive definite* if
$$\langle Ax, x \rangle > 0 \qquad x \in \mathbb{R}^d \setminus \{0\}$$
and *negative definite* if the above inequality is reversed. It is *nonnegative definite* or *nonpositive definite* if
$$\langle Ax, x \rangle \geq 0 \qquad \text{or} \qquad \langle Ax, x \rangle \leq 0$$
for all $x \in \mathbb{R}^d$, respectively.

A.1.4   Linear systems

A.1.4.1   The *linear system*
$$a_{1,1}x_1 + a_{1,2}x_2 + \cdots + a_{1,d}x_d = b_1,$$
$$a_{2,1}x_1 + a_{2,2}x_2 + \cdots + a_{2,d}x_d = b_2,$$
$$\vdots$$
$$a_{d,1}x_1 + a_{d,2}x_2 + \cdots + a_{d,d}x_d = b_d,$$
is written in vector notation as $Ax = b$. It possesses a unique solution $x = A^{-1}b$ if and only if A is nonsingular.

A.1.4.2   If a square matrix A is singular then there exists a nonzero solution to the *homogeneous* linear system $Ax = 0$. The *kernel* of A, denoted by $\ker A$, is the set of all $x \in \mathbb{R}^d$ such that $Ax = 0$. If A is nonsingular then $\ker A = \{0\}$, otherwise $\ker A$ is a subspace of $\mathbb{R}^d$ of dimension ≥ 1.

Recall that a $d \times d$ matrix A is a linear transformation mapping $\mathbb{R}^d$ into itself. If A is nonsingular ($\Leftrightarrow \det A \neq 0 \Leftrightarrow \ker A = \{0\}$) then this mapping is an *isomorphism* (in other words, it has a well-defined and unique inverse linear transformation A^{-1}, acting on the image $A\mathbb{R}^d := \{Ax : x \in \mathbb{R}^d\}$ and mapping it back to $\mathbb{R}^d$) *on* $\mathbb{R}^d$ (i.e., the image $A\mathbb{R}^d$ is all of $\mathbb{R}^d$). However, if A is singular ($\Leftrightarrow \det A = 0 \Leftrightarrow \dim \ker A \geq 1$) then $A\mathbb{R}^d$ is a proper vector subspace of $\mathbb{R}^d$ and $\dim(A\mathbb{R}^d) = d - \dim \ker A \leq d - 1$.

A.1.4.3   The practical solution of linear systems such as the above can be performed by *Gaussian elimination*. We commence by subtracting from the ℓth equation, $\ell = 2, 3, \ldots, d$, the product of the first equation by the real number $a_{\ell,1}/a_{1,1}$. This does not change the solution x of the linear system, while setting zeros in the first column, except in the first equation, and replacing the system by
$$a_{1,1}x_1 + a_{1,2}x_2 + \cdots + a_{1,d}x_d = b_1,$$
$$\tilde{a}_{2,2}x_2 + \cdots + \tilde{a}_{2,d} = \tilde{b}_2,$$
$$\vdots$$
$$\tilde{a}_{d,2}x_2 + \cdots + \tilde{a}_{d,d} = \tilde{b}_d,$$

where

$$\tilde{a}_{\ell,j} = a_{\ell,j} - \frac{a_{1,1}}{a_{\ell,1}} a_{1,j}, \quad j = 2, 3, \ldots, d, \qquad \tilde{b}_\ell = b_\ell - \frac{a_{1,1}}{a_{\ell,1}} b_1, \qquad \ell = 2, 3, \ldots, d.$$

The unknown x_1 does not feature in equations $2, 3, \ldots, d$ because it has been *eliminated;* the latter thereby constitute a set of $d - 1$ equations in $d - 1$ unknowns. Continuing this process by induction results in an *upper triangular* linear system:

$$\begin{aligned}
a_{1,1}^{(1)} x_1 + a_{1,2}^{(1)} x_2 + a_{1,3}^{(1)} x_3 + \cdots + a_{1,d}^{(1)} x_d &= b_1^{(1)}, \\
a_{2,2}^{(2)} x_2 + a_{2,3}^{(2)} x_3 + \cdots + a_{2,d}^{(2)} x_d &= b_2^{(2)}, \\
a_{3,3}^{(3)} x_3 + \cdots + a_{3,d}^{(3)} x_d &= b_3^{(3)} \\
&\vdots \\
a_{d,d}^{(d)} x_d &= b_d^{(d)},
\end{aligned}$$

where $a_{1,j}^{(1)} = a_{1,j}$, $a_{2,j}^{(2)} = \tilde{a}_{2,j}$, $a_{3,j}^{(3)} = a_{3,j}^{(2)} - \left(a_{2,2}^{(2)} / a_{3,2}^{(2)} \right) a_{2,j}^{(2)}$ etc.

The upper triangular system is solved successively from the bottom:

$$x_d = \frac{1}{a_{d,d}^{(d)}} b_d^{(d)},$$

$$x_{d-1} = \frac{1}{a_{d-1,d-1}^{(d-1)}} \left[b_{d-1}^{(d-1)} - a_{d-1,d}^{(d)} x_d \right],$$

$$\vdots$$

$$x_1 = \frac{1}{a_{1,1}^{(1)}} \left[b_1^{(1)} - \sum_{j=2}^{d} a_{1,j}^{(1)} x_j \right].$$

The whole procedure depends on the *pivots* $a_{\ell,\ell}^{(\ell)}$ being nonzero, otherwise it cannot be carried out in this fashion. It is perfectly possible for a pivot to vanish even if A is nonsingular and (with few important exceptions) it is impractical to determine whether a pivot vanishes by inspecting the elements of A. Moreover, if some pivots are exceedingly small (in modulus), even if none vanishes, large rounding errors can be introduced by computer arithmetic, thereby destroying the precision of Gaussian elimination and rendering it unusable.

A.1.4.4   A standard means of preventing pivots becoming small is *column pivoting.* Instead of eliminating the ℓth equation at the ℓth stage, we first search for $m \in \{\ell, \ell + 1, \ldots, d\}$ such that $|a_{m,\ell}^{(\ell)}| \geq |a_{i,\ell}^{(\ell)}|$ for all $i = \ell, \ell + 1, \ldots, d$, interchange the ℓth and the ith equations and only then eliminate.

A.1.4.5   An alternative formulation of Gaussian elimination is by means of the *LU factorization* $A = LU$, where the $d \times d$ matrices L and U are *lower* and *upper* *triangular*, respectively, and all diagonal elements of L equal unity,

$$L = \begin{bmatrix} 1 & 0 & \cdots & & \cdots & 0 \\ \ell_{2,1} & 1 & 0 & & & \vdots \\ \vdots & \ddots & \ddots & & \ddots & \vdots \\ \ell_{d-1,1} & \cdots & \ell_{d-1,d-2} & 1 & & 0 \\ \ell_{d,1} & \cdots & \ell_{d,d-2} & \ell_{d,d-1} & 1 \end{bmatrix} \quad \text{and} \quad U = \begin{bmatrix} u_{1,1} & u_{1,2} & \cdots & & \cdots & u_{1,d} \\ 0 & u_{2,2} & u_{2,3} & & & \vdots \\ \vdots & \ddots & \ddots & \ddots & & \vdots \\ 0 & \cdots & 0 & & u_{d-1,d-1} & u_{d-1,d} \\ 0 & \cdots & & \cdots & 0 & u_{d,d} \end{bmatrix}.$$

Provided that an LU factorization of A is known, we replace the linear system $A\boldsymbol{x} = \boldsymbol{b}$ by the two systems $L\boldsymbol{y} = \boldsymbol{b}$ and $U\boldsymbol{x} = \boldsymbol{y}$. The first is lower triangular, hence

$$y_1 = b_1,$$
$$y_2 = b_2 - \ell_{2,1}y_1,$$
$$\vdots$$
$$y_d = b_d - \sum_{j=1}^{d-1} \ell_{d,j}y_j,$$

while the second is upper triangular – and we have already seen in A.1.4.3 how to solve upper triangular linear systems.

The practical evaluation of an LU factorization can be done explicitly, by letting, consecutively for $k = 1, 2, \ldots, d$,

$$u_{k,j} = a_{k,j} - \sum_{i=1}^{j-1} \ell_{k,i}u_{i,j}, \qquad j = k, k+1, \ldots, d,$$

$$\ell_{j,k} = \frac{1}{u_{k,k}} \left(a_{j,k} - \sum_{i=1}^{k-1} \ell_{j,i}u_{i,k} \right), \qquad j = k+1, k+2, \ldots, d.$$

As always, empty sums are nil and empty number ranges are disregarded.

LU factorization, like Gaussian elimination, is prone to failure due to small values of the pivots $u_{1,1}, u_{2,2}, \ldots, u_{d,d}$ and the remedy – column pivoting – is identical. After all, LU factorization is but a recasting of Gaussian elimination into a form that is more convenient for various applications.

A.1.4.6   A positive definite matrix A can be factorized into the product $A = LL^T$, where L is lower triangular with $\ell_{j,j} > 0$, $j = 1, 2, \ldots, d$; this is the *Cholesky factorization*. It can be evaluated explicitly via

$$\ell_{k,j} = \frac{1}{\ell_{j,j}} \left(a_{k,j} - \sum_{i=1}^{j-1} \ell_{k,i}\ell_{j,i} \right), \qquad j = 1, 2, \ldots, k-1,$$

$$\ell_{k,k} = \left(a_{k,k} - \sum_{i=1}^{k-1} \ell_{k,i}^2 \right)^{1/2}$$

for $k = 1, 2, \ldots, d$.

Having evaluated a Cholesky factorization, the linear system $A\boldsymbol{x} = \boldsymbol{b}$ can be solved by a sequential evaluation of two tridiagonal systems, firstly $L\boldsymbol{y} = \boldsymbol{b}$ and then $L^T\boldsymbol{x} = \boldsymbol{y}$.

An advantage of Cholesky factorization is that it requires half the storage and half the computational cost of an LU factorization.

A.1.5 Eigenvalues and eigenvectors

A.1.5.1 We say that $\lambda \in \mathbb{C}$ is an *eigenvalue* of the $d \times d$ matrix A if $\det(A - \lambda I) = 0$. The set of all eigenvalues of a square matrix A is called the *spectrum* and denoted by $\sigma(A)$.

Each $d \times d$ matrix has exactly d eigenvalues. All the eigenvalues of a symmetric matrix are real, all the eigenvalues of a skew-symmetric matrix are pure imaginary and, in general, the eigenvalues of a real matrix are either real or form complex conjugate pairs.

If all the eigenvalues of a symmetric matrix are positive then it is positive definite. A similarly worded statement extends to negative, nonnegative and nonpositive matrices.

We say that an eigenvalue is of *algebraic multiplicity* $r \geq 1$ if it is a zero of multiplicity r of the *characteristic polynomial* $p(z) = \det(A - zI)$; in other words, if

$$p(\lambda) = \frac{\mathrm{d}p(\lambda)}{\mathrm{d}z} = \cdots = \frac{\mathrm{d}^{r-1}p(\lambda)}{\mathrm{d}z^{r-1}} = 0, \qquad \frac{\mathrm{d}^r p(\lambda)}{\mathrm{d}z^r} \neq 0.$$

An eigenvalue of algebraic multiplicity one is said to be *distinct*.

A.1.5.2 The *spectral radius* of A is a nonnegative number

$$\rho(A) = \max\{|\lambda| \, : \, \lambda \in \sigma(A)\}.$$

It always obeys the inequality

$$\rho(A) \leq \|A\|$$

(where $\| \cdot \|$ is the Euclidean norm) but $\rho(A) = \|A\|$ for a normal matrix A. It is possible to express the Euclidean norm of a general square matrix A in the form

$$\|A\| = [\rho(A^T A)]^{1/2}.$$

A.1.5.3 If $\lambda \in \sigma(A)$ it follows that $\dim \ker (A - \lambda I) \geq 1$, therefore there are nonzero vectors in the *eigenspace* $\ker (A - \lambda I)$. Each such vector is called an *eigenvector* of A corresponding to the eigenvalue λ. An alternative formulation is that $\boldsymbol{v} \in \mathbb{R}^d \setminus \{\boldsymbol{0}\}$ is an eigenvector of A, corresponding to $\lambda \in \sigma(A)$, if $A\boldsymbol{v} = \lambda\boldsymbol{v}$. Note that even if A is real, its eigenvectors – like its eigenvalues – may be complex.

The *geometric multiplicity* of λ is the dimension of its eigenspace and it is always true that

$$1 \leq \text{geometric multiplicity} \leq \text{algebraic multiplicity}.$$

If the geometric and algebraic multiplicities are equal for all its eigenvalues, A is said to have a *complete set of eigenvectors*. Since different eigenspaces are linearly independent and the sum of algebraic multiplicities is always d, a matrix possessing a complete set of eigenvectors provides a basis of $\mathbb{R}^d$ formed from its eigenvectors – specifically, the assembly of all bases of its eigenspaces.

If all the eigenvalues of A are distinct then it has a complete set of eigenvectors. A normal matrix also shares this feature and, moreover, it always has an orthogonal basis of eigenvectors.

A.1.5.4   If a $d \times d$ matrix A has a complete set of eigenvectors then it possesses the *spectral factorization*

$$A = VDV^{-1}.$$

Here D is a diagonal matrix and $d_{\ell,\ell} = \lambda_\ell$, $\sigma(A) = \{\lambda_1, \lambda_2, \ldots, \lambda_d\}$, the ℓth column of the $d \times d$ matrix V is an eigenvector in the eigenspace of λ_ℓ and the columns of V are selected so that $\det V \neq 0$; in other words so that the columns form a basis of $\mathbb{R}^d$. This is possible according to A.1.5.3.

If A is normal, it is possible to normalize its eigenvectors (specifically, by letting them be of unit Euclidean norm) so that V is an orthogonal matrix.

Let two $d \times d$ matrices A and B share a complete set of eigenvectors then $A = VD_AV^{-1}$, $B = VD_BV^{-1}$, say. Since diagonal matrices always commute,

$$AB = (VD_AV^{-1})(VD_BV^{-1}) = (VD_A)(V^{-1}V)(D_BV^{-1}) = V(D_AD_B)V^{-1}$$
$$= V(D_BD_A)V^{-1} = (VD_B)(V^{-1}V)(D_AV^{-1}) = (VD_BV^{-1})(VD_AV^{-1}) = BA$$

and the matrices A and B commute as well.

A.1.5.5   Let

$$f(z) = \sum_{k=0}^{\infty} f_k z^k, \qquad z \in \mathbb{C},$$

be an arbitrary power series that converges for all $|z| \leq \rho(A)$, where A is a $d \times d$ matrix. The matrix function

$$f(A) := \sum_{k=0}^{\infty} f_k A^k$$

then converges. Moreover, if $\lambda \in \sigma(A)$ and v is in the eigenspace of λ then $f(A)v = f(\lambda)v$. In particular,

$$\sigma(f(A)) = \{f(\lambda_j) : \lambda_j \in \sigma(A), \quad j = 1, 2, \ldots, d\}.$$

If A has a spectral factorization $A = VDV^{-1}$ then $f(A)$ factorizes as follows

$$f(A) = Vf(D)V^{-1}.$$

A.1.5.6   Every $d \times d$ matrix A possesses a *Jordan factorization*

$$A = W\Lambda W^{-1},$$

where $\det W \neq 0$ and the $d \times d$ matrix Λ can be written in block form as

$$
\Lambda = \begin{bmatrix}
\Lambda_1 & O & \cdots & O \\
O & \Lambda_2 & \ddots & \vdots \\
\vdots & \ddots & \ddots & O \\
O & \cdots & O & \Lambda_s
\end{bmatrix}.
$$

Here $\lambda_1, \lambda_2, \ldots, \lambda_s \in \sigma(A)$ and the kth *Jordan block* is

$$
\Lambda_k = \begin{bmatrix}
\lambda_k & 1 & 0 & \cdots & 0 \\
0 & \lambda_k & 1 & \ddots & \vdots \\
\vdots & \ddots & \ddots & \ddots & 0 \\
\vdots & & \ddots & \lambda_k & 1 \\
0 & \cdots & \cdots & 0 & \lambda_k
\end{bmatrix}, \qquad k = 1, 2, \ldots, s.
$$

Bibliography

Halmos, P.R. (1958), *Finite-Dimensional Vector Spaces,* van Nostrand–Reinhold, Princeton, NJ.

Lang, S. (1986), *Introduction to Linear Algebra* (2nd ed.), Springer-Verlag, New York.

Strang, G. (1988), *Linear Algebra and its Applications* (3rd ed.), Harcourt Brace Jovanovich, San Diego, CA.

A.2   Analysis

A.2.1   Introduction to functional analysis

A.2.1.1   A *linear space* is an arbitrary collection of objects closed under addition and multiplication by a scalar. In other words, $\mathbb{V}$ is a linear space over the field $\mathbb{F}$ (*a scalar field*) if there exist operations $+ : \mathbb{V} \times \mathbb{V} \to \mathbb{V}$ and $\cdot : \mathbb{F} \times \mathbb{V} \to \mathbb{V}$, consistent with the following axioms (for clarity we denote the elements of $\mathbb{V}$ as vectors):

(1)   *Commutativity:* $x + y = y + x$ for every $x, y \in \mathbb{V}$.

(2)   *Existence of zero:*   There exists a unique element $0 \in \mathbb{V}$ such that $x + 0 = x$ for every $x \in \mathbb{V}$.

(3)   *Existence of inverse:*   For every $x \in \mathbb{V}$ there exists a unique element $-x \in \mathbb{V}$ such that $x + (-x) = 0$.

(4)   *Associativity:* $(x + y) + z = x + (y + z)$ for every $x, y, z \in \mathbb{V}$ [*These four axioms mean that* $\mathbb{V}$ *is an abelian group with respect to addition*].

(5)   *Interchange of multiplication:* $\alpha(\beta x) = (\alpha\beta)x$ for every $\alpha, \beta \in \mathbb{F},\ x \in \mathbb{V}$.

(6) *Action of unity:* $1\boldsymbol{x} = \boldsymbol{x}$ for every $\boldsymbol{x} \in \mathbb{V}$, where 1 is the unit element of $\mathbb{F}$.

(7) *Distributivity:* $\alpha(\boldsymbol{x} + \boldsymbol{y}) = \alpha\boldsymbol{x} + \alpha\boldsymbol{y}$ and $(\alpha + \beta)\boldsymbol{x} = \alpha\boldsymbol{x} + \beta\boldsymbol{x}$ for every $\alpha, \beta \in \mathbb{F}$, $\boldsymbol{x}, \boldsymbol{y} \in \mathbb{V}$.

If $\mathbb{V}_1, \mathbb{V}_2$ are both linear spaces over the same field $\mathbb{F}$ and $\mathbb{V}_1 \subseteq \mathbb{V}_2$, the space $\mathbb{V}_1$ is said to be a *subspace* of $\mathbb{V}_2$.

A.2.1.2   We say that a linear space is of *dimension* $d < \infty$ if it has a *basis* of d linearly independent elements (cf. A.1.1.2). If no such basis exists, the linear space is said to be of infinite dimension.

A.2.1.3   We assume for simplicity that $\mathbb{F} = \mathbb{R}$, although generalization to the complex field presents no difficulty whatsoever.

A familiar example of a linear space is the d-dimensional vector space $\mathbb{R}^d$ (cf. A.1.1.1). Another example is the set $\mathbb{P}_\nu$ of all polynomials of degree $\leq \nu$ with real coefficients. It is not difficult to verify that $\dim \mathbb{P}_\nu = \nu + 1$.

More interesting examples of linear spaces are the set $C[0, 1]$ of all continuous real functions in the interval $[0, 1]$, the set of all real power series with a positive radius of convergence at the origin and the set of all *Fourier expansions* (that is, linear combinations of 1 and $\cos nx, \sin nx$ for $n = 1, 2, \ldots$). All these spaces are infinite-dimensional.

A.2.1.4   Inner products and norms over linear spaces are defined exactly as in A.1.3.1 and A.1.3.3 respectively.

A linear space equipped with a norm is called a *normed space*. If a normed space $\mathbb{V}$ is *closed* (that is, all Cauchy sequences converge in $\mathbb{V}$), it is said to be a *Banach space*.

An important example of a normed space is provided when a function is measured by a *p-norm*. The latter is defined by

$$\|f\|_p = \left(\int_\Omega |f(\boldsymbol{x})|^p \, \mathrm{d}\boldsymbol{x} \right)^{1/p}, \qquad 1 \leq p < \infty, \qquad f \in \mathbb{V},$$

$$\|f\|_\infty = \sup_{\boldsymbol{x} \in \Omega} |f(\boldsymbol{x})|,$$

where Ω is the domain of definition of the functions (in general multivariate). A normed space equipped with the p-norm is denoted by $L_p(\Omega)$. If Ω is a closed set, $L_p(\Omega)$ is a Banach space.

A.2.1.5   A closed linear space equipped with an inner product is called a *Hilbert space*. An important example is provided by the inner product

$$\langle f, g \rangle = \int_\Omega f(\boldsymbol{x}) g(\boldsymbol{x}) \, \mathrm{d}\boldsymbol{x}, \qquad f, g \in \mathbb{V},$$

where Ω is a closed set (if the space is over complex numbers, rather than reals, $g(\boldsymbol{x})$ needs to be replaced by $\overline{g(\boldsymbol{x})}$). It induces the *Euclidean* norm (the 2-norm)

$$\|f\| = \left(\int_\Omega |f(\boldsymbol{x})|^2 \, \mathrm{d}\boldsymbol{x} \right)^{1/2}, \qquad f \in \mathbb{V}.$$

A.2.1.6   The Hilbert space $L_2(\Omega)$ is said to be *separable* if it has either a finite or a countable orthogonal basis, $\{\varphi_i\}$, say. (In infinite-dimensional spaces the set $\{\varphi_i\}$ is a basis if each element lies in the *closure* of linear combinations from $\{\varphi_i\}$. The closure is, of course, defined by the underlying norm.)

The space $L_2(\Omega)$ (denoted also by $L(\Omega)$ or $L[\Omega]$) is separable with a countable basis.

A.2.1.7   Let $\mathbb{V}$ be a Hilbert space. Two elements $f, g \in \mathbb{V}$ are *orthogonal* if $\langle f, g \rangle = 0$. If

$$\langle f, g \rangle = 0, \qquad g \in \mathbb{V}_1,$$

where $\mathbb{V}_1$ is a subspace of the Hilbert space $\mathbb{V}$ and $f \in \mathbb{V}$, then f is said to be *orthogonal* to $\mathbb{V}_1$.

A.2.1.8   A mapping from a linear space $\mathbb{V}_1$ to a linear space $\mathbb{V}_2$ is called an *operator*. An operator $\mathcal{T}$ is *linear* if

$$\mathcal{T}(\boldsymbol{x} + \boldsymbol{y}) = \mathcal{T}\boldsymbol{x} + \mathcal{T}\boldsymbol{y}, \qquad \mathcal{T}(\alpha\boldsymbol{x}) = \alpha\mathcal{T}\boldsymbol{x}, \qquad \alpha \in \mathbb{R}, \quad \boldsymbol{x}, \boldsymbol{y} \in \mathbb{V}.$$

Let $\mathcal{T}$ be a linear operator from a Banach space $\mathbb{V}$ to itself. The *norm* of $\mathcal{T}$ is defined as

$$\|\mathcal{T}\| = \sup_{\substack{\boldsymbol{x} \in \mathbb{V} \\ \boldsymbol{x} \neq \boldsymbol{0}}} \frac{\|\mathcal{T}\boldsymbol{x}\|}{\|\boldsymbol{x}\|} = \sup_{\substack{\boldsymbol{x} \in \mathbb{V} \\ \|\boldsymbol{x}\|=1}} \|\mathcal{T}\boldsymbol{x}\|.$$

It is always true that

$$\|\mathcal{T}\boldsymbol{x}\| \leq \|\mathcal{T}\| \times \|\boldsymbol{x}\|, \qquad \boldsymbol{x} \in \mathbb{V},$$

and

$$\|\mathcal{T}\mathcal{S}\| \leq \|\mathcal{T}\| \times \|\mathcal{S}\|,$$

where both $\mathcal{T}$ and $\mathcal{S}$ are linear operators from $\mathbb{V}$ to itself.

A.2.1.9   The *domain* of a linear operator $\mathcal{T} : \mathbb{V}_1 \to \mathbb{V}_2$ is the Banach space $\mathbb{V}_1$, while its *range* is

$$\mathcal{T}\mathbb{V}_1 = \{\mathcal{T}\boldsymbol{x} : \boldsymbol{x} \in \mathbb{V}_1\} \subseteq \mathbb{V}_2.$$

- If $\mathcal{T}\mathbb{V}_1 = \mathbb{V}_2$, the operator $\mathcal{T}$ is said to map $\mathbb{V}_1$ *onto* $\mathbb{V}_2$.

- If for every $\boldsymbol{y} \in \mathcal{T}\mathbb{V}_1$ there exists a unique $\boldsymbol{x} \in \mathbb{V}_1$ such that $\mathcal{T}\boldsymbol{x} = \boldsymbol{y}$ then $\mathcal{T}$ is said to be an *isomorphism* (or an *injection*) and the linear operator $\mathcal{T}^{-1}\boldsymbol{y} = \boldsymbol{x}$ is the *inverse* of $\mathcal{T}$.

- If $\mathcal{T}$ is an isomorphism and $\mathcal{T}\mathbb{V}_1 = \mathbb{V}_2$ then it is said to be an *isomorphism onto* $\mathbb{V}_2$ or a *bijection*.

A.2.1.10   A mapping from a linear space to the reals (or to the complex numbers) is called a *functional*. A functional L is *linear* if $L(\boldsymbol{x} + \boldsymbol{y}) = L\boldsymbol{x} + L\boldsymbol{y}$ and $L(\alpha\boldsymbol{x}) = \alpha L\boldsymbol{x}$ for all $\boldsymbol{x}, \boldsymbol{y} \in \mathbb{V}$ and scalar α.

A.2.2 Approximation theory

A.2.2.1 Denote by $\mathbb{P}_\nu$ the set of all polynomials with real coefficients of degree $\leq \nu$.

Given $\nu + 1$ distinct points $\xi_0, \xi_1, \ldots, \xi_\nu$ and $f_0, f_1, \ldots, f_\nu$, there exists a unique $p \in \mathbb{P}_\nu$ such that

$$p(\xi_\ell) = f_\ell, \qquad \ell = 0, 1, \ldots, \nu.$$

It is called the *interpolation polynomial*.

A.2.2.2 Suppose that $f_\ell = f(\xi_\ell)$, $\ell = 0, 1, \ldots, \nu$, where f is a $\nu + 1$ times differentiable function. Let $a = \min_{i=0,1,\ldots,\nu} \xi_i$ and $b = \max_{i=0,1,\ldots,\nu} \xi_i$. Then for every $x \in [a, b]$ there exists $\eta = \eta(x) \in [a, b]$ such that

$$p(x) - f(x) = \frac{1}{(\nu + 1)!} f^{(\nu+1)}(\eta) \prod_{k=0}^{\nu} (x - \xi_k).$$

[*This is an extension of the familiar Taylor remainder formula and it reduces to the latter if $\xi_0, \xi_1, \ldots, \xi_\nu \to \xi^* \in (a, b)$.*]

A.2.2.3 An obvious way of evaluating an interpolation polynomial is by solving the *interpolation equations*. Let $p(x) = \sum_{k=0}^{\nu} p_k x^k$. The interpolation conditions can be written as

$$\sum_{k=0}^{\nu} p_k \xi_\ell^k = f_\ell, \qquad \ell = 0, 1, \ldots, \nu,$$

and this is a linear system with a nonsingular Vandermonde matrix (cf. A.1.2.5).

An explicit means of writing down the interpolation polynomial p is provided by the *Lagrange interpolation formula*

$$p(x) = \sum_{k=0}^{\nu} p_k(x) f_k, \qquad x \in \mathbb{R},$$

where each *Lagrange polynomial* $p_k \in \mathbb{P}_\nu$ is defined by

$$p_k(x) = \prod_{\substack{j=0 \\ j \neq k}}^{\nu} \frac{x - x_j}{x_k - x_j}, \qquad k = 0, 1, \ldots, \nu, \quad x \in \mathbb{R}.$$

A.2.2.4 An alternative method of evaluating the interpolation polynomial is the *Newton formula*

$$p(x) = \sum_{k=0}^{\nu} f[\xi_0, \xi_1, \ldots, \xi_k] \prod_{j=0}^{k-1} (x - \xi_k), \qquad x \in \mathbb{R},$$

where the definition of the *divided differences* $f[\xi_{i_0}, \xi_{i_1}, \ldots, \xi_{i_k}]$ is given by recursion,

$$f[\xi_i] = f_i, \qquad i = 0, 1, \ldots, \nu,$$

$$f[\xi_{i_0}, \xi_{i_1}, \ldots, \xi_{i_k}] = \frac{f[\xi_{i_1}, \xi_{i_2}, \ldots, \xi_{i_k}] - f[\xi_{i_0}, \xi_{i_1}, \ldots, \xi_{i_{k-1}}]}{\xi_{i_k} - \xi_{i_0}},$$

where $i_0, i_1, \ldots, i_k \in \{0, 1, \ldots, \nu\}$ are pairwise distinct.

An equivalent definition of divided differences is that $f[\xi_0, \xi_1, \ldots, \xi_\nu]$ is the coefficient of x^ν in the interpolation polynomial p.

A.2.2.5 The practical evaluation of $f[\xi_0, \xi_1, \ldots, \xi_k]$, $k = 0, 1, \ldots, \nu$ is done in a recursive fashion and it employs a *table of divided differences*, shown below. Only the underlined divided differences are required for the Newton interpolation formula.

The cost is $\mathcal{O}(\nu^2)$ operations.

An added advantage of the above procedure is that only 2ν numbers need be stored, provided that overwriting is used.

A.2.2.6 Let L be a linear functional from the linear space $C^{\nu+1}[a, b]$ of functions that are defined in the interval $[a, b]$ and possess $\nu + 1$ continuous derivatives there. Let us suppose that

$$L \int_a^b f(x, \xi) \, \mathrm{d}\xi = \int_a^b Lf(x, \xi) \, \mathrm{d}\xi$$

for any bivariate function f such that $f(\cdot, \xi), f(x, \cdot) \in C^{\nu+1}[a, b]$ and that L *annihilates* all polynomials of degree $\leq \nu$,

$$Lp = 0, \qquad p \in \mathbb{P}_\nu.$$

The *Peano kernel* of L is the function

$$k(\xi) := L[(x - \xi)_+^\nu], \qquad \xi \in [a, b],$$

where

$$t_+^m := \begin{cases} t^m, & t \geq 0, \\ 0, & t < 0. \end{cases}$$

The *Peano kernel theorem* states that

$$Lf = \frac{1}{\nu!} \int_a^b k(\xi) f^{(\nu+1)}(\xi) \, \mathrm{d}\xi, \qquad f \in C^{\nu+1}[a, b].$$

The following bounds on the magnitude of Lf can be deduced from the Peano kernel theorem:

$$|Lf| \leq \frac{1}{\nu!} \|k\|_1 \times \|f^{(\nu+1)}\|_\infty;$$

$$|Lf| \leq \frac{1}{\nu!} \|k\|_\infty \times \|f^{(\nu+1)}\|_1;$$

$$|Lf| \leq \frac{1}{\nu!} \|k\|_2 \times \|f^{(\nu+1)}\|_2,$$

where $\| \cdot \|_1$, $\| \cdot \|_2$ and $\| \cdot \|_\infty$ denote the 1-norm, the 2-norm and the ∞-norm, respectively (A.2.1.4).

A.2.2.7 In practice, the main application of the Peano kernel theorem is in estimating approximation errors. For example, suppose that we wish to approximate

$$\int_0^1 f(\xi) \, \mathrm{d}\xi \approx \tfrac{1}{6} \left[f(0) + 4f(\tfrac{1}{2}) + f(1) \right].$$

Letting

$$Lf = \int_0^1 f(\xi) \, \mathrm{d}\xi - \tfrac{1}{6} \left[f(0) + 4f(\tfrac{1}{2}) + f(1) \right]$$

we verify that L annihilates $\mathbb{P}_3$, therefore $\nu = 3$. The Peano kernel is

$$k(\xi) = L[(x-\xi)_+^3] = \int_\xi^1 (x-\xi)^3 \, \mathrm{d}x - \tfrac{1}{6} \left[4(\tfrac{1}{2} - \xi)_+^3 + (1-\xi)^3 \right]$$

$$= \begin{cases} -\frac{1}{12}\xi^3(2 - 3\xi), & 0 \leq \xi \leq \tfrac{1}{2}, \\ -\frac{1}{12}(1-\xi)^3(3\xi - 1), & \tfrac{1}{2} \leq \xi \leq 1. \end{cases}$$

Therefore

$$\|k\|_1 = \frac{1}{480}, \qquad \|k\|_2 = \frac{\sqrt{14}}{1344}, \qquad \|k\|_\infty = \frac{1}{192}$$

and we derive the following upper bounds on the error,

$$|Lf| \leq \frac{1}{1152}\|f^{(iv)}\|_1, \quad \frac{\sqrt{14}}{8064}\|f^{(iv)}\|_2, \quad \frac{1}{2880}\|f^{(iv)}\|_\infty, \qquad f \in C^4[0,1].$$

A.2.3 Ordinary differential equations

A.2.3.1 Let the function $\boldsymbol{f} : [t_0, t_0 + a] \times \mathbb{U} \to \mathbb{R}^d$, where $\mathbb{U} \subseteq \mathbb{R}^d$, be continuous in the cylinder

$$\mathbb{S} = \{(t, \boldsymbol{x}) : t \in [t_0, t_0 + a], \ \boldsymbol{x} \in \mathbb{R}^d, \ \|\boldsymbol{x} - \boldsymbol{y}_0\| \leq b\}$$

where $a, b > 0$ and the vector norm $\| \cdot \|$ is given. Then, according to the *Peano theorem* (not to be confused with the Peano kernel theorem), the ordinary differential equation

$$\boldsymbol{y}' = \boldsymbol{f}(t, \boldsymbol{y}), \quad t \in [t_0, t_0 + \alpha], \qquad \boldsymbol{y}(t_0) = \boldsymbol{y}_0 \in \mathbb{R}^d,$$

where

$$\alpha = \min\left\{a, \frac{b}{\mu}\right\} \qquad \text{and} \qquad \mu = \sup_{(t,\boldsymbol{x})\in\mathbb{S}} \|\boldsymbol{f}(t,\boldsymbol{x})\|,$$

possesses at least one solution.

A.2.3.2   We employ the same notation as in A.2.3.1.

A function $\boldsymbol{f} : [t_0, t_0 + a] \times \mathbb{U} \to \mathbb{R}^d$, where $\mathbb{U} \subseteq \mathbb{R}^d$, is said to be *Lipschitz continuous* (with respect to a vector norm $\|\cdot\|$ acting on $\mathbb{R}^d$) if there exists a number $\lambda \geq 0$, a *Lipschitz constant*, such that

$$\|\boldsymbol{f}(t,\boldsymbol{x}) - \boldsymbol{f}(t,\boldsymbol{y})\| \leq \lambda\|\boldsymbol{x} - \boldsymbol{y}\|, \qquad \boldsymbol{x}, \boldsymbol{y} \in \mathbb{S}.$$

The *Picard–Lindelöf theorem* states that, subject to both continuity and Lipschitz continuity of $\boldsymbol{f}$ in the cylinder $\mathbb{S}$, the ordinary differential equation has a unique solution in $[t_0, t_0 + \alpha]$.

If $\boldsymbol{f}$ is smoothly differentiable in $\mathbb{S}$ then we may set

$$\lambda = \max_{(t,\boldsymbol{x})\in\mathbb{S}} \left\| \frac{\partial \boldsymbol{f}(t,\boldsymbol{x})}{\partial \boldsymbol{x}} \right\|,$$

therefore smooth differentiability is sufficient for existence and uniqueness of the solution.

A.2.3.3   The linear system

$$\boldsymbol{y}' = A\boldsymbol{y}, \quad t \geq t_0, \qquad \boldsymbol{y}(t_0) = \boldsymbol{y}_0,$$

always has a unique solution.

Suppose that the $d \times d$ matrix A possesses the spectral factorization $A = VDV^{-1}$ (cf. A.1.5.4). Then there exist vectors $\boldsymbol{\alpha}_1, \boldsymbol{\alpha}_2, \ldots, \boldsymbol{\alpha}_d \in \mathbb{R}^d$ such that

$$\boldsymbol{y}(t) = \sum_{\ell=1}^{d} \mathrm{e}^{\lambda_\ell(t-t_0)}\boldsymbol{\alpha}_\ell, \qquad t \geq t_0,$$

where $\lambda_1, \lambda_2, \ldots, \lambda_d$ are the eigenvalues of A.

Bibliography

Birkhoff, G. and Rota, G.-C. (1978), *Ordinary Differential Equations* (3rd ed.), Wiley, New York.

Bollobás, B. (1990), *Linear Analysis: An Introductory Course,* Cambridge University Press, Cambridge.

Boyce, W.E. and DiPrima, R.C. (1992), *Elementary Differential Equations* (5th ed.), Wiley, New York.

Davis, P.J. (1975), *Interpolation and Approximation,* Dover, New York.

Powell, M.J.D. (1981), *Approximation Theory and Methods,* Cambridge University Press, Cambridge.

Rudin, W. (1991), *Functional Analysis* (2nd ed.), McGraw–Hill, New York.

Index